ENVIRONMENTAL TOXICOLOGY

Environmental Toxicology is a comprehensive introductory textbook designed for undergraduate and graduate students of this subject.

The text is arranged in four tiers and covers most aspects of environmental toxicology, from the molecular to the ecosystem level. Early chapters deal with basic and advanced concepts, methods, and approaches for environmental toxicology. The next tier of chapters discusses the environmental toxicology of individual substances or groups of substances. The third tier of chapters addresses complex issues that incorporate and integrate many of the concepts, approaches, and substances covered in the first two tiers. The fourth part includes chapters on risk assessment, rehabilitation, and regulatory toxicology. A final chapter dicusses areas of study for current and future emphasis.

Throughout the book concise case studies from Europe, the United Kingdom, and North America illustrate the issues. Each chapter has a comprehensive list of references and further reading, as well as student exercises that are designed to reinforce the subject matter.

There is an extensive glossary and a list of abbreviations and acronyms. *Environmental Toxicology* is primarily a textbook for undergraduate and graduate students in environmental toxicology, environmental chemistry, ecotoxicology, applied ecology, environmental management, and risk assessment. It will also be valuable for specialists in ecology, environmental science, and chemistry, for example, practitioners in the metals and energy industries and in agriculture.

David A. Wright is a professor at the Center for Environmental Science at the University of Maryland and Director of the Chesapeake Bay Ambient Toxicity Program for the state of Maryland. Professor Wright has published more than 100 journal articles primarily on the physiology of ionic regulation and the uptake, toxicology, and physiology of trace metals in aquatic organisms. In recent years he has developed an interest in the dispersion and control of non-indigenous species. He has served on numerous review panels at the state and federal level and has testified in many court cases and hearings concerned with environmental pollution. He holds a DSc degree from the University of Newcastle upon Tyne.

Pamela Welbourn is a professor at Queen's University, previously a professor at Trent University, and former director of the Institute for Environmental Studies and a professor at the University of Toronto. Professor Welbourn has published more than 150 articles in scientific journals including *Nature, Environmental Science and Technology*, the *Canadian Journal of Fisheries and Aquatic Sciences*, and *Environmental Toxicology and Chemistry* and has contributed to ten scholarly books on aspects of the environmental toxicology of inorganic substances. She has served on numerous panels and boards in Canada and the United States, as well as on various public advisory committees. She has also had experience testifying as an expert witness in cases involving environmental contamination.

CAMBRIDGE ENVIRONMENTAL CHEMISTRY SERIES

Series editors:
P. G. C. Campbell, *Institut National de la Recherche Scientifique, Université du Québec, Canada*
R. M. Harrison, *School of Chemistry, University of Birmingham, England*
S. J. de Mora, *International Atomic Energy Agency – Marine Environment Laboratory, Monaco*

Other books in the series:
A. C. Chamberlain   *Radioactive Aerosols*
M. Cresser and A. Edwards   *Acidification of Freshwaters*
M. Cresser, K. Killham, and A. Edwards   *Soil Chemistry and Its Applications*
R. M. Harrison and S. J. de Mora   *Introductory Chemistry for the Environmental Sciences* Second Edition
S. J. de Mora   *Tributyltin: Case Study of an Environmental Contaminant*
T. D. Jickells and J. E. Rae   *Biogeochemistry of Intertidal Sediments*
S. J. de Mora, S. Demers, and M. Vernet   *The Effects of UV Radiation in the Marine Environment*

Environmental toxicology

DAVID A. WRIGHT, PhD, DSc
University of Maryland

PAMELA WELBOURN, PhD
Queen's University

CAMBRIDGE
UNIVERSITY PRESS

PUBLISHED BY THE PRESS SYNDICATE OF THE UNIVERSITY OF CAMBRIDGE
The Pitt Building, Trumpington Street, Cambridge, United Kingdom

CAMBRIDGE UNIVERSITY PRESS
The Edinburgh Building, Cambridge CB2 2RU, UK
40 West 20th Street, New York, NY 10011-4211, USA
477 Williamstown Road, Port Melbourne, VIC 3207, Australia
Ruiz de Alarcón 13, 28014 Madrid, Spain
Dock House, The Waterfront, Cape Town 8001, South Africa

http://www.cambridge.org

© Cambridge University Press 2002

First published 2002

Printed in the United Kingdom at the University Press, Cambridge

Typefaces Times New Roman 10.75/13.5 pt. and Univers   *System* QuarkXPress   [BTS]

A catalog record for this book is available from the British Library.

Library of Congress Cataloging in Publication Data
Wright, David A., 1948–
Environmental toxicology / David A. Wright, Pamela Welbourn.
p.   cm. – (Cambridge environmental chemistry series; 11)
Includes bibliographical references and index.
ISBN 0-521-58151-6 – ISBN 0-521-58860-X (pb.)
1. Environmental toxicology.   I. Welbourn, Pamela, 1935–   II. Title.   III. Series.
RA1226 .W75 2001
615.9′02 – dc21   2001018486

ISBN   0 521 58151 6   hardback
ISBN   0 521 58860 X   paperback

This book is dedicated to Rex Welbourn and Lee Ann Wright

for all their support and understanding.

When you can measure what you are speaking about, and express it in numbers, you know something about it; but when you cannot measure it, when you cannot express it in numbers, your knowledge is of a meager and unsatisfactory kind: it may be the beginning of knowledge, but you have scarcely, in your thoughts, advanced to the stage of science.

Thompson, William (Lord Kelvin). *Popular Lectures and Addresses* (1841–4).

Nowadays, people know the price of everything and the value of nothing.

Oscar Wilde, Definition of a Cynic. *Lady Windermere's Fan* (1892).

Contents

○ ○

Foreword

○ ○

Environmental Toxicology is a welcome addition to the Cambridge University Press Environmental Chemistry Series. The inclusion of a textbook on toxicology in a series devoted to environmental chemistry might, at first glance, appear surprising. However, as will become evident to the reader, the authors have approached their topic in a truly interdisciplinary manner, with environmental chemistry playing a prominent role in their analysis.

Environmental toxicology is a young and dynamic science, as pointed out by the authors in their welcome historical perspective of its development over the past 30+ years. One of the inevitable consequences of the rapid evolution of this area of science has been the scarcity of useful textbooks. Several multiauthored books have appeared in recent years, usually consisting of specialised chapters written by researchers familiar with a specific area of environmental toxicology. The individual chapters in such volumes are often very useful as state-of-the-art reviews, but links between and among chapters are difficult to establish. *Environmental Toxicology* breaks with this trend and offers a broad and coherent vision of the field, as developed by two senior researchers who have been active in this area of research since its inception in the 1970s.

In keeping with the aims of the Environmental Chemistry Series, this book is designed for use in courses offered to senior undergraduates and to graduate students. As university professors, the authors have used much of the material in their own courses, and thus, in a certain sense, the overall approach has already been tested and refined in the classroom. In choosing illustrative examples to include in their treatise, Wright and Welbourn have taken pains to maintain an international perspective – their frequent use of examples from the United Kingdom, Europe, the United States, Canada, and elsewhere should prove invaluable to readers seeking to learn from the scientific and regulatory experience of their global neighbours.

Professor Peter G. C. Campbell
Université du Quebec INRS
INRS-Eau CP 7500
Rue Einstein
Ste. Foy, Quebec
Canada

Preface

○ ○

This book is intended for use as a general text for courses given to intermediate undergraduate students with some basic background in chemistry, biology, and ecology. Graduate students with backgrounds in such traditional disciplines as chemistry, geography, or engineering, who are beginning studies that require an understanding of environmental toxicology, will also find the text useful. Additional readings, beyond those cited in the text, have been provided for those students who wish to take the subject matter further.

In common with many university and school texts, the original idea for this book grew from a course that the authors designed and presented. This began in 1989. Since that time, we have modified the material for use in different courses, in both the United States and Canada. Also since that time in environmental toxicology, existing approaches have evolved and new ones have been introduced. Technological advances, particularly in computers and in analytical chemistry and its applications, have facilitated progress. Beginning in the 1970s, but notably over the past decade, a number of excellent essay collections, as well as various texts addressing environmental toxicology, aquatic toxicology, ecotoxicology, and related topics, have been published. We have attempted to incorporate information on most of the significant items of progress, while providing the core and accepted components of the science, and to convey the enthusiasm that we have experienced, and continue to experience, over the subject area.

Whether there has been progress in the fundamental understanding and theory of the multidisciplinary subject known as ecotoxicology or environmental toxicology is less easy to determine. A lot depends on progress in other disciplines, some of them still young, notably ecology.

Paraphrasing Schuurmann and Markert (1998), ecotoxicology aims to characterise, understand, and predict deleterious effects of chemicals on biological systems. Various definitions have been provided for the term *ecotoxicology*, but in essence the subject involves the study of sources, pathways, transformations, and effects of potentially harmful chemicals in the environment, including not only their effects on individuals and populations of organisms but also their effects at the ecosystem level. The decision was made for the present text to use the more general term *environmental toxicology* in the title, while attempting in the main text, wherever it was deemed appropriate, to distinguish between this and the more specifically defined *ecotoxicology*, in Truhaut's (1975) sense.

In 1980, during the Aquatic Toxicology and Hazard Assessment Symposium, organized by the American Society for Testing and Materials, Macek stated, "There are unquestionably much more aquatic toxicity data on many more chemicals. However, there has been no real 'qualitative growth' in the science. No new and better questions are being asked; there are few new theories and precious little in the way of new scientific truths which have led to a better understanding of unifying concepts in the science" (Boudou and Ribeyre, 1989).

This somewhat gloomy statement concerning aquatic toxicology may well still be true in 2001. In our opinion, however, there is sufficient information that is genuinely new and original to stimulate the writing of a basic text that includes some of the still incomplete and controversial components of the science.

This book has been organised in an hierarchical manner, generally progressing from the simple to the complex. Following some discussion of the social context from which the science developed, early chapters look at the "tools of the trade", with definitions, methods, and approaches. The sources, behaviour, fate, and effects of individual contaminants are then treated, as inorganic, organic, and radioactive substances. Some relatively simple case studies have been provided where appropriate to illustrate these earlier chapters.

It will be noted that, for the most part, categories such as air pollution and water pollution have not been used as main headings for chapters or sections. This reflects our attitude that even though these compartments of the environment have value for regulatory purposes and possibly for policy formulation, they are often quite artificial in terms of an ecological approach to the science.

A number of complex issues were selected for the later chapters, with two major objectives in mind. One was to provide vehicles to integrate a number of the principles of methodology and approaches, and the characteristics of contaminants, which had already been described. The other was to illustrate the nature of real-world issues, in which contaminants do not exist in isolation from other contaminants, pre-existing conditions, or the natural complexity and variability of the ecosystem.

Chapters on risk assessment and rehabilitation draw on some earlier and by now familiar examples, and regulatory toxicology is addressed by incorporating hazard and risk assessment with reviews of some of the state-of-the-art regulatory approaches. The objective here is to consider some of the philosophy and approaches underlying the regulation of toxic substances and not to provide comprehensive coverage of statutory environmental regulation.

David A. Wright. University of Maryland, Center for Environmental Science, Chesapeake Biological Laboratory, Solomons, Maryland

Pamela Welbourn. Trent University, Peterborough, Ontario, Canada, and Queen's University, Kingston, Ontario, Canada

Abbreviations

○ ○

AAF   2-acetylaminofluorene

Ar   aryl hydrocarbon (receptor)

AHH   aryl hydrocarbon hydroxylase

ALA   aminolaevulinic acid

ALAD   aminolaevulinic acid dehydratase

ARNT   Ah receptor nuclear translocator

ASP   amnesic shellfish poisoning

ASTM   American Society for Testing and Materials

ATP   adenosine triphosphate

AVLS   atomic vapour laser separation

AVS   acid volatile sulphide (see glossary)

BSCF   biota-sediment concentration factor

CCME   Canadian Council of Ministers of the Environment (formerly CREM)

CFP   ciguatera fish poisoning

CTV   critical toxicity value

CYP1A1 and CYP1A2   subfamilies of the CYP1 gene family of P450 enzymes responsible for transformation of xenobiotics and endogenous substrates (see glossary, *cytochrome P 450*)

CWS   Canadian Wildlife Service

DDD   1,1-dichloro-2,2-bis(*p*-chlorophenyl) ethane

DDE   1,1-dichloro-2,2-bis(*p*-chlorophenyl) ethylene

DDT   1,1,1-trichloro-2,2-bis(*p*-chlorophenyl) ethane

DMRP   Dredged Material Research Programme

DMSO   dimethyl sulphoxide

DSP   diarrhetic shellfish poisoning

2,4-D   2,4-dichlorophenoxyacetic acid

EDTA   ethylenediaminotetraacetic acid

EEV   estimated exposure value

EF   enrichment factor

ELA   Experimental Lakes Area

ENEV   estimated no effects value

ER   endoplasmic reticulum

EROD   ethoxyresorufin-o-deethylase

ETS   electron transport system

FISH   fluorescence in situ hybridisation

GC-MS   gas chromatography-mass spectrometry

GSSG   glutathione disulphide

GUS   Groundwater ubiquity score (United Kingdom); defined as
$(1 \text{ g soil } t_{1/2}) \cdot (4 - (1 \text{ g } K_{oc}))$

HAB   harmful algal bloom

HPLC   high-pressure liquid chromatography

pH   (negative logarithm of) hydrogen ion concentration

IARC   International Agency for Research on Cancer

ICP-MS   inductively coupled plasma-mass spectrometry

ICRP   International Commission on Radiological Protection

IQ   intelligence quotient

ISE   ion selective electrode

K_a dissociation constant for weak acid (see glossary)

kDa   kilodaltons

K_{ow} octanol: water partition coefficient (see glossary)

LAS   linear alkylbenzene sulphonate

LLIR   low-level ionising radiation

LTE   linear transfer energy (see glossary)

LULU   locally unwanted land use

NAD(H)   nicotinamide adenine dinucleotide (reduced form)

NADP(H)   nicotinamide adenine dinucleotide phosphate (reduced form)

NIMBY   not in my backyard

NIMTO   not in my term of office

NTA   nitrilotriacetic acid

NOAA   National Oceanographic and Atmospheric Administration
(United States)

NSP   neurotoxic shellfish poisoning

OPEC   Organisation of Petroleum Exporting Countries

OSHA   Occupational Safety and Health Administration (United States)

PAH   polycyclic aromatic hydrocarbon

PAN   peroxyacetyl nitrate

PCB   polychlorinated biphenyl

PMR   premanufacturing registration

PPAR   peroxisome proliferase-activated receptor

PSP   paralytic shellfish poisoning

RAIN   Reversing Acidification in Norway

RAR   retinoid receptor

RXR   retinoic acid receptor

SEM   simultaneously extracted metals (used in association with acid volatile sulfides, AVS)

SERF   Shoreline Environmental Research Facility

SETAC   Society for Environmental Toxicology and Chemistry

SOD   superoxide dismutase

STP   sewage treatment plant

TBT   tributyltin

3,4,5-T   3,4,5-trichlorophenoxyacetic acid

2,3,7,8,TCDD   2,3,7,8-tetrachlorodibenzodioxin

TOC   total organic carbon

UDG   glucoronosyl transferase

UNSCEAR   United Nations Scientific Committee on the Effects of Atomic Radiation

U.S. EPA   United States Environmental Protection Agency

WHAM   Windermere Humic Acid Model

Acknowledgements

The authors are grateful to the following colleagues and students who contributed to this volume in various ways: Tom Adams, Carol Andews, Joel Baker, Gord Balch, Allyson Bissing, Canadian Environmental Law Association (Kathleen Cooper, Lisa McShane, and Paul Muldoon), Thomas Clarkson, Peter Dillon, Susan Dreier, Catherine Eimers, Hayla Evans, Mary Haasch, Landis Hare, Holger Hintelmann, Thomas Hutchinson, Maggie Julian, Allan Kuja, David Lasenby, David McLaughlin, Kenneth Nicholls, David Richardson, Eric Sager, Rajesh Seth, Douglas Spry, David Vanderweele, Chip Weseloh.

Particular thanks are due to the following for their special contributions, such as painstaking review of certain sections: Dianne Corcoran, R. Douglas Evans, Robert Loney, Donald Mackay, Sheila Macfie, Ann MacNeille, Lisa McShane, Diane Malley, Christopher Metcalfe, Macy Nelson, Robert Prairie, David Schindler, Elizabeth Sinclair, Judith Wilson. Robert Loney is thanked for his drafting of some of the figures, and Guri Roesijadi is thanked for the juxtaposition of the two quotes in Chapter 13.

Above all, this book is a testament to the patience of three people; Peter Campbell, who edited the whole text and made *many* helpful suggestions; Linda Rogers, who produced the typescript and collated the references; and Fran Younger, who drafted most of the figures.

This project benefitted in part from financial support from Trent University.

1

○ ○

The emergence of environmental toxicology as science

1.1   The context

This textbook is predominantly about the science, that is to say the observable, verifiable science, of environmental toxicology. The introductory quotation from Lord Kelvin emphasises the need for quantification in scientific studies. But studies of the development of most branches of science reveal linkages among intellectual curiosity, technological advances, and an awareness of human related problems, usually of social or of economic importance. Philosophically, it can be argued that quantification is not in itself sufficient to define and describe many environmental problems. The rich fabric of ideas, ideology, and technology that has led up to the current scientific discipline of environmental toxicology has been described and discussed by many authors in a variety of disciplines. Lord Ashby, referring in 1978 to protection of the environment, has aptly referred to the "quickening of the public conscience in most industrial countries" over the preceding 10 years. In the same series of essays, Ashby made a strong case for the position that values intrinsic to the environment cannot reasonably be measured, particularly in the context of monetary values. The quotation from Wilde, also quoted at the beginning of this book, although not made in the context of environment, captures this concept.

Environmental toxicology is only one component of the broader set of topics frequently referred to as environmental science and environmental studies. Other important aspects of environmental science are conservation of species, habitats and ecosystems, protection of endangered species, and various levels of management for water, soil, wildlife, and fisheries.

In common with other branches of environmental science, in environmental toxicology there are reciprocal relationships among scientific investigations and social problems. The influence of increased social awareness on the development of environmental toxicology is difficult to measure objectively. Although the main emphasis of this text is on the scientific and technological aspects of environmental toxicology (i.e., the things that we can measure), we begin by including some fairly general ideas concerning the social context in which the science has developed. To omit the social context completely would limit the educational value of the science, since the media and the publications of various environmentally concerned social groups have surely influenced some of the knowledge base that a modern student brings to class.

Many students of environmental science are sufficiently young that they accept the science of environmental toxicology as a recognised and important discipline, so that most of the readers of this text will probably not have had the opportunity to witness or understand the remarkable rapidity of its development.

As Schuurmann and Markert (1998) pointed out, "The need to achieve results of practical relevance for politics and society made the development of ecotoxicology rapid indeed". Most authorities would date its beginning in the mid- to late 1960s. As stated by van der Heijden et al. (in Finger, 1992), "What we call environmental problems today were not defined as such before 1965". It is notable that, even though advances in environmental toxicology have resulted in part from "need", the same time span has seen great advances in technology, which have also stimulated progress. Technology has provided tools to facilitate the rapid growth of the science. As environmental toxicology has evolved, fundamental scientific advances in related fields such as ecology and geochemistry have been accelerated, partly in response to the need to better understand the fundamental processes upon which environmental toxicants are acting. In these and other aspects, the past few decades have been exceptionally exciting for the environmental toxicologist.

Most texts of environmental toxicology or ecotoxicology refer to the publication of Rachel Carson's *Silent Spring* (1962) as a landmark in the public's awareness of potential damage to human and environmental health from man-made toxic substances. Carson's book could be described as bringing about "the ignition of public concern". These are Lord Ashby's words, which he uses to describe the process of raising public awareness about an environmental problem. He refers not only to the past but also to the current process. According to Rodricks (1992), Carson's book "almost single-handedly created modern society's fears about synthetic chemicals in the environment and, among other things, fostered renewed interest in the science of toxicology". Certainly the consolidation of academic and related pursuits into the study of toxic substances in the environment dates from about the same time as the publication of *Silent Spring*. Prior to the 1960s, there were no coordinated programmes in research, in education or in regulation that systematically addressed toxic substances in the environment. The considerable progress that has been made in all these areas during the past four decades bears further examination.

1.2   The historical background: Classical toxicology, ecotoxicology, and environmental toxicology

As indicated already, the systematic study of the effect of toxic substances on ecosystems is essentially a twentieth-century phenomenon. But, although the science of environmental toxicology is a relatively new one, it has its roots in the study of human or classical toxicology, which has a longer and somewhat more sinister history. A review of Shakespeare's plays as well as other historic literature reveals that deliberate poisoning has long been regarded as a means of solving personal and political problems. The first students of toxicology were physicians and alchemists whose jobs involved the treatment of poisoning cases. Perhaps the best known is the Swiss physician Paracelsus (1499–1541) who first formulated the notion of dose-response, including the idea that chemicals, which had a therapeutic value in low doses, were toxic at higher doses. He was also responsible for a treatise, *On the Miner's Sickness and Other Diseases of Miners*, which was published posthumously in 1567 and represents one of the first works on occupational toxicology. This theme was expanded to a variety of different occupational hazards in the classic work of Bernadino Ramazzini, *Discourse on the Diseases of Workers* (1700).

The modern science of toxicology can be traced back to the Spanish physician Mattieu Orfila, who published a comprehensive treatise on the toxicity of natural agents, which articulated many of the basic components of the discipline we know today, including the relation between toxic symptoms (pathology) and chemical content of tissues as determined by analysis, mechanisms of eliminating poisons from the body and treatment with antidotes (Orfila, 1815). The work of physiologist Claude Bernard (1813–78) introduced a more mechanistic approach to toxicology through controlled experiments on animals in the laboratory.

During the latter half of the nineteenth century, along with the rapid expansion of the chemical manufacturing industry, the science of toxicology proceeded rapidly in many countries both as an appendix to and an offshoot of pharmacology. Chemicals being tested were raw and refined natural products and synthetic compounds of both medical and industrial origin. The principal focus was human health, and much of the concern arose from the obvious effects of occupational exposure to chemicals. However, in both Europe and North America, an awareness that increasing industrialisation was having an adverse effect on species other than *Homo sapiens* was beginning to take shape. Arsenic had been used as a crop treatment since the 1850s, and as early as the 1870s there were signs that arsenic from pesticides and lead from lead shot were causing wildlife deaths.

One of the first indications that government was taking an interest in the problem was the creation of a royal commission in Britain to examine the effects of toxic chemicals in industrial wastewater. Although a report from the commission included some short-term bioassays (see Chapter 2), standardised tests designed to quantify environmental toxicity were still a long way off. In 1924 Carpenter published the first in a series of papers on the effect of trace metals from acid mine drainage on fish, and in 1951 Doudoroff and co-workers advocated the use of standardised fish assays for testing effluent toxicity. Anderson suggested in 1944 the use of the ostracod crustacean *Daphnia magna* as a standard test organism.

The term *ecotoxicology* was first used by Truhaut as recently as 1969 and focuses primarily on the toxic effects of chemicals and radiation on levels of biological organisation from individual organisms to communities. His definition for ecotoxicology (Truhaut, 1977) was "the branch of Toxicology concerned with the study of toxic effects, caused by natural or synthetic pollutants, to the constituents of ecosystems, animal (including human), vegetable and microbial, in an integral context". Inherent in this concept is the investigation of how and to what degree biota are exposed to these toxic elements. As a corollary, ecotoxicology also involves the study of the manner in which chemicals and various energy forms are released into the environment, their transport, and transformation. Butler (1978) cites Truhaut's 1975 identification of the ways in which ecotoxicology differs from classical, as follows:

> In ecotoxicology, any assessment of the ultimate effect of an environmental pollutant must involve a four-part process of
>
> release into the environment;
> transport into biota, with or without chemical transformation;
> exposure of one or more target organisms and
> response of individual organisms, populations or communities.

Classical toxicology has been primarily concerned with the toxic effects of chemicals and radiation on levels of biological organisation from the subcellular to the individual organisms, but with a primary focus on humans. Where other species have been used, their role has been as human surrogates to a greater or lesser degree. Improvements in statistical methods and epidemiology have, in many instances, extended the science to the population level, but once again the emphasis has been on the human population.

It is fair to say that the modern science of environmental toxicology embraces the disciplines of both classical toxicology and ecotoxicology. It is but a short step from a subcellular or physiological assay on a rodent used as a human surrogate to a similar test performed on a mink, ferret, bird, or fish seen as an integral part of a particular ecosystem. Additionally, it is recognised that an understanding of human chemical exposure, which is an integral part of classical toxicology, often involves food chain studies of other species. The passage of chemical contaminants through food chains with consequential damaging effects at different trophic levels is a primary focus of ecotoxicology. Therefore, the distinction among ecotoxicology and environmental toxicology becomes blurred.

Wildlife toxicology has been identified by some as a particular branch of environmental toxicology. As already indicated, some of the earliest examples of non-human targets for anthropogenic contaminants were indeed drawn from effects on wild animals or free-ranging livestock. Unplanned "experiments" on the effects of pest control products and industrial releases on whole populations in the field, while regrettable and even tragic at a humane level, provided some of the earliest cause-effect data in environmental toxicology. But is there a fundamental scientific rationale for making a distinction between wildlife toxicology and other types of environmental toxicology? To single out this branch of environmental toxicology, based as it is on the particular group of receptors that humans have chosen to single out as "wildlife", may be useful in the context of management and policy formulation, but the fundamental scientific approaches of environmental toxicology are the same for wildlife as for other forms of life.

In all endeavours in toxicology for which the immediate target is nonhuman, the need for some appreciation and consideration of ecology, itself a relatively young science, is crucial. Yet our understanding of ecology is still very limited. As pointed out by Moriarty (1988), "The underlying science on which ecotoxicology rests is far less developed than is the scientific basis of [classical] toxicology". Moriarty warns against "a disinclination to support research once the immediate practical problem has been resolved". This opinion is particularly relevant to what is referred to as applied ecotoxicology, where concrete solutions and "fixes" are expected to be found for real-world problems. Because of the urgency, real or perceived, of certain problems, emphasis may be necessary on some immediate solution. Yet without pursuit of the more fundamental aspects of the problem, two types of penalty can result. Intellectual pursuit is thwarted, and in the absence of an improved understanding of the issue, similar environmental problems are likely to occur in the future.

Terms and their definitions can be a source of confusion for the student as for the practitioner. For the present text, we decided to use the term *environmental toxicology* in the title, while attempting in the main text, wherever it was deemed appropriate, to bridge this subject area to the more specifically defined *ecotoxicology*, in Truhaut's sense.

Figure 1.1 is a scheme that encompasses most of the elements comprising our definition of environmental toxicology, including the dissemination of scientific information to the public and to those responsible for formulating and enacting regulatory legislation. The scheme incorporates the principal components of what has become known as a risk assessment (i.e., exposure assessment and hazard assessment). This concept is dealt with in more detail in Chapter 10 but may be very simply expressed in the form of the question: What are the chances of a potentially toxic agent (A) reaching a particular receptor (B), and what degree of damage will result from A meeting B? Despite this empirical definition, it is clear that public perception of risk has a clear bearing on how priorities are ordered and regulations enacted.

1.3   Social aspects: The environmental movement

Social and political concerns that have accompanied the growth of environmental toxicology may be complementary to or even part of the scientific approach; others appear to be at odds with the objectivity that characterises science. The so-called Environmental Movement is deeply involved with social and political activities. The Environmental Movement is normally understood to include the activities, both in written works and in political activism, that began in the mid to late 1960s

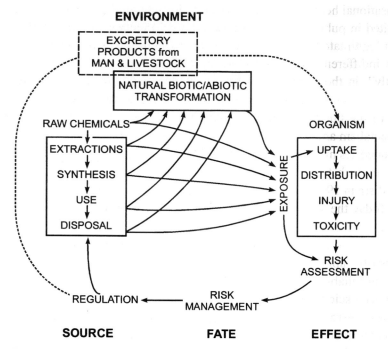

Figure 1.1 Principal components of environmental toxicology.

and continues into the present time, under the general rubric of protecting the environment from damage by human activities. The level of organisation ranges from groups that have been formed to address single issues on a local scale, usually with modest budgets and a great deal of voluntary work, to international groups that have large budgets and are run by professional advocates. The issues range from saving a specific wetland to attempting to stop all nuclear power generation. The strategies range from distributing information at Town Hall types of meetings through highly publicised demonstrations designed to attract media attention to actual violence and destruction. Common to all groups is the goal of raising of awareness, which ideally means educating the audience. However, environmental issues frequently evoke emotional responses, which are likely to be at odds with objective science but which the activist may decide to exploit as part of his or her strategy.

Early in the history of the movement, the dramatisation of selected facts and events by activists was probably necessary to make points and raise awareness. This was often done at arm's length from practising scientists, many of whom felt that their role was to provide information but not to become politically involved.

Ashby (1978) (see Section 1.9) pointed out a number of instances in which objective and calm reporting of an issue by a technical or scientific practitioner failed to result in any action, whereas a certain compromise with objectivity, resulting in sensational headlines (e.g., "Lake Erie is dead"), with intensive media attention, resulted in public response followed by political action. He asked the question: "Is it legitimate to dramatise some potential environmental hazard in order to overcome indifference among the public?" His own answer, in 1978, was that "reluctantly", in the social climate of that time, there was no alternative but to concede to the techniques of the mass media. The media's job is not only to *inform* members of the public but also to *alert* them. What is the relationship between environmentalism and science, between the environmentalist and the scientist? Are the two categories mutually exclusive? Ashby, himself a scientist, apparently would not resort to such techniques, yet he is sympathetic. He articulated the issue of sensationalism in 1978 as an ethical problem for the would-be protector of the environment. Today the same ethical problems exist; techniques used by environmentalists range from calm investigation to rampant activism. Environmental scientists' opinions of their own roles range from active participation in advocacy groups to the "hands off" approach in which the scientist's responsibility ends at the point of producing reliable data.

Yet between scientists and environmentalists there are logically some reciprocal activities. There is at least circumstantial evidence that, on the one hand, social and political awareness have stimulated or pushed forward scientific investigations of environmental problems, whereas, on the other, scientific findings have clearly made possible advances in environmentalism. In reality, the various practitioners of science and environmentalism are spread along a continuum, but there is in practice, as expressed by Paehlke (1989), "a great gulf between environmentalism as a political movement and much of the scientific community. Environmentalists have been hostile to much of what science has offered society. Likewise, many scientists deeply resent the claims and style of environmentalists".

The publication in 1962 of the biologist Rachel Carson's *Silent Spring* has already been referred to in the context of both public awareness about chemicals in the environment and renewal of interest in the science of toxicology. It is instructive at this point to examine what she did and how she was viewed in her time as well as more recently.

Carson wrote of the dangers of pesticides and using a dramatic rather than scientific idiom, alerted the public to the damage that pesticides were doing to the environment, particularly to humans and wildlife. She is generally credited with founding the Environmental Movement. Her work was followed by a number of other works of the movement, not all of which are of relevance to environmental toxicology, including *The Closing Circle* (Commoner, 1971), *The Population Bomb* (Ehrlich, 1975), *The Limits to Growth* (Meadows et al., 1979), and *Small Is Beautiful* (Schumacher, 1973). These works, only a few of many written during the earlier years of the Environmental Movement, are not reviewed in the present text but are provided to illustrate the climate of thinking at that time.

Carson herself did not pretend that her book was a thoroughly objective study – she felt passionately about her subject, as indicated by the titles of some chapters: "Elixirs of Death", "Needless Havoc", and "And No Birds Sing". Predictably, Carson's work was extremely controversial when it was written. It has been subjected to many analyses since then. The types of criticism and analysis that Carson's book has received have varied over time: some came from parts of the established scientific community, and many were predominantly defensive, particularly from the alleged culprits of pesticide pollution. More recently however, a different, non-defensive, type of analysis has been made of Carson's and similar works; attempts have been made to identify the scientific strengths and weaknesses of the literature and the activities of it and more generally of the Environmental Movement.

The recent critiques of Carson's and other works of the Environmental Movement tend to address the scientific validity of the claims, the accuracy of predicted outcomes, and, more generally, the role of environmental groups in the 1990s. Charles Rubin, now sceptical about the goals of the movement, attempted in *The Green Crusade* (1994) to determine the source of the ideas that made up what "we all know" about environmental concerns. He reinforced awareness of the "two prongs" of the environmental movement: the study of ecology and related science, juxtaposed with recognition of environmental degradation. He provided a very detailed evaluation of Carson's work. Although many praised Carson for exposing the damage wrought to human and ecological systems by chemical pesticides, as well as for her "impeccable" scientific credentials, even those who praised it admitted that, for example, it was not a "fair and impartial appraisal of all the evidence". It addition to reviewing the criticisms made at the time, by those in the chemical industry that she attacked and as well by disinterested scientists, Rubin presented us with many examples of more recent criticisms. These included the extremely selective manner in which she used the available literature to support her thesis and the flawed conclusions that she drew from much of the literature that she cited in support of her cause. But this type of analysis does not change the impact that "this remarkable woman" made through her moral stance and the more subtle political messages that have been recognised in her work.

Attitudes change and may become moderated, and when this happens, compromises tend to be made by all interested parties. Ideally, this is reflected in improved understanding among the parties and a maturing of attitudes. The scientific along with the industrial community understands that, in the context of their respective interests as well as from some sense of responsibility for stewardship of the environment, they need to become involved in the process of environmental protection. Some environmental groups have already recognised the fact that their role too can change. Now that terms like *ecology* and *toxicology* are familiar to the general public, and now that environmental protection is more generally accepted as a real need for the sustainability of the ecosystems of the Earth, perhaps environmentalists can be equally effective when educating the public by providing more objective and less emotional scenarios. As expressed in *Eco-sanity* by Bast et al. (1994), "a more mature environmental movement would consistently rely on reason and deliberation, not emotion and fear. It would be considerate of the rights and expectations of those affected by its proposals. It would accept the necessity of compromise sometimes". Some environmental groups already cooperate with other communities, namely academic scientists and industrialists. Others have remained in a more confrontational mode. Although discussion of this subject must be limited in the present text, the list of further readings in Section 1.9 is recommended, providing for more detailed examination of this important topic.

Initially the focus of the movement was mainly on the North American continent. Soon lobby groups and other special interest groups began to come together in many parts of the world. Some of these had their origins in, or connections with, the more "traditional" conservation and natural history groups. For example, the Sierra Club, which originated in the nineteenth century as a conservationist group, now addresses issues such as human health concerns of industrial workers. Originally such groups maintained a low profile among the general public. According to Paehlke (1989), "Without forgetting its origins, the movement as a whole has travelled from a perspective rooted in the wilderness, mountains and forests to a point far nearer to the immediate concerns of urban North Americans."

Table 1.1 lists some of the major existing environmental groups, with some basic information about the groups. They were selected to include those groups that have been active in the field of environmental toxicology, as well as to illustrate the range of the respective strategies and expertise that is represented in the movement. The founding dates of many of these organisations coincide closely with the emergence of the movement.

1.4   Social aspects: Regulation

Fully 2,500 years ago, Athens had a law requiring refuse disposal outside the city's boundaries. This was long before any cause-effect relationships between harmful substances and disease were clearly understood.

Scientific and social considerations of environmental toxicology have been followed closely and accompanied by the growth of government departments, ministries, and legislation, all dealing with various aspects of the study and management of environmental quality. During the 1960s and 1970s, many government departments and ministries were formed, either de novo or evolving out of existing units, in most of the developed countries. Environmental legislation grew concurrently. A brief overview of the process of regulation and its links with other social aspects of environmental awareness is included in this Introduction, and regulatory toxicology is covered in more detail in Chapter 12.

With the general perception of environmental deterioration that began in the 1960s and flourished in the 1970s, environmental problems came to be viewed by the public, largely through the success of the Environmental Movement, as a pattern of damage related mainly to human economic activity. Several dramatic accidents in those years, such as oil spills, provided concrete evidence that

Table 1.1. Some environmental non-government organisations (ENGOs) with activities related to environmental toxicology

Organisation/acronym	Date and place of origin	General goal(s)	Strategies	Membership and supporters or donors[a]	Examples of projects involving environmental toxicology	Budget
Canadian Environmental Law Association (CELA)	1970, Toronto, Canada	To "give voice to the environment in policy and laws"; A "legal aid" approach to environmental protection	Advocacy for comprehensive laws, standards, etc., to protect environmental quality in Canada, especially Ontario; support and in some cases help to found other groups	Approximately 350	Toxics caucus works extensively on developing member groups' positions on the control of toxic substances and products of biotechnology; CELA is representing a client who lives adjacent to a corporation that is discharging fluoride fumes: a SLAPP (strategic law suit against participation); CELA intervened successfully concerning PCB releases by Quebec Hydro; CELA as counsel for Community Health Centre concerning removal of lead-contaminated soil	CAN $800,000 per annum including Ontario Legal Aid Plan core funding and additional project funding
EarthFirst!	1991, United Kingdom – now active in	A network of radical ecoactivists; no	Direct action: conventional "green"	No members in conventional sense; no paid	No details available: toxic substances component currently	No budget in the conventional sense

	Founded	Purpose	Activities/Methods	Members/Workers	Focus/Projects	Budget
	13 countries	central office, no decision making body; their slogan "No compromise in defence of the earth"	campaigning is not enough; fight is against "ecological destruction [through] the domination of the majority by elite groups"	workers	not well advanced	
Environmental Defense Fund (EDF)	1967, Long Island, NY, originally by volunteers	To achieve environmental goals through legal action and education	Use of scientific evidence in court; use of economic incentives to solve environmental problems; participation in environmental education	300,000 (in 1997) members; staff of 163 including 60 full-time professionals	First project was attempt to stop DDT spraying in Long Island	U.S. $25.7 million (1994)
Energy Probe Research Foundation, with divisions: Energy Probe, Probe International, Environment Probe, Consumer Policy Institute	1970, founded in Canada as a team of Pollution Probe; separated from Pollution Probe in 1980	To provide business, government and the public with information on energy, environmental, and related issues	Publication of papers and books, newspaper articles, and mailings to households	Supported by 18,000 Canadians; main foundation has staff of 9 persons and many volunteers	Various issues related to radioactive waste and health effects of nuclear power plants; working to limit acid gas emissions from fossil fueled power plants	General fund turned over approximately CAN $2 million per annum in each of 1997 and 1998
Friends of the Earth International (FOE), a federation of 54	1971	Research and advocacy	FOE Canada is a campaign advocacy	FOE International, nearly 1	Not applicable	Not applicable

Table 1.1. *(continued)*

Organisation/acronym	Date and place of origin	General goal(s)	Strategies	Membership and supporters or donors[a]	Examples of projects involving environmental toxicology	Budget
independent member groups in the world			organisation	million members in 1991		
Greenpeace	1971; Vancouver, Canada; involved 12 protesters sailing a boat into the U.S. atomic test zone off Amchitka, Alaska	Protection of the environment	Nonviolent direct action to protest and attract media attention to issues	5 million supporters in 58 countries; as of 1993, Greenpeace reported more than 1,330 staff in 43 offices in 30 countries	Campaigns against toxic waste, acid rain, ocean pollution	
Pollution Probe	1969, Toronto, Canada	To work for enforceable environmental policies; to identify critical effective solutions for environmental problems	Works in partnership with stakeholders; works through education and outreach	10,000 donors and 14 full-time staff members	Were involved in banning DDT, limiting phosphates in detergents, and in helping to launch the Canadian Coalition on Acid Rain. Involved in agreements for hospitals to voluntarily reduce their use of mercury	CAN $1,600
The Sierra Club	1892, founded in the United States; active	Original goal was conservation of wilderness	To influence public policy and raise environmental	Superfund which	Involved in a number of issues including the US	

12

	in Canada since 1969	areas; now expanded to include toxic chemical contamination and other issues beyond conservation	awareness		requires polluters to pay for cleanup of abandoned waste	
World Wildlife Fund (WWF)/also called World Wide Fund for Nature International, an international network of 25 organisations	1961	To harness public opinion and educate the world about the necessity for conservation; mission revised in the 1990s, see next column	Strategy revised in 1990s to also include promotion of the concept of reducing wasteful consumption and pollution	WWF "almost 4 million members" in 1991	Details not available, but see WWF Canada	WWF earned U.S. $200 million in revenue in 1991
World Wildlife Fund Canada, of which the Wildlife Toxicology Fund (Canada) was part	World Wildlife Fund Canada started in 1967, part of World Wildlife Fund; Wildlife Toxicology Fund (Canada) operated from 1985 to 1997	As above: wildlife toxicology was added in 1985, to identify the impacts of pollution on wildlife	Wildlife Toxicology Fund (Canada) from 1985 to 1997 set up to fund research, requiring matching funding from other agencies	WWF Canada 300 true (voting) members; many other donors and supporters	Funded research projects included a number of studies on acidification and wildlife; effects of various pesticides on aquatic invertebrates, amphibians; effects of various metals on wildlife and a number of studies of biomonitors	From 1985 to 1997, WW toxicology fund spent a total of CAN $5 million on research projects

[a] Membership is difficult to define. Strictly speaking, members of an organisation should have a role in its activities, such as voting rights at meetings. Many of the environmental groups count as members every person who has donated or in any way supported it.

unregulated growth and development was likely to cause ecological degradation. In cases of environmental pollution, the link was perceived as being very close to industrial activity. Parallel arguments by environmentalists concerning the use of resources, even in the absence of pollution, added to the demand for regulation of activities that seemed to threaten not only ecosystems but also the social condition of humans.

By the early 1970s, politicians in most developed countries had been responsive to pressure from the various environmental groups, some of which were professional lobbyists, as well as from the general public. As an example, in 1969, Canada's Prime Minister Pierre Trudeau stated in the debate on the Speech from the Throne (cited in Roberts, 1990):

> We intend to tackle the problems of environment, not only in the Northern regions but everywhere in Canada, by directing our efforts mainly to the two major sources of pollution: urbanization and the invasion of modern technology. The present [Liberal] government is firmly determined that no such acts of madness [challenging not only his own species but also the whole life of our planet] will be allowed to go on indefinitely, at least in Canada.

President Kennedy set up a Council on Environmental Quality. On the first Earth Day, 1970, the Council wrote:

> A chorus of concern for the environment is sweeping the country. It reaches to the regional, national and international environmental problems. It embraces pollution of the earth's air and water, noise and waste and the threatened disappearance of whole species of plant and animal life. (cited in Van der Heijdden et al., 1992)

During the late 1960s and early to mid 1970s, some existing laws including the British Common Law served as tools for environmental protection and suits for damages. Beginning the late 1960s, amendments that increased the power of governments to designate substances as "deleterious" were made to existing acts, and a great deal of new environmental legislation was enacted. By the time the United Nations Conference on the Human Environment was held in Stockholm in 1972, most developed countries had instituted departments or ministries of the environment, with growing staffs of civil servants and many new environmental laws. The environment had become a powerful part of the political platform.

Funding was provided from the public purse for academics and government scientists to conduct research on pollution, as well as pollution control. Federal expenditure on pollution control alone in the United States, according to estimates from the Council for Environmental Quality, grew from about $0.8 million in 1969 to about $4.2 million in 1974. By 1986 the U.S. Environmental Protection Agency (U.S. EPA) was spending $8.1 million annually on risk impact assessments associated with the Clean Air Act, the Clean Water Act, and other regulatory programs while recognising that such expenditures represented only a small fraction of the national cost of these programs.

By the 1980s, many of the local or point source releases of pollutants were understood and were regulated. As understanding of cause and effect was improved, and as technology in engineering and analytical chemistry grew in tandem with this understanding, many of the worst point source emissions and discharges into air and water were under reasonable control through regulations, control orders, and permits. Controversies continued and violations and accidents happened, but for the most part, one could concede that the "first generation" of environmental regulation was beginning to show benefits. For new endeavours, the cost of ongoing pollution control or environmental clean-up at the time of decommissioning was of necessity built in at the planning stage. Legislation concerning environmental impact assessment for certain types of projects was beginning to emerge.

By the 1980s, a new kind of environmental problem related to toxic substances arose, demanding an altogether more difficult approach to regulation. Diffuse pollution, particularly long-range transport of atmospheric pollutants, exemplified by acid deposition, was identified as a serious environmental issue. In Europe, particularly in the Nordic countries, and in parts of North America, particularly the northeastern states and the southeastern Canadian provinces, "acid rain" assumed scientific and political prominence. The transboundary movement of the airborne contaminants introduced a new type of regulatory problem. Pollutants released in one jurisdiction were being deposited in a different jurisdiction, often in a different country. The Stockholm Conference had adopted a "declaration on the human environment", with 109 resolutions of principles to guide the development of international law, but the problem of transboundary pollution was not identified at that time. Control of these diffuse emissions would often result in severe economic hardship, even closure of whole plants, for the "donor" industries; therefore, prior to any agreement to control emissions by the "donor", governments in the receiving jurisdiction were challenged to provide concrete proof of the transport and effects of diffuse airborne pollutants. During the late 1970s and well into the 1980s, there was a great deal of scientific and political activity on airborne pollutants and their ecological and health effects. Some outstanding research on aquatic ecosystems, including advances in fundamental limnological research, grew from this issue. Some improvements in air quality were eventually forced at international levels, in Europe and in North America, and the responses in ecosystems are still being assessed. The acid rain issue no longer makes headlines nor does it create international incidents. Government funding for research has declined. Yet the jury is still out on the current risk to ecosystems from acidic deposition.

A tendency has emerged over the past two or three decades for the establishment of international treaties of "good will" among nations, concerning environmental protection. Regrettably these treaties, "gentlemen's agreements" signed by politicians with no expert understanding of the technological or economic implications, are frequently unrealistic and unenforceable.

1.5   Education in environmental toxicology

The emerging scientific interest in and social concern for environmental toxicology was logically followed by the opening of programmes, departments, and even schools in universities and other postsecondary institutions. These originated from many different roots: some grew out of programmes in ecology, marine biology, or limnology; others, from geography; others, from geochemistry; others, from engineering; whereas others still had roots in public health departments. Some of the programmes deal exclusively with toxicology, and others embrace environmental science in a broader context. Yet others describe their offerings as environmental studies. The latter is understood to include any or all of social science, law, and political science, as well as the so-called hard sciences.

The older universities have tended to adapt or add to existing programmes or courses of study to include environmental sciences, whereas the universities that grew in the 1960s and later often have autonomous departments or faculties of environmental studies. Models for the programmes of study are as varied in their content and presentation as are their respective histories, and will not be pursued further in this text. Environmental toxicology, in common with other branches of environmental science, is considered to still be a hybrid science, often referred to as "interdisciplinary". In contrast to the more traditional disciplines, which have a body of literature and a core of knowledge that is rather similar throughout most educational institutions, at the present time, there is no such undisputed core for environmental toxicology. Partly for this reason, there is also some controversy concerning the stage in a curriculum at which it should be introduced.

It is our particular belief that, ideally, the postsecondary student of environmental toxicology should bring to the subject area some basic education in chemistry and biology, especially ecology, with the prerequisites such as basic physics and mathematics that these other subjects require. For this reason, the subject is normally introduced into the higher years of undergraduate programmes, or at the postgraduate level. Other academics would disagree, and introductory overviews of environmental toxicology have been successfully taught in a number of universities. Although these courses may provide the student with a broad general sense of the topic, and may well stimulate his or her interests, in our opinion they will not provide the in-depth scientific understanding that can lead to specialisation. In our preferred model, we would provide one or more courses at the senior undergraduate level; then most of the specialist programmes in environmental toxicology would be offered at the postgraduate level, either by incorporating research or as technical training.

1.6   The role of technology

In addition to social and economic influences on the development of environmental toxicology since the late 1960s, technical advances have also enhanced the science over the same time period. Analytical chemistry has become at once more sophisticated yet simplified for the user. Over the time period that we have been considering, rather rapid advances have been made in techniques for analysis for chemicals. In particular, the change from "wet chemistry" to spectroscopic and similar methods, accompanied by certain electronic and other devices resulting in streamlining and automation of the actual instruments, have facilitated the processing of large numbers of samples. The simplification of operation has made possible the use of analytical methods by scientists, such as ecologists and biologists, outside the exclusive realm of analytical chemistry.

The analytical chemist's improved ability to reliably identify and detect very low concentrations of chemical substances has enhanced toxicological research. This applies to certain situations in which there are sublethal toxic effects and also to certain carcinogens. Subtle changes may be related to very low exposures to or residues of chemicals, providing the potential for informed environmental criteria. On the other hand, some chemicals may be shown to have no effect at very low concentrations, if there is a true threshold of response. Either way, the technology may well lead to improved regulation of chemicals in the environment.

Improved accuracy and precision of measurement, combined with attaining finer and finer detection limits for inorganic and organic chemicals, has nevertheless produced some unexpected results. On the one hand, it has become apparent that many contaminants are far more widely dispersed in the environment, albeit at very low concentrations, than was previously imagined. Trace quantities of manufactured chemicals, even certain pesticides that have been banned and are no longer used, continue to be detected in abiotic and biotic components of remote areas such as the arctic and antarctic regions. This detection has been used to claim that humans have polluted the entire planet, which may be true, depending upon one's definition of pollution. On the other hand, we can now "measure more than we can understand", meaning that we do not know if these traces of substances that can now be detected analytically have any biological significance. In the purest sense, the very fact that the substances are "out there" is evidence of human impact on the ecosystem, yet until the real risk from these low concentrations is understood, excessive concern might divert attention and finite resources from more acute problems. This only emphasises once again the need for careful scientific research into cause and effect and mechanisms for the action of environmental toxicants.

The electronic revolution has arguably had its main impact through the development of computers. Advances in computers, both in hardware and in software, have at once simplified and increased the complexity of environmental toxicology, both in research and in more routine tasks. The same is, of course, true for almost every area of scholarly and technical endeavour. The computer has made possible the processing of large data sets such as those resulting from field collections that hitherto would have been insurmountable by manual methods. As pointed out earlier, the computer has streamlined the recording and processing of data from machines that are used for chemical analysis. The computer has enabled many more people than were hitherto able to use mathematical models for research and for assessment of toxic chemicals.

There are additional areas where technological advances from other disciplines have worked in tandem with toxicology. Some have even become accepted as part of toxicology. The price of these and other technological advances is the increased burden of maintaining, repairing, and staying current with a rapidly changing world of equipment. Other costs are literally monetary concerns. Equipment breaks down or becomes obsolete with alarming rapidity.

The younger generation of toxicologists probably takes both the benefits and costs of technology more or less for granted. Indeed, there is really very little choice in these matters. It is salutary, however, to remember that, in common with the other components of environmental toxicology, many relevant changes in technology have been very impressive, very rapid, and very recent.

The role of technology in approaches to toxicology is revisited in Section 4.8.

1.7   Questions

1. List the ways in which, according to Truhaut, ecotoxicology differs from classical toxicology.

2. Using diagrams as required, outline the essential elements of environmental toxicology.

3. Indicate the approximate time and the main events that mark the beginnings of the Environmental Movement.

4. Not all environmental activists are alike. Provide some examples of different groups that have been involved in issues related to toxic substances. Indicate the goal(s), strategies, and examples of issues that have been addressed by at least three of the groups that you name.

5. Environmentalists have been criticised for their lack of scientific objectivity and for dramatising some potential environmental hazards. Discuss these criticisms and provide your own opinion, either based on facts or on principles, concerning the legitimacy and the role of environmental activists.

6. Indicate the approximate time and the main events leading up to the recognition of environmental toxicology as a science in its own right.

7. Are there demonstrated connections between the Environmental Movement and the emergence of the science of environmental toxicology? If so, identify them. If not, indicate possible connections.

8. Provide examples of technological advances that appear to have enabled the advancement of the science of environmental toxicology. Within the limits of your own information at this time, give your opinion regarding

the relative importance of these technological advances and original ideas in the context of environmental toxicology.

1.8   References

Anderson, B. G. (1944) The toxicity thresholds of various substances found in *Daphnia magna*, *Sewage Works Journal*, **16**, 1156–65.

Ashby, E. (1978) *Reconciling Man with the Environment*, Stanford University Press, Stanford, CA.

Bast, J. L., Hill, P. J., and Rue, R. C. (1994) *Eco-sanity: A common sense guide to environmentalism*. Madison Books, Lanham, MD.

Boudou, A., and Ribeyre, F. (1989) *Aquatic Ecotoxicology: Fundamental Concepts and Methodologies*, CRC Press, Boca Raton, FL.

Butler, G. (1978) *Principles of Ecotoxicology, SCOPE 12*, Wiley and Sons, New York.

Carpenter, K. E. (1924) A study of the fauna of rivers polluted by lead mining in the Aberystwyth district of Cardiganshire, *Annals of Applied Biology*, **11**, 1–23.

Carson, R. (1962) *Silent Spring*, Houghton Mifflin, Boston.

Commoner, B. (1971) *The Closing Circle*, Bantam Books, New York.

Doudoroff, P., Anderson, B. G., Burdick, G. E., Galtsoff, P. S., Hart, W. B., Patrick, R., Strong, E. R., Surber, E. W., and Van Horn, W. M. (1951) Bio-assay methods for the evaluation of acute toxicity of industrial wastes to fish, *Sewage and Industrial Wastes*, **23**, 1380–97.

Ehrlich, P. (1975) *The Population Bomb*, Rivercity Press, Rivercity, MA.

Finger, M. (1992) *Research in Social Movements, Conflicts and Change: The Green Movement Worldwide*, Jai Press, Greenwich, CT.

Meadows, D. H. (1979) *The Limits to Growth*, Universe Books, New York.

Moriarty, F. (1988) *Ecotoxicology: The Study of Pollutants in Ecosystems*, Academic Press, San Diego.

Orfila, M. J. B. (1815) *Traites des Poisons Tires des Regnes Mineral, Vegetal et Animal ou, Toxicologic Generale Consideree sous les Rapports de la Physiologic, de la Pathologie et de la Medecine Legale*, Crochard, Paris.

Pachlke, R. C. (1989) *Environmentalism and the Future of Progressive Politics*, Yale University Press, New Haven and London.

Paracelsus, P. A. T. B. von Hohenheim (1567) *On the Miner's Sickness and Other Diseases* (trans.). Dillingen.

Ramazzini, B. (1700) (*Discourse on the Diseases of Workers.*) De Morbis Artificum Diatriba: Typis Antonii Capponic.

Roberts, J. (1990) Meeting the environmental challenge. In *Towards a Just Society: The Trudeau Years*, eds. Axworthy, T. S., and Trudeau, P. E. Penguin Books, Markham, Ontario.

Rodricks, J. V. (1992) *Calculated Risks: The Toxicity and Human Health Risks of Chemicals in Our Environment*, Cambridge University Press, Cambridge.

Rubin, C. T. (1994) *The Green Crusade: Rethinking the Roots of Environmentalism*, Maxwell Macmillan International, The Free Press, New York.

Schumacher, E. F. (1973) *Small Is Beautiful*, Vintage Press, London.

Schuurmann, G., and Markert, B. (1998) *Ecotoxicology. Ecological Fundamentals, Chemical Exposure, and Biological Effects*, John Wiley and Sons and Spektrum Akademischer Verlag, copublication, New York.

Truhaut, R. (1975) Ecotoxicology – A new branch of toxicology. In *Ecological Toxicology Research*, eds. McIntyre, A. D., and Mills, C. F., p. 323, Proc. NATA Science Comm. Conf. Mt. Gabriel, Quebec, May 6–10, 1974. Plenum Press, New York.

(1977) Ecotoxicology: Objectives, principles and perspectives, *Ecotoxicology and Environmental Safety*, **1**, 151–73.

Van der Heijdden, H.-A., Koopmans, R., and Guigni, M. G. (1992) The West European environmental movements. In *Research in Social Movements, Conflicts and Change: The Green Movement Worldwide*, ed. Finger, M., pp. 1–19. Jai Press, Greenwich, CT.

1.9   Further reading

Social concerns for the state of the environment: Ashby, E. 1978. *Reconciling Man with the Environment*, Stanford University Press, Stanford, CA.

The Environmental Movement: Rubin, C. T. 1994. *The Green Crusade: Rethinking Environmentalism*, Maxwell Macmillan International, The Free Press, New York.

The Environmental Movement: Paehlke, R. C. 1989. *Environmentalism and the Future of Progressive Politics*, Yale University Press, New Haven and London.

2

○ ○

The science of environmental toxicology: Concepts and definitions

2.1 The development of environmental toxicology

2.1.1 An historical perspective on the science of environmental toxicology

In Chapter 1, we discussed some of the historical, social, and economic background that accompanied the development of environmental toxicology. Here, we address some of the scientific and technical aspects of the subject.

The connection between community living and human health problems has been recognised for many centuries, long before any cause-effect relationship was clearly understood. It was probably in the earliest centres of human civilisation that the first pollution control laws were enacted. Fully 2,500 years ago, Athens had a law requiring refuse disposal outside the city boundaries. Today, the disposal of "unwanted" products of humans, including sewage, domestic refuse, and industrial waste continues to occupy a large proportion of the resources of human communities, and many of the attendant problems continue to challenge environmental scientists, public health professionals, and engineers.

The development of methods for determining the impact of man on the environment has advanced on several different fronts. When assessing damage to an ecosystem, we are chiefly concerned with the route(s) of exposure of biota to toxic agents and the degree of effect. The latter will depend on the inherent toxicity of the chemical(s), their availability to sensitive segments of the ecosystem, and their persistence in the biosphere.

Toxic chemicals enter the environment through a great variety of human activities including the mining, smelting, refining, manufacture, use, and disposal of products. Natural products of organisms themselves may reach levels toxic to portions of the ecosystem (Figure 2.1).

2.1.2 An evolutionary perspective on environmental toxicology

When considering the impact on the environment of toxic chemicals and other agents developed by humans, differentiating between the anthropogenic component and products of natural biological and geological processes is not always easy.

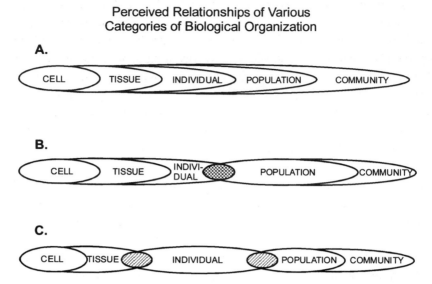

Perceived Relationships of Various
Categories of Biological Organization

A.

CELL TISSUE INDIVIDUAL POPULATION COMMUNITY

B.

CELL TISSUE INDIVI-DUAL POPULATION COMMUNITY

C.

CELL TISSUE INDIVIDUAL POPULATION COMMUNITY

Figure 2.1 A conceptual scheme describing the relationship between toxic chemical effects at different levels of biological organization. Shaded areas represent areas of uncertainty. For example, difficulties may exist in relating changes seen at the subcellular or physiological level to the survivability of an individual organism. Further difficulties may exist in extrapolating toxic effects on an individual (usually in laboratory assays) to effects seen at the population or community level.

An appreciation of how organisms have evolved both before and after human influence is important in understanding not only the extent of human impact but also the mechanisms through which organisms regulate their chemical environment. In this context, the metabolism of toxic chemicals may be seen as an extension of homeostatic mechanisms that maintain normal physiological processes. Where such mechanisms result in the elimination of a toxic chemical from its site of influence or the formation of a nontoxic by-product, we refer to the process as detoxification.

Most normal physiological functions of living organisms rely on a relatively few elements that are ubiquitous in the biosphere. Life probably originated in a strongly reducing atmosphere containing simple gaseous compounds of carbon, nitrogen, oxygen, and hydrogen. Although some bacteria retain an obligatory anaerobic metabolism, oxygen has become a constant and stable component (ca. 21%) of the Earth's atmosphere. Initially, a significant proportion of the oxygen in the atmosphere was probably formed by the photolytic effect of solar ultraviolet energy on water in the upper atmosphere, although photosynthesis is now by far the largest source of atmospheric oxygen.

Every year 146 billion tons of carbon combine with 234 billion tons of water to release over 400 billion tons of oxygen. Almost all of this oxygen is used by organisms to oxidise organic food molecules to carbon dioxide. The unique property of carbon atoms to combine with each other and a variety of other atoms creates the potential for vast numbers of organic chemicals, several million of which have been identified by chemists. Hundreds of thousands of synthesised organic chemicals have been added to the list, although the manufacture of organic chemicals only started at the beginning of the nineteenth century. Organic chemistry took a quantum leap forward in 1858 with the development of valency theory through independent publications by Couper and by Kekulé (Leicester, 1956). In 1865 Kekulé first described the structure of the benzene ring, and by 1880 over 10,000 organic chemicals were in production. It may be argued, then, that in fewer than 200 years humans have unleashed on the Earth a huge array of chemicals to which biota have had little time to adapt. Certainly none would argue that humans continue to have a dramatic effect on their environment. However, the adaptation of organisms to foreign chemicals long predates human influence.

Humans have contributed to toxic organic pollution ever since they deposited uneaten food and untreated excreta from themselves and their domestic animals within their operational living area and that of their neighbours. Early use of fire for heat and cooking would have made a modest contribution to airborne pollution, and probably within the last 10,000 years primitive smelting began. Many of these activities amounted to nothing more than rearranging and concentrating chemicals that were already part of the biosphere.

Organisms have adapted to many chemical and physical changes in their environment over several hundred million years and have evolved mechanisms to regulate and protect themselves from chemicals that may damage them. One legacy of life having evolved in a marine environment is that body fluids of even terrestrial and freshwater animals resemble salt water in several respects. Further, it is not surprising that the major ionic constituents of seawater (chloride, sodium, potassium, calcium, and magnesium) perform a variety of vital functions including maintenance of skeletal and cellular integrity, water balance, and neuromuscular function. Several enzymes require calcium and magnesium as cofactors, and many more are supported by a number of trace elements such as copper, zinc, and iron, which may be an integral part of the enzyme molecule or may act as electron receptors. For a physiologically useful element such as copper, the "window" between the amount required for normal physiological function and that which is toxic may be quite small (see Section 6.1 and Figure 6.1). Therefore, we can reasonably assume that regulatory mechanisms exist for physiologically useful trace metals just as they do for major electrolytes. However, the substrate concentrations involved and their scale of operation are much smaller. Consequently, these regulatory mechanisms are more difficult to characterise than the transepithelial pumps responsible for major ionic regulation.

When we refer to detoxification mechanisms, we may, therefore, be speaking about regulatory mechanisms that may serve a detoxifying function if the substrate (toxicant) has reached a potentially harmful level. Specific metal-binding proteins such as metallothioneins (see also Sections 4.3.2, 4.4.2, 6.2.6, and 6.6.2) probably both regulate and detoxify nutrient metals such as zinc and copper, yet apparently perform a solely detoxifying function for metals that play no physiological role such as mercury. Whether or not organisms were exposed to potentially toxic concentrations of these metals prior to human mining and smelting activity is a matter of speculation. There are, however, several instances of plant and animal populations existing in environments naturally enriched with trace metals. Grasses growing in the Zambian copper belt represent a well-known example. Therefore, concluding that some members of a population were preadapted for some degree of metal detoxification when elevated metal levels were introduced into the environment seems reasonable. Following man's influence, there are many examples of rapid development of metal-tolerant races of organisms through selection of detoxification and/or metal exclusion mechanisms (Mulvey and Diamond, 1991).

An even stronger case can be made for preadaptation to the metabolism and detoxification of toxic organic compounds. The following chapters contain several examples of relatively nonspecific mechanisms capable of handling a broad spectrum of both endogenous and exogenous organic compounds. The latter may include many of the novel compounds and natural compounds currently released from anthropogenic sources. Novel or newly synthesised compounds are commonly given the name xenobiotics (from the Greek word *xenos* meaning stranger). With some notable exceptions, newly synthesised compounds are often sufficiently similar to natural products that they mimic their toxic action and share aspects of their metabolism. Where waste products from human activity are similar or identical to naturally occurring chemicals, it is important from a regulatory standpoint to differentiate between contaminant and background levels.

2.2   Assessment of toxicity

To assess the potentially detrimental effects of a chemical (or other agent) on biota, it is necessary to establish a reproducible quantitative relationship between chemical exposure and some measure of damage to the organism or group of organisms under investigation.

Most current environmental toxicological data come from controlled laboratory assays and generally involve single chemicals and very small populations of test organisms. Even in a standard toxicity bioassay, an attempt is made to simulate what would happen in a large population by observing perhaps only 30 (10 organisms $\times$ 3 replicates) per exposure or treatment. Hypotheses concerning the mechanism of toxic action of a specific chemical or group of chemicals are typically tested at the cellular or subcellular level on the basis of which investigators often make assumptions about how a particular biochemical or physiological change may affect the overall fitness of the organism.

In linking the whole organism response to that of a population or community, we are also making assumptions about how toxic responses of individuals may be reflected at higher levels of biological organisation. Such an approach is embodied in the "nested" scheme shown in Figure 2.1A where cellular effects imply effects at all levels of biological organisation. Clearly, this represents an enormous oversimplification. Two particular areas of uncertainty are represented by the shaded areas in Figures 2.1B and 2.1C. For example, a suborganismal response such as an alteration in enzyme activity or immune response may represent a healthy reaction to a stress; therefore, finding a quantitative relationship between such a response and the "fitness" of an organism may be difficult (see discussion of biochemical markers in Section 4.4). Extrapolation from the individual to the population or community (Figure 2.1C) presents a different set of problems. For example, the disruptive effects of humans on the ecosystem must be viewed within the context of natural physical and chemical variations within the environment, such as climatalogical effects. In addition, organisms may become adapted to man-made pollution in the same way that they may adapt to numerous other environmental variables such as temperature, salinity, and food and oxygen availability.

A fundamental aspect of toxicological investigation is the relationship between the amount of chemical exposure and the degree of toxic response. This dose-response relationship is usually explored using a toxicity bioassay, the principal features of which follow.

2.2.1 The dose-response

The relationship between chemical exposure and toxicity is fundamental to toxicological investigation. Typically, the relationship is characterised from the relationship between two variables: dose and response. A dose is a measure of the amount of chemical taken up or ingested by the organism and may be quantified or estimated in a number of different ways (Table 2.1). From Table 2.1, we can see that dose may be broadly or narrowly defined. In toxicity bioassays involving terrestrial vertebrates, dose may be precisely circumscribed by injecting an exact amount of chemical into the organism. In several environmental toxicological studies, however, dose is implied from a measure of chemical concentration in the exposure medium (e.g., air, water, or sediment) and a knowledge of total exposure time. Together, these two parameters provide a measure of exposure rather than dose, and, even though the term dose-response is commonly used to describe laboratory-based bioassays, what we are often determining is an exposure-response relationship. The term exposure has great relevance to a field setting, where estimates of ambient chemical concentration and exposure time provide the basis for risk estimation when coupled with data from laboratory and other controlled exposures.

Toxicity bioassays are performed by exposing a representative population of organisms to a range of chemical concentrations and recording responses, or end-points, over a period of time. Responses may be an all-or-none (quantal) phenomena such as mortality, or they may be graded effects such as growth or reproductive performance (fecundity). An end-point may be any quantifiable response that can

Table 2.1. Concept of toxic dose: Ways by which toxic chemical dose[a] is measured or estimated in different applications

A measured amount of chemical may be injected into the tissue/extracellular space of the test organism (e.g., premarketing test for therapeutic agent).

Chemical may be administered as part of a controlled dietary ration (e.g., therapeutic agent/food additive/pesticide testing).

Chemical may be administered as a one-time dose in the form of a skin patch for cutaneous testing (e.g., testing therapeutic, cosmetic agents).

Ingested chemical may be analysed at target tissue or in surrogate tissue/excretory product having predetermined relationship with receptor site/tissue: blood, urine, hair (e.g., therapeutic agent/pesticide testing, industrial exposure studies).

In small organisms, whole body burden may be measured, although important information on tissue distribution is lost (e.g., skeletal chemical composition may be very different from soft tissues and may be misleading if included).

Dose may be inferred from the product of exposure concentration and time. This estimation is important in many macrophytic plant assays where the toxicant is a gas that may not bioaccumulate (see also ppm hr). It is common in aquatic toxicity testing.

Lethal body burden (critical body residue) is measured at the time of death. It may be related to a laboratory bioassay or be a result of intoxication in the field.

[a] See Section 8.2.2 for a discussion of ionizing radiation dose.

be related to chemical dose or exposure and may include changes in enzyme activity, tissue chemistry, pathology, or even behavioural changes.

The population response may be viewed at any one test concentration over time or at any prescribed time (t) over the concentration range. In either case, a quantal response such as mortality usually follows a normal distribution. For example, if percent mortality at time (t) is plotted against toxicant concentration on an arithmetic scale, a sigmoid curve results. In Figure 2.2, this is superimposed on a mortality frequency plot. The Gaussian curve essentially reflects a normally distributed population response with early mortalities among more sensitive individuals and prolonged survival of the most resistant organisms.

For statistical reasons we are most interested in the centre of the response curve, specifically the point at which the mortality of the test population is 50%. If a vertical line is drawn through this point, it intersects the abscissa at a concentration termed the LC (Lethal Concentration)$_{50}$. This is the toxicant concentration, which kills 50% of the test population at time (t). As such, it represents an estimate of the concentration that would cause 50% mortality of the infinitely large population from which the test population is taken. Such an estimate is strengthened by the use of larger test populations and appropriate replication of the assay. Preferably, such an assay should be conducted at least in triplicate, although it is recognised that large-scale replication presents logistical problems. In Figure 2.2, the mortality frequency curve follows the typical bell-shaped (Gaussian) distribu-

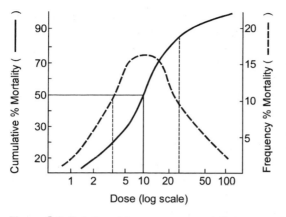

Figure 2.2 Relationship among cumulative percentage mortality, frequency percent mortality, and dose.

tion describing the overall population response. The LC_{50} represents the mean response of the population. In a normally distributed test population, 68.3% of the population is included within ± 1 standard deviation (SD) about the mean (or median). This is the region of the curve between 16 and 84% (Figure 2.2). The mean $\pm 2SD$ includes 95.5% of the test population, and the mean $\pm 3SD$, 99.7% of the test population. Various mathematical devices have been used to straighten out the response curve so that this central portion of the curve can be examined in a standardised manner. The most commonly used transformation is the probit transformation where percentage mortality is plotted on a probability scale on the y-axis versus log chemical concentration on the x-axis (hence the logarithmic range of chemical concentrations used for the test). The plot is based on the normal equivalent deviate or NED (equivalent to the standard deviation about the mean), initially proposed by Gaddum (1933) and then modified to the probit unit by Bliss (1934a, 1934b) by adding 5 to avoid negative numbers (Figure 2.3). The typical plot encompasses 6 probit units and results in a mortality curve that theoretically never passes through 0 or 100% mortality.

In addition to determining the LC_{50}, it is also desirable to obtain values for the 95% confidence limits and the slope of the line. Methods for calculating these parameters are described by Litchfield and Wilcoxon (1949) and Finney (1971). Probably the first and most commonly used method for determining the LC_{50} was the probit transformation for the analysis of quantal assay data, which, as we have seen, generates a sigmoid curve when plotted against the log of concentration. Empirical studies, however, suggested that probit transformation was not always the optimal method to use, and Finney (1971) showed that the method could be laborious and time-consuming. Subsequently, Gelber et al. (1985) proposed the logistic function (Hamilton et al., 1977), the angular arc-sine transformation of the percent mortalities, and a combination of the moving average interpolation with the latter as competitors to the probit transformation. Not until the 1970s did a

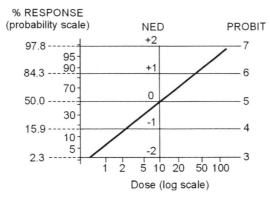

% RESPONSE
(probability scale)

Figure 2.3 The development of probits from the log probability dose
response relationship. NED ≡ normal equivalent deviations (numerically
equivalent to standard deviations).

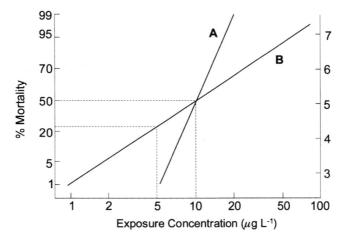

Figure 2.4 Dose response for two chemicals A and B having the same LC$_{50}$
but different slope functions.

nonparametric technique, the trimmed Spearman-Karber method, come into widespread usage. The method produced fail-safe LC$_{50}$ estimates but was not applicable in its original form if the bioassay yielded test concentrations with no partial
kills. The method could not be used to generate confidence limits.

Calculated slopes can provide useful additional information on the mode of toxic
action. Figure 2.4 shows probit-transformed response curves for two chemicals A
and B. In the case shown, the LC$_{50}$s are identical, although the response for A
exhibits a much steeper slope than that for B. The steep slope may signify a high
rate of absorption of chemical A and indicates a rapid increase in response over a
relatively narrow concentration range. By contrast, the flat response curve for B
suggests a slower absorption or, perhaps, a higher excretion or detoxification rate.

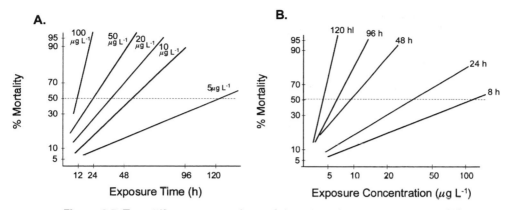

Figure 2.5 Two different expressions of the same dose response data. (A) Mortality at different toxicant concentrations is recorded over time. (B) Mortality at set time intervals is recorded for different toxicant concentrations.

Notwithstanding the fact that LC_{50}s for A and B are both $10 \, \mu g \, L^{-1}$, the steeper slope for A might, at first sight, suggest a greater degree of toxicity associated with this chemical. However, in environmental toxicology, we are increasingly concerned with toxic effects of low concentrations of chemicals (i.e., those less than the LC_{50}). In the lower left portion of the graph, we note that, at concentrations half of the LC_{50} ($5 \, \mu g \, L^{-1}$), chemical A kills less than 1% of the test population whereas the mortality associated with chemical B is still as high as 20%.

All the foregoing methods for calculating LC_{50} tend to agree very well, but there is much less agreement at the upper and lower ends of the curve (e.g., LC_{10} and LC_{90}). A fundamental weakness of the LC_{50} is that, in focusing on the statistically more robust centre of the curve, where confidence limits are relatively narrow compared with the mean, potentially useful information is lost at the two ends of the curve. This is particularly true at the lower end, where we are interested in the lowest concentration of a chemical that elicits a toxic effect. This is known as the threshold concentration.

Information from acute toxicity tests can be expressed in one of two basic ways. Cumulative mortality may be plotted over (log) time for different chemical concentrations. Alternatively, log concentration is used as the x-axis, and plots are made of differential mortality at different time intervals. In Figure 2.5, the same data have been plotted both ways. Time-based plots, favoured particularly by European laboratories, occasionally illustrate instances where toxicity apparently ceases after an initial die-off (Figure 2.6). The point of inflection may occur anywhere along the curve and indicates that the chemical is only toxic to a part of the test population at that concentration. In some rare data sets, mortality may proceed at a different rate following an inflection point. An example is the split probit plot (Figure 2.7), which is probably indicative of a change in mode of toxic action by the chemical. The concentration-based plot (Figure 2.5B) illustrates how the LC_{50}

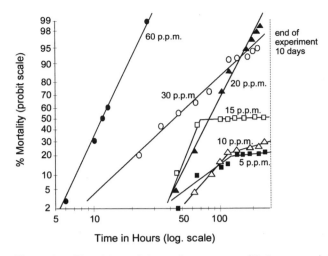

Figure 2.6 Fluoride toxicity to brown trout (*Salmo trutta*) fry showing curtailment of mortality at lower F⁻ concentrations. From Wright (1977).

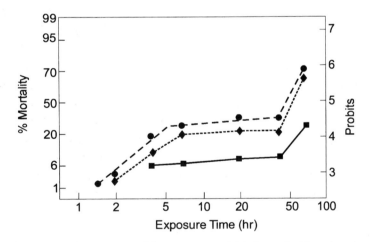

Figure 2.7 The split probit response, which is probably related to different toxic actions varying in intensity.

becomes progressively smaller at longer exposure time, although the incremental decrease in LC_{50} diminishes as the time of the test is extended. When viewed in this way, it becomes clear that there is a point where longer exposure does not lead to any further change in LC_{50}. This concept is illustrated in Figure 2.8, where toxicant concentration is plotted against the time to reach 50% mortality. This time is referred to as the median lethal time. The concentration at which the curves become asymptotic to the time axis is referred to as the threshold, incipient LC_{50}, or incipient lethal level.

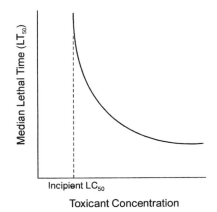

Incipient LC$_{50}$

Toxicant Concentration

Figure 2.8 Relationship between median lethal time and dose, showing derivation of incipient LC$_{50}$.

2.2.2 The acute toxicity bioassay

The most commonly used test is the so-called acute toxicity bioassay wherein cumulative mortality is recorded over a 48- or 96-hr period. Sprague (1973) has advocated more frequent observations at the beginning of the test, although, typically, mortalities are recorded at 24-hr intervals. The time course for such an assay has been arbitrarily chosen for the convenience of the investigator and has been accepted as one of a number of "standard practices" designed to assist comparability of data between chemicals, between organisms, and between investigators. If no background information is available on the probable toxicity of the chemical to be tested, an initial "range-finding" test is performed using a wide range of concentrations, which may cover several orders of magnitude. As a result of preliminary data, this test is reduced to a more narrow, logarithmic, range of concentrations. Usually, at least four chemical concentrations plus a (chemical-free) control are included in the definitive test.

In an aquatic assay, organisms may remain in the same medium throughout the test, in which case it is termed static. Flow-through tests are performed using metering devices (dosers) designed to deliver an appropriate range of chemical concentrations on a once-through basis. The flow-through test better maintains the integrity of the chemical concentrations where uptake, degradation, evaporation, and adsorption to container surfaces may lead to the depletion of the test chemical in the exposure medium and prevents build-up of waste products from the test organisms. A compromise is the static-renewal test, where some or all of the test medium is replenished periodically.

2.2.3 Subacute (chronic) toxicity assays

A variety of approaches has been used to investigate subacute toxicity. This term is used here to describe toxic effects at concentrations less than the acute LC$_{50}$.

Mortality may still be used as an end-point but may involve longer exposures than 96 hr. In this regard, the term *chronic toxicity* is often used. However, in many texts, the term *chronic* is conventionally reserved for a group of sublethal bioassays involving graded end-points such as level of biochemical activity, fecundity, or growth.

Although the sensitivity of a bioassay may be augmented by lengthening the exposure time, the use of mortality as an end-point remains a somewhat crude measure of toxicity. Considerable improvements in sensitivity may be achieved by employing the most susceptible stages of the life-cycle (e.g., embryos and larvae). However, the most sensitive bioassays have usually involved long-term (chronic) exposures and a variety of subtle end-points. In the scientific community, a good deal of debate over the selection of these end-points has occurred, particularly in terms of how they relate to mortality. Several dose-response studies deal with mechanistic aspects of toxic action where the linkage between the end-point and the health of the organism is not fully established (see Sections 4.4.1 and 4.4.2). Nevertheless, most sublethal assays involve parameters clearly linked to productivity such as growth and reproduction. Although the terms acute and chronic have been used in association with lethal and sublethal end-points, respectively, the time-based distinction between these two categories of test has become blurred. Chronic assays, which were originally designed to cover the complete life-cycle of the organism, or a substantial proportion of it, have been largely replaced by shorter tests, sometimes no more than a few weeks in duration.

For example, for some species of fish, so-called egg-to-egg tests might last more than a year, and have been superseded by early life stage tests, which go from the egg or embryo through to the larval or juvenile stage. Such tests may include multiple end-points such as the delay of onset and/or completion of hatch, percentage abnormal embryos, delays in swimming/feeding activity, abnormal swimming behaviour, abnormal or retarded metamorphosis to juveniles, and reduction in growth rate.

In the last two decades, increased efforts have been made to expand sublethal tests to a broad spectrum of organisms from a variety of aquatic and terrestrial habitats. Some of these assays are summarised in Table 2.2. In the aquatic environment, invertebrate assays figure prominently. One reason for this emphasis is the relatively short life-cycle of mysids such as *Mysidopsis bahia* and ostracods such as *Daphnia magna*. Another reason is the improvement in conditioning and spawning techniques facilitating the availability of gametes for much of the year. This availability has enabled the development of fertilisation tests for sea urchins and bivalve molluscs as well as single gamete (sperm) assays.

In keeping with the objective of characterising threshold levels of toxicity as far as possible, a primary objective of the sublethal (chronic) bioassay is the determination of what is called the maximum acceptable toxic concentration (MATC). The MATC has been defined as the geometric mean of two other values: the no observed effect concentration (NOEC) and the lowest observed effect concentration (LOEC). The NOEC is the highest concentration having a response not significantly different from the control, and the LOEC is the lowest concentration that shows significant difference from the control. Sometimes these concentrations are referred to as the no observed effect level (NOEL) and the lowest observed effect level (LOEL). In most cases, it is possible to calculate a median effective concentration or EC_{50}. This is analogous to the LC_{50} and is the chemical concentration causing a sublethal response in 50% of the test population. In each case, the significance (or otherwise) of the response is measured by comparison with the mean control value but is counted in a quantal way (e.g., an organism is either scored as normal or abnormal).

Another useful way of expressing graded sublethal results is in terms of percentage inhibition concentration (IC_p). For example, a chemical concentration causing a 30% inhibition of hatch or growth rate (relate to controls) is referred to as the IC_{30}. The IC_p has the advantage of flexibility insofar as degrees of response can be compared.

2.2.4   The relationship between acute and chronic toxicity

Because of the relative ease and economy with which acute toxicity data can be obtained, this type of information is much more commonly and widely available than chronic data. Although the gap is being closed with more abbreviated "chronic" assays, such as the early life stage test in the aquatic environment and a variety of screening tests for mutagenicity and teratogenicity, a great deal of attention is still given to determining the relationship between short-term and long-term toxic effects. The acute-to-chronic ratio (ACR), the ratio of acute LC_{50} to a measure of chronic toxicity (e.g., MATC),

$$ACR = \frac{Acute\ LC_{50}}{MATC}$$

has been used to extrapolate between different species and different chemicals where chronic toxicity data are unavailable.

Thus, an acute LC_{50} divided by an ACR determined for a similar species will provide an estimate of chronic toxicity. A high value for this ratio indicates that the chronic toxicity of a chemical may not be satisfactorily determined from its acute toxicity and may indicate two quite different modes of toxic action in both the short term and the long term. Kenaga (1982) found acute-to-chronic toxicity ratios varying from 1 to over 18,000, although for 93% of the chemicals surveyed the mean ratio was about 25. The ACR represents a somewhat crude means of estimating chronic toxicity in the absence of more appropriate data, but it probably has its place in environmental toxicology as long as acute toxicity tests form the basis of toxicological assessment. The aim of this procedure is to provide a regulatory tool for affording maximum protection for the largest number of species without spending precious resources performing toxicity tests on very large numbers of organisms.

Table 2.2. Bioassays in common use in environmental toxicology

Test species	Test/conditions	End-point
	Aquatic – Freshwater	
Teleosts		
Fathead minnows (*Pimephales promelas*)	48/96h acute bioassay	Mortality, % hatch, swim-up time, % normal larvae, growth
	21–32 d early life stage (20–25°C)	
	Modified as sediment-related test in United States	
Bluegill (*Lepomis machrochirus*)	48/96h acute bioassay	Mortality, % hatch, swim-up time, % normal larvae, growth
	28 d early life stage (28°C)	
Brook trout (*Salvelinus fontinalis*)	48/96h acute bioassay	Mortality, % hatch, swim-up time, % normal larvae, growth
	28 d early life stage (28°C)	
Rainbow trout (*Oncorhynchus mykiss*)	48/96h acute bioassay	Mortality, % hatch, swim-up time, % normal larvae, growth
	30–60 d early life stage	
Lake trout (*Salvelinus namaykush*)	60–72 d early life stage	% hatch, swim-up time, % normal larvae, mortality, growth
Amphibians (*Xenopus laevis Rana* spp.)	96 h embryonic exposure (FETAX test)	Malformation of head, gut, skeleton (teratogenicity)
	Sediment – Freshwater/marine	
Amphipod		
Hyallella azteca	Freshwater	
	8 parts water: 1 part sediment	
	10 d assay	Mortality, growth
	28 d assay	Mortality, growth
Leptocheirus plumulosus	Brackish water (5–20‰)	
	8 parts water: 1 part sediment	
	10 d assay	Mortality, growth
	28 d assay	Mortality, growth
Ampelisca abdita	Marine (>25‰)	
	8 parts water: 1 part sediment	
	10 d assay	Mortality, growth
	28 d assay	Mortality, growth
Polychaetes		
Neanthea arenaceodentata	Marine	
	8 parts water: 1 part sediment	
	96 h assay (20°C)	Mortality, growth
Crustaceans – Ostracods		
Ceriodaphnia dubia	48 h acute bioassay	Mortality, mean no. and survival of larvae from 3 broods
	10–14 d fecundity per surviving female (20°C)	
Daphnia magna,	48 h acute bioassay	Mortality, mean no. and survival of larvae from 3 broods
D. pulex	7–10 d fecundity per surviving female (20°C)	
	21–28 d life-cycle test	Survival and brood size of F0. No. of F1 generation
Microalgae		
Selenastrum capricornutum	96 h biomass production from concentrated innoculum	Cell counts, dry weight, chlorophyll a, ^{14}C assimilation

34

Saline water		
Teleosts		
Sheepshead minnows (*Cyprinodon variegatus*)	48/96h acute bioassay	Larval mortality
	7 d growth assay	Larval growth
	30 d life-cycle test (25–30°C)	Hatch, embryo/larval
		Development, growth, mortality
Mummichogs (*Fundulus heteroclitus*)	48/96h acute bioassay (20–25°C)	Larval mortality
Turbot (*Scophthalmus maximus*)	48 h acute bioassay (15°C)	Yolk sac, larval mortality
Crustaceans		
Grass shrimp (*Palaemonetes pugio*)	48/96h acute bioassay (20°C)	Larval mortality
Gulf shrimp (*Mysidopsis bahia*)	7 d fecundity test (27°C)	No. of eggs per female
	28 d life-cycle test (27°C)	Growth, survival of F0, no. of F1 per female productive day
Acartia tonsa (copepod)	96h acute bioassay (20°C)	Adult mortality
Eurytemora affinis (copepod)	14 d life-cycle test (starting with young naupliar larvae) 20°C	Mortality of F0, no. and survival of F1 generation
	Brackish water conditions (5–15‰)	
	96h acute bioassay (20–25°C)	Adult mortality
	10–14 d life-cycle test (starting with young naupliar larvae)	Mortality of F0, no. and survival of F1 generation
Bivalve molluscs		
Pacific oyster (*Crassostrea gigas*)	48 h embryo/larval assay	Survival to the shelled (D-hinge) stage
	Exposure begins with fertilized eggs (20°C)	
Echinoderms		
Atlantic purple sea urchin (*Arbacia punctulata*)	Short-term (10–20 min) sperm cell exposure (10–15°C)	Fertilisation of unexposed eggs
Sediments/aquatic		
Dipteran larvae		
Chironomus tentans	fresh – brackish water 10 d assay 28 d partial life-cycle (from 2nd/3rd instar)	Larval mortality
		Adult emergence from pupa
Terrestrial		
Birds		
Mallard (*Anas platyrhynchos*)	Acute oral; single dose followed by >14 d observations	Mortality (necropsy); feeding behaviour, other signs of intoxication; LD_{50} normalised to body weight
Northern bobwhite (*Colinus virginianus*) various avian species	Dietary LC_{50} (toxicant mixed with daily diet); c.10 d–10 weeks before egg laying	Mortality, no. of eggs, thickness of egg shell, hatching
Mammals		
Rat (*Rattus norvegicus*)	Acute oral; single dose followed by >14 d observations	Mortality (necropsy); feeding behaviour, other signs of intoxication; LD_{50} normalised to body weight
Rats and wild rodent (e.g., voles, *Microtus* spp.)	14–30 d LC_{50} (toxicant mixed with daily diet)	Mortality, breeding success

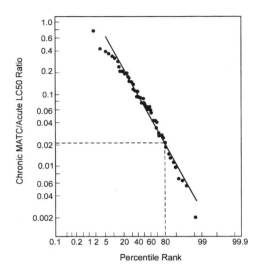

Figure 2.9 Probability plot relating application factors (MATC/Acute LC$_{50}$) to relative chemical toxicity. Data computed from the most sensitive couplet of three species, *Daphnia magna, Oncorhyncus mykiss*, and *Pimephales promelas* (Giesy and Graney, 1989).

One drawback of this approach is the variability in ACR seen between different species exposed to the same chemical. This drawback is due, in part, to the different mode of toxic action on dissimilar taxonomic groups. In many instances, a compromise is reached by pooling data from a small group of sensitive species. In the state of Michigan, for example, toxicity data from three species have been used as the basis of a probability plot relating MATC : acute LC$_{50}$ ratios, or application factors (AF) to relative chemical toxicity characterised in terms of percentile rank (Figure 2.9). In this figure, data points are computed from the most sensitive couplet of three species: *Daphnia magna*, rainbow trout (*Oncorhynchus mykiss*), and fathead minnows (*Pimephales promelas*). The objective is to determine an AF that can be satisfactorily used to protect a predetermined percentage (in this case 80%) of species through regulatory action. From Figure 2.9, an AF of 0.022 applied to acute toxic data would estimate a protective chronic toxicity value for 80% of compounds used to compile the data set. A similar approach is used to protect 80% of a broad species range. An advantage of the method is that it can define the degree of certainty and probability of protection desired when estimating safe or chronic, no-effect concentrations from acute toxicity data.

Mayer and co-workers (Mayer et al., 1994; Lee et al., 1995) have used regression techniques to extrapolate time, concentration, and mortality data from acute toxicity bioassays to estimate long-term effects. Computer software available from the U.S. Environmental Protection Agency laboratory in Gulf Breeze, Florida, essentially uses a three-dimensional approach to arrive at a no-effect concentration (NEC), which is operationally defined as the LC$_{0.01}$ from a probit curve at t_∞. Oper-

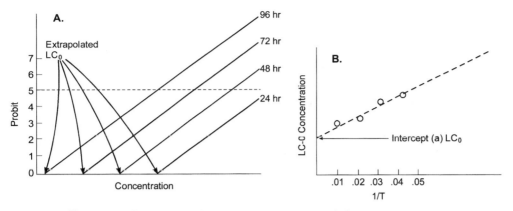

Figure 2.10 Derivation of no-effect concentration (A) by regression of dose response data at different times to determine LC_0 ($\equiv LC_{0.01}$) and (B) plot of LC_0 vs. $1/T$ to determine LC_0 @ t_∞ ($\equiv y$ NEC) (Giesy and Graney, 1989).

ationally, the program performs two separate regressions: initially determining the LC_0 at each time interval (Figure 2.10A) and then plotting these LC_0 values against reciprocal time to extrapolate to a hypothetical infinite exposure time (Figure 2.10B). NEC values derived from acute toxicity data for several fish species have been shown to correlate highly with corresponding MATCs from more time-consuming chronic bioassays (Mayer et al., 1994).

A useful variant of the acute-to-chronic ratio is the chronicity index (Hayes, 1975), which uses data obtained from dose response assays using vertebrates such as mammals and birds. Here, the chemical dose is accurately known, whether given intravenously or inserted into the stomach with a food bolus, and is commonly measured in milligrams per kilogram. Acute toxicity may be measured following a single, 1-day dose or may be assessed as a result of repeated doses up to 90 days. In both cases, the toxic end-point is expressed as the rate of dose (i.e., $mg\,kg^{-1}\,day^{-1}$). The chronicity index is the ratio of acute to chronic toxicity data expressed in terms of rate and is designed to detect cumulative toxic effects. Lack of cumulative toxicity is signified by a chronicity index of 1.

When toxic exposure is transient, difficulties may sometimes occur in characterising dose response in the environment in a satisfactory manner. Such a situation may arise in the case of a spill, when highly toxic concentrations may exist locally for a short period of time before they are dissipated. Exposure times and concentration may vary, and it is useful to have a flexible means of expressing toxicity. A typical example is an oil spill where the term parts per million ($mg\,L^{-1}$) hour has been used to characterise total petroleum hydrocarbon toxicity. Such a concept assumes that a toxic "dose" may be approximated by the product of concentration and time of exposure, as shown in Figure 2.11. Any number of rectangles may be subtended by the curve, each representing the same degree of toxicity (i.e., short exposure to high toxicant concentration or long exposure to low chemical levels). This approach may be applied to a variety of dose-response

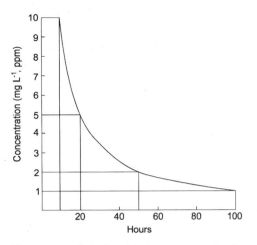

Figure 2.11 Relationship between toxic chemical concentration and exposure time. Subtended rectangles hypothetically have same toxicity value, although boundary conditions apply (see Section 2.2.4).

relationships, although its utility is clearly limited at either end of the curve. For example, at the high concentration/short exposure time end of the curve, toxicity will become limited by the kinetics of transepithelial transport of the chemicals or saturation kinetics at the receptor site (see Section 2.3.6), and at the low concentration/long exposure time end the boundary will be represented by the toxic threshold.

2.2.5   Statistical considerations

Statistics play two principal roles in environmental toxicology. One is in the process of hypothesis testing (i.e., determining the difference between two populations with some predetermined level of confidence). The other can be defined as predictive in nature (e.g., determining the outcome of chemical-biological interaction) and often involves the influence of one or more independent variables on one or more measures of effect [dependent variable(s)].

Statistical methods associated with bioassays are concerned with differences among a few very narrowly defined test populations [e.g., those exposed to different chemical concentrations (treatments) and differences between treatments and controls]. They are also concerned with the relationship between test populations and the larger population from which they were taken. Measurement of treatment/treatment and treatment/control differences requires calculations that take into account the variance associated with the observed end-points (e.g., mortality). Differences between or among treatments can be measured with varying degrees of precision, depending on the numbers of organisms and replicates used. Increases in both of these parameters will decrease the variance associated with end-point measurement and will increase the precision with which treatment differences can be determined. Precision is characterised in this case in terms of a confidence limit.

For example, most tests of population (treatment) differences allow a 5% error. This difference means that there is one chance in 20 of reaching the wrong conclusion based on the data obtained. Even though such an error can be lessened through increased replication of each treatment, there are clearly practical limits to the numbers of replicates that can be handled in an assay. For treatment populations to reflect accurately the overall population from which they were taken, care must be taken to avoid inappropriate clumping of treatments either within a laboratory assay or a field sampling program. This is called pseudoreplication. There are many different examples of this. One example is the selection of organisms for chemical body burden analysis from only one of the replicates. Likewise, replication of tissue sample analyses from a single individual is not true replication within the context of a bioassay. A good overview of pseudoreplication is provided by Hurlbert (1984).

For both lethal and sublethal data, control values need to be taken into account. Where control mortality is low, simple subtraction of control values from other treatments will suffice. However, organisms that spawn large numbers of offspring [e.g., *Arbacia* (sea urchin); *Crassostrea* (oyster); *Morone saxatilis* (striped bass)] often sustain large mortalities of embryos and larvae, even under control conditions, perhaps 30% over a 96-h period. In such cases, adjustment of a treatment value relative to a control is made using the formula of Abbott (1925):

$$\text{Adjusted \% response} = \frac{\text{Test \% response} - \text{Control \% response}}{100 - \text{Control \% response}} \times 100$$

For a 70% observed test response and a 30% control response, the adjusted response becomes

$$\frac{70 - 30}{100 - 30} \times 100 = 57\%$$

Often, sublethal (chronic) data are not normally distributed (as determined by a Chi-Square or Shapiro-Wilks test) or homogeneous (e.g., Bartlett's test). In such cases, results may be analysed using a nonparametric procedure such as Steel's multiple rank comparison test (for equal size replicates) or Wilcoxon's rank sum test (unequal replicates). Normally distributed data can be analysed using Dunnett's test. In certain instances, nonnormally distributed data may be transformed to a normal distribution using a mathematical device that has the effect of lessening differences between data points. One such transformation is the arcsine transformation, which still has a role where statistical treatment demands normal distribution of data. One example is multiple regression analysis, which may be appropriate where more than one variable is being tested. For example, if toxicity due to metal exposure is being examined at different metal concentrations over time at different temperature and salinities, metal concentration, exposure time, temperature, and salinity are all regarded as independent variables and the toxic

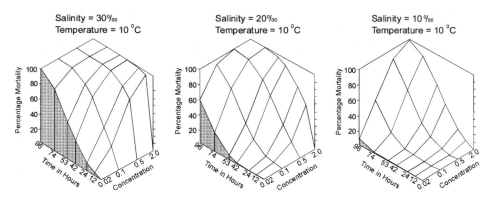

Figure 2.12 Response surfaces describing toxicity of trimethyltin to hermit crab (*Uca pugilator*) larvae at different temperature/salinity combinations. Data from Wright and Roosenburg (1982).

end-point, say, mortality, is the dependent variable. In such an instance, the regression analysis is performed on arcsine-transformed results, which are then back-transformed for presentation in a regression equation, or as response surfaces. Figure 2.12 shows response surfaces describing the mortality of hermit crab larvae (*Uca pugilator*) as a function of trimethyltin concentration and exposure time in the different salinities (Wright and Roosenburg, 1982). A comprehensive account of appropriate statistical methods and procedures is given by Newman (1995).

Statistical treatments of acute toxicity tests. Historically, the LC_{50} that forms the basis of the acute toxicity test has been determined using either the parametric, moving average, or nonparametric technique. The first is based on obtaining a known parametric form for the transformed concentration (dose) mortality relationship, whereas the second is based on numerical interpolation where a moving mortality rate is obtained by calculating the weighted sum for each concentration, and the third uses the data to generate an empirical curve that is based on the symmetry of the tolerance distribution. These techniques are discussed in greater detail in Section 2.2.1.

Statistical treatments for chronic bioassays. For many years, the use of the analysis of variance (ANOVA) F test and Dunnett's procedure has formed the basis for the statistical assessment of chronic toxicity tests. Gelber et al. (1985) outlined the rationale behind the use of these procedures and highlighted some of the potential difficulties associated with these tests.

A review of the literature reveals no clear consensus on the most appropriate statistical procedures to use with chronic toxicity data. The American Society for Testing and Materials (1992) issued a guideline for toxicity tests with *Ceriodaphnia dubia*, which stated that several methods may be suitable. Recommended procedures for reproductive data included analysis of variance, t-tests, various multiple comparison tests, regression analysis, and determination of the test concentration associated with a specified difference from control. The latter is purported to represent a biologically "important" value. Opinions differ as to what this biologically important value should be. The U.S. Environmental Protection Agency recommends the use of an IC_{25} (i.e., toxicant concentration causing 25% reduction in reproductive output relative to controls), although such a figure must be regarded as somewhat arbitrary pending more detailed information on the reproductive strategy of the test species concerned. The ASTM (1992) guideline also described methods for detecting outlying data values and for transforming response data to obtain homogeneous variances. Nonparametric methods were recommended for heterogeneous data.

An increasing concern on the part of many toxicologists is the overuse and misuse of multiple comparison procedures in chronic toxicity tests. Although many tests are highly appropriate in, say, field situations where the toxicities of various sites are compared to a control or reference site, they are inherently poorly suited to analysing data from experiments where treatments form a progressive series (e.g., the serial dilution of a toxicant in a bioassay). A characteristic of Dunnett's test, for example, is that its relative immunity to false positives in a dose-response situation is obtained at a cost of reduced sensitivity (i.e., more false negative readings). Nevertheless, this procedure is part of a battery of tests recommended by the U.S. EPA for data analysis of chronic assays of effluent toxicity and is incorporated into several computer software packages serving this purpose.

Today, researchers are faced with the dilemma of choosing the most appropriate statistical approaches to test data from acute and chronic bioassays in light of an expanding range of statistical tests and computer programs. Software packages (such as SAS®, Toxstat®, ToxCalx, and TOXEDO) present a number of different alternative methods for the statistical analysis of data sets, and an application using SAS/AF software has been recently developed to estimate the sample size, power, and minimum detectable differences between treatments for various statistical hypotheses and tests of differences between means and proportions.

TIME RESPONSE (TIME-TO-DEATH) MODELS
An increasingly attractive alternative to the dose-response approach is the time response or time-to-death model, wherein the mortality of test organisms is followed for a period that typically extends beyond the duration of an acute bioassay. The approach is similar to that illustrated in Figure 2.5B, except that the time scale of the assay may embrace most or all of the life-cycle of the organism. As such, the assay may be defined as chronic but utilizes more information than the chronic assays described previously. Where observations are made over the complete life-cycle, the organism may grow through different stages (embryo, larva, juvenile, adult). Developmental data from different stages, together with information on mortality and reproductive performance, are collectively known as vital rates (see

Caswell, 1996) and form the basis of demographic models of population structure (Section 4.5.3).

The life table integrates all the stage-specific developmental data and can be used as the basis for a set of analyses variously called survival analysis or failure time analysis. The advantage of these methods over dose-response relationships is that they may incorporate survival times for all invididuals within each treatment group rather than select an arbitrary timed observation such as percentage killed by 96 hr. Thus, more information is utilized and can result in better estimates of mortality rates and covariates.

A critical parameter in this regard is the cumulative mortality distribution function, $F(t)$, which has a value of zero at $t = 0$ and 1 at 100% mortality. The inverse of this is the survival function, $S(t)$, which is equal to $1 - F(t)$. The hazard function, $(h)t$, describes the probability of percentage survival at time t and is mathematically related to the survival function as follows (Dixon and Newman, 1991):

$$h(t) = -d \log S(t) / dt = \frac{-1}{S(t)} \cdot \frac{dS(t)}{dt}$$

The hazard function is useful in determining whether toxicity is constant over time, or whether it accelerates or decelerates as exposure continues. As such, it may determine the choice of the model that best fits the results of a time-to-death assay. Typical regression analyses, accompanied by ANOVA, which assume normal distribution of data and a variety of different distributions have been used. For example, variants of the Weibull distribution have been used to accommodate time-to-death data that are skewed either toward a preponderance of deaths earlier in the assay or to accelerated mortality as the assay proceeds. If no specific distribution can be satisfactorily assigned to the results, a semiparametric proportional hazard model, the Cox proportional hazard model, is commonly employed. If the underlying time-to-death relationship can be satisfactorily described by a specific distribution model, results can be treated using parametric methods such as proportional hazard or accelerated failure time. The latter model can incorporate covariants such as may modify toxicity as the assay progresses. Likewise, the model can accommodate comparisons of two or more different strains of organism, which may have differential tolerance of the toxicant. Several computerized procedures are available for time-to-death data analysis. A typical example is the SAS PROC LIFEREG program, which describes the relationship between failure time (time-to-death) and covariates in the following form:

$$\ln(t_i) = f(x_i + de_i)$$

where t_i = time-to-death of individual I, x_i = vector of p predictor variables (x_{i1}, x_{i2}, ..., x_{ip}), $f(x_i)$ = linear combination of covariates ($\beta_0 + \beta_1 x_{i1} + \beta_2 x_{i2}, + ... + \beta_p x_{ip}$), d = scale parameter, e_i = error term for individual I.

In this analysis, individuals that have not died by the end of the assay are taken into account through censoring. Censoring is a general term that describes a common phenomenon in time-to-death studies (i.e., organisms not reaching the end-point of interest at the time of record). A variety of statistical methods for handling such data is available in the literature (e.g., Cox and Oakes, 1984). The reader is directed to accounts of time-response models in the context of environmental toxicology by Dixon and Newman (1991) and Newman and Dixon (1996).

Generalised linear models. Many toxicologists are moving towards the generalised linear models (GLM) approach, which produces more accurate lethal concentration and effects concentration estimates and confidence intervals. Rather than using regression analysis on data that have been transformed to fit a preconceived notion of how a test population might respond, the GLM approach accurately reflects the true nature of the data no matter how they are distributed. For example, 0 and 100% mortality are true end members of the response curve instead of an expanding logarithmic scale that never reaches absolute values at either end. Nyholm et al. (1992) pointed out that the tolerance distribution that results in the sigmoid curve (Figure 2.2) is particularly inappropriate for microbial toxicity tests where the response is typically a continuous variable and is much more suited to a GLM approach. An important advantage of the GLM method is that it provides a much more accurate picture of the response probabilities associated with all levels of toxicant concentration, not just around the 50% mark. The method supplies confidence intervals not only for the LC value but also for the true toxicant concentration that would give rise to a specified response. Kerr and Meador (1996) outlined several advantages of the GLM approach and concluded empirically that any LC value below LC_{10} calculated from their data set would be statistically indistinguishable from the control. They suggested that the method may be useful in determining the lowest observed effect concentration in chronic toxicity tests and proposed that the LOEC would occur where the horizontal line from the upper 95% confidence interval for zero concentration intersects the right 95% band of the model. For their model (Figure 2.13), this occurs at $16\,mg\,L^{-1}$, which becomes the LOEC.

As already mentioned, one of the principal advantages of using a GLM approach is that it allows researchers to focus on information imparted by toxicity data over the whole concentration range without having to fit the data into a preconceived distribution model. This is particularly important at the low end of the concentration range where the focus is on the threshold of toxic effect. GLM methods are also appropriate for time-to-death data described in the preceding section.

2.2.6 Comparative bioassays

Most of the features of the toxicity bioassay described previously are incorporated into a broad range of different types of assay covering a variety of media and circumstances. Bioassays are integral parts of several regulatory programs ranging

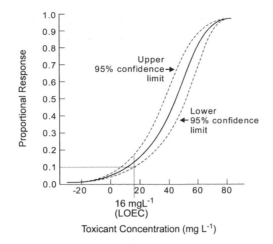

Figure 2.13 Generalised linear model approach for assessing lowest observed effect concentration (based on Kerr and Meador, 1996). The horizontal line intersecting the upper 95% confidence limit associated with zero concentration also intersects the lower confidence limit associated with the $16\,mg\,L^{-1}$ (the LOEC). Note that the response curve encroaches into "negative dose" territory, not uncommon for such empirical curves. Although the LOEC coincides with the LC_{10} for this data set, this may vary considerably with other empirical dose-response relationships.

from premarketing registration of new products to testing of water quality. Water quality assays have included a multitude of studies from university and government laboratories of known compounds, which have been used to produce water quality criteria. Increasingly, however, bioassays are being used to test groundwater, runoff, and effluent where the actual chemical composition of the medium may be unknown. In these circumstances, a range of "doses" is created by serial dilution of the environmental sample with clean (nontoxic) water, and toxicity is related to the degree of dilution. Assays of this nature are generally conducted in parallel with comprehensive chemical analyses of the sample, and both chemical and toxicological data are used to arrive at decisions on effluent treatment options. An extension of this approach is the toxicity identification estimate (TIE), also called toxicity reduction estimation (TRE), which involves testing and retesting an effluent before and after a series of chemical extractions and/or pH adjustments designed to isolate and identify the toxic fraction (U.S. EPA, 1991). In a review of this approach also known as whole effluent toxicity testing, Chapman (2000) makes the point that such tests identify hazard, not risk (see Chapter 10), and as such should only be regarded as the preliminary phase of a risk assessment process.

Birds and mammals. Toxicity bioassays on birds and mammals are often important components of preregistration testing of a variety of chemicals such as agrochemicals, therapeutic agents, and food additives. In these tests, the range of chemical exposure is created by serial dilution of an injected dose or by mixing

Table 2.3. Examples of types of end-points in use for plant assays

Nonlethal
Chlorophyll-a concentration
^{14}C uptake
Cellular effect (e.g., membrane permeability)
Growth (rate of increase of weight, volume, cell density, etc.)
Rate of change in leaf biomass
Sexual reproduction
Abundance (number of individuals)
Overwintering
Population growth rate
Acute, lethal
Mortality

Modified from Swanson et al. (1991).

different amounts of chemical with food. A more precise means of delivering an oral dose is used in the LD_{50} test, where a food + toxicant bolus is inserted into the stomach of the animal. Often, such tests are conducted on the basis of a one-time dose, normalised to body weight, followed by a 14-day observation period during which animals are fed uncontaminated diets. Mortality, growth (weight), and feeding behaviour are all used as end-points in these tests. A typical example is the bob-white quail test. It is common for such assays to form part of a tiered approach to regulatory toxicity testing. For example, if no adverse effects result from a dose of $2\,mg\,kg^{-1}$ in the foregoing bob-white quail assay, no further oral tests are performed. This dose also represents the cut-off dose for the standard cutaneous assay, where the chemical product is applied in the form of a patch applied to the shaved skin of the test species, usually a rabbit. Essentially, the assay has the same design as the LD_{50} test; a one-time, weight-normalised dose followed by a 14-day observation period. Depending on the proposed use of a particular product, genotoxicity assays may also form part of the chemical preregistration procedure. Examples of these are given in the following section.

Plant assays. Photosynthetic organisms can be used in acute or chronic assays of potentially toxic substances or effluents. Acute and chronic tests have been developed using various types of aquatic or terrestrial vascular plants as well as unicellular or multicellular algae. Tests using vascular plants frequently employ nonlethal end-points, such as root elongation of seedlings, and effects on development or physiological responses such as carbon assimilation rates. Table 2.3 provides a more comprehensive list of the types of end-points that have been used in plant testing.

Recently, tissue cultures (cell suspension or callous culture) have been proposed as test systems (Reporter et al., 1991; Christopher and Bird, 1992); using tissue cultures should avoid some of the variability that occurs and the long time periods required for whole plant tests, but may not provide such realistic results in terms of the environmental effects.

Higher plants can also be used for mutagenicity tests, which are normally done with microorganisms or higher animals. A system for the detection of gene mutations has been devised using the stamens hairs of spiderwort (*Tradescantia*). For chromosomal aberrations, the micronucleus of the spiderwort, along with root tips of bean, *Vicia faba*, have been employed. According to Grant (1994), "higher plants are now recognised as excellent indicators of cytogenic and mutagenic effects of environmental chemicals".

Unicellular plants, such as algae and blue-green bacteria, cultured in static or flow-through systems, provide population assays, since all the cells in the test are essentially identical and undifferentiated and have conveniently short generation (mitotic) times. For the most part, only a few species of algae have been employed in routine testing; of these, the green algae (Chlorophyta) *Selenastrum capricornutum* and *Chlorella vulgaris* are the most frequently cited. Parameters that can be used include cell density (direct counting or particle analysis), measurement of production through carbon fixation, or chlorophyll-*a* production (see Nyholm, 1985). The latter pigment, being common to all green plants, including the various groups of different coloured algae, is a particularly useful parameter, because it can be used in the field or laboratory, and in mixed- or single-species collections, at least in a comparative way, as a universal measure of plant production.

Some caution is needed in selecting the type of test to use when designing bioassays with algae. Although the physiological parameters such as carbon fixation rates can provide rapid results (generally hours, rather than days), measurements of growth by cell number may prove more sensitive. For *Selenastrum capricornutum*, short-term tests with endpoints such as CO_2 fixation rates or oxygen generation were less sensitive, for cadmium and simazine respectively, than population growth tests, Figure 2.14 (Versteeg, 1990). A more recent aquatic bioassay utilises clonal lines of Sago pondweed *Potamogeton pectinatus* and relies on respirometry under light and dark conditions to determine net photosynthesis rates. LC_{50} values are calculated from linear regression of log concentration and photosynthetic response expressed as a percentage of control (undosed) plants (Fleming et al., 1995).

Table 2.4 lists some of the ways in which plant responses can be used in testing and assessment. For the purposes of the present chapter, phytotoxicity will be emphasised. Table 2.5 provides information on several of the most widely used phytotoxicity tests. The application of other types of plant responses, such as their application as sentinels and monitors in environmental toxicology, are discussed in Section 4.4.3.

Plant tests often produce rapid results, and the procedures are often less expensive than, for example, rainbow trout assays, but their use should not be based simply on convenience. Selection of the type of organism and the type of test obvi-

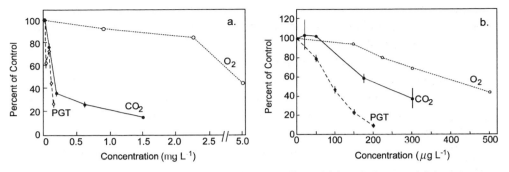

Figure 2.14 Three types of tests for the effect of (a) cadmium and (b) simazine on the green alga *Selenastrum capricornutum*. Based on Versteeg (1990). PGT = population growth tests; CO_2 = carbon fixation test; O_2 = oxygen generation test.

Table 2.4. Classes of environmental assessment analyses using plant responses

Class	Purpose	Example end-points
Biotransformation	Determine influence of plants on chemical fate of environmental pollutants	Change in chemical concentration
Food chain uptake	Establish the amounts and concentrations of toxic chemicals that enter food chains via plant uptake	Chemical concentration
Phytotoxicity	Evaluate the toxicity and hazard posed by environmental pollutants to the growth and survival of plants	Death, discolouration, reduced growth
Sentinel	Monitor the presence and concentration of toxic chemicals in the environment by observing toxicity symptoms displayed by plants	Death, discolouration, reduced growth
Surrogate	Use inexpensive, socially acceptable plant tests as a substitute for an animal or human assay	Chromosome aberrations

Source: Fletcher (1991).

ously should be geared to the questions of concern; furthermore, the relevance and potential for extrapolation to other plants as well as to the more general situation from plant responses need to be considered. This latter caution is not, of course, unique to plant assays. In all toxicity testing, the choice of organisms and type of test has to be made based upon a number of practical considerations, as well as concern for realism. In human or classical toxicology, we find the oft-repeated question of how useful rat (or guinea pig) studies are when considering the poten-

Table 2.5. Plant phytotoxicity tests use by regulatory agencies

Test	Response measured to chemical treatment	Agencies requiring limited use[a]
Enzyme assay	Enzyme activity	None
Process measurement	Magnitude of a process: photosynthesis, respiration, etc.	None
Tissue culture growth	Changes in fresh or dry weight	None
Seed germination (soil solution)	Percent of seeds that germinate	EPA (1985)[b] (50)[c] FDA (1981)[b] (25)[c] OECD
Root elongation (soil solution)	Length of root growth during fixed time period	EPA (1985)[b] (50)[c] FDA (1981)[b] (25)[c] OECD
Seedling growth (soil solution or foliar application)	Changes in height, fresh weight, and dry weight	EPA (1985)[b] (50)[c] FDA (1981)[b] (25)[c] OECD
Life-cycle	Changes in height, fresh weight, dry weight, flower number, and seed number	None

[a] EPA, Environmental Protection Agency; FDA, Food and Drug Administration; OECD, Organization for Economic Co-operation and Development.
[b] Year when the agency started requiring plant testing as a concern for potential hazard posed by chemicals to nontarget vegetation.
[c] A rough approximation of the current annual number of agency-requested data sets used for hazard assessment.
Source: Fletcher (1991).

tial toxicity of substances to humans. In many ways, the choice of test organisms in environmental toxicology is immensely more complex.

Comparative studies have been made, addressing the response to a particular chemical, among different types of plants or, more rarely, between the response of plants and higher organisms. Fletcher (1990, 1991) reviewed the literature on phytotoxicology with the objective of making comparisons among plant species and between algae and higher plants. He compared the variability related to taxonomic differences with that resulting from differences in test conditions and found large differences among taxa, above the level of genus. He concluded that taxonomic variation among plants in terms of the effects of chemicals had a greater influence than did the testing condition.

In the context of the relevance of algae as test organisms, Fletcher (1991) found that in 20% of the time, algal tests might not detect chemicals that elicit a response in vascular plants. This relative lack of sensitivity among algae may be explained in part by the fact that higher plants have more complex life-cycles than algae, and that a number of physiological phenomena in higher plants (e.g., development, leaf abscission) have no equivalent in algae. Comparing algal and higher plant tests for

21 herbicides, Garten (1990) showed, using *Selenastrum capricornutum* in 96-hr tests, that there was only a 50% chance of identifying herbicide levels that would reduce the biomass of terrestrial plants.

A few examples that compare plant responses with those of other organisms exist. Gobas et al. (1991) compared the response of guppies (*Poecilia reticulata*) with the aquatic macrophtye Eurasian water milfoil (*Myriophyllum spicatum*) to a series of chlorinated hydrocarbons. They concluded from their study that, since the toxicokinetics in both types of organisms involve passive transport and are related to the octanol-water partition coefficient, the acute lethality would be similar in plants and fish. Such generalisation would not be possible for inorganic toxicants or organics for which uptake could not be readily modelled. Normally, a more detailed consideration, based on the problem of concern, would be needed before plants could be used as surrogates for higher organisms.

Plants constitute more than 99.9% of the biomass on Earth (Peterson et al., 1996) and have intrinsic value of themselves. In the context of environmental toxicology, the eradication of plant communities can result in modified habitats and changes in trophic structure of ecosystems that may have more impact on higher trophic levels than the direct effects of a toxic substance. Thus, plants may be used in conjunction with animals in batteries of tests, in attempts to represent various trophic levels. Many regulatory agencies already require a plant and two higher trophic-level organisms to be used in screening "new" chemicals. The approach of incorporating several trophic levels in one test, as in microcosms and field enclosure tests, is even more realistic (see Section 4.7.1).

2.2.7 Sediment toxicity assays

By the late 1960s, concern over the deterioration of water quality in the Great Lakes prompted the Federal Water Quality Administration (the predecessor of the U.S. Environmental Protection Agency) to request the U.S. Army Corps of Engineers to initiate studies on the chemical characteristics of harbours around the perimeter of the Great Lakes system. Initially, harbour sediments were analysed using methods that were developed to characterise municipal and industrial wastes rather than sediments. In 1970, the Rivers and Harbors Act of 1890, which was originally written to protect access to navigable waters in the United States, was amended to include specific provisions to begin a "comprehensive evaluation of the environmental effects of dredged material" through the Dredged Material Research Program (DMRP). Under this program, the first effects-based testing approach was developed; it relies on the use of bioassays to integrate the potential effects of all the contaminants present in a complex sediment matrix.

The first published study of the potential toxicity of field-collected sediments was performed by Gannon and Beeton (1971). Experiments were specifically designed to test the effects of dredged material on freshwater benthic organisms. The amphipod *Pontoporeia affinis* was exposed in 48-h acute bioassays to five sediments collected from different sites. The tests were designed to assess sediment selectivity (behaviour) and viability (survival). Although survival data indicated a graded response to contaminated sediments, *P. affinis* clearly preferred a sediment particulate profile similar to their natural sediments. Over the succeeding 20 years or so, it became apparent that substrate preferences, salinity tolerance, and ease of culturing/availability were all-important considerations in the selection of test organisms. Candidate species for freshwater assays have included the burrowing mayfly *Hexagenia limbata*, the ostracod *Daphnia magna*, and the isopod *Asellus communis* (Giesy and Hoke, 1990). Early suggested species for bioassays in saline environments included the bivalves *Yoldia limatula* and *Mytilus* sp., the infaunal polychaetes *Neanthes* sp. and *Nereis* sp., and the crustaceans *Mysidopsis* sp. and *Callinectes sapidus*. Variants on the use of whole sediments have included sediment leachates and pore-water extracts/dilutions obtained through squeezing and centrifugation. Such techniques had the effect of transforming sediment toxicity tests into standard assays that could be carried out in aquatic media, yet they suffered the disadvantage of removing many of the unique characteristics of the sediments that affected sediment chemical bioavailability.

In most current sediment toxicity assays, there has been an increasing focus on the use of small infaunal (usually amphipod) crustaceans, which have proven to be among the most sensitive phyla to sediment contamination. Several different species have been considered depending on such attributes as substrate preference, salinity tolerance, and ease of culture. Both acute and chronic toxicity have been examined, and most protocols now include a standard 10-day bioassay. Fecundity and reburial behaviour have also been considered as end-points. Commonly used species include *Ampelisca abdita* (marine), *Repoxynius abronius*, *Leptocheirus plumulosus* (marine/estuarine), and *Hyalella azteca* (brackish/freshwater).

Along with the focus on standardized infaunal bioassays has come a reaffirmation of the idea that measures of ecosystem integrity should include indicators at the population and community level. Consequently, ecological risk assessments for sediment-associated chemical contaminants are universally based on a "weight-of-evidence" approach consisting of what is called a sediment quality triad (Figure 2.15) (Chapman, 1986). This triad uses a combination of sediment chemistry, macrobenthic community data, and results from single-species bioassays to assess the sediment condition. Such an approach has proven cost effective for regulatory purposes. However, in specific studies of sediments contaminated by potentially carcinogenic compounds (see Section 4.4.2), histopathology of benthic organisms (fish) has often been included in the investigation.

2.3   Toxicity at the molecular level

A major concern of environmental toxicologists in recent decades has been the effect of chemicals and other agents on the normal development of organisms and their offspring. By the 1960s, it was clear that reproductive performance of many

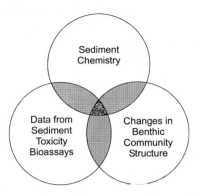

Figure 2.15 Components of a sediment quality triad (Chapman, 1986).

wild animals was being compromised by chlorinated pesticides and other similar compounds. Since then, research has elucidated the mechanistic basis of many of these phenomena and has identified major problems such as the disruption of endocrine and other chemical signals (see Section 7.5) and impairment of genetic transcription. Damage to the DNA molecule will result in abnormal somatic growth and the development of neoplasms (cancer).

Where DNA damage occurs in germinal tissue (sperm and eggs), degraded genetic material may be passed on to offspring. This, in turn, may result, directly, in decreased survival or fitness of these offspring. Changes in the genotype of succeeding generations are also likely to lead to subtle shifts in the organism's relationship to a variety of environmental parameters not necessarily toxicological in nature (see Section 13.5).

Thirty years ago, chemical carcinogenicity was largely viewed within the context of occupational exposure and was, therefore, seen principally as the purview of classical or clinical toxicology. This attitude is no longer the case. The identification of numerous surface water, groundwater, and sediment contaminants (e.g., solvents, pesticides, combustion, and chlorination products) as carcinogens has placed such studies squarely within the realm of environmental toxicology. For example, in 1998, the U.S. Environmental Protection Agency issued a report stating that 7% of coastal sediments in the United States were heavily contaminated with chemicals, many of which were carcinogenic in nature. Many studies of chronically polluted sites have focussed on neoplasm development in fish from these sites; in some instances, carcinogenic models using fish have been examined for their applicability to higher vertebrates.

Although regulations governing the handling and release of most toxic chemicals are formulated as a result of information from assays such as those described in Sections 2.2.1–2.2.7, cancer-causing agents (carcinogens) are treated differently. The concept of toxicity for carcinogens is founded on a knowledge of their mode of toxic action (i.e., mechanistic considerations). These modes shape our perception of dose-response and, in particular, toxic threshold. However, before this is

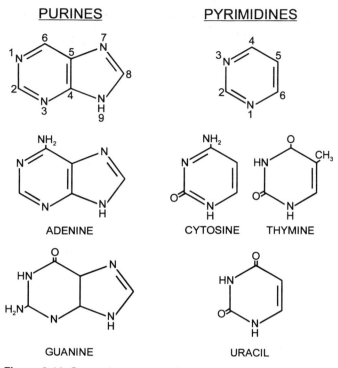

Figure 2.16 General structure of purines, pyrimides, and component bases of RNA and DNA.

discussed, it is important to understand the principal factors that initiate and influence the development of cancer.

2.3.1   Carcinogenesis

DNA LESIONS

The origin of most cancers or neoplasms can be traced to damage (lesions) in the DNA molecule leading to faulty transcription of the genetic signals that regulate the normal growth of somatic tissue. If scrambled genetic information is passed to germinal tissue, chromosomes may be damaged and normal development of offspring will be compromised.

DNA contains two pyrimidine bases (cytosine and thymine) and two purine bases (guanine and adenine). Each base in combination with the monosaccharide 2-deoxy-β-ribofuranose and phosphate forms a nucleotide. Each cytosine-based nucleotide on one strand of DNA is paired with a guanine-based nucleotide on the opposing strand; each thymine-based nucleotide is paired with an adenine-based nucleotide. RNA has a similar structure except that uracil replaces thymine, and the constituent monosaccharide is β-D-ribofuranose. The general purine and pyrimidine structure is shown in Figure 2.16, together with individual bases.

Damage to DNA caused by chemicals or other agents such as UV light or radioactivity takes a variety of forms. Chemical lesions usually result from the

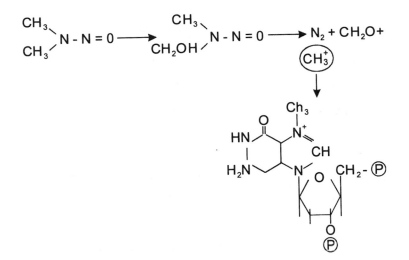

Figure 2.17 Alkylation of N-7 position of guanine by dimethylnitrosamine.

formation of covalent bonds (adducts) between a carcinogen and one of the bases. Sites of action differ depending on the carcinogen involved, and these in turn affect the type and severity of the resulting lesion.

Due to the electron-rich (nucleophilic) nature of the N-7 position of guanine, this site is particularly vulnerable and is subject to attack by alkylating agents such as dimethylnitrosamine (Figure 2.17), epoxides of aflatoxin, and benzo-a-pyrene. The latter also forms a covalent bond with the exocyclic N of guanine. Activated acetylaminofluorene bonds at the C-8 position of guanine. Some planar cyclic compounds insert themselves between the bases of the DNA without forming a covalent bond. This process is called intercalation and may result in distortion of the DNA molecule leading to interference with replication and transcription. Examples of intercalating molecules include acriflavine and 9-aminoacridine.

MUTAGENESIS
Gene mutations may be caused by breaks, omissions, and translocations of chromosome pieces (clastogenesis); uneven distribution of chromosomes following mitosis (aneuploidy); or alterations of DNA known as point mutations. Point mutations may result from base substitutions or from the addition or deletion of base pairs. Numerical changes in base pairs will alter the sequence of base triplets (codons), which form the basis of the transcription process. Such frameshift mutations may, therefore, lead to major changes in protein structure.

The progression from ingestion of a potentially carcinogenic compound to the development of a cancerous condition involves an enormous array of chemical reactions which may advance or retard neoplasia. Many of these chemical transformations are interactive, and the product may be thought of as the net result of a complex of additive, synergistic, and antagonistic events. This complexity, and the fact that some reactions set the stage for further reactions, virtually eliminates the possibility of a single carcinogenic molecule causing a malignant neoplasia (i.e., a one-hit model where the dose response curve passes through zero). Existing multihit models, although closer to reality, must obviously represent gross simplifications of all the components that contribute to the threshold situation referred to earlier.

Some of these complicating factors are outlined next.

Multidrug resistance. Multidrug resistance (MDR) describes a phenomenon whereby cells become resistant to a variety of unrelated compounds, including xenobiotic chemicals with carcinogenic properties. The term multixenobiotic resistance has been used in some studies. MDR has been described in both prokaryotes and eukaryotes and results from the activity of a 170-kDa transmembrane protein capable of extruding chemicals from cells using an energy-dependent, saturable process. In eukaryotes and some prokaryotes, the protein has been identified as P-glycoprotein. It has a low substrate specificity, and the principal common property of potential substrates appears to be a moderate degree of specificity. p-Glycoprotein is encoded by a small family of genes that differ in number from species to species. In humans, two genes have been identified and in other mammalian species there are up to five genes. p-Glycoprotein is induced by exposure to a variety of endogenous and xenobiotic compounds including estrogen, progesterone, aflatoxin, sodium arsenite, and cadmium as well as X-ray irradiation and heat shock. p-Glycoprotein is particularly highly expressed in late tumour stages, although how this equates with its normal function is poorly understood.

Chemical activation. Some chemicals must be activated to become carcinogenic. Usually the mixed function oxidase (MFO) (cytochrome P450) system is involved. Details of this enzyme system and associated reactions are given in Chapter 7. Carcinogens are produced as by-products in the metabolism of other, relatively benign compounds. Examples of chemical activation include phase I epoxidations such as those of benzo(a)pyrene, aflatoxin B_1, and vinyl chloride (Section 7.8.3). Dichloroethane, a laboratory solvent, may be activated by phase II conjugation with glutathione to haloethyl-S-glutathione, an unstable triangular electrophile capable of conjugation with DNA. Activation of acetylaminofluorine (AAF) to the powerful electrophilic nitrenium ion occurs through a phase I hydroxylation and phase II formation of an unstable sulphonate intermediate.

MFO activity may result in the formation of reactive compounds capable of direct interaction with DNA or the formation of reactive oxygen intermediates (ROIs). ROIs may also result from metabolism of compounds that are not themselves carcinogenic. During normal respiration, oxygen accepts four electrons in a stepwise manner in becoming reduced to water. This progressive univalent reduction of molecular oxygen results in the production of such ROIs as the superoxide radical ($O_2^-\bullet$), hydrogen peroxide, and the hydroxyl radical (Figure 2.18). The

$$O_2 \xrightarrow{e^-} O_2^- \cdot \xrightarrow[2H^+]{e^-} HOOH \xrightarrow[H^+]{e^-} H_2O + OH \xrightarrow[H^+]{e^-} H_2O$$

Figure 2.18 Four-stage oxidation involving production of reactive oxygen intermediates.

superoxide anion free radical ($O_2^- \cdot$) is formed by the transfer of a single electron to oxygen. It may be converted to hydrogen peroxide (HOOH) spontaneously or through catalysis by superoxide dismutase (SOD). Cleavage of HOOH to the hydroxyl ion (OH^-) and the superreactive hydroxyl radical $OH \cdot$ is catalysed by a number of trace metals in their reduced form (e.g., Cu^+, Fe^{2+}, Ni^{2+}, and Mn^{2+}).

HOOH is also the direct or indirect by-product of several other enzyme reactions (e.g., monoamine oxidase). Most, but not all HOOH production occurs in peroxisomes (see Section 7.5.6), catalysed by oxidases that remove two electrons from substrates and transfer then to O_2. HOOH is also produced by the oxidation of fatty acids.

ROIs generated through normal metabolism or induced by environmental contaminants (e.g., metal ions, pesticides, air pollutants, radiation) can cause significant metabolic dysfunctions. For example, superoxide radicals can cause peroxidation of membrane lipids, resulting in loss of membrane integrity and the inactivation of membrane-bound enzymes. Although not highly reactive, HOOH can inhibit some metalloenzymes, including superoxide dismutase, and is capable of rapid penetration of cell membranes thereby creating toxic effects at several different subcellular locations. Its most important cytotoxic effect is its role in producing hydroxyl radicals, which can indiscriminately attack and damage every type of macromolecule in living cells including lipids, proteins, and DNA.

Carcinogens are categorised as initiators, promoters, progressors, or complete carcinogens.

Initiation is caused by damage to cellular DNA that results in a mutation. If the mutation is not repaired, it becomes permanent, although the cell is nonmalignant and may remain in the organism for several years. However, there is evidence to suggest that such premalignant cells may be selectively excised through a process of programmed cell death, or apoptosis. It is now believed that initiation is a common occurrence and as such may not be a limiting step in the development of neoplasia.

Promotion is the process of increased replication (hyperplasia) of initiated cells leading to the production of a precancerous condition. Several promoters have been identified, including several inducers of cytochrome P450, such as phenobarbitol, butylated hydroxytoluene, 2,3,7,8 TCDD, polychlorinated biphenyls (PCBs), and several chlorinated pesticides (DDT, aldrin, dieldrin, chlordane, and others). In contrast to the situation for initiators, for promoters there is no evidence that they act directly on DNA. Although the mode of action of promoters is poorly understood, an important component seems to be their stimulation of cell division, which propagates the numbers of mutated cells. The end product is a preneoplastic lesion, which is regarded as a reversible condition insofar as removal of the promoter will halt the process or reverse the condition through apoptosis. Work with phorbol esters has indicated that promotion is influenced by the ageing process (Van Duuren et al., 1975) and by dietary (Cohen and Kendall, 1991) and hormonal factors (Sivak, 1979; Carter et al., 1988).

Although the promotion process is reversible, it often leads to a series of irreversible complex genetic changes known as *progression*. Progression can be a spontaneous process, but it is also enhanced by a variety of chemicals including 2,5,2′,5′-tetrachlorobiphenyl, hydroxyurea, arsenic salts, asbestos, benzene, and derivatives (Pitot and Dragan, 1996). These progressor agents differ from promoters in that they have genotoxic properties. A characteristic of progression is irreversible changes in gene expression, notably that of oncogenes. The continued presence of promoters following the onset of progression will result in an increased number of neoplastic lesions.

Examples of initiators, promoters, and progressors are given in Table 2.6. Chemicals that are capable of performing all these functions independently of other agents are called *complete carcinogens*. *Cocarcinogenicity* is a term that seems to be disappearing from modern usage, describing the potentiation of a carcinogen by a noncarcinogen. A common example of this is the potentiation of polycyclic aromatic hydrocarbons (PAHs), the main cancer-causing agents in tobacco, by catechols, another tobacco by-product. As more is learned about the process of carcinogenesis, the distinction between cocarcinogen and promoter becomes increasingly blurred, and the likelihood is that terms such as *promoter* and *progressor* will also evolve as more is learned about the mechanistics of carcinogenicity.

In Table 2.6, a subset of chemicals, identified with an asterisk, has no specific genotoxic action (i.e., they do not interact with DNA and have no apparent mutagenic activity). Little is known of the exact role(s) of these epigenetic carcinogens, although in view of their chemical diversity, it is likely that they may have several different modes of activity. Some have been shown to alter genetic transcription through a variety of receptor pathways (see Sections 2.3.6 and 7.5.2).

DNA repair. Mechanisms for DNA repair may be broadly categorised into excision-repair and direct repair. Excision-repair involves the removal of a sequence of several nucleotides including the damaged portion of the DNA strand and the insertion of the correct nucleotide sequence using the opposing, undamaged DNA stand as a template. Direct repair involves a reversal of the event which caused a DNA lesion. For example, the methylation of the O^6 position of guanine causes it to be misread as adenine, thereby resulting in a cytosine $\rightarrow$ thymine transition. The enzyme O^6-methylguanine-DNA methyltransferase reverses this process and restores the correct base-pairing (Shevell et al., 1990). There are

Table 2.6. Examples of different modes of action of chemical carcinogens[a]

Initiators	Promoters	Progressors
Dimethylnitrosamine	2,3,7,8 TCDD*	Benzene
2-Acetylaminofluorene*	Cholic acid	Arsenic salts
Benzo(a)pyrene	Polychlorinated biphenyls*	
Dimethylbenz(a)anthracene	Dichlorodiphenyl trichloroethane (DDT)	
	Peroxisome proliferators (see Section 7.5.6)	
	Dieldrin*	

[a] Chemicals with asterisks (*) show no direct genotoxic activity (i.e., they are epigenetic).

at least four genes (ada, AlkA, AlkB, and aidB) known to exercise control over this DNA repair system.

Cell death. Exposure to toxic chemicals results in two types of cell death: necrosis and apoptosis.

Necrosis is a form of mass cell death caused by high doses of toxic chemicals or severe hypoxia. It is an uncontrolled process characterised by the swelling of cells and organelles, random disintegration of DNA, acute inflammation of cell clusters, and secondary scarring. The time course of necrosis is several hours to several days.

Apoptosis is a highly regulated, energy-dependent homeostatic process that takes the form of a programmed excision of ageing, damaged, or preneoplastic cells. It is, therefore, responsible for maintaining the normal morphometric and functional integrity of tissues as well as acting as a means of preventing the onset of neoplasia. It is a controlled process involving individual cells rather than clumps of cells (as in necrosis) and acts on a faster time scale (<10 h) than necrosis. DNA is cleaved in an orderly manner, first in large, then smaller fragments. This process is mediated by members of the caspase family of protease enzymes (Enari et al., 1998). The cells shrink without the collapse of the tissue, and there is no associated inflammation or scarring. Apoptosis may be induced by low levels of chemicals that may cause necrosis in higher concentration [e.g., 1,1-dichloroethylene (Reynolds et al., 1984) and dimethylnitrosamine (Ledda-Columbano et al., 1991)]. Apoptosis may also be stimulated by small doses of ionising radiation, cancer therapeutic agents, changes in hormone balance, and hyperthermia.

Increasing evidence demonstrates that apoptosis may be induced in response to DNA damage, through the mediation of the p53 tumour-suppresser gene, although there is clearly a delicate balance between doses of chemicals that may induce apoptosis and those that may stimulate counteractive effects and necrosis. Tumour promoters may reverse apoptosis (Bellomo et al., 1992), and oxidative stress and ROIs have been shown to stimulate the opposing effects of abnormal cell growth on one hand and apoptosis on the other.

2.3.2   Genotoxicity assays

Although there is no direct evidence that mutations result in carcinogenicity, the indirect evidence for this is very strong, and screening for mutagenicity is a valuable tool in assessing cancer-causing potential. The most widely used assay is the Ames test (Ames et al., 1975), which uses a mutant strain of *Salmonella typhimurium* incapable of synthesising histidine and unable to grow in a histidine-free medium. Reversion to a histidine-producing wild type provides a means of quantifying mutagenic activity. The Ames *Salmonella* assay is very versatile and has proven effective over a range of situations and media including water, gaseous emissions, tobacco smoke, marine sediments, fuels, and food. It possesses some limitations as a means of forecasting carcinogenicity. For example, epigenic carcinogens act in a nongenotoxic manner and would not be expected to give a positive result in an Ames test. Examples include arsenic, chlordane, and dioxins. Studies comparing *Salmonella* mutagenicity with tumour induction in rodents indicate a 70–80% coincidence between bacterial and rodent assays (Tennant and Ashby, 1991; Gold et al., 1993). This correlation improves to >90% with the exclusive selection of probable carcinogens identified through structure activity relationships (SAR) (see also Section 5.5), although it is recognised that forecasting carcinogenicity using the *Salmonella* assay is more reliable for some groups of compounds than for others. One inherent problem is that of time scale. The *Salmonella* test provides a rapid (ca. 2 days), convenient means of quantifying mutagenic potential. However, the development of a neoplastic condition is clearly a much longer process, often several years, and is subject to numerous checks and balances. Thus, the correlation is somewhat analogous to judging the outcome of a long and complex journey based on the first few steps. Nevertheless, the scientific community generally regards the Ames test as an effective screening assay for an important correlate of carcinogenicity. In accepting this model, the extrapolation is made from prokaryotes to eukaryotes. However, much smaller phylogenic jumps still cause interpretive problems for carcinogenicity data. For example, it is not always straightforward to predict human carcinogenicity from experiments on laboratory test animals such as rats and mice, even after normalising for body size. Tests on rodents result in an approximately 500-fold overestimate of human cancer risk from vinyl chloride, and there are many other mismatches of such data to a greater or lesser degree. Nevertheless, very often, such assays are all we have, and there is sufficient confidence in them to support a large body of regulatory legislation involving chemical registration and discharge.

Typically, full carcinogenicity bioassays involve 50 individual male and female rats and the same numbers of mice. Six-week-old animals are exposed for a 2-year period, with a further, postexposure observation period of up to 6 months. Each test costs between U.S. $0.5 million and $1.0 million. To reduce reliance on such tests and to minimise, or at least better understand, phylogenetic extrapolations, most human mutagenicity testing is now performed using mammalian cell preparations. These include in vitro assays of mammalian cell lines, including human somatic cells, and in vivo tests on human and other mammalian lymphocytes. The latter tests may be followed by in vitro culturing of mutant cells or amplification of altered DNA using the polymerase chain reaction to provide sufficient material for molecular analysis.

2.3.3 Chromosome studies

Chromosomal aberrations have been detected by microscopic examination of a variety of cells, notably lymphocytes, from test animals and humans exposed to mutagens. Physical derangement of chromosomes (clastogenesis) takes a number of forms including breaks, omissions, and abnormal combinations of chromosomes or fragments. Sister chromatid exchange has been correlated with DNA strand breakage through chemical intercalation (Pommier et al., 1985) and is used as an assay for several different carcinogens. Many of these observations are carried out during mitosis when the chromosomes may be easily visualised with the use of differential stains. Discernibility of omissions, inversions, and exchanges has been greatly enhanced by the use of fluorescent probes, a technique known as fluorescence in situ hybridisation (acronym FISH).

Aneuploidy is a condition wherein chromosomes become unevenly distributed during cell division in either meiosis or mitosis. It is caused by damage to the meiotic or mitotic spindle, or to the point of attachment between the chromosome and the spindle. Known or suspected aneugens include X-rays, cadmium, chloral hydrate, and pyrimethamine (Natarajan, 1993). Aneuploidy may also arise from hereditary conditions such as Down's syndrome.

2.3.4 The concept of threshold toxicity

Characterisation of the dose or ambient chemical concentration at which toxic effects first appear (threshold) is a fundamental objective of toxicology, but one that has been approached in a variety of ways and is still the subject of controversy. The concept of threshold is embodied in a number of population-based empirical measures of toxicity (e.g., LOEC, NOEC). However, other models dealing with this concept usually adopt a more mechanistic approach. As more information is gathered on how toxic chemicals exert their effect at the molecular level, a variety of mechanistic models have been proposed. Most of these models move us away from the notion of the one-hit model, which was originally proposed to describe carcinogenicity, to multihit models that incorporate a value for the critical number of hits that must occur to elicit a toxic response. Such models may incorporate putative single or multiple receptors and form the basis of the linearised multistage (LMS) model which is used by the U.S. Environmental Protection Agency to determine life-time cancer risk. Embodied in this approach is the

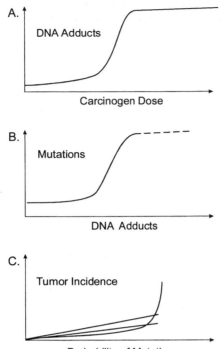

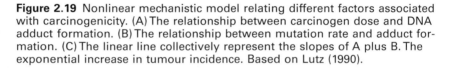

Figure 2.19 Nonlinear mechanistic model relating different factors associated with carcinogenicity. (A) The relationship between carcinogen dose and DNA adduct formation. (B) The relationship between mutation rate and adduct formation. (C) The linear line collectively represent the slopes of A plus B. The exponential increase in tumour incidence. Based on Lutz (1990).

notion that eukaryotic organisms have a whole arsenal of defence mechanisms to cope with an anticipated range of noxious environmental stimuli. The toxic threshold is, therefore, perceived as the point at which there is a preponderance of damaging overcompensatory (repair) effects. It may be seen as a gradual shift in balance or a more precipitous "spillover" effect.

Evidence from mammalian bioassays indicates a strongly sigmoid relationship between carcinogen dose (e.g., 2-acetylaminofluorene) and DNA adduct formation. Mutation rate, in turn, is nonlinearly related to adduct formation and shows a steep increase after adducts reach a certain concentration. Dose-response relationships associated with chemical carcinogenesis have been examined by Lutz (1990), who concluded that the sigmoid curves relating dose to adduct formation and adduct formation to mutagenesis would result in an exponential increase in tumour incidence as a function of probable mutations (Figure 2.19). It is, therefore, clear that evidence for a mechanistic basis for threshold toxicity is beginning to accumulate.

A more recent extension of this concept is the demonstration of an active response to low-level doses of toxic chemicals, which actually stimulate cell growth and repair. This may happen to the extent that, at low levels of toxicant exposure, normal (i.e., nonneoplastic) cellular proliferation may even exceed control values. This is called hormesis.

2.3.5 Hormesis

Hormesis is a stimulatory effect of low levels of toxic agent on an organism. It may be expressed as an enhancement of some measure of physiological or reproductive fitness relative to control values and is superseded by inhibitory effects at increasing levels of toxicant exposure. Hormetic effects have been shown to result from exposure to trace metals and organic chemicals and have been reported from both plants and animals. Additionally, hormetic effects of radiation on animals, particularly humans, have been extensively studied.

Mechanistically, hormesis has been explained in two principal ways. A more specific, yet less common, occurrence stems from the fact that several inorganic contaminants such as trace metals, while toxic at high levels, are essential to the survival of both plants and animals at very low levels. Examples of essential metals include arsenic, cobalt, copper, iron, manganese, molybdenum, selenium, and zinc. In situations where an organism is deficient in one of these metals, a low dose may cause a beneficial effect. A less specific yet probably more broadly applicable explanation for hormesis is a transient overcorrection by detoxification mechanisms responding to inhibitory challenges well within the organism's zone of tolerance.

Chemicals known to stimulate nonneoplastic hepatocellular generation include allyl alcohol, bromotrichloromethane, carbon tetrachloride, chloroform, and ethylene dibromide. Hormesis has been interpreted as evidence for the functional separation of the "damage" component, and the "repair" component, and, although it is assumed that these in turn have subcomponents, this provides support for looking at cancer development as a stochastic two-stage process.

Radiation hormesis describes the stimulatory and arguably beneficial effects of low-level ionising radiation, generally doses less than $10\,mSv\,day^{-1}$. The concept that low levels of radiation might actually be beneficial is still radical and far from proven. However, various studies have shown that cells and organisms adapt to low-level ionising radiation (LLIR) (see Chapter 8) by the stimulation of repair mechanisms. Holzman (1995) described a radiation-induced adaptive response wherein rat lung epithelial cells exposed to low-level alpha radiation increased production of the p53 tumour suppressor protein. Several reports have been made of radiation-induced protein induction. Although some have speculated that this induction may serve to maintain optimal protein levels for cell repair and function (Boothman et al., 1989), further beneficial effects remain speculative at present.

2.3.6 Receptors

From a regulatory standpoint most legislation is based on toxicological information gathered at the whole organism or tissue level. However, from a mechanistic

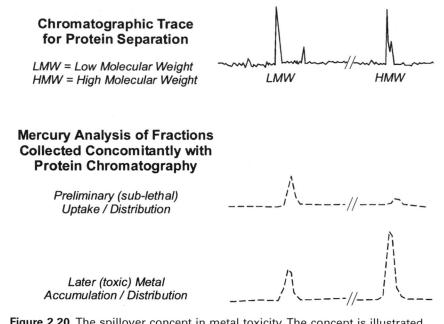

**Chromatographic Trace
for Protein Separation**

LMW = Low Molecular Weight
HMW = High Molecular Weight

LMW   *HMW*

**Mercury Analysis of Fractions
Collected Concomitantly with
Protein Chromatography**

*Preliminary (sub-lethal)
Uptake / Distribution*

*Later (toxic) Metal
Accumulation / Distribution*

Figure 2.20 The spillover concept in metal toxicity. The concept is illustrated by the saturation by a toxic metal of a low-molecular-weight, protective protein (e.g., metallothionein) and the subsequent association of the metal with a high-molecular-weight protein (e.g., enzyme).

point of view, we recognise that such information is often at best correlative and may tell us little about the mode of toxic action. We may document unusually high exposure to and bioaccumulation of potentially toxic chemicals and may be able to arrive at estimates of toxic dose [e.g., through the lethal body burden (LBB) approach described in Chapter 3]. Yet, in this approach, we recognise that such a dose represents only a crude indication that defence mechanisms are overloaded. In fact, much of the tissue or body burden of a toxic chemical or mixture of chemicals often results from its storage in a benign form (e.g., bioaccumulation of hydrophobic nonpolar organics in neutral storage lipids or storage of trace metals associated with metallothioneins). When chemicals move from benign storage to potentially damaging receptor sites, toxicity occurs. The actual toxic lesion may result from only a tiny proportion of the ingested toxicant acting on a narrowly circumscribed receptor. Sometimes, such a situation can arise as a spillover effect following saturation of available sites on protective proteins such as metallothioneins. Such an effect may be traced at the molecular level as a shift in metal associated with a low-molecular-weight protein to a high-molecular-weight protein such as an enzyme (Figure 2.20).

Receptors are defined as normal body constituents that are chemically altered by a toxicant resulting in injury and toxicity. Usually receptors are proteins, although they can also be nucleic acids or lipids. For example, alkylating agents such as *N*-nitroso compounds are activated by the mixed function oxidase (MFO)

Table 2.7. Inhibition of metabolic enzymes by xenobiotics

Toxicants	Effect
As	Inhibition of pyruvate dehydrogenase, α-ketoglutarate dehydrogenase, blockage of mitochondrial electron transport, oxidative phosphorylation.
Sb^{3+}	Competitive inhibition of fructose-6-phosphate. The hydrated species $Sb^{3+} (H_2O)_3$ is a structural analog of fructose-6-phosphate and binds to substrate binding site on phosphofructokinase. Inhibition of glycolysis, partial blockage of energy production leading to loss of cell function and cell death.
Pentachlorophenol, 2-*sec*-butyl-4,6-dinitrophenol (Dinoseb)	Uncouples electron transport system (ETS) from control by oxidative phosphorylation. ETS reverses, resulting in ATP hydrolysis and heat generation.
Rotenone	Binds reversibly with NADH dehydrogenase resulting in lack of flow of reduced electrons and inhibition of energy production.
2,4-D	Inhibition of succinate dehydrogenase and cytochrome-c reductase. Damage to bioenergetic function of mitochondria.

system in the liver to carbonium ions which react spontaneously with N-, O-, and S-containing functional groups and with nonesterified O groups in phosphates. The N-7 of the purine DNA base guanine is particularly vulnerable to electrophilic attack (Figure 2.17), and methylation may result in the pyrazine ring being broken, thereby distorting the base or resulting in the excision of the purine with consequent mutagenic potential. Several chemicals may cause DNA lesions through covalent bonding with purine and pyrimidine bases. Examples are given in Chapter 7. For nucleic acid receptors, the end-point is mutation. In the case of protein receptors, toxic chemicals elicit a variety of responses depending on the function of the protein and the part it plays in the survival of the cell. Depending on the constituent amino acids, a number of different functional groups (e.g., amino, carboxyl, hydroxyl, sulphydryl) may be affected, leading to protein denaturation or blockage of an active site for the binding of endogenous substrates. In the latter case, the protein may be an enzyme and/or a transporter molecule. Enzymic function often confers a high degree of specificity on the receptor, and the bond formed with the toxicant is often reversible.

Where toxicants act competitively with endogenous substrates, the response may be dependent on the relative proportions of the substrate, antagonist, and available active sites or ligands. Enzyme-substrate kinetics may be diagnostic in determining the competitive or noncompetitive nature of the interaction (Section 3.5.5). Where the enzyme is catabolic in nature, occupation of the active site by a toxicant acting as an analog of the normal substrate may result in a build-up of that substrate or a failure in transmembrane movement of the substrate. For example,

Table 2.8. Receptors capable of handling organic compounds

Receptor	Characteristics
Aromatic (aryl) hydrocarbon receptor (AhR)	High-affinity aromatic (aryl) hydrocarbon-binding protein forms a complex with nuclear translator (ARNT) prior to DNA binding (Figure 7.20) and transcription of genes responsible for the production of various metabolic enzymes [e.g., glutathione-s-transferase, glucuronyl transferase, CYP1A1, CYP1A2 (see Section 7.8.3)].
Thyroid receptor (TR)	Thyroid hormone receptors are associated with the plasma membranes, mitochondrial membranes, and nuclear membranes in cells of hormone-responsive tissues. The principal activators are thyroxine and triiodothyronine. Activation of nuclear receptors results in the increased formation of RNA and subsequent elevated levels of protein synthesis and enzyme activity.
Retinoid receptors (RAR); Retinoic acid Receptor (RXR)	Retinoids are important for reproduction and development. They inhibit β-hydroxysteroid dehydrogenase, which converts estradiol to estrone. The peroxisome proliferator receptor, the RAR, and the thyroid receptor are capable of association with the RXR, resulting in the formation of a heterodimer.
Peroxisome proliferator–activated receptor (PPAR)	Peroxisomes are subcellular organelles in most plant and animal cells, performing various metabolic functions including cholesterol and steroid metabolism, β-oxidation of fatty acids, and hydrogen peroxide–derived respiration. Although endogenous activators of PPAR are fatty acids and prostaglandins, several industrial and pharmaceutical products have demonstrated affinity for the receptor and, therefore, peroxisome proliferating properties.
Estrogen receptor (ER)	This member of the thyroid/retinoic acid/steroid hormone receptor superfamily exhibits much structural homology with other receptors. Although its primary activator is the major female sex hormone 17β-estradiol, the ER is capable of interacting with a broad range of chemicals, including pesticides, PCBs, industrial phenols, and phthalates – the so-called xenoestrogens. Endocrine disruption through the mediation of ER is highly complex. Complications include cross-reactivity with other receptors and the fact that ER will interact with both estrogen agonists and antiestrogens.
Invertebrate receptors – ecdysteroid receptor, farnesoid receptor	Invertebrates possess unique enzymes such as the molting hormone (ecdysone) and the juvenile hormone (methyl farnesoate). Although these hormones bear little structural resemblance to the vertebrate hormone estrogen, they show a significant degree of homology with the retinoid receptor (RAR) and are capable of forming functional heterodimers with RXR.

an elevated level of aminolaevulinic acid (ALA) is regarded as diagnostic of lead intoxication through its inhibition of ALA-dehydratase. Lead may also displace calcium from binding sites on the intracellular protein calmodulin. Effects of various toxicants on enzymes involved with energy utilisation are shown in Table 2.7. In some cases, inhibition is achieved through occupation of the active site of the enzyme. In other cases, the effect is caused by the displacement of an integral component of the enzyme molecule. For example, trace metals such as copper and zinc are components of several enzymes, and their displacement by mercury may denature those enzymes. For some enzymes, the bond formed with the toxicant may be very strong and essentially irreversible. An example is the bond formed between an acetylcholine analogue such as paraoxon and the enzyme acetyl-cholinesterase. Here the loss of the nitrophenol group from the paraoxon increases the basicity of the serine hydroxyl group on the enzyme through an interaction with an adjacent imidazole nitrogen. This facilitates a phosphorylation, which is extremely difficult to reverse and functionally incapacitates the enzyme (Figure 7.13).

Intracellular receptors that regulate gene expression belong to a family of receptor proteins that apparently follow a similar phylogenetic lineage and that interact with a variety of endogenous substrates such as thyroid hormones, sex hormones (e.g., estrogen), and corticosteroids. The xenobiotic (Ah) receptor (see Section 7.5.2) may be a relative of these. These receptors are found intracellularly because their substrates are sufficiently lipid-soluble to cross the plasma membrane, and some, such as the estrogen receptor, are found in the nucleus itself. In the case of these receptors, the xenobiotic itself may be the substrate (e.g., Ah receptor) or may mimic the normal endogenous substrate. These circumstances would still fall within our initial (toxicological) definition of receptor if they led to an inappropriate function of the gene regulator, such as the induction of transsexual characteristics. Examples of different receptors capable of handling endogenous and exogenous organic compounds are shown in Table 2.8.

2.4 Questions

1. How is the term *dose* used in environmental toxicology? Describe the different ways dose has been related to toxicity in all branches of the subject.

2. What is an LC_{50}? Describe how toxicologists measure it, including some of the statistical models that have been used to arrive at this parameter. What are the advantages and disadvantages of this means of assessing toxicity?

3. Chronic or subacute toxicity bioassays have often been shown to be more sensitive than acute toxicity tests. Using examples from a broad range of assays, describe the types of end-points commonly employed in these tests. Give two examples of strategies employed to extrapolate from acute to chronic toxicity end-points.

4. Describe two different ways of arriving at a lowest observed effect concentration and a no (observed) effect concentration. What are the advantages of a generalised linear model approach compared with preceding population response distribution models?

5. Compare and contrast toxicity bioassays in plants and animals in (a) aquatic and (b) terrestrial environments.

6. Describe the major components of carcinogenesis, including factors that may advance or retard the process. Give an account of three assays that have been used as direct or correlative evidence of carcinogenesis.

7. How is the concept of threshold toxicity dealt with in population-based dose-response models compared with more mechanistic models? Give two explanations for hormesis. Discuss the pros and cons of current dose-response models in describing and quantifying this phenomenon.

8. The term *receptor* has acquired a broad array of definitions in environmental toxicology. At the molecular and submolecular level, give an account of the toxicological significance of different kinds of receptors, illustrating your answer with specific examples.

2.5   References

Abbott, W. S. (1925) A method of computing the effectiveness of an insecticide, *Journal of Economic Entomology*, **18**, 265–7.

American Society for Testing and Materials (ASTM). (1992) Standard guide for conducting three-brood renewal toxicity rests with *Ceriodaphnia dubia*, Designation E 1295–89. Annual Book of ASTM Standards, Vol. 11.04.669–684, ASTM, Philadelphia.

Ames, B., McCann, J., and Yamasaki, E. (1975) Methods for detecting carcinogens and mutagens with the *Salmonella*/mammalian microsome mutagenicity test, *Mutation Research*, **31**, 347–64.

Bellomo, G., Perotti, M., Taddei, F., Mirabelli, F., Finardi, G., Nicotera, P., and Orrenius, S. (1992) Tumor necrosis factor-inf induces apoptosis in mammary adenocarcinoma cells by an increase of intranuclear free Ca^{2+} concentration and DNA fragmentation, *Cancer Research*, **52**, 1342–6.

Bliss, C. I. (1934a) The method of probits, *Science*, **79**, 38–9.

(1934b) The method of probits – A correction!, *Science*, **79**, 409–10.

Boothman, D., Bouvard, I., and Hughes, E. N. (1989) Identification and characterization of X-ray-induced proteins in human cells, *Cancer Research*, **49**, 2871–8.

Carter, J. H., Carter, H. W., and Meade, J. (1988) Adrenal regulation of mammary tumorigenesis in female Sprague-Dawley rats: Incidence, latency, and yield of mammary tumors, *Cancer Research*, **48**, 3801–7.

Caswell, H. (1996) Demography Meets Ecotoxicology: Untangling the Population Level Effects of Toxic Substances. In *Ecotoxicology. A Hierarchial Treatment*, eds. Newman, M. C., and Jagoe, C. H., pp. 255–92, Lewis, Boca Raton, FL.

Chapman, P. M. (1986) Sediment quality criteria from the sediment quality triad: An example, *Environmental Toxicology and Chemistry*, **5**, 957–64.

(2000) Whole effluent toxicity testing – Usefulness of level of protection, and risk assessment, *Environmental Toxicology and Chemistry*, **19**, 3–13.

Christopher, S. V., and Bird, K. T. (1992) The effects of herbicides on development of *Myriophyllum spicatum* L. cultured in vitro, *Journal of Environmental Quality*, **21**, 203–7.

Cohen, S. M., and Kendall, M. E. (1991) Genetic errors, cell proliferation and carcinogenesis, *Cancer Research*, **51**, 6493–505.

Cox, D. R., and Oakes, D. (1984) *Analysis of Survival Data*, Chapman and Hall, London.

Dixon, P. M., and Newman, M. C. (1991) Analyzing Toxicity Data Using Statistical Models for Time-to-Death: An Introduction. In *Metal Ecotoxicology: Concepts and Applications*, eds. Newman, M. C., and McIntosh, A. W., pp. 207–42, Lewis, Boca Raton, FL.

Enari, M., Sakahira, H., Yokoyama, H., Okaura, K., Iwamatsce, A., and Nagata, S. (1998) A caspase-activated DNase that degrades DNA during apoptosis and its inhibitor ICAD, *Nature*, **391**, 43–50.

Finney, D. J. (1971) *Probit Analysis*, Cambridge University Press, Cambridge.

Fleming, W. J., Ailstock, M. S., and Momot, J. J. (1995) Net Photosynthesis and Respiration in Sago Pondweed (*Potamogeton pectinatus*) Exposed to Herbicides. In *Environmental Toxicology and Risk Assessment*, STP 1218, eds. Hughes, J., Biddinger, G., and Moses, E., American Society for Testing and Materials, Philadelphia.

Fletcher, J. S. (1990) Use of Algae Versus Vascular Plants to Test for Chemical Toxicity. In *Plants for Toxicity Assessment*, ASTM STP 1091, eds. Wang, W., Gorsuch, J. W., and Lower, W. R., pp. 33–9, American Society for Testing and Materials, Philadelphia.

Fletcher, J. (1991) Keynote Speech: A Brief Overview of Plant Toxicity Testing. In *Plants for Toxicity Assessment*, vol. 2, ASTM STP 1091, eds. Gorsuch, J. W., Lower, W. R., Wang, W., and Lewis, M. A., pp. 5–11, American Society for Testing and Materials, Philadelphia.

Gaddum, J. H. (1933) Reports on biological standards. III. Methods of biological assay depending on quantal response. Medical Research Council Special Report Series 183, HMSO, London.

Gannon, J. E., and Beeton, A. M. (1971) Procedures for determining the effects of dredged sediment on biota-benthos viability and sediment selectivity tests, *Journal Water Pollution Control Federation*, **43**, 393–8.

Garten, J., and Garten, C. T., Jr. (1990) Multispecies Methods of Testing for Toxicity: Use of the Rhizobium-Legume Symbiosis in Nitrogen Fixation and Correlations Between Responses by Algae and Terrestrial Plants. In *Plants for Toxicity Assessment*, ASTM STP 1091, eds. Wang, W., Gorsuch, J. R., and Lower, W. R., pp. 69–84, American Society for Testing and Materials, Philadelphia.

Gelber, R. D., Lavin, P. T., Mehta, C. R., and Schoenfeld, D. A. (1985) Statistical Analysis. In *Fundamentals of Aquatic Toxicology*, eds. Rand, G. M., and Petrocelli, S. R., pp. 110–23, Hemisphere, Washington, New York, and London.

Giesy, J. G., and Hoke, R. A. (1990) Freshwater Sediment Quality Criteria: Toxicity Bioassessment. In *Sediments: Chemistry and Toxicity of In-Place Pollutants*, eds. Baudo, R., Giesy, J. P., and Mantau, H., pp. 265–348, Lewis, Ann Arbor, MI.

Giesy, J. P., and Graney, R. L. (1989) Recent developments in the intercomparisons of acute and chronic bioassays and bioindicators, *Hydrobiologia*, **188/189**, 21–60.

Gobas, F. A. P. C., Lovett-Doust, L., and Haffner, G. D. (1991) A Comparative Study of the Bioconcentration and Toxicity of Chlorinated Hydrocarbons in Aquatic Macrophytes and Fish. In *Plants for Toxicity Assessment*, vol. 2, ASTM STP 1115, eds. Gorsuch, J. W., Lower, W. R., Wang, W., and Lewis, M. A., pp. 178–93, American Society for Testing and Materials, Philadelphia.

Gold, L. S., Stone, T. H., Stern, B. R., and Bernstein, L. (1993) Comparison of trafet organs of carcinogenicity for mutagenic and non-mutagenic chemicals, *Mutation Research*, **286**, 75–100.

Grant, W. F. (1994) The present status of higher plant bioassays for the detection of environmental mutagens, *Mutation Research*, **310**, 175–85.

Hamilton, M. A., Russo, R. C., and Thurston, R. V. (1977) Trimmed Spearman-Karber method for estimating median lethal concentrations in toxicity bioassays, *Environmental Science and Technology*, **7**, 714–19.

Hayes, W. J., Jr. (1975) *Toxicology of Pesticides*, Williams and Wilkins, Baltimore.

Holzman, D. (1995) Hormesis: Fact or fiction, *Journal of Nuclear Medicine*, **36**, 13–16.

Hurlbert, S. H. (1984) Pseudoreplication and the design of ecological field experiments, *Ecological Monographs*, **54**, 187–211.

Kenaga, E. E. (1982) Predictability of chronic toxicity from acute toxicity of chemicals in fish and aquatic invertebrates, *Environmental Toxicology and Chemistry*, **1**, 347–58.

Kerr, D. R., and Meador, J. P. (1996) Modelling dose response using generalized linear models, *Environmental Toxicology and Chemistry*, **15**, 395–401.

Ledda-Columbano, G. M., Coni, P., Curto, M., Giacomini, L., Faa, G., Oliverio, S., Piacentini, M., and Columbano, A. (1991) Induction of two different modes of cell death, apoptosis and necrosis, in rat liver after a single dose of thioacetamide, *American Journal of Pathology*, **139**, 1099–109.

Lee, G., Ellersieck, M. R., Mayer, F. L., and Krause, G. F. (1995) Predicting chronic lethality of chemicals to fishes from acute toxicity test data: multifactor probit analysis, *Environmental Toxicology and Chemistry*, **14**, 345–9.

Leicester, H. M. (1956) *Historical Background of Chemistry*, Dover, New York.

Litchfield, J. T., and Wilcoxon, F. (1949) A simplified method of evaluating dose-effect experiments, *Journal of Pharmacology and Experimental Therapeutics*, **96**, 99–113.

Lutz, W. K. (1990) Dose-Response Relationships in Chemical Carcinogenesis: From DNA Adducts to Tumor Incidence. In *Biological Reactive Intermediates IV*, ed. Witmer, C. M., Plenum Press, New York.

Mayer, F. L., Krause, G. F., Buckler, D. R., Ellersieck, M. R., and Lee, G. (1994) Predicting chronic lethality of chemicals to fishes from acute toxicity test data: Concepts and linear regression analysis, *Environmental Toxicology and Chemistry*, **13**, 671–8.

Mulvey, M., and Diamond, S. A. (1991) Genetic Factors and Tolerance Acquisition in Populations Exposed to Metals and Metalloids. In *Metal Ecotoxicology: Concepts and Applications*, eds. Newman, M. C., and McIntosh, A. W., pp. 301–21, Lewis, Chelsea, MI.

Natarajan, A. T. (1993) An overview of the results of testing of known or suspected aneugens using mammalian cells *in vitro*, *Mutation Research*, **287**, 47–56.

Newman, M. C. (1995) *Quantitative Methods in Aquatic Ecotoxicology*, Lewis, Boca Raton, FL.

Newman, M. C., and Dixon, P. M. (1996) Ecologically Meaningful Estimates of Lethal Effects in Individuals. In *Ecotoxicology: A Hierarchial Treatment*, eds. Newman, M. C., and Jagoe, C. H., pp. 225–53, Lewis, Boca Raton, FL.

Nyholm, N. (1985) Response variable in algal growth inhibition tests: biomass or growth rate?, *Water Research*, **19**, 273–9.

Nyholm, N., Sorenson, P. S., and Kusk, K. O. (1992) Statistical treatment of data from microbial toxicity tests, *Environmental Toxicology and Chemistry*, **11**, 157–67.

Peterson, H. G., Nyholm, N., Nelson, M., Powell, R., Huang, P. M., and Scroggins, R. (1996) Development of aquatic plant bioassays for rapid screening and interpretive risk assessments of metal mining liquid waste water, *Water Science and Technology*, **33**, 155–61.

Pitot, III, H. C., and Dragan, Y. P. (1996) Chemical Carcinogenesis, Chapter 8. In *Casarett & Doull's Toxicology: The Basic Science of Poisons*, ed. Klaassen, C. D., pp. 201–67, McGraw-Hill, New York.

Pommier, Y., Zwelling, L. A., Kao-Shan, C. S., Whang-Peng, J., and Bradley, M. O. (1985) Correlations between intercalator-induced DNA strand breaks and sister chromatid exchanges, mutations and cytotoxicity in Chinese hamster cells, *Cancer Research*, **45**, 3143.

Reporter, M., Robideauz, M., Wickster, P., Wagner, J., and Kapustka, L. (1991) Ecotoxicological assessment of toluene and cadmium using plant cell cultures. In *Plants for Toxicity*

Assessment, vol. 2, ASTM STP 1115, eds. Gorsuch, J. W., Lower, W. R., Wang, W., and Lewis, M. A., pp. 240–9, American Society for Testing and Materials, Philadelphia.

Reynolds, E. S., Kanz, M. F., Chieco, P., and Moslen, M. T. (1984) 1,1-Dichloroethylene: An apoptotic hepatotoxin?, *Environmental Health Perspectives*, **57**, 313–20.

Shevell, D. E., Friedman, B. M., and Walker, G. C. (1990) Resistance to alkylation damage in *Escherichia coli*: Role of the Ada protein in induction of the adaptive response, *Mutation Research*, **233**, 53–72.

Sivak, A. (1979) Carcinogenesis, *Biochemical and Biophysical Research Communications*, **560**, 67–89.

Sprague, J. B. (1973) The ABCs of pollutant bioassay with fish. In *Biological Methods for the Assessment of Water Quality*, STP 528, eds. Cairns, J. J., and Dickson, K. L., pp. 6–30, American Society for Testing and Materials, Philadelphia.

Swanson, S. M., Rickard, C. P., Freemark, K. E., and MacQuarrie, P. (1991) Testing for pesticide toxicity to aquatic plants: Recommendations for test species. In *Plants for Toxicity Assessment*, vol. 2, ASTM STP 1115, eds. Gorsuch, J. W., Lower, W. R., Wang, W., and Lewis, M. A., pp. 77–97, American Society for Testing and Materials, Philadelphia.

Tennant, R. W., and Ashby, J. (1991) Classification according to chemical structure, mutagenicity to Salmonella and level of carcinogenicity of a further 39 chemicals tested for carcinogenicity by the U.S. National Toxicology Program, *Mutation Research*, **257**, 209–27.

U.S. Environmental Protection Agency (U.S. EPA). (1991) *Methods for Measuring the Acute Toxicity of Effluents and Receiving Waters to Freshwater and Marine Organisms*, 4th ed., ed. C. I. Weber, EPA-600/4-90/027, U.S. EPA, Washington, DC.

Van Duuren, B. L., Sivak, A., and Katz, C. (1975) The effect of aging and interval between primary and secondary treatment in two-stage carcinogenesis on mouse skin, *Cancer Research*, **35**, 502–5.

Versteeg, D. J. (1990) Comparison of short- and long-term toxicity test results for the green alga, *Selenastrum capricornutum*. In *Plants for Toxicity Assessment*, ASTM STP 1091, eds. Wang, W., Gorsuch, J. W., and Lower, W. R., pp. 40–8, American Society for Testing and Materials, Philadelphia.

Wright, D. A. (1977) Toxicity of fluoride to brown trout fry *Salmo trutta*, *Environmental Pollution*, **12**, 57–62.

Wright, D. A., and Roosenburg, W. H. (1982) Trimelthyltin toxicity to larval *Uca pugilator*: Effects of temperature and salinity, *Archives of Environmental Contamination and Toxicology*, **11**, 491–5.

3

○ ○

Routes and kinetics of toxicant uptake

3.1 General considerations

To gain access to an organism, toxicants must usually penetrate at least one layer
of cells. A variety of rate-limiting uptake processes may be conveniently studied
using unicellular organisms such as algal cells, and much has been learned about
the kinetics of chemical bioaccumulation from such simple systems. However,
when cells are formed into single or complex membranes, the situation rapidly
becomes more complicated. Transepithelial chemical movement is now studied
by comparing the external medium with a defined bodily fluid such as blood, which
serves as a vector of the toxicant to internal tissues where it may be stored or metabolized. The plasma membrane of a single cell consists of a lipid bilayer containing
various embedded protein molecules. Some proteins play a role in transepithelial
chemical transport. Specialized carrier proteins may mediate the transport of a
variety of chemicals including ions, sugars, amino acids, and several xenobiotics,
and in doing so they perform a range of functions encompassing nutrition, homeostasis, excretion, and detoxification.

Other models for transepithelial chemical movement include channels, which
may have specific voltage or chemical triggers controlling their opening and
closing, and pores operating on the basis of molecular size exclusion. Many nonpolar xenobiotic chemicals move through the lipid portion of the plasma membrane by passive diffusion, a process that may be limited by the plasma membrane's
relative lipid solubility.

Configuration of cells into complex membranes or epithelia often confers "ori-
entation" to those cells. Cells within an epithelium may move chemicals in a specific direction. For example, ions that may enter a cell through channels at the outer
or apical surface of an epithelium may be transported from the cell to the blood
by a pump located at the inner basolateral membrane. Recent evidence suggests
that intracellular transporter proteins may be responsible for the translocation of
trace metals such as zinc across epithelial cells.

Some pathways responsible for transport of physiological ions across a membrane are shown in Figure 3.1. Where specific mechanisms are indicated, any of

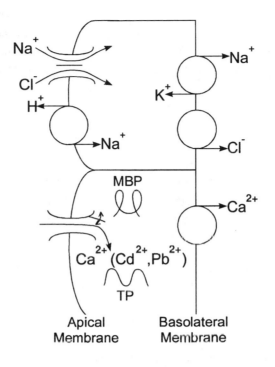

 Channels may be separate-but-linked cation/anion channels or specific ion (e.g., Ca^{2+}) channels. May be blocked by various chemical agents. Channels may be controlled by gating mechanisms bestowing ionic specificity.

Ionic pumps may be electroneutral ionic exchange mechanisms or single ion electrogenic pumps. Both pumps and specific ion channels may be subject to competitive or non competitive inhibition.

 Intracellular proteins may function as transporter proteins (TP) capable of delivering metal ions to membrane pumps (ᴧᴧ) or metal binding proteins (MBP), which may compete with transporter proteins and enzymes. These may serve a detoxifying function for trace metals.

Figure 3.1 Examples of some mechanisms responsible for transepithelial transport. Pumps and channels for physiological electrolytes may accommodate nonphysiological trace metals.

these may be subject to competitive or noncompetitive inhibition by a variety of toxic agents such as trace metals. In some instances, chemical binding to proteins may create concentration gradients between the external medium and cytoplasm or between the external medium and body fluids, which will cause nutrients, electrolytes, and chemical toxicants to diffuse into the organism.

3.2   Route of toxicant uptake

In terrestrial animals, the three principal routes of uptake of xenobiotic chemicals are the skin, the lungs, and the gastrointestinal tract. Respiratory and digestive epithelia are, of course, specialised to permit selective molecular uptake and are, therefore, more vulnerable to the entry of toxic chemicals. In aquatic animals, gills are modified for both respiratory exchange and the regulation of electrolytes and may permit entry of dissolved substances through diffusion or ionic regulatory pathways. In plants, too, root epithelia are specialized for the uptake of water and electrolytes and present a mode of chemical accumulation, which is in marked contrast to aerial deposition on leaves.

3.2.1   Skin

The skin of terrestrial animals acts as a protective barrier to isolate them from their environment, and it was originally thought to be impervious to environmental toxicants. However, it is now regarded as an important route of entry for several xenobiotics, although uptake is slower than through other epithelia. In mammals, the skin consists of three layers: the outer, nonvascularized layer, the epidermis; the middle, highly vascularized layer, the dermis; and the inner layer, hypodermis, which consists of connective and adipose tissue. The primary barrier to chemical absorption is the outer layer of the epidermis, which contains a dense lipid and keratin matrix. Organic solvents may cross this layer by solubilizing constituent lipids. Particularly effective are small molecules with some lipid and aqueous solubility such as dimethyl sulfoxide (DMSO), chloroform, and methanol. These may increase the permeability of the skin to other compounds, which may be absorbed as solutes. For example, DDT powder has poor skin permeability but is readily absorbed when dissolved in kerosene.

A specific skin pathology associated with chemical exposure is the development of chloracne resulting from contact with halogenated hydrocarbons. Polychlorinated biphenyls and tetrachlorodibenzodioxin (TCDD) have been particularly implicated in this regard, and the condition is regarded as diagnostic of exposure to these types of compounds. Chloracne is characterized by hyperkeratosis and progressive degeneration of the sebaceous glands. The epidermis is a significant source of both phase I and phase II enzymes capable of metabolizing organic xenobiotic chemicals (see Section 7.8.2 for a description of these enzymes). The TCDD-inducible enzyme P4501B1 has been reported from human skin keratinocytes (Sutter et al., 1994) and many other P450s are also present. Several phase II enzymes are also found in mammalian epidermal cells including uridine diphosphate (UDP)-glucuronosyltransferase, β-glucuronidase, and sulphatases (Section 7.8.5).

Pathological skin changes may also be caused by ultraviolet (UV) light from solar radiation or man-made sources. UV light is subdivided into UVA (wavelength 320–400 nm), UVB (290–320 nm), and UVC (<290 nm). Maximal absorbance of UV light by DNA and protein is at wavelengths less than 300 nm. Although the penetrative power of UV light increases exponentially at higher wavelengths, it does little or no damage, and its toxicity becomes a trade-off between light penetration and specific molecular absorption. Chronic UV exposure results in a range of preneoplastic and neoplastic skin conditions, although the development of these cancers varies greatly among individuals and is highly dependent on skin pigmentation.

3.2.2 Lungs

The mammalian respiratory system consists of three regions: the nasopharyngeal canal, the tracheobranchial region, and the pulmonary region. The first two regions are protective in nature and are responsible for filtering larger particles, trapping them in mucous secreted by goblet cells and transporting them by ciliated epithelia to the oral cavity where they are swallowed or expelled with the sputum. Polar compounds such as sulphuric acid (H_2SO_4) and hydrochloric acid (Cl) are highly water soluble and are readily absorbed by the aqueous mucous lining of the nose, throat, trachea, and bronchii.

Dense particles have greater momentum in the upper respiratory tract and tend to collide with the walls of the tract particularly where the airways bend. Smaller particles (i.e., those less than 1 μm) are carried to the lower respiratory tract, the pulmonary region. This region consists of increasingly subdivided bifurcating lobes ending in terminal branchioles, which are small tubes approximately 0.5–1 mm in diameter. These further subdivide into alveolar ducts and alveoli, which are tiny sacs (150–350 μm) that comprise about 80% of the total lung capacity of 5.7 L. The total surface area of the alveoli is about 120 m^2 compared with the skin surface area of 1.75 m^2.

Toxicants absorbed by the lungs are usually gases, although several major diseases of the lung have been caused by particulates, notably aluminosilicate microfibrils such as asbestos. Lead (Pb) is also introduced as inhaled particulates.

Absorption of gases by the lungs is highly dependent on their water solubility. At equilibrium, the ratio of chemical concentration in blood to its concentration in air is known as the blood:gas partition coefficient. According to Henry's law, the amount of gas dissolved in a liquid is proportional to its partial pressure in the gaseous phase until the point of saturation in the liquid is reached. With highly soluble gases such as chloroform, equilibration with the blood takes longer than with relatively insoluble gases, and the rate of absorption is limited by the respiration rate. Gases with low blood:gas partition coefficients, such as ozone, rapidly equilibrate with the blood, and in this case the loading of the blood is limited by the rate of blood flow. At rest, the total volume of air moved in and out of the lung (tidal volume) is about 500 cm^3. Resting adults average 16 breaths per minute, which means that the total ventilation rate is approximately 8 L min^{-1}. This may increase sevenfold during exercise, which also elevates the heart rate and blood circulation. Thus, exercise greatly increases absorption of gases of both high and low water solubility. Carbon monoxide presents a special case: Although its water solubility is low, its high affinity for hemoglobin causes a diffusion gradient favoring rapid blood absorption.

Poorly water soluble gases such as ozone (O_3) and nitrogen dioxide (NO_2) penetrate deep into the lungs wherein they exert toxic action through the formation of unstable free radicals such as superoxide, hydroxyl ions, and possibly singlet oxygen. These are produced as a result of mixed function oxidase activity (Section 2.3.1). MFO activity has been implicated in promoting the toxicity of the powerful pneumotoxicant paraquat, which cycles between its oxidized and reduced forms with the prodigious production of active oxygen metabolites.

The presence of many invasive chemicals and particles in the lungs stimulates chemotactic and phagocytic activity in a variety of specialized protective cells such as macrophages. Phagocytic cells activated in this way release potent oxidants, notably hydrogen peroxide, which while aiding the phagocytic process may exacerbate the potential for tissue damage through elevated oxidant activity. In addition to its oxidizing potential, NO_2 may cause tissue damage through the formation of nitric and nitrous acids with alveolar fluid.

Several diseases of the respiratory system have been linked to the inhalation of metals and metal compounds in the form of fumes and vapors. Respiratory absorption of cadmium (Cd) may represent up to 30% of total Cd ingestion. Cigarette smoke contains Cd, and heavy smoking may double overall Cd uptake. Cadmium oxide (CdO) fumes are produced during the manufacture of battery electrodes, glass, and semiconductors and in cadmium electroplating. They consist of fine particles capable of reaching the alveoli. Symptoms include edema, fibrosis of alveolar membranes, and emphysema. Nickel carbonyl [Ni(CO)$_4$] is a volatile liquid used in nickel refining and plating. It is acutely toxic with a variety of symptoms ranging from nausea, chest pains, and cyanosis to respiratory failure, cerebral edema, and death. Long-term exposure in refineries to particulate nickel compounds such as nickel monoxide (NiO) and nickel subsulfide (Ni$_2$S$_3$) has been epidemiologically linked to lung and nasal cancer. Inhalation of particulate hexavalent chromium compounds such as chromate (CrO^{2-}_4) and bichromate ($Cr_2O_7^{2-}$) during chromium smelting and plating operations causes bronchial irritation after short-term exposure and pulmonary cancers following chronic exposure.

Aerial exposure to beryllium (Be) dust occurs during the manufacture of ceramics and alloys and during the extraction of the metal from its ore. Be dust particles enter the alveoli where they cause granulomas, a condition known as berylliosis. This progresses to interstitial fibrosis. The International Agency for Research on Cancer (IARC) listed Be as a human carcinogen in 1994.

3.2.3   Gills

Gills are the primary respiratory organs of aquatic animals and are typified by an unusually thin epithelium, which serves to minimize the distance between the oxygenated external medium and the internal body fluids. They range from simple external epithelial appendages to elaborate structures containing numerous folds or lamellae (Figure 3.2) enclosed in a branchial chamber that is ventilated by a continuous stream of water. In most aquatic organisms, the remainder of the external epithelium is relatively impermeable.

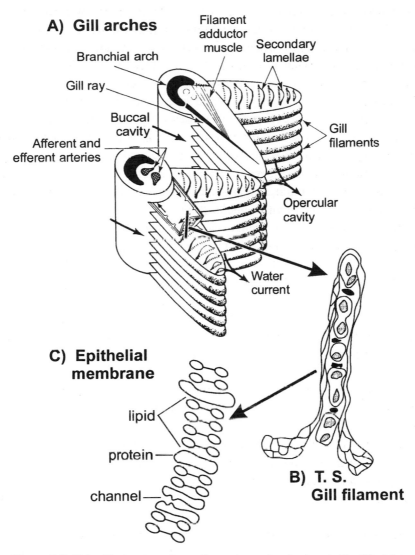

A) Gill arches

Branchial arch

Gill ray

Buccal cavity

Afferent and efferent arteries

Filament adductor muscle

Secondary lamellae

Gill filaments

Opercular cavity

Water current

C) Epithelial membrane

lipid

protein

channel

B) T. S. Gill filament

Figure 3.2 Fish gill structure from tissue to molecular level. Modified from Hughes and Shelton (1962).

Like lungs, gills are designed to present a large surface area for respiration and are also the primary sites for the exchange of dissolved chemicals. As such, they contribute to important functions such as ionic and osmotic regulation. In freshwater animals, electrolytes are actively taken up by specialised ion-transporting cells in gill epithelia. In fish, these are called chloride cells. Excess water accumulated through osmosis is excreted through a kidney or kidney-like organ. Marine fish drink seawater to compensate for osmotic water loss, and the gut epithelium is responsible for the excretion of excess electrolytes.

In gills, exchange of gases and other solutes is facilitated by a countercurrent flow wherein the circulation of blood through the lamellae runs counter to the flow of water past the gill surface. Laminar flow of water creates a boundary layer immediately adjacent to the gill surface, which is further modified by the presence of mucus. Many of the chemical interactions occurring at the gill surface are dictated by the chemistry of this boundary layer.

Gill surfaces are composed of one or more layers of epithelial cells (Figure 3.2B) joined by tight junctions, which may still allow extracellular solvent and solute movement. The lipid bilayer that composes the outer membrane of these cells (Figure 3.2C) includes phospholipids, glycolipids, and cholesterol. The hydrophilic, outer ends of these compounds are rich in nitrogen-containing bases, which function as ligands for the attachment of cations, especially trace metals. Embedded in the membrane are proteins, which may traverse the whole of the outer membrane of the epithelial cells. These may function as ion channels or pumps (see Figure 3.1), which may also transport trace metals in addition to physiological electrolytes. The hydrogen ion (H^+) and several metals have been shown to compete with the calcium ion (Ca^{2+}) for ligands at gill surfaces. Calcium is responsible for maintaining the integrity of cellular junctions in gill epithelia, and its depletion may increase permeability to water and electrolytes.

The gills of many organisms also contain mucus-secreting cells, which are largely protective in nature. In fish, the mucus has a rich glycoprotein content and forms a polyanionic matrix capable of binding several trace metals. The physiological effects of aluminium (Al) at the surface of gills have been the subject of extensive studies because of its toxicity to freshwater fish. Al speciation is highly pH-dependent with hydroxo-complexes predominant at higher pHs. Such conditions are thought to exist at the gill boundary layer where aluminium hydroxide [$Al(OH)_3$] precipitation may take place at the gill surface. Aluminium binding to functional groups such as phosphatidylchloline and phosphatidylserine may also occur. Aluminium exposure has been associated with loss of sodium ion (Na^+) and cloride ion (Cl^+) through the gill epithelium. Iron (Fe) may act in a similar manner (i.e., the formation of insoluble ferric compounds which physically alter epithelial permeability).

3.2.4 Digestive system

The absorption of toxicants through the digestive tract represents a major uptake route in multicellular animals. Molecular uptake is, after all, the major function of the gastrointestinal system, which is superbly adapted for this purpose. The surface area of the human small intestine ($240 m^2$), for example, is approximately twice that of the lung. The digestive tract may be considered as an extension of the external medium, and all its contents may be regarded as being outside the body. Chemical uptake can occur along the whole length of the digestive system including the mouth and rectum, although in mammals most absorption occurs in the small intestine. The gastrointestinal tract possesses specialized transport systems for certain

The Henderson - Hasselbach Equations

For weak acids (HA)

$$HA \rightleftharpoons H^+ + A^-$$

$$K_a = \frac{(H^+)\,(A^-)}{(HA)}$$

$$\log K_a = \log\left[\frac{(H^+)\,(A^-)}{(HA)}\right]$$

$$\log K_a = \log(H^+) + \log\frac{(A^-)}{(HA)}$$

$$-\log(H^+) = -\log K_a + \log\frac{(A^-)}{(HA)}$$

$$pH = pK_a + \log\frac{(A^-)}{(HA)}$$

For weak bases (HB⁺)

$$HB^+ \rightleftharpoons H^+ + B$$

$$K_a = \frac{(H^+)\,(B)}{(HB^+)}$$

$$\log K_a = \log\left[\frac{(H^+)\,(B)}{(HB^+)}\right]$$

$$\log K_a = \log(H^+) + \log\frac{(B)}{(HB^+)}$$

$$-\log(H^+) = -\log K_a + \log\frac{(B)}{(HB^+)}$$

$$pH = pK_a + \log\frac{(B)}{(HB^+)}$$

Figure 3.3 Henderson-Hasselbach equations – the relationship between dissociation constant (pK_a) and pH for weak acids and weak bases.

nutrients such as amino acids, fatty acids, carbohydrates, iron, sodium, and calcium, although most chemicals are absorbed through diffusion. In the gastrointestinal tract, chemicals are in an aqueous medium with variable pH, and their entry may be heavily influenced by pH. If a chemical is an organic acid or base, it is most readily absorbed as the nonpolar or un-ionised form of the acid or base conjugate pair. Organic acids, AH, dissociate to produce anions A⁻, and organic bases combine with hydrogen ions (H⁺) to form cations BH⁺. The extent of dissociation is determined by the pH and the acid dissociation constant, K_a, according to the Henderson-Hasselbach equations (Figure 3.3). For a weak acid such as benzoic acid with a $pK_a = 4$, the ratio of non-ionised:ionised form in the stomach (pH 2) will be 100 and will favor absorption. The equivalent ratio in the small intestine (pH 6) is 0.01 and, therefore, does not favor absorption. Nevertheless, the enormous surface area of the small intestine still facilitates the absorption of substantial quantities of benzoic acid.

Some particles several nanometers in diameter can be absorbed from the gastrointestinal tract by pinocytosis and are carried to the circulatory system via the lymphatic system.

3.2.5   Toxicant uptake by plants

Toxicant chemicals enter plants through foliar uptake or root uptake and can bind to waxes on leaf surfaces. The rate of movement and target sites depends on physiocochemical properties of the toxicant (Henry's law constant, hydrophilicity) and environmental conditions including temperature, humidity, exposure to UV light, and height of the leaf boundary layer.

Stomatal uptake of pollutants in terrestrial vascular plants can occur in the gaseous form and through both wet and dry deposition for sulphur dioxide (SO_2), carbon dioxide (CO_2), carbon monoxide (CO), oxides of nitrogen (NO_x), ammonia (NH_3), hydrogen peroxide (H_2O_2), hydrofluoric acid (HF), and some semivolatile organics including PCBs, DDT, and lindane (see Chapter 7). Particulate deposition onto leaf surfaces brings other particulate sorbed pollutants into contact with plant surfaces, including heavy metals, radionuclides, polynuclear aromatic hydrocarbons (PAHs), and sulphur (often associated with metals), which become bound to leaf cuticles (Veijalainen, 1988). In addition, clogging of stomatal pores by fine particulates can adversely affect plant CO_2 and water regulation. Rainfall interacts with vegetation surfaces modifying their original composition, leaching elements and ions released by plant organs, and washing out powders and aerosols deposited on the leaf surfaces. These substances may be transferred to the soil where some of them become available to the root systems of plants (e.g., metals). Throughfall and stemflow are the most important vectors of pollutants and nutrients from the atmosphere to the soil.

Pollutants first come into contact with the epidermal layer, the interface between the inside and outside of the plant. Ozone affects the cuticle by vigorously reacting with components such as hydroxy fatty acids therein, stripping away leaf waxes. However, cuticular transpiration is not affected by O_3. Ozone enters leaves via stomata and is received at the apoplastic space where its concentration decreases, indicating some degree of detoxification. Atmospheric photooxidation products such as peroxyacetyl nitrate (PAN) also enter plants via stomata, moving into the substomatal space and dissolving in the extracellular fluid of cells. PANs are transported by the transpiration stream to developing areas of leaves, damaging sulphydryl groups of proteins and unsaturated double bonds of lipids. About 90% SO_2 is absorbed by stomata. Entry of SO_2 into leaves is abnormal (sulphur normally enters via the roots). Ammonia is codeposited with SO_2.

A summary of routes of uptake of various toxicants by plants is provided in Table 3.1.

3.3   Uptake at the tissue and cellular level

In rare cases, chemicals may exert toxic action without entering the tissues. For example, excess mucus production by fish may occur in reaction to the presence of aluminium at the gill epithelial surface, thereby inhibiting gaseous exchange. However, generally speaking the toxicant needs to cross both aqueous and lipid compartments of the integument. The first barrier encountered is the plasma membrane of the epithelial cells lining the integument. This is a typical biomembrane about 80Å thick consisting of a lipid biolayer and a variety of proteins, some of which may function as transporters. Water and small organic molecules may diffuse across the membrane through small pores (0.2–0.4 nm diameter), although these pores are relatively few in number. The nonpolar regions of the lipids and proteins

Table 3.1. Uptake routes for various toxicants in plants

Toxicant	Routes of entry of toxicants to plants
SO_2	90% absorbed by stomata.
NH_3/NH_4	50% wet/50% dry deposition found in most European countries. Wet deposited N taken up by roots. NH_3 entering leaf stomata dissolves in leaf to form NH_4.
CO	Stomatal uptake incorporated into CO_2 uptake routes (Wellburn, 1994).
F	Toxicant occurs as particulate or gaseous HF. Enters old or weathered leaves via stomata, dissolves in water surrounding mesophyll cell, and is transported in transpiration stream to leaf tips and margins causing necrosis.
H_2O_2	Route of entry is principally gaseous in stomata. Toxicant is received in a localised place called the apoplastic space in aqueous matrix of cell walls.
Semivolatile organics	The pattern of uptake depends on Henry's law constant (relating to volatility) and octanol:water partition coefficient (relating to lipid solubility). Mono- to tetrachlorobiphenyls are transferred to plants in gaseous form. Higher chlorinated PCBs sorbed to particulates and aerosols are transferred to plants by depositional processes and revolatilisation as gases in hot summer months. Stomatal entry and solubilisation in epicutical waxes are important entry routes of many semivolatile organics and PCBs.
Bromacil, 2,4-D (herbicides)	High water solubility and low Henry's law constants dictate root uptake from soil and transfer to leaves in transpiration stream (Paterson et al., 1994).
Mercury	Toxicant transported to plants in free gaseous form (as well as particulate bound) where it is oxidized and removed by wet and dry deposition processes to ionic and particulate species (Hacon et al., 1995).
Trace metals from smelting activities	Toxicant principally taken up from soil by roots but also through bark into wood or through leaves and transported via phloem to xylem.
Radionuclides (e.g., [134, 137]Cs, ^{210}Pb, [210]Po)	Absence of cuticle in lichens permits favourable uptake, particularly in foliose forms (e.g., *Xanthonia*) and species with large flat thalli (e.g., *Peltigera*). Lichens lack root system and uptake is principally from the atmosphere (ombrotrophic) by deposition and mechanical trapping (e.g., Pyatt and Beaumont, 1989).

consist of the long hydrocarbon chains of the lipids and the nonpolar side chains of the constituent amino acids of the proteins. These comprise the bulk of the membrane. Hydrated polar heads of the phospholipids and polar amino acid side chains of the protein constitute the surface of the membrane (Figure 3.2C) and provide compatible interaction with aqueous compartments on either side. Hydrogen bonding in these aqueous phases provides support for the main structure of the membrane.

3.3.1 Toxicokinetics

DIFFUSION

Toxicants that penetrate the nonpolar portion of the membrane do so by a process of passive diffusion. Nonpolar organic compounds are highly lipid soluble and cross lipid membranes easily. This tendency to move from the aqueous to the lipid phase (hydrophobicity) is conveniently modelled by the octanol:water partition coefficient. Passage of nonpolar chemicals through the lipid portion of the membrane is principally limited by the molecular size and lipid solubility of the compound. Other limitations include the thickness and surface area of the membrane and the overall diffusive process is defined by Fick's law:

$$D = \frac{[C_0/C_i] \times s \times A}{\mathrm{MW}^{1/2} \times d} \tag{3.1}$$

where D is the diffusion rate, C_0/C_i is the concentration gradient, MW is the molecular weight of the diffusing chemical, s is its solubility in the membrane, and A and d are the membrane surface area and thickness, respectively.

In respiratory epithelia such as gills and lungs, where A may be very large and d very small, the solubility of the chemical in the membrane may become much less important, and in terrestrial animals exposed to gaseous compounds, s is better represented by blood solubility, which would limit absorption through respiratory epithelia.

COMPARTMENTAL MODELS

As we have seen, toxicant uptake can occur through direct absorption from ambient media such as air or water or through the ingestion of contaminated food. After ingestion, toxic chemicals follow a variety of pathways. They may enter short-term storage (i.e., tissues where they have a limited residence time), enter long-term tissue storage, or undergo rapid elimination or metabolism.

Interest in the uptake and retention of toxic substances stems from concern over their toxicity to the organism and their potential for passing toxicants through different trophic levels. The kinetics of toxic chemical transfer through food webs are further mentioned in later chapters (Sections 4.5.1, 4.5.2, and 5.4.1).

Here we describe some of the models that have been developed to describe the kinetics of chemical bioaccumulation by an individual organism. Models tend to be either empirical or mechanistic in nature, although many integrate these two approaches. Typical are the compartmental models developed initially by pharmacologists.

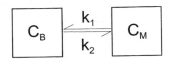

Figure 3.4 Single-compartment model. C_B, C_M = toxicant concentrations in body and external medium, respectively. k_1, k_2 = rate constants for uptake and elimination.

3.3.2 Single-compartment model

Bioaccumulation of a toxicant by an organism can be written as a mass balance equation wherein the net rate of accumulation is the difference between the uptake and loss (elimination) rates:

$$\frac{dC}{dt}B = k_1 C_M - k_2 C_B \tag{3.2}$$

where C_B is the toxicant concentration in the organism, C_M is the concentration in the ambient medium, and k_1 and k_2 are rate constants for uptake and loss, respectively (Figure 3.4).

Integration of (3.2) gives

$$C_B = \frac{k_1}{k_2} C_M \cdot (1 - e^{k_2 t}) \tag{3.3}$$

where C_B is the toxicant concentration in the organism at time t.

At low concentrations of toxicant, uptake is a first-order process wherein the rate of accumulation is proportional to the external concentration (Figure 3.5). As the toxicant concentration increases, a steady state is reached wherein the rate of uptake approaches the loss rate and the concentration in the organism reaches a plateau. At this point,

$$k_1 C_M = k_2 C_B \tag{3.4}$$

and

$$\frac{dC_B}{dt} = k_1 C_M - k_2 C_B = 0 \tag{3.5}$$

The system is now in a state of zero-order kinetics.

Following termination of toxicant exposure, the rate of uptake, $k_1 C_M$, falls to zero, and the elimination term becomes

$$\frac{dC_B}{dt} = -k_2 C_B \tag{3.6}$$

If this is integrated,

$$C_B = C_0 e^{-k_2 t} \tag{3.7}$$

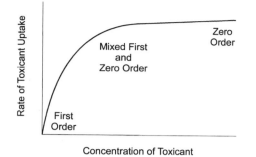

Figure 3.5 Rate of toxicant uptake versus toxicant concentration as described by first-order and zero-order kinetics.

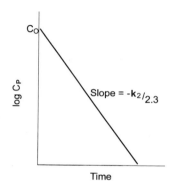

Figure 3.6 Kinetics of toxicant elimination from a single (plasma) compartment. C_P = toxicant concentration in plasma; toxicant concentration in plasma at time zero (i.e., $t = 0$). k_2 = elimination constant.

where C_0 is the concentration in the organism at the beginning of the elimination process. The logarithmic form of this equation,

$$\log C_B = \log C_0 - \frac{k_2 t}{2.3} \tag{3.8}$$

plots as a straight line with a slope of $-k_2/2.3$ (Figure 3.6).

The conventional way of expressing this information is in the form of biological half-life ($t_{1/2}$). This is the time taken for an organism to clear half of the toxicant content of its body. In a single-compartment system, it is the time for C_0 to be reduced by half (i.e., at $t_{1/2}$, $C = C_0/2$). Thus,

$$\ln \frac{C_0}{2} = \ln C_0 - k t_{1/2} \tag{3.9}$$

and

$$t_{1/2} = \frac{\ln 2}{k_2} = \frac{0.693}{k_2} \tag{3.10}$$

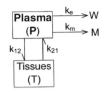

Figure 3.7 Two-compartment model. W = waste (excretion); M = metabolism; k_{12} = rate constant plasma → tissues; k_{21} = rate constant tissues → plasma; k_e = rate constant for excretion; k_m = rate constant for metabolism.

In multicelled organisms, the single-compartment model/system is a special case that is only approximated under certain circumstances, for example, (a) where the toxicant remains unchanged in the circulatory fluid and is not (or only very slowly) taken up by the tissues or (b) where the toxicant freely diffuses throughout the blood and tissues without any rate-limiting diffusional barrier.

3.3.3 Two-compartment model

Often the time-course of toxicant concentration in blood plasma following its rapid introduction (e.g., by injection) is a curve rather than a straight line. Such a curve is the result of toxicant distribution into more than one compartment. The simplest case is the two-compartment system where the toxicant distributes rapidly into the plasma and tissues, but its excretion and metabolic transformation (collectively termed elimination) proceed more slowly. The differential equation for this model is

$$\frac{dC_P}{dt} = k_{21}C_T - k_{12}C_P - k_2C_P \tag{3.11}$$

where C_P and C_1 are toxicant concentrations in plasma and tissues, respectively; $k_2 = [k_e + k_m]$; and k_e, k_m, k_{12}, and k_{21} are, respectively, the first-order rate constants for elimination, metabolic transformation, distribution from plasma to tissues, and distribution from tissues to plasma. A schematic representation of this model is shown in Figure 3.7.

Integration gives

$$C_P = Ae^{-\alpha t} + Be^{-\beta t} \tag{3.12}$$

where A, B, α, and β are complex constants of k_{12}, k_{21}, and k_2; in other words,

$$A = C_0 \cdot (\alpha - k_{12})/\alpha - \beta \tag{3.13}$$

$$B = C_0 \cdot (k_{12} - \beta)/\alpha - \beta \tag{3.14}$$

and

$$\alpha + \beta = k_{12} + k_{21} + k_2 \tag{3.15}$$

A semi-logarithmic plot of C_P versus time gives a biphasic curve with two linear portions, a steep distribution phase with slope α and a more shallow elimination

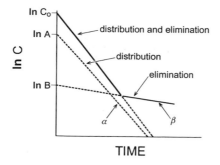

Figure 3.8 Kinetics of toxicant elimination from a two-compartment system. As in Figure 3.6. A, B = ordinate intercepts for distribution and elimination phases; α, β = slopes of distribution and elimination phases.

phase with slope β (Figure 3.8). In the integrated curve, the initial portion represents both distribution and elimination, whereas the terminal portion represents elimination only. The elimination curve may be extrapolated to $t = 0$, and this can be separated from the initial portion of the curve to enable characterisation of the distribution phase only. A and B are the ordinate intercepts of the distribution and elimination phases, respectively. Thus, α, β, A, and B may be derived graphically, and from these values k_{12}, k_{21}, and k_2 may be determined as follows:

$$k_{21} = \frac{A\beta + B\alpha}{A + B} \qquad (3.16)$$

$$k_{12} = \alpha + \beta - k_{21}(-k_2) \qquad (3.17)$$

$$k_2 = \frac{\alpha\beta}{k_{21}} \qquad (3.18)$$

Although it might be considered that the characteristics of these elimination curves would be governed by gross molecular properties such as size, shape, and electrical charge, sometimes relatively minor changes in molecular configuration may have significant effects on the elimination kinetics. Figure 3.9 illustrates the effect of altering the position of a chlorine atom on a chlorinated aniline on its elimination from zebra fish, *Brachydanio rerio* (see Table 5.3). A shift in the chlorine from the ortho to the meta position significantly extends the distribution phase, suggesting a larger volume of distribution (see following discussion) relative to the ortho form (indicated by the dashed line in Figure 3.9).

In some cases, several different tissues (each with different rate constants) might be substituted for the single-tissue compartment described earlier. If, for any tissue, the rate constant from blood to tissue (k_{12}) = the rate constant from tissue to blood (k_{21}), that tissue can be functionally combined with blood as a single compartment. Beginning with a single source of toxicant available to the organism, the relative changes in toxicant dose occurring in compartments P (plasma), T (tissues), M (metabolised toxicant), and W (excreted toxicant) are shown in Figure 3.10.

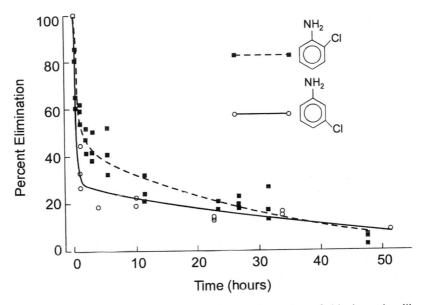

Figure 3.9 Effect of change in molecular configuration of chlorinated aniline on elimination from the zebra fish, *Brachydanio rerio* (after Kalsch et al., 1991).

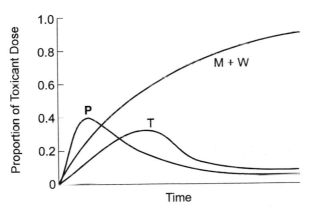

Figure 3.10 Relative changes in a single toxicant dose in different body compartments during uptake. P = plasma compartment; T = tissue compartment; M + W = sum of metabolism and waste compartments.

Changes in the various rate constants have a marked effect on the shape of these curves. In classical pharmacokinetic studies, much can be learned by monitoring the central (blood) compartment in a human or a laboratory animal. Figure 3.11 shows the consequences of changing different parameters in a toxicokinetic study. The ability of a chemical to exchange between the blood compartment and one or more tissue compartments owes much to its physical and chemical characteristics. A large, relatively hydrophilic molecule is much more likely to remain in the bloodstream than small, hydrophobic molecules, which will be rapidly distributed to tissue lipids. This phenomenon is addressed through a concept known as volume of distribution.

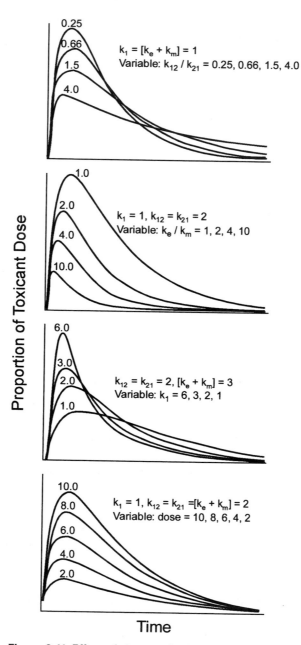

Figure 3.11 Effect of changes in dose and different rate constants on uptake curve seen in central (plasma) compartment. See Figure 3.7 for description of compartments and rate constants. Modified from Notari (1987).

3.3.4   Volume of distribution

If a known dose of toxicant is injected into the blood of an organism, we should be able to estimate the blood volume occupied by the toxicant by measuring the toxicant concentration after equilibration in the bloodstream, but before distribution to other tissues. For a one-compartment system, the volume of distribution, V_d, can be calculated from the equation

$$V_d = \frac{\text{Dose}}{C_0} \qquad (3.19)$$

where C_0 is the extrapolated toxicant concentration at time zero.

For a two-compartment system

$$V_d = \frac{\text{Dose}}{\beta \times \text{AUC}} \qquad (3.20)$$

where the area under the curve (AUC) from $t = 0$ to $t = \infty$ is expressed by

$$\text{AUC} = \frac{A}{\alpha} + \frac{B}{\beta} \qquad (3.21)$$

The volume of distribution is a hypothetical measure because it may describe a volume much greater than the actual volume of the organism. As such, it is more accurately described as the apparent volume of distribution. A high V_d is seen where a toxicant is strongly bound to protein or lipid in tissues. Conversely, a toxicant with poor binding properties will display a low V_d.

Much of the existing information on toxicokinetics has been derived from pharmacological studies where toxicant sources are usually carefully circumscribed in terms of dose. In the field of environmental toxicology, in addition to resolving differences between different toxicant sinks such as tissue deposition and products of elimination, a good deal of recent attention has been devoted to differentiating between different toxicant sources, which are often poorly delineated. For example, food chain studies are often concerned with the degree to which organisms ingest toxicants from food as opposed to other sources and different food sources (e.g., plant, animal, seafood). In some respects, the models referred to in Section 4.6.2 are attempts to expand toxicokinetics from the single organism to the ecosystem level. Whereas the fish bioaccumulation model referred to in Section 4.6.2 follows a similar approach to that outlined in this chapter, the models designed to quantify broader environmental "compartments" and fluxes represent means of achieving similar goals, but on a different scale. Where large compartments are being measured (e.g., water, sediment, plankton, benthos), toxicant concentration may be analysed directly by taking environmental samples, and the dynamics of the system (equivalent to rate constants) may be estimated by a variety of methods including isotope fluxes, fugacity estimates, and estimates based on following chemicals in various media over time.

3.3.5 Transporter-mediated transport

To understand the mechanistic aspects of toxicant uptake, it is important to know whether specific transport mechanisms are involved. Some toxicants are transported across membranes through their association with transporter proteins. Where no energy is involved in the process, it is termed facilitated diffusion. Where energy is required, the process is referred to as active transport. Both facilitated diffusion and active transport display saturation characteristics reflecting the fact that the toxicant is reversibly bound to a finite number of carrier proteins in the membrane.

The kinetics of a saturable transmembrane carrier system can be explained by analogy to enzyme substrate dynamics. We assume that the system has a finite number of carrier proteins available for chemical (substrate) transport.

> Total molar concentration of protein carrier, P_0
> Molar concentration of unoccupied carrier, P
> Molar concentration of free chemical substrate, S
> Molar concentration of substrate carrier complex, PS

The dissociation constant for substrate chemical interaction is the Michaelis constant, K_m.

$$K_m = \frac{(P) \times (S)}{(PS)} = \frac{(P_0 - PS) \times (S)}{(PS)} \tag{3.22}$$

$$K_m(PS) = (P_0 - PS) \times (S) \tag{3.23}$$

Thus,

$$PS = \frac{(P_0) \times (S)}{K_m + (S)} \tag{3.24}$$

The velocity or rate of formation of PS is $v = k \times (PS)$, and the rate is

$$v = \frac{k \times (P_0) \times (S)}{K_m + (S)} \tag{3.25}$$

When the carrier reaches saturation, $v = k \times (P_0)$. This can be described as the maximum rate, v_{max}, at which the system will operate and

$$v = \frac{v_{max} \times (S)}{K_m + (S)} \tag{3.26}$$

where v_{max}, S, and v may all be determined experimentally, and K_m is the substrate concentration at which the carrier is half saturated. This may be derived graphically from plots such as the Michaelis-Menten (Figure 3.12) or Scatchard plots (Figure 3.13). The K_m value derived from these plots provides a measure of the affinity of the carrier for the chemical (substrate). A smaller K_m value indicates a higher carrier-substrate affinity.

Where toxicants inhibit carrier-mediated transport of an endogenous substrate, they may do so in either a competitive or noncompetitive manner. Competition results in a change in value of K_m with respect to the normal substrate. Such a sit-

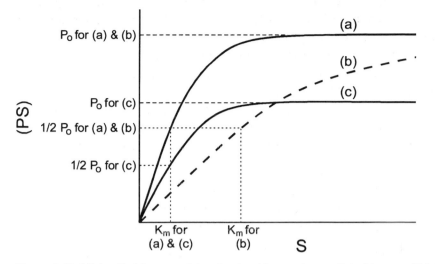

Figure 3.12 Michaelis-Menten plots of saturable, carrier-mediated transepithial transport. P_0 represents the total number of available sites on the carrier protein for three different systems (a), (b), and (c). (a) and (b) have the same carrying capacity. (a) and (c) have the same substrate affinity as determined by the Michaelis constant K_m, which describes the substrate, S, concentration at which the carrier is half saturated. PS is the carrier-substrate complex.

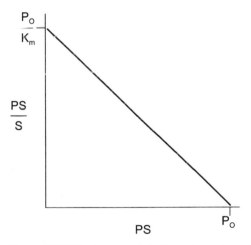

Figure 3.13 Scatchard plot of carrier substrate kinetics. Parameters are the same as in Figure 3.12.

uation would cause a reversion from curve a (Figure 3.12) to curve b wherein the total number of carrier sites (P_0) remains unaltered. Note that a lowered affinity for the substrate is reflected by a higher K_m value. Noncompetitive inhibition, on the other hand, is characterised by a reduction in the overall number of available carriers. This reduction results in a reduced value for the upper asymptote of the saturation curve (i.e., a lower maximum flux rate), although K_m is not affected (Figure 3.12, curve c). Trace metals such as lead and cadmium may gain access to

A. Model of human P-glycoprotein based on amino acid sequencing

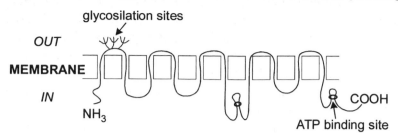

B. Schematic of P-glycoprotein acting as an efflux pump for drugs

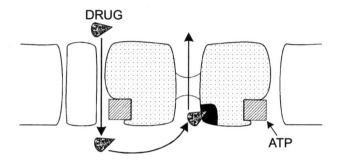

Figure 3.14 Structural configuration model (A) and functional schematic (B) of the mammalian multidrug resistance protein P-glycoprotein.

Ca ATPases present on the basolateral membranes (serosal side) of epithelial membranes via voltage-gated Ca^{2+} channels on the apical (mucosal) side of the membrane. In some cases, cadmium has been shown to have a higher affinity for the Ca^{2+} carrier than calcium itself. The herbicide paraquat has its primary site of action on the mammalian lung and moves from blood into lung tissue via an active transport mechanism. The MDR protein P-glycoprotein (Figure 3.14) is a further example of an ATP-dependent toxicant transporter.

3.3.6   Lethal body burden (critical body residue)

Although toxicokinetic information is important in understanding chemical toxicity, it has only limited relevance to environmental toxicology. Part of the problem is that most toxicity bioassays with application to field situations are based on the chemical concentration in the exposure medium rather than dose. Such bioassays may not, in fact, reflect the true route of environmental exposure (e.g., toxicant uptake from food and particulates may be more important than the accumulation of soluble toxicant).

For several chemicals, their relative toxicity may be dominated by their physical characteristics. For example, narcosis caused by organic chemicals with similar modes of action is related to their hydrophobicity (Könemann, 1981; Veith et al.,

1983). This relation does not necessarily reflect their inherent toxicity. The differences in the slope of the response curves seen in Figure 2.4 may reflect differences in toxic action or bioaccumulation, but we cannot differentiate between them using such a test. A toxic chemical of lower potency but rapid bioaccumulatory potential may elicit toxic symptoms faster than an ultimately more toxic compound that is taken up more slowly. Even though physicochemical properties are important in assessing relative chemical toxicity (see Section 5.5), important information may be lost unless dose is considered.

These considerations have led to an exploration of the concept of lethal body burden, also referred to as critical body residue, as a basis for determining aquatic toxicity (see also Section 2.3.6). Such a concept is analogous to dosimetry, which represents an integral part of mammalian toxicology. In aquatic toxicology, the LBB is defined as the molar concentration in the organism at the time of death. It was initially derived theoretically (Friant and Henry, 1985), but more recently it has been supported by experimental data from a variety of lipophilic organic compounds. Data indicate that chemicals with higher K_{OW} tend to have higher LBBs in organisms with higher fat content (Lassiter and Hallam, 1990; van Wezel et al., 1995; Geyer et al., 1994), although such a relation may not hold for more hydrophilic compounds (van den Heuvel et al., 1991). This relation suggests that neutral "storage" fats probably have a sparing effect on the toxicity of highly lipophilic organic compounds.

The importance of the LBB approach is that it represents an attempt to equate toxicity to actual tissue concentration in the organisms being examined. Very often, in a field situation, knowledge of chemical exposure history (i.e., time and external concentration) is poor, and tissue chemical concentration is the only reliable information available to the toxicologist. Models that use toxicokinetic principles but use body burden, or total toxicant content, as the basis for toxicity assessment have obvious advantages in environmental toxicology.

The LBB has been used in conjunction with a first-order toxicokinetics model to estimate LC_{50} values from exposure information. For example, in a situation where chemical uptake is principally from food, for a fixed exposure time t and chemical concentration c, survival can be expressed as

$$S(t,c) = \frac{e^{-\mu t}}{1+\left(\dfrac{c}{LC_{50}(t)}\right)^b} \tag{3.27}$$

where $S(t,c)$ is the survival time t (days) at external concentration c ($\mu g\,g\,dry\,wt^{-1}$), μ is the natural mortality rate (day^{-1}), $LC_{50}(t)$ is the LC_{50} as a function of t ($\mu g\,g$ dry wt food^{-1}), and b is the slope of the survival function.

Toxicant uptake may be described by the single compartment uptake-elimination equation:

$$D(t,c) = (k_1/k_2) = (1 - e^{-k_2 t}) \tag{3.28}$$

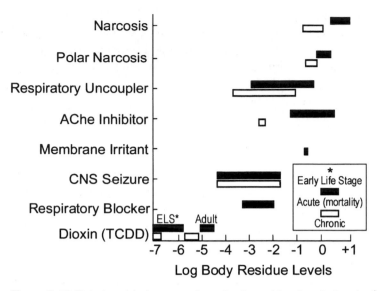

Figure 3.15 Relationship between $\log_{10}$ body residue levels (mg kg^{-1}) and various types of sublethal toxic end-points. Modified from McCarty and Mackay (1993).

where $D(t,c)$ is the chemical concentration (μg g dry wt^{-1}) in the organism following an exposure of t days to b constant concentration c in food (μg g dry wt^{-1}), k_1 is the uptake rate constant (day^{-1}), and k_2 is the elimination constant (day^{-1}).

In this case, the relationship between time and toxicity may be expressed as

$$LC_{50}(t) = \frac{LC_{50\infty}}{1 - e^{-k_2 t}} \tag{3.29}$$

and

$$LC_{50\infty} = LBB(k_2/k_1) \tag{3.30}$$

where LBB is the lethal body burden (μg g dry wt^{-1}), $LC_{50\infty}$ is the ultimate LC_{50} value (μg g dry wt^{-1}), and $LC_{50}(t)$ is the LC_{50} (μg g dry wt^{-1}) following t days of exposure.

In this model, the value for LC_{50} will decrease with increasing exposure time until the chemical concentration in the organism reaches equilibrium with the concentration in the food and LC_{50} reaches the ultimate value.

Alternatively, the LBB can be estimated as the product of the threshold LC_{50} and the bioconcentration factor (Hickie et al., 1995). At present, the LBB approach has been chiefly applied to acute end-points, either mortality or narcosis. Fewer data are available concerning its applicability to chronic toxicity. Figure 3.15 summarises LBBs associated with various acute and chronic toxicity end-points in fish.

SAMPLE PROBLEMS

A. Following the administration of a single toxicant dose to a 5-kg animal, the toxicant becomes completely distributed throughout the bloodstream and achieves a concentration of $20\,mg\,L^{-1}$. At this time, there has been no biotransformation of the toxicant or significant flux into the tissues, although 15% of the toxicant has been excreted unchanged. The toxicant is assumed to have a volume of distribution equivalent to 20% vol/wt. What is the amount in grams of the original dose?

> SOLUTION. The apparent volume of distribution is $0.2 \times 5\,kg = 1\,L$. If it is assumed that the toxicant remains in the blood compartment, the mass of toxicant required to achieve a concentration of $20\,mg\,L^{-1}$ must be $20\,mg$. However, 15% has already been excreted. Therefore, the original dose must have been $20 \times 1.15 = 23\,mg$.

B. Toxicant concentration in the plasma of an organism declines according to the following time-course:

Time (h)	Plasma concentration (mg)
0	10
0.5	6.2
1.0	4.1
1.5	3.3
2.0	2.4
3.0	1.9
4.0	1.4
5.0	1.3
7.0	1.0

Calculate the values of k_2 ($= k_e + k_m$), k_{12}, and k_{21}, where k_e, k_m, k_{12}, and k_{21} are, respectively, the first-order rate constants for elimination, metabolic transformation, distribution from plasma to tissues, and distribution from tissues to plasma.

> SOLUTION. Plasma toxicant concentration is plotted semi-logarithmically against time as shown in Figure 3.16 (crosses). A straight line is then drawn through the later portion of the curve and interpolated back to the ordinate. This line represents the elimination phase and intercepts the ordinate at B. Values for the distribution phase can be obtained graphically by plotting the deviation of the initial part of the curve from the straight-line elimination plot (e.g., after 1 h the curve is exactly $2\,mg\,L^{-1}$ above the line). These separate plots form the steeper line (circles), which represents the distribution phase. This intercepts the ordinate at A. The slopes of the distribution and elimination phases are α and β, respectively. Values for A, B, α, and β can

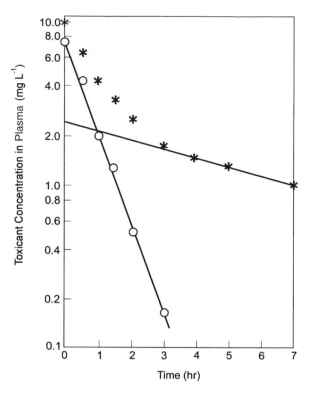

Figure 3.16 Graphical solution to determine first-order rate constants for [elimination and metabolic transformation], distribution from plasma to tissues and from tissues to plasma.

all be calculated from the graph and are as follows: $A = 7.7\,\mathrm{mg\,L^{-1}}$, $B = 2.4\,\mathrm{mg\,L^{-1}}$, $\alpha = 2.2\,\mathrm{h^{-1}}$, $\beta = 0.2\,\mathrm{h^{-1}}$.

These values can all be substituted in Equations 3.16, 3.17, and 3.18 to give the following values: $k_{21} = 0.68\,\mathrm{h^{-1}}$, $k_{12} = 1.07\,\mathrm{h^{-1}}$, and $k_2 = 0.65\,\mathrm{h^{-1}}$.

3.4   Questions

1. Describe the various ways by which toxicants are transported across epithelia. What are the limiting factors influencing transepithelial transport?

2. What is the relationship between the pH of a salt of a weak acid or base and its degree of ionization? What are the implications of this for solute uptake by the gastrointestinal tract?

3. How does the structure of the mammalian respiratory system affect the nature of toxic particulates and gases taken up by this route? Give a short account of some of the respiratory diseases resulting from the inhalation of metals and metal compounds.

4. Write an essay on toxicant uptake by plants.

5. Define the following terms: first-order kinetics, zero-order kinetics, two-compartment model, apparent volume of distribution, Michaelis constant.

6. Describe the general form of the Michaelis-Menten equation, explaining the meaning of each of the terms in the equation. How would you design an experiment to determine the K_m for a toxicant (x) transported into a single-compartment "system" by a carrier-mediated mechanism? How would a competing molecule (y) be likely to affect the value of K_m for x? How would you design an experiment to determine the relationship between x and y in this respect?

7. How has the concept of critical body residue been used as a basis for determining aquatic toxicity?

3.5 References

Friant, S. L., and Henry, L. (1985) Relationship between toxicity of certain organic compounds and other concentrations in tissues of aquatic organisms: A perspective, *Chemosphere*, **14**, 1897–907.

Geyer, H. J., Scheunert, I., Brüggemann, R., Matthies, M., Steinberg, C. E., Zitko, V., Kettrup, A., and Garrison, W. (1994) The relevance of aquatic organisms' lipid content to the toxicity of lipiophilic chemicals: Toxicity of lindane to different fish species, *Ecotoxicology and Environmental Safety*, **28**, 53–70.

Gibaldi, M. (1990) *Biopharmaceutics and Clinical Pharmacokinetics*, Williams and Wilkins, Baltimore.

Hacon, S., Artaxo, P., Gerab, F., Yamasoe, M. A., Campos, R. C., Conti, L. F., and de Lacerda, L. (1995) Atmospheric mercury and trace elements in the region of Alta Floresta in the Amazon Basin, *Water, Air, and Soil Pollution*, **80**, 273–83.

Hickie, B. E., McCarty, L. S., and Dixon, D. G. (1995) A residue-based toxicokinetic model for pulse-exposure toxicity in aquatic systems, *Environmental Toxicology and Chemistry*, **14**, 2187–97.

Hughes, G. M., and Shelton, G. (1962) Respiratory mechanisms and their nervous control in fish, *Advances in Comparative Physiology and Biochemistry*, **1**, 275–364.

Kalsch, W., Nagel, R., and Urich, K. (1991) Uptake, elimination, and bioconcentration of ten anilines in zebrafish (*Brachydanio rerio*), *Chemosphere*, **22**, 351–63.

Könemann, H. (1981) Quantitative structure-activity relationships in fish toxicity studies. Part I. Relationships for 50 industrial pollutants, *Toxicology*, **19**, 209–21.

Lassiter, R. R., and Hallam, T. G. (1990) Survival of the fattest: Implications for acute effects of lipophilic chemicals on aquatic populations, *Environmental Toxicology and Chemistry*, **9**, 585–95.

McCarty, L. S., and Mackay, D. (1993) Enhancing ecotoxicological modeling and assessment: Body residues and modes of toxic action, *Environmental Science and Technology*, **27**, 1719–28.

Notari, R. E. (1987) *Biopharmaceutics & Clinical Pharmacokinetics: An Introduction.* 4th ed. Dekker, New York.

Paterson, S., Mackay, D., and McFarlane, C. A. (1994) A model of organic chemical uptake by plants from soil and the atmosphere, *Environmental Science and Technology*, **28**, 2259–66.

Pyatt, F. B., and Beaumont, E. H. (1989) Three years after Chernobyl: Some gamma radioactivity values from Barra, Outer Hebrides, UK, *The Environmentalist*, **9**, 213–17.

Sutter, T. R., Tang, Y. M., and Hayes, C. L. (1994) Complete DNA sequence of human dioxin-inducible mRNA identifies a new gene subfamily of cytochrome P-450 that maps to chromosome 2, *Journal of Biological Chemistry*, **269**, 13092–9.

van den Heuvel, M. R., McCarty, L. S., Lanno, R. P., Hickie, D. E., and Dixon, D. G. (1991) Effect of total body lipid on the toxicity and toxicokinetics of pentachlorophenol in the rainbow trout (*Oncorhynchus mykiss*), *Aquatic Toxicology*, **20**, 235–52.

van Wezel, A. P., deVries Dieuuke, A. M., Kostense, S., Sijm, D. T. H. M., and Opperhuizen, A. (1995) Intraspecies variation in lethal body burdens of narcotic compounds, *Aquatic Toxicology*, **33**, 325–42.

Veijalainen, H. (1988) The applicability of peat and needle analysis in heavy metal deposition surveys, *Water, Air, and Soil Pollution*, **107**, 367–91.

Veith, G. C., Dall, D. J., and Brooke, L. T. (1983) Structure-activity relationships for the fathead minnor *Pinephales promelas*: Narcotic industrial chemicals, *Canadian Journal of Fisheries and Aquatic Sciences*, **40**, 734–48.

Wellburn, A. (1994). *Air Pollution and Climate Change – The Biological Impact*, 2nd ed., Longman, London.

3.6   Further reading

Gibaldi, M. 1990. *Biopharmaceutics and Clinical Pharmacokinetics*, Williams and Wilkins, Baltimore.

4

○ ○

Methodological approaches

4.1   Introduction

In Chapter 2, attention was drawn to some of the distinctions between classical toxicology and environmental or ecotoxicology, as well as to the fact that the two types of toxicology share considerable common ground. For example, concepts such as acute and chronic toxicity and thresholds, which were developed for classical toxicology, have been applied to environmental toxicology (see Sections 2.2.2, 2.2.3, and 2.2.4). Classical toxicology has relied mainly on evidence from controlled exposure of individual organisms or from epidemiological approaches including retrospective case studies, and more recently it has seen the development of more generic tests on cell lines and microorganisms. Environmental toxicology includes comparable types of testing but by its very nature has to go beyond the responses of individual organisms or populations.

The development of methods for determining the impact of man on the environment has advanced on several different fronts. The term *ecotoxicology*, first used by Truhaut as recently as 1969 (see Chapter 1), encompasses the study of all levels of biological organisation described in Figure 2.1. At lower levels of biological organisation (subcellular to individual), the approaches used have much in common with human toxicology. Indeed, the distinction between the two disciplines may be somewhat arbitrary, particularly because so many aspects of environmental toxicology have implications for human health. A clearer distinction between the two disciplines is seen in studies at the population and community levels where methods for determining the impact of man on ecosystems have matured in parallel with a better understanding of ecology. Ecology is itself a young science. Concepts such as trophic layers, food chains, and food webs were introduced in 1927 in the book *Animal Ecology* by British ecologist Charles Elton (1900–91), who helped to establish ecology as a quantitative science with dynamic properties. Incorporation of the effects of environmental pollution into the science of ecology is one of the most fundamental challenges confronting toxicologists today.

Because of the nature of the differences between classical toxicology on one hand and environmental or ecotoxicology on the other, different methodological approaches have varying degrees of applicability between the two branches. Rand (1995) made (for aquatic toxicology) a succinct summary of the differences between mammalian and ecotoxicology (Table 4.1). His final point states that for mammalian toxicology "Test methods are well developed, their usefulness and limits well understood", whereas for ecotoxicology, "Many commonly used test methods are relatively new and some are formalised (standardised). However, their usefulness in predicting field impacts and protecting natural ecosystems is often uncertain". This statement sums up many of the concerns that the approaches to be described in the present chapter may help to address.

This chapter deals with some of the ways in which methods have been developed, modified, and applied in the study of toxic substances in the environment. In the main, it addresses the questions and problems that are specific to environmental toxicology. Not surprisingly, these approaches show considerable departure from the classical or traditional approaches to toxicology. As indicated by Rand, many of these methods and approaches are still relatively new. This too should come as no surprise. The over-all impression that emerges from these discussions is one of considerable variability in needs, opinions, and reliability. However, along with the unavoidable uncertainty and incompleteness, this young science typically provides some exciting opportunities for real progress in basic science and its application toward environmental protection.

Environmental toxicology has involved a wide range of chemical, physical, and biological approaches, and there is little doubt that many important advances in the study of environmental toxicology have been facilitated, some might say driven, by technical developments such as improved analytical capabilities and the availability of more powerful computers. This aspect was touched upon already in Chapter 1. For example, prior to the availability of sophisticated computers, certain types of complex information and large data sets simply could not be handled in any practical manner. This advantage has relevance to a number of disciplines, not the least to ecology and related studies, especially those involving collections of field data. Data from large-scale surveys can now be manipulated and subjected to statistical tests in ways that reveal relationships that cannot reliably be deduced from experimental studies in the laboratory. Computers have also facilitated various types of mathematical modelling, some of which are now in routine use in environmental toxicology. Research continues in the development and application of models to environmental toxicology.

In the course of the present chapter, the theoretical as well as the practical aspects of various approaches will be introduced, discussed, and evaluated. We consider biological indicators and biomarkers, various types of ecological indicators, and modelling, all in the context of environmental or ecotoxicology. There is also some consideration of scale, or level, of biological organisation (from subcellular to ecosystem). At the end of this chapter, we make an appraisal of standardised versus other means of assessing toxic damage and provide some guidance on the appropriate choices of methods and approaches.

Table 4.1. Differences between classical toxicology and ecotoxicology

Mammalian toxicology	Ecotoxicology
Objective is to protect *humans* from exposure to toxic substances at concentrations that are actually or potentially harmful.	**Objective** is to protect populations and communities of *many diverse species* from exposure to toxic substances at concentrations that are actually or potentially harmful.
Subject of investigation cannot be subjected to experimentation; therefore, it must always rely on *animal models* (e.g., rat, guinea pig).	**Species of concern** can be subjected to *direct experimentation*, although choice of indicator or sensitive species may be uncertain.
Species of interest (human) is known; thus, the degree of extrapolation is fairly certain.	**Not able to identify all the test species of concern**; thus, the degree of extrapolation is uncertain. Furthermore, organism responses may be different in complex systems from those in laboratory test conditions.
Test organisms are warm-blooded, and body temperature is independent of environmental temperature; thus, *toxicity is predictable in the context of temperature effects*.	**Test organisms** (particularly aquatic) live in variable environment, and, except for birds and mammals, body temperature varies with environmental temperature. Thus, *toxicity in the context of temperature is less predictable*.
The dose of a chemical can be measured directly, and route(s) of administration can be controlled. The *absorbed dose* should be made through tissue dosimitry, but in fact many estimates are based on an external or exposure dose.	**The external or exposure dose and the length of exposure** is known indirectly from measurements in the water, sediment, food, etc.; the *absorbed dose* can be determined experimentally through bioconcentration/bioaccumulation studies.
Mechanisms of toxic action are fairly well studied through *extensive basic research*.	**Mechanisms of toxic action and structure-activity relationships** have recently been emphasized through basic research; otherwise, emphasis has been on *measuring effects and generating threshold concentration data*, addressing regulatory needs.
Test methods are well-developed; their *usefulness and their limits are well understood*.	**Commonly used test methods** are relatively new, and some are standardised, but *their usefulness in the context of ecosystems is often uncertain*.

Modified from Rand (1995).

4.2   The general concepts and principles for biological indicators

The general theory or principle that underlies the use of biological indicators is that organisms respond to some specific condition, or change in a condition, of the environment in which the biological system exists, and that the biological response can be measured.

The dictionary definition of an indicator is "person, thing, that points out, especially a recording instrument attached to apparatus, etc." The word originates from the Latin *dicare*, to make known. Biological indicators can, therefore, be seen as a means of making known a condition or state. It is instructive to consider the relative usefulness and application of biological as compared with chemical measurements in environmental toxicology. In the pursuit of studying and understanding environmental toxicology, the protection of the biological system(s) should be of primary concern. In keeping with this precept, the concept of a biological indicator includes the idea that the biological response is in some way or ways more desirable or useful than direct physical or chemical measurements alone. Furthermore, a biological indicator is normally expected to integrate over time the effects of a perturbation, whereas chemical measurements are typically taken at points in time. A measurement taken at a single point in time could "miss" some short-term but serious perturbation of the ecosystem such as a chemical spill.

Ecological indicators are biological responses that indicate the condition of the ecosystem, but an ecological indicator is not necessarily a measurement at the ecosystem level. For example, the concentration of chlorophyll-*a* in lake water at midsummer is a biological measurement of the phytoplankton community, yet it is used as an indicator of the overall condition of the lake ecosystem in the context of phosphorus concentration. High concentrations of phosphorus lead to cultural eutrophication (Case Study 4.5 and Section 6.10), which evokes responses of many trophic levels in the lake. To assess the extent of eutrophication, one could measure phosphorus concentrations directly or use some fish population or community parameter. But for routine assessment, the chlorophyll-*a* determination is a reliable and relatively simple indicator. In this instance, the mechanism is understood. In essence, the indicator "works" because the relationship between phosphorus concentration and chlorphyll-*a* is predictable within an acceptable range of uncertainty (Dillon and Rigler, 1974). However, all indicators are by no means as satisfactory.

For the ideal biological indicator, a number of primary criteria should be satisfied. The response should be quantifiable, specific to the perturbation, observable both in the natural environment and in experimental conditions, reproducible and relevant to the overall functioning of the system.

The choice of biological responses that occur as a result of chemical perturbations is theoretically unlimited, yet in practical terms there are many limitations on the choice of a biological indicator. Obviously death, expressed as mortality (see

Table 4.2. Changes in flower colour of fire weed (Epilobium angustifolium) in relation to radioactivity

Petals	Sepals	Filament	Capsule
Normal forms			
Light pink	Clear red	White to pale pink at base	Red above, pale green beneath
Magenta	Dark magenta	Very pale pink, turning white	Reddish above, greenish beneath
Abnormal forms			
Pale rose pink	Cerise	Pure white	Red above, paler beneath
Intense magenta	Purple, darker than petals	Pure white	Pale pink
Pale magenta	Reddish purple	Pure white	Paler than normal
Very pale rosy pink	Clear rose red	White, pink at base	Dusky pink

Selected and modified from Shacklette (1964).

Sections 2.2.1 and 2.2.2) is a biological response, and it can be argued that the acute lethal bioassay is the ultimate biological indicator! But because death may be caused by many conditions, or combinations of conditions, as an indicator, particularly in the ecosystem, mortality is not generally very useful. Not only does it lack subtlety, but it also lacks specificity: in many instances, particularly under natural conditions, cause and effect of death cannot be scientifically linked. Even so, lethal effects in field situations have, particularly in the decades prior to the 1980s, alerted the public and the scientific community to the presence of toxic substances in an environment. Such retrospective, often accidental, case studies are clearly not the most desirable means of studying environmental toxicology: A prospective or predictive approach is preferable.

A useful biological indicator would be a nonlethal response, consistent with the primary criteria listed earlier. An extensive variation in the flower colour of *Epilobium angustifolium* (fireweed) was observed by Shacklette (1964) when this species was growing over uranium deposits. The radioactivity caused mutational changes in the colour of various flower parts (Table 4.2). Morphological and colour changes related to elements in the substrate have been used extensively by trained geobotanists to detect mineralisation in soils and rocks. Metal ions can affect flower colour, particularly through their chemical reactions with anthocyanin pigments. In the nickel-accumulating plant *Hybanthus floribundus* in Western Australia, Severne (1974) noted that colour variations from white to deep blue were to some extent related to the nickel content of the soil. Perhaps more familiar is the colour change of the hydrangea flower from red to blue when salts of iron or aluminium are added to soil, a device known by generations of gardeners. The mechanism for the colour change is related to the complexation of anthocyanin oxonium salts with iron or aluminium. These types of indicator are nonlethal, easily recognisable, and fairly substance-specific: The element caused the pigment to develop or become chemically modified. But they are still subject to the natural variability of the colour, are somewhat subjective, and could be considered as qualitative indicators. Better still would be a colour change that could be quantified, as, for example, if intensity of colour increased in some manner that could be related to the concentration of an element or other chemical substance in the substrate.

The use of ecological indicators in ecotoxicology has been advocated for a number of practical issues, including baseline assessment, environmental impact assessment, routine monitoring and rehabilitation or recovery after damage. The ideal indicator should, in addition to the primary criteria, provide some "early warning" of change. In order to do this, one might look for a response that is not only nonlethal, but which does not involve a process that is crucial to the system, yet is readily measurable and specific to the perturbation of interest.

That having been said, it is immediately apparent that a potential conflict among some of the criteria has been identified. If the indicator is reliable and of significance to the ecosystem, it is likely also to be a crucial or vital component of the system, and by the time it has shown a response, it may be "too late" because the whole system may already be damaged. On the other hand, if the response is specific to the perturbation (normally the chemical substance), and readily measured, it may be of limited relevance to the functioning of the biological system. The latter criticism, which is of considerable concern, has been levelled particularly at the use of biochemical markers (called biomarkers by some authorities, but see our definition, Section 4.4). This criticism is discussed in Section 4.4.1 for biochemical markers.

Biological response to a change in the environment (the change in the environment variously referred to as perturbation or stress) can occur and be measured at many levels, from the subcellular to the whole ecosystem. One can cite the production of specific proteins called heat shock proteins, which is a conveniently measured biochemical response to unfavourable temperature. At the other end of the scale, one can cite the alteration of the trophic structure of a forest resulting from the complete elimination of one tree species due to a plant disease. In between, one could cite the example of a change in the relative numbers of phytoplankton species resulting from acidification of a lake. Each of these examples illustrates a relationship between the cause (the perturbation) and the response. Each of these types of biological responses can be measured. The selection of the most appropriate biological or ecological indicator, if indeed a decision is made to take this route, will almost inevitably be a compromise among scientific, practical, and economic concerns. In Section 4.9 the bases for such choices are discussed.

Development of biological monitoring from theoretical considerations into a practical tool has occupied many years of effort. Probably one of the earliest, and certainly one of the most commonly cited, examples is the application of the canary's sensitivity to carbon monoxide and the use of the miner's canary as an early warning of the gas reaching potentially toxic concentrations. The canary was a practical and direct indicator and also illustrates the concept of a surrogate. The canary is a surrogate for the human, and it also satisfies many of the criteria listed earlier for indicators. As an aside, note that instruments that measure the concentration of carbon monoxide, in combination with knowledge of the toxicological threshold for damage, have now supplanted the biological indicator for assessing the condition of the environment in mines, but in earlier times, the canary was more sensitive and reliable than were the available chemical measurements.

Much more recently, with technical improvements in our ability to measure subcellular biochemical changes or molecular responses, attempts have been made to "calibrate" biochemical responses for specific chemical or physical perturbations. These biochemical and molecular responses are referred to as biochemical markers or biomarkers (Landis and Yu, 1999) and, as such, are a particular category of indicators. Caution is advised, however, in the use of the term *biomarkers*, since some authors have used it in a broader sense, even as synonymous with biological indicators. In the present text, we make a compromise and use the term *biochemical markers* for indicators at the molecular or biochemical level.

Since the subject of biomarkers has received a great deal of attention in recent years, and since it is still fairly controversial, we present a fairly extensive discussion (Section 4.4) of biochemical markers in the context of the biological scale for indicators.

Before proceeding to the subject of biological scale, however, some clarification of terms is required.

DEFINITIONS OF BIOLOGICAL INDICATOR AND BIOLOGICAL MONITOR
Two types of biological response can be clearly distinguished in terms of their application in toxicology. These are (a) the attainment of an end-point such as those discussed in Chapter 2 and (b) the accumulation in tissue of a chemical substance.

The first can be perceived as an expression of "sensitivity" whereby the organism has a limited range of tolerance to an environmental factor and will show some response when that range is exceeded. This tolerance is exemplified by the miner's canary, which dies when it is exposed to a certain concentration of carbon monoxide. Many other examples can be cited: certain species of lichens are absent from areas where sulphur dioxide concentrations exceed their tolerance for this air pollutant; fish develop tumours when exposed to contaminated lake sediments; sludge worms flourish in sediments of polluted waters, replacing other forms of benthic invertebrates that require higher concentrations of oxygen; organisms produce mixed function oxidases in response to certain types of organochlorine compounds (see Sections 4.4.2 and 7.8.2 and Table 4.5).

The second type of biological response involves the accumulation of a chemical compound or element in one or more tissues of an organism, ideally without symptoms over a wide range of exposure concentration. The processes involved, known as bioconcentration and bioaccumulation, occur very commonly and, in some instances, can be shown to have a quantitative relationship to exposure (Figure 4.1).

The literature uses the terms *biological indicator* and *biological monitor* more or less interchangeably. The present text makes a distinction, which, although by no means is universally accepted, conveniently distinguishes between the sensitive indicator and the accumulator organism. Such a distinction is quite fundamental to our discussion of approaches.

We propose to use the term *biological indicator* for all types of responses, subcellular to whole system, other than accumulations of substances. Thus, good indicator organisms, in our sense, are those previously identified as typically rather sensitive to specific types of insult, or narrow in their range of tolerance for a chemical or physical condition.

We propose to use biological monitor (monitor, from the Latin *monere*, to warn) for the accumulator type of organism. Measuring the burdens of potentially harmful substances in biotic components of ecosystems or in plants and animal tissues, in theory, not only provides an integrated measure of the substances delivered over time but also, in the case of living organisms, distinguishes the available from the inert portions of environmental contaminants (i.e., provides information on dose). This is an attractive concept, and a dose-uptake relationship is frequently assumed but is, in fact, rarely demonstrated scientifically. It is technically difficult to properly calibrate organisms for dose-accumulation. It is generally true, however, that organisms in systems where a given contaminant is present at high concentrations will have higher tissue concentrations of the substance than the same species of organism from reference or noncontaminated areas (i.e., there are clear qualitative differences that can be applied in a comparative manner to illustrate spatial or temporal variation).

If native organisms are chosen as monitors, a choice that would seem to be favoured for the purpose of realism in assessment, a number of assumptions obviously need to be made. Arguably, the most important of these is the choice of a reference (sometimes called control) site or sites, resembling the test site in every way, except for the perturbation, and having a comparable population of the same species of monitor organisms. It is assumed, but is difficult to substantiate, that the exposed and reference populations are genetically and physiologically identical. And yet it is not unreasonable to suggest that there may have been genetic or phenotypic adaptations in exposed as compared to reference populations. This is examined further in Section 4.3, which addresses tolerance.

To overcome the potentially confounding factor of genetic differences, much routine biological monitoring is done by "planting" standard organisms into the sites of interest, which would normally include both potentially contaminated and reference or control sites. The standard organisms may be cultured under standard conditions in the laboratory or, more frequently, collected from a "clean" site where there is a large supply.

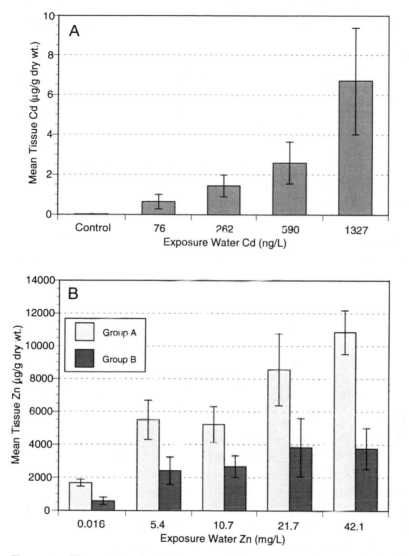

Figure 4.1 The relationship between metal exposure and accumulation by stream populations of caddis fly (*Hydropsyche*) larvae. Both panels show the concentration of metal in the larvae of caddis fly (*Hydropsyche*), whole body, exposed to metal in solution for 48 days, over a range of exposure concentrations. Panel A shows the tissue concentrations of cadmium, which clearly increases with exposure over the range provided. Panel B shows a similar pattern for zinc, but with two different populations of *Hydropsyche betteni* larvae. Both were collected from the same water course; however, Group B was collected from a distinct site upstream of Group A. The site from which Group A was collected has an unknown history but is thought to have been metal contaminated at one time. Group B is considered to have been exposed to less metal; therefore, it has a history of living in a "clean" site with respect to metals. The concentrations of zinc were higher for the organisms in Group B, even though the general pattern of the relationship between exposure and accumulation is very similar for respective populations. Modified from Balch, Evans, and Welbourn (1999, unpublished).

A COMPARISON OF THE PROPERTIES OF BIOLOGICAL INDICATORS AND
BIOLOGICAL MONITORS

Table 4.3 provides a comprehensive list of criteria that are considered desirable for
biological indicators or monitors, respectively. Clearly, in many respects, the
criteria for good indicators are the opposite of those for good monitor species. It
should be noted, however, that rarely is the "ideal" indicator or monitor found that
satisfies all of the criteria.

4.3   Tolerance and resistance to potentially toxic substances

4.3.1   Some conundrums related to tolerance in the context of environmental assessment

Several authors have pointed out that the fact that a perturbation (or stress) has
occurred may cause an unusual or abnormal response, yet this response is not of
itself immediately "harmful" to the organism, indeed it may confer an advantage.
The responses of this type that have been cited most frequently are tolerance and
resistance to potentially toxic substances. It is now well established that organisms
from polluted sites or from sites with natural geological anomalies frequently show
increased tolerance to one or more of the contaminants that were elevated in their
environment. The tolerance may be phenotypic or genotypic.

Terhivuo et al. (1994), in a study in Finland of a lead smelter, found that the
numbers, species, and biomass of earthworms all decreased with proximity to the
smelter. But in the laboratory, the worms from the smelter region took up less lead
in polluted soil, suggesting that they were able to regulate lead uptake. Worms from
control soil, on the other hand, took up high levels of lead in experiments with
polluted soil. The authors of this study pointed out that since worms have different tolerances to lead and different lead uptake patterns, there are major implications involving the choice of species (or ecotypes) for biomonitoring.

Klerks and Weis (1987) reviewed genetic adaptation to metals in aquatic
organisms and found that metal tolerance does occur widely, but not necessarily
with a genetic basis. They warn that the evolution of resistance to environmental
pollutants warrants taking that possibility into account when evaluating bioassays
and designing monitoring programmes.

There is another type of consideration for which the possession of tolerance by
certain organisms has relevance. Some scientists have argued that far from being
a barrier to environmental monitoring, the acquisition or the property of being tolerant may of itself be a useful indicator of past exposure to a potentially toxic substance. Thus, if there has been a history of contamination in a particular area, the
indigenous organisms may have developed tolerance to one or more contaminants,
and this of itself may be used as an indicator. Even if the contamination were of a
transient or occasional nature, such that there remained no measurable or obvious

Table 4.3. Properties of biological monitors and indicators

Indicators	Monitors
Accumulation of a contaminant in tissue, even if it occurs, is not normally of concern when using a species as an indicator.	The monitor should accumulate contaminant in tissue, and the tissue concentrations must be above analytical detection limits.
The indicator can be a specialised or even rare species.	The monitor species should be relatively widespread and common.
The indicator species should be of limited range so that it can represent local conditions.	Monitor species should be sedentary/nonmigratory, sessile, or of limited mobility.
The indicator species should be readily identifiable without specialist skills.	The monitor species should be readily identifiable without specialist skills.
The indicator species should (arguably) be of ecological relevance to the system and ideally endemic to area.	The ecological significance of the species to a system is not of concern; a species can even be "planted" for the duration of a test.
Indicator species should be sensitive to low levels of contaminant (ideally they should provide an "early warning" of exposure).	Monitor species should be tolerant to a reasonably wide range of concentrations of the contaminant.
The indicator should show a measurable response in a reasonably short timeframe.	The monitor should show no physiological or other adverse response over the range of contaminant exposure.
The indicator species can be a specialist with very limited ranges of tolerance for various physical and chemical conditions.	The monitor species should be tolerant of a broad range of physical and chemical conditions and should be amenable to transplantation.
The indicator species should be easily cultured or, at least, maintained in the laboratory and should be amenable to laboratory experimentation.	The monitor species should be easily cultured or at least maintained in the laboratory and should be amenable to laboratory experimentation.
Contaminant sensitivity should be specific to one contaminant or family of contaminants (i.e., can be used to establish cause-effect relationships).	Ideally, it should be possible to demonstrate a statistically reliable relationship between contaminant concentrations in the tissue of the biological monitor and concentrations in environment.

residues of harmful chemicals on a site, there could, nevertheless, be a "memory" of the contaminant as seen in the existence of a tolerant population if a population at the site has adapted to tolerate the contaminant. This concept will be revisited in later parts of the present chapter, particularly in a discussion of metallothionein (Section 4.4.2).

Before moving on to a more general scientific consideration of the mechanisms of tolerance, we can conveniently introduce one more issue at this point. This issue concerns the validity of using "standard" test organisms for bioassays, for establishing criteria, or as indicators for field situations. It is obvious even to a casual observer that site-specific conditions are "normal" for a given site and affect the flora and fauna at that site. However, these site-specific conditions may be outside the range of what is typically considered normal. For example, if the site is highly mineralised and has naturally high concentrations of cadmium and zinc, then the flora and fauna are likely to be more tolerant to cadmium and zinc than those from sites with "background" concentrations of the same metals. Normally, environmental quality criteria or bioassays depend upon tests, typically laboratory tests, using standard laboratory organisms. The question then becomes, should sites with abnormally high natural concentrations of potentially toxic substances be assessed using the same criteria in terms of indicators, monitors or bioassays, as more normal sites? The resolution of this question is clearly of considerable significance in a regulatory context. A case study of such an instance as it relates to a mine site is provided at the end of this chapter (Case Study 4.1).

The mechanism(s) by which the organisms tolerate the substance(s) are treated in more detail in Section 4.3.2, but, briefly, they may be of the type known as exclusion, whereby the substance does not enter the cell or is secreted rapidly. Alternatively, the organism may actually accumulate the substance, perhaps attaining internal concentrations that exceed those that would harm nontolerant forms, but retain it in nonsensitive sites or organs inside the organism, effectively isolating it from the machinery of the cell that would normally be impaired. Another type of nonharmful accumulation is achieved by sequestering the potentially toxic substance into a chemical form that is not biologically available.

Any or all of these phenomena have the potential to affect and possibly to invalidate the use of a species or ecotype as an indicator or in a monitoring system. The limitations follow.

For indicators. If the selected test organism is tolerant to higher concentrations of a potentially toxic substance than a normal organism, it may thrive or be dominant at sites that are contaminated so that, arguably, it will not be a true indicator of the original or ideal condition of the site. Further, if an unusually tolerant species or a tolerant form of a species were to be used in laboratory bioassays, the result, in terms of determining the potential toxicity of a site or a chemical, could be misleading. Finally, if a site or region has naturally high concentrations of a potentially toxic substance, it is problematic to make a decision whether one should use an indigenous, tolerant indicator or test species rather than a standard one. And if, in principle, the former type of indicator or test species is deemed appropriate, then how should the organisms be selected in a regulatory context?

For biomonitors. If the monitoring (test) organism is an "excluder", it will tend to underestimate the concentrations of available chemical in the environment or the test medium; conversely, if it is an accumulator, the opposite problem will arise.

Because there is more than one type of mechanism by which a population or species can tolerate a potentially harmful substance, one cannot make generalised rules concerning the suitability of species for their use as indicators or monitors. Site-specific considerations, where possible, are to be advocated.

4.3.2 Selection for tolerance, mechanisms of tolerance, and potential practical applications of the phenomenon

The terms *tolerance* and *resistance* have been used to mean different things by different authors in the literature. The National Research Council of Canada (NRCC, 1985) defines both tolerance and resistance as "the ability of an organism to exhibit decreased response to a chemical relative to that shown on a previous occasion". NRCC distinguishes the two by defining tolerance as the change that is within the normal adaptive range of the organism and can be sustained indefinitely. Resistance, in contrast, implies that the magnitude of the chemical change lies outside the normal range, and that negative effects of that stressor will eventually be manifested in the organism. Other authors make quite different distinctions. For example, Foster (1982) uses tolerance for situations when a species that is normally metal sensitive in the absence of metal pollution develops a "race" that can tolerate higher than normal metal concentrations. This race would have evolved as a distinct population of the normally sensitive species. There may be a series of races with different tolerances, corresponding to environmental conditions (Turner, 1969). According to Foster (1982), resistance is a more general term implying interspecies but not necessarily intraspecies comparisons (i.e., when typical members of entire species can grow without ill effects in the presence of elevated concentrations of potentially toxic substances). For the purposes of the present text, none of these distinctions were deemed useful, so no distinction will be made between the terms tolerance and resistance. The term tolerance is preferred. Normally it is assumed that there is a genetic basis for tolerance, whether the trait is phenotypic or genotypic.

We use the term tolerance to mean the ability to withstand exposure to abnormally high concentrations of substances (elements or compounds) that would otherwise cause adverse biological effects and that can be sustained indefinitely. Many organisms, both plant and animal, have been shown to be tolerant to unusually high concentrations of potentially harmful substances in their environment. However, all definitions and examples of tolerance are, in fact, relative, not absolute, and this fact may lead to confusion in the literature, even after removing the complication of the term resistance.

The most commonly documented examples of tolerance occur *within* a species, when ecotypes or strains of a given species exhibit an ability to withstand concentrations of a given contaminant that are definitely harmful to the original strain (sometimes called the wild type). This is believed to result from the selection of preadapted individuals in the original population, which then reproduce under conditions adverse to the wild type. After reproducing, the individuals become a select, tolerant (or strictly speaking, *more* tolerant) population. The selection pressure may be from environmental conditions, such as organic contamination of sediments selecting populations of bacteria that can tolerate high concentrations of the organic substance or grasses that grow on a mine waste site that has very high concentrations of certain metals. This type of selection has been referred to as evolution in action. Over time, if the selection pressure persists, one can envisage entire species developing tolerance. Sites that have naturally elevated concentrations of potentially toxic chemicals, such as soils over serpentine rocks, with very high concentrations of nickel and chromium, have very distinctive flora, including certain examples where the entire species is tolerant to these metals. One can speculate that the substrate selected ecotypes, which over time have undergone divergent evolution and speciation.

It is useful to note at this point that the phenomenon of resistance has also been described in the context of bacterial response to antibiotics and of insect and fungal pests to pesticides. In all such examples, the mechanism is similar. Exposure of a population to a toxic substance selects those individuals that are inherently able to survive at concentrations that kill most of the population. These individuals then produce progeny, and, with time, the population is dominated by the "new" tolerant or resistant forms. Thus, the phenomenon can be explained in terms of Darwinian evolution. In other instances, entire species are tolerant to certain potentially toxic chemicals, yet no description explains the selection pressure that brought this about. Populations or whole species particularly of microorganisms have shown tolerance to certain xenobiotic substances, yet there has been no obvious prior selection pressure.

Selection pressure can be exerted deliberately, under laboratory or other experimental conditions, as it is for traits other than tolerance to toxic substances. This process is sometimes referred to as artificial selection and has an important place in applied ecology. For example, the selection of tolerant ecotypes can facilitate rehabilitation of contaminated sites, as discussed in Chapter 11.

Two different views of the manner in which selection for chemical tolerance might develop in the environment follow. Even though they may seem almost contradictory at first sight, on further examination they can be seen to operate on premises that, although somewhat different, are by no means mutually exclusive.

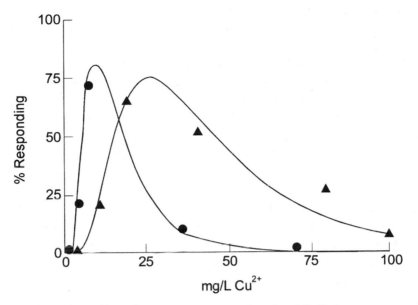

Figure 4.2 The effect of copper on two species of caddis fly larvae (*Hydropsyche*). *H. saxonica* (●) is a specialist having more restricted habitat; *H. augustipennis* (▲) is a generalist having broader habitat preference. The recorded end-point is damage to anal papillae. From Petersen (1986).

Tolerance in specialists versus generalists. Empirically, specialists are defined as species found at only a limited number of habitats, whereas generalists are found over a broad range of habitats. Stable environments with abundant resources would favour the selection of specialists. Selection pressure for generalists would be higher in more unstable environments with limited resources.

At the molecular level, a generalist would be seen as having a greater number of options available to counter disrupting influences including chemical toxicity. This increased versatility is, in many cases, probably related to greater genetic heterogeneity compared with specialist species and is manifest as a larger diversity of enzymes, sequestering proteins and other agents involved with the detoxification process. Petersen (1986) provided evidence showing how this concept might apply to trace metal toxicity. Two species of *Hydropsyche* were chosen for copper toxicity experiments using damage to the anal papillae (osmoregulatory organs) as an end-point. The two species were *H. angustipennis*, a generalist found at a large number of different sites in the Kolbäcksån River system, and *H. saxonica*, a specialist found at relatively few locations in the river. Figure 4.2 compares the percentage of each species showing anal papilla damage over a range of copper concentrations. The narrower curve shown for the specialist (*H. saxonica*) indicates a greater sensitivity and smaller (stenotoxic) tolerance range for this species, whereas the generalist *H. angustipennis* has a much broader (eurytoxic) tolerance range.

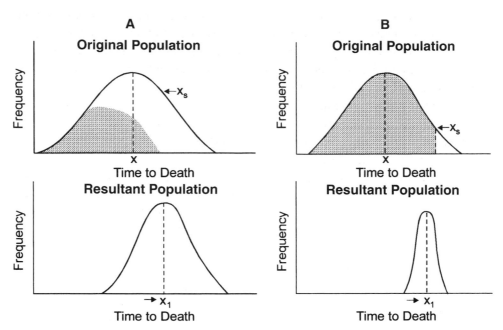

Figure 4.3 Change in response (mortality) frequency in a hypothetical population exposed to selection pressure. Response is shown in the form of (A) 50% mortality and (B) 90% mortality (shaded areas). The population is assumed to have a typical range of genetic variation and rare mutations associated with a few loci that are responsible for tolerance. Modified from Mulvey and Diamond (1991).

The survivor effect. Toxic chemicals may exert selection pressure at both population and community level. At the population level, both short- and long-term effects may be observed. Even within the confines of a short-term bioassay lasting only a few days, it is not uncommon to find incomplete mortality expressed as a truncated dose-response curve and indicating the indefinite survival of a certain percentage of the test population. These survivors represent the more resistant end of the phenotype which, if allowed to reproduce, may form the basis for a more chemically tolerant population over the long term. Thus, tolerant characteristics may be incorporated into the genotype. A good account of this phenomenon is given by Mulvey and Diamond (1991), who emphasised that the impact of the selection process is determined by the degree of selection pressure (Figure 4.3). Following 90% mortality, only resistant genotypes survive, and the response to selection is significantly larger than if the initial mortality had been only 50%. The response to selection (R) is defined as the difference between the mean tolerance before selection (x) and the mean tolerance after selection (x_1). Hypothetical mortalities at the 50 and 90% level are shown as the shaded portions in Figures 4.3B and 4.3C. Thus, the result of intensive selective pressure may be a discreet population of tolerant individuals having lower genetic diversity and a relatively reduced tolerance range. The same concept can be applied to communities. There is evidence that communities may increase their tolerance to chemical toxicants through a transition from less to more tolerant species. Blanck et al. (1988) showed that a periphyton community subject to toxicant selection pressure became restructured in a way that increased its overall tolerance to chemical pollutants, a phenomenon called pollution-induced community tolerance.

This tolerance represents a different interpretation from the specialist/ generalist approach, which equates higher chemical tolerance with greater genetic diversity. By contrast, the survivor concept assumes that tolerance is conferred by mutations that occur at a few highly specific loci.

Several studies have indicated that one or only a few genes are implicated in differential metal tolerance (e.g., Shirley and Sibley, 1999), and the involvement of few genes may explain the rapid development of genetic tolerance in some species (Sullivan et al., 1984; Shirley and Sibley, 1999).

MECHANISMS OF TOLERANCE
Organisms have demonstrated tolerance to a contaminant in several ways, including:

1. Prevention of uptake;
2. Uptake followed by storage of contaminant in physically isolated structures (concretions, special inclusions, vacuoles) or in biologically inactive forms (e.g., for metals complexed with metallothioneins and phytochelatins);
3. Degradation of the chemical, externally or internally, into less harmful components;
4. Uptake followed by removal from the cell by excretion or secretion;
5. Avoidance (applicable to motile organisms only).

Example of studies on different types of tolerance are provided in Table 4.4.

An example of exclusion is illustrated by Russel and Morris' (1970) study of brown algae isolated from copper-treated structures in the marine environment. Subsequent laboratory studies comparing the tolerant isolate and nontolerant strains of the species suggested that the tolerant races produced extracellular organic compounds, which sequestered copper thus preventing its uptake. In this example, there was not an unequivocal demonstration that the mechanism described was responsible for the demonstrated tolerance. Similarly, Butler et al. (1980) described the production of an organic metal-binding substance by a metal-tolerant green alga. Again, absolute proof of the tolerance mechanism was lacking. This illustrates a problem that is frequently encountered: Potential mechanisms for tolerance can be hypothesised, but absolute proof that tolerance is conferred by a particular mechanism is rarely obtained.

Plant cell walls have the capacity to bind metals in negatively charged sites. Several studies on metal-tolerant grasses (see Table 4.4) have demonstrated the potential for cell walls of tolerant grasses to bind metal and thus exclude it from

Table 4.4. Examples of different types of mechanisms for tolerance

Chemical	Organism(s)	Mechanism	Reference
Copper	Marine macro-algae *Ectocarpus* (+ one other brown alga)	Prevention of uptake by external production of chelators	Russel and Morris (1970)
Zinc	Terrestrial grass, *Agrostis tenuis*	Metal binding to cell wall prevents uptake.	Turner and Marshall (1972)
Lead	Terrestrial grass, *Anthoxanthum odoratum*	Cell walls appear to protect tolerant strain: intact cells are unaffected by Pb, which harmed unprotected protoplasts.	Poulter et al. (1985)
Cadmium	Marine bivalve, *Mytilus edulis*	Contaminant stored in lysozomes; granular concretions are stored in the kidney.	George (1983)
Zinc	Vascular plants	Citrate and malate sequester Zn and Cd in the vacuole.	Mathys (1980)
Cadmium	Mayfly larvae	Metallothionein induced by Cd correlated with increased Cd tolerance.	Aoki et al. (1989)
Cadmium	White suckers, *Catostomus commersoni*	Metallothionein induced by Cd correlated with increased Cd tolerance.	Klaverkamp and Duncan (1987)
Cadmium	Freshwater alga/protozoan, *Euglena gracilis*	Main mechanism for Cd tolerance is the sequestering of Cd by cysteine-rich polypeptides.	Wikfors et al. (1991)
Selenium	Terrestrial herb, *Astragalus* spp.	Se sequestered inside the cell into selenomethionine, an amino acid that is biologically inactive.	Trelease et al. (1960)
Copper and lead	Aquatic isopod, *Asellus meridianus*	Tolerant strains compared with nontolerant strains of same species indicated tolerance associated with accumulation of copper or lead respectively in the hepatopancreas from food; this difference does not occur when metal supplied in water. Field collections of tolerant animals showed markedly more metal in the hepatopancreas than did nontolerant strains.	Brown (1977)
Metals	Collembola, *Folsomonia fimetaroides* and *Isotomiella minor*	Study shows ability to avoid metals gives advantage of one species over another in metal-polluted soils. Both collembola feed on fungi *F. fimetaroides* showed higher preference for metal-tolerant fungi than did *I. minor* when they were offered fungi of differing metal tolerance. *F. fimetaroides* avoided the polluted substrate, but *I. minor* did not.	Tranvik and Eijsackers (1989)
Hydrocarbons	*Flavobacterium putida*, strain DS-711	Halo-tolerant strain was also tolerant to a number of organic solvents including benzene. In culture, *F. putida* had greater ability to degrade alkanes than did control strains.	Moriya and Horikoshi (1993)
Phenolic compounds	Sulphate-reducing bacteria	Degradative ability of sediment bacteria can be related to previous exposure patterns to phenol. For organisms from industrially polluted sediments, rates of degradation for phenol and *p*-cresol were faster than for those from nonindustrialised sites. Substrate mineralisation shown to be attributable to sulphate-reducing bacteria.	Mort and Dean-Ross (1994)

the interior of the cell. Recent short-term studies on a wallless mutant of the normally walled alga *Chlamydomonas reinhardtii* have shown that the wallless strain is indeed more sensitive to metals than is the walled form (Macfie et al., 1994). Metal binding to the cell wall of the walled strain has not been consistently related to the relative tolerance, and this relation is expected: cell-wall binding is unlikely to confer tolerance over the long term because eventually the cell wall will reach equilibrium with the external medium. At this point, it will no longer protect the interior of the cell.

Intracellular detoxification of metals has been demonstrated for a variety of plants and animals. This detoxification may occur through physical isolation of the metal in concretions as well as by incorporation into various organic molecules, effectively protecting sensitive intracellular sites from exposure. Such biochemical detoxification is included in some of the examples discussed in Section 4.4.1. Certain biochemical markers are, in fact, related to detoxification mechanisms. Probably the best known but still incompletely understood of biochemical markers that detoxify and also reflect tolerance are probably the metallothioneins (MTs). Animal metallothioneins and plant metallothioneins, the latter usually called phytochelatins, are small proteins rich in sulphur-containing amino acids (thiols), which effectively bind many trace metals. Indeed, the presence of high levels of phytochelatins in plant tissues has been used as an indicator of metal stress (see Section 4.4.2). For nonessential metals such as cadmium[1], it is now generally agreed that MT induction is indeed a detoxification mechanism, and in some studies, as shown in Table 4.4, it has also been related to tolerance. However, when essential elements such as copper and zinc are considered, MT may have both a regulatory and a detoxifying role.

In common with the mechanisms that involve physical means of avoiding toxicity to sensitive sites inside the cell, some biochemical mechanisms of tolerance can result in an abnormally high accumulation of potentially toxic contaminants. One fascinating example of this phenomenon is selenium (Se)-accumulating species of plants in the genus *Astragalus*. Several species of the genus are able to accumulate very high concentrations of selenium, through the formation of an amino acid seleneomethionine (Trelease et al., 1960), which protects the cell from Se toxicity. Nontolerant species of *Astragalus* do not produce this amino acid; in these species the Se is incorporated into essential amino acids; therefore, their metabolism is harmed by Se. Interestingly enough, the Se accumulator plants, which may contain up to several thousand parts per million of Se, are toxic to herbivores that graze on them, giving symptoms of Se toxicity in horses and cattle. Hence, the consequence of this detoxification, although advantageous to the population of plants, can be seen as a potential disadvantage to the ecosystem as a whole. The Se accumulators are also tolerant of high concentrations of Se in the

---

[1] Even though cadmium is generally regarded as a nonessential element, it has been shown to substitute for Zn in algae as a micronutrient (Section 6.6.1) (Price and Morel, 1990).

soil in which they grow, even to the extent of having a requirement for high Se. One would have to question the utility of such plants for use as either indicators or monitors. On the other hand, if their specific ecological and physiological characteristics were known, a case could be made that they would be ideal indicators of the available Se in soil.

Tolerance to some of the persistent organic substances (xenobiotics, see Chapter 7) has received less attention to date than tolerance to metals. The history of the literature on metal tolerance starting in the late 1950s shows that the phenomenon has attracted researchers not only for its toxicological significance, but also for its relevance as "an evolutionary paradigm" (Antonovics, 1975). Concern for organic contaminants is in general much more recent. The examples in Table 4.4 indicate that tolerance to organic substances involves the ability to break down complex organic molecules into harmless or less harmful products (i.e., differs mechanistically from most of the processes known for metal tolerance). In the case of metal tolerance, an exception can be seen in the microbial transformation of mercury II to metalic mercury. To date, the best studied examples of tolerance to organic toxicants are microbial processes. Tolerance to persistent organic substances, such as PCBs, has been demonstrated in a number of microorganisms. If the mechanism by which an organism is able to tolerate the potentially toxic organic substance is through degradation of the substance into benign forms, in a process referred to as bioremediation, the organisms can literally "clean up" the contaminated environment, a phenomenon that obviously has potential for rehabilitation and will be considered again in more detail in Chapter 11.

COST OF TOLERANCE

It must be stressed that, in some cases, toxicant adaptation may be developed at some cost to the tolerant population. For example, enhanced metal tolerance in trace-metal-selected populations may be linked to other traits that confer detrimental ecological consequences such as diminished reproduction and growth and a reduction in ability to forage, avoid predators, or resist infection (Mulvey and Diamond, 1991; Posthuma and Van Straalen, 1993).

4.4 Biological scale

In the introductory sections of this chapter, the principles of the various approaches to the use of indicators and monitors were addressed. The following section provides a more detailed description of some of these approaches and includes a discussion of scale. Biological changes that result from exposure to chemicals can be used in biological assessment to indicate chemical contamination. Biological changes, sometimes termed biomarkers, may occur at different levels of biological organisation, ranging from the suborganismal level (biochemical, physiological, or histological) to the levels of community and ecosystem. As in other topics that have been considered in this text, one encounters the issue of definitions of terms. The term *biomarker* has been used variously, from the wide range of meaning used in the current paragraph, to the limitation of the term for changes or responses at the biochemical level.

To avoid confusion, while recognising that the student will almost certainly encounter various different uses of the term *biomarker*, we are recommending the practice of simply qualifying the terms used, with *biochemical indicator* or biochemical marker being used for the subcellular responses, whereas similarly descriptive and nonambiguous terms such as *morphological indicator*, *population level indicator*, *surrogate*, and *community level response* can be used as appropriate.

4.4.1   Principles and properties of biochemical markers/biochemical indicators

Toxicologists have developed an extensive set of biochemical and physiological tests that can be used in field situations as indicators of toxic chemical exposure and toxic stress. The biochemical indicator or biochemical marker is really a special case of a biological indicator. Response to toxic substances as well as other perturbations such as temperature changes, at the biochemical or other subcellular level, has recently received a great deal of attention from environmental toxicologists, and a book of essays published in 1990 by McCarthy and Shugart illustrates a fairly wide range of these markers for terrestrial and aquatic systems.

Biochemical markers of chemical exposure include enzymes, proteins, and other macromolecules associated with physiological function. Molecular markers commonly used in monitoring programs include stress proteins (also known as heat shock proteins), cytochrome P450 mono oxygenases, phase II conjugating enzymes, heme porphyrin systems, metallothioneins, and responses associated with reactive oxidants. More generalised responses often considered biochemical markers include a variety of immune responses and physiological/cellular responses associated with energy utilisation. Table 4.5 shows a range of biochemical markers. Some of these are discussed in more detail later.

The use of responses at the biochemical level to potentially toxic substances, or to other insults, may have a number of advantages over responses at more complex levels of organisation. Contaminant effects at the organism or population levels are preceded by biochemical reactions in individuals. Not surprisingly, biochemical changes are frequently detectable before any overt effects such as morphological or behavioural changes occur. Thus, as indicators of harm, they may provide more sensitive, and/or earlier warnings than are seen through higher levels of organisation.

As indicated earlier, several different definitions of biochemical markers have been published, but there is general agreement that the following characteristics are important for biochemical indicators:

- A particular biochemical response can be observed in a wide range of organisms;

Table 4.5. Examples of some commonly used biochemical markers

Stressor	Response	Specificity (yes/no)	Implications for health of organism
Photochemical oxidants (e.g., sulfur dioxide, ozone, nitrous oxides), acid precipitation, peroxides	Production of antioxidant enzymes, (e.g., super oxide dismutase, peroxidase, catalase) or nonenzymatic antioxidants (e.g., glutathione, ascorbic acids, α-tocopherol or vitamin E, β-carotenes)	No	Free radical generation through membrane receptor perturbations, MFO activity regulated enzymically and nonenzymically. • Overproduction of free radicals can damage DNA, membrane lipids, and proteins • Suppression of free radical production may inhibit immune response
Chlorinated and aromatic hydrocarbons (e.g., PAHs)	Production of cytochrome P450 monooxygenase enzymes [e.g., EROD (ethoxyresorufin-o-deethylase), AHH (aryl hydrocarbon hydroxylase)]	Yes	Oxidize organic compounds and convert them to more polar (soluble) forms. May result in carcinogenic intermediates.
Trace metals, organic chemicals	Phase II (conjugative) Enzymes (e.g., glutathione transferase)	No	Increase solubility and elimination rates of foreign compounds.
Lead	ALAD (erythrocyte o-amino laevulinic acid dehydratase)	Yes	Alteration in ALAD activity. May disrupt iron/cytochrome metabolism.
Organophosphates, carbamates, some pyrethroid pesticides	Inhibition of acetylcholinesterase (AChE) activity by attacking OH groups on enzyme	Yes	Hyperactivity of neurotransmitter. Loss of neuromuscular function.
Some class B and borderline metals[a] (Ag, Cd, Cu, Hg, Zn)	Metallothionein synthesis	Limited	Regulate distribution of metal in tissues (detoxifying function). Free radical scavengers.
Physical and chemical perturbations (e.g., salinity, osmotic changes, trace metals, anoxia, heat/cold shock, xenobiotics)	Stress protein synthesis, also called heat shock proteins (e.g., hsp90, ubiquitin)	No	Involved with cellular repair.
Metals, organics	Altered immune function • Macrophage activity	No	Macrophage activity may be impaired due to previous exposure to stressors.
PAHs, PCBs	Tissue pathology • Neoplasms epithelial and renal lesions	No	Tissue lesions may result from direct contact with contaminants.
Ethynylestradiol alkylphenols chlorinated pesticides, PCBs, phthalates	Disruption of reproductive endocrine system	No	Proliferation of female genital tract in rats. Unnatural production of yolk protein vitellogenin (e.g., in male fish transcription of estrogen-responsive reporter genes).

[a] Nieboer and Richardson (1980) Δβ ranking.

- The response should be specific to a particular contaminant, or family of contaminants;
- Laboratory results can be related to responses in the field;
- Biochemical response is related to or correlated with decreased or impaired function.

The fourth principle is the one that gives most cause for disagreement or concern, in that most environmental toxicologists agree that biochemical markers often provide good evidence of toxic chemical exposure. There is less agreement on the degree to which the biochemical and physiological parameters observed represent pathological conditions that compromise the overall fitness of the organism. Several different approaches have been adopted to clarify this perception.

a. Biochemical markers in field-collected organisms have been compared with organisms exposed to specific chemicals or groups of chemicals under laboratory conditions. Where laboratory organisms have been exposed to a range of chemical concentrations as part of a dose-response study, these results may then provide a benchmark for determining the degree of exposure and/or pathology of field specimens.

b. Biochemical markers have been measured in organisms collected from different points along a pollution gradient. Obviously such an approach demands parallel measurement of chemicals either in the ambient environment or in the organisms themselves. Selection of control or reference sites requires careful consideration to ensure uniformity of physicochemical conditions apart from the degree of chemical contamination. In rare circumstances, where baseline data exist, it may be possible to monitor biochemical marker data at a specific site before, during, and after a pollution event.

This approach has been particularly useful in ascribing a characteristic biochemical marker to a specific chemical group thereby imparting what might be termed a "forensic" property to the test.

c. A weight-of-evidence strategy is bolstered by a correlation between different biochemical markers. This approach, often associated with approach b, adopts the notion that the correlation of two or more biochemical markers or multiple "hits" in organisms from specific sites probably indicates increased likelihood of risk. Even if these biochemical markers have little clear functional linkage, such a correlation would indicate that several different metabolic systems have been affected.

4.4.2 Some of the more commonly used groups of biochemical markers

ENZYMES

Environmental chemicals influence enzyme levels and activity in a variety of ways. Even though altered enzyme activity may give a general indication of the effects of trace metals, biochemical markers that can give more specific information on a particular pathological condition and, if possible, a specific causative agent are sought.

A good example of such a biochemical marker is the inhibition of M-amino-laevulinic acid dehydratase (ALAD) by lead. This effect has proved to be a useful biomarker for this metal (see Table 4.5 and Section 6.5).

Where pesticides are formulated specifically to interact with a particular enzyme, alterations in the activity of that enzyme in field-collected biota may represent a very effective biomarker. For example, measurement of cholinesterase activity in nontarget species is a useful means of assessing environmental contamination by organophosphate and carbamate pesticides that are primarily anti-cholinesterases (Grue et al., 1991).

Other measures of enzyme activity may indicate nonspecific forms of toxic damage. For example, significant activity of liver enzymes such as glutamic pyruvic transaminase in serum may be a sign of general liver damage in vertebrates.

Several enzymes that are used as biochemical markers may actually be induced by exposure to specific groups of environmental chemicals and may be involved in their metabolism and detoxification. Some examples follow.

MONOOXYGENASE (MIXED FUNCTION OXIDASE) ENZYME

One of the first cellular responses produced in response to exposure to certain organic chemicals is the enhanced production of enzymes collectively known as the mixed function oxidase system. This system, which is found in most plants and animals, alters the toxicity of nonpolar organic chemicals, such as polychlorinated biphenyls, through structural transformation, converting them into more polar forms that can be metabolised and degraded. Base-line levels of MFO enzymes reflect the processing of endogenous nonpolar compounds produced through normal metabolism. Two enzymes that are increasingly becoming incorporated into field monitoring for xenobiotic chemicals are aryl hydrocarbon hydroxylase (AHH) and ethoxyresorufin-o-deethylase (EROD) (Table 4.5). In several studies, ex-pression of these enzymes has shown a close correlation with organic chemical exposure, with cytochrome P450 induction, and with each other. Studies of EROD activity have become incorporated into regulatory bioassays associated with eval-uation of effluents from paper and pulp mill discharges into freshwater systems, and AHH measurement is often an integral part of the investigation of petroleum hydrocarbon toxicity to marine organisms following major oil spills. There are varying reports of the time-course of AHH induction and decline relative to chem-ical exposure. Studies of English sole (*Parophrys vitals*) exposed to a discrete dose of PAH showed a fairly rapid decline in hepatic AHH activity as the PAH was metabolised, indicating that elevated AHH levels in field-caught specimens might indicate recent PAH exposure (Collier and Varanasi, 1991). However, other evi-dence from the sheepshead (*Archosargus probatocephalus*) suggests more sus-tained elevation of tissue AHH levels following PCB exposure (James and Little,

Figure 4.4 Map of the United States showing sites where fish have exhibited significantly elevated aryl hydrocarbon hydroxylase levels relative to background conditions (after Varanasi et al. 1992).

1981). The U.S. National Status and Trends program run by the National Oceanographic and Atmospheric Administration (NOAA) has documented significant elevations in hepatic AHH activity in fish from contaminated sites around the coast of North America relative to specimens taken from pristine areas (Figure 4.4).

PHASE II (CONJUGATIVE) ENZYMES (GLUTATHIONE TRANSFERASE)
Glutathione functions as an antioxidant, mitigates the effects of reactive oxygen species, and detoxifies reactive xenobiotic metabolites through conjugation reactions. It exists primarily in the reduced form (glutathione, GSH), although about 5% is normally present in cells as an oxidised form (glutathione disulphide, GSSG). Hepatic GSH levels have been shown to be altered following exposure to chemicals that are potential substrates or that form reactive oxygen intermediates. It has been suggested recently that, in addition to organic chemicals, trace metals may also bind to glutathione in a manner that may protect the organism. Several investigations have now documented increases in fish hepatic GSH following exposure to elevated metals and PAH, either under laboratory conditions or as complex mixtures in contaminated sediments or effluents.

Glutathione transferases (GST) are a family of enzymes that function as catalysts for the conjugation of various electrophilic compounds (e.g., epoxides of PAH). They increase the availability of lipophilic toxicants to phase I enzymes (i.e., monooxygenases) by serving as carrier proteins.

Reduced glutathione is a tripeptide compound, which plays two contrasting roles in detoxification – as an intermediate in phase II metabolism via GST (see earlier discussion) and as an important antioxidant. Other enzymes important in neutralising, removing, or scavenging reactive oxidative intermediates may also be used as indicators of ROI activity. These include catalases, ascorbate peroxidase and other peroxidases capable of removing hydrogen peroxide, and superoxide dismutases that scavenge the superoxide anion (see Section 2.3.1).

Both phase I (MFO) and phase II enzyme activity form part of a comprehensive study of biomarkers in Puget Sound fish, which has illustrated a hierarchy of toxic chemical effects at different levels of biological organisation (Case Study 4.2). Effects at various levels of biological organisation are also addressed in Section 4.4.6.

BIOMARKERS OF DNA DAMAGE AND REPAIR

As described in Chapter 2 (Section 2.3.1), the development of carcinogenicity apparently results from an imbalance between DNA damage and repair. In a healthy organism, repair mechanisms generally keep pace with the incidence of DNA damage; however, any process that enhances damage or slows repair will increase the risk of carcinogenicity. Covalent bonding between DNA and potential carcinogens results in the formation of adducts, which are examples of primary DNA damage and which may be detected using very sensitive techniques such as ^{32}P postlabelling. Many oxidatively damaged DNA bases can be detected in tissue and body fluids using a suite of high-pressure liquid chromatography (HPLC), gas chromatography-mass spectroscopy (GC-MS), fluorescence, tritium (^{3}H), and ^{32}P labelling and immunochemical techniques.

Physical breakage of DNA strands can be determined with the use of a dye that fluoresces at much greater intensity with double-stranded than with single-stranded DNA. Fluorescence is measured following alkali-induced unwinding, which occurs at single-strand breakage points in the DNA molecule. DNA breakage may also result in chromosomal damage, which may be detected as increasing numbers of micronuclei in specific cells. The incidence of micronuclei in fish erythrocytes is employed as a biomarker in the National Status and Trends Program conducted by the U.S. National Oceanic and Atmospheric Administration around the U.S. coast.

METALLOTHIONEINS

Metallothioneins, a family of low-molecular-weight, cysteine-rich metal-binding proteins, are commonly used biochemical indicators of metal exposure in animals. Synthesis of these proteins is induced by exposure to metals in groups IB and IIB of the periodic table. Metallothioneins are thought to function in the storage, transport, and detoxification of both essential [e.g., copper (Cu) and zinc (Zn)] and nonessential [e.g., lead (Pb) and cadmium (Cd)] metals. Metallothionein levels in fish and aquatic invertebrates have been used to indicate metal pollution in the aquatic environment and to demonstrate metal acclimation to heavy metal toxicity. For example, fish from lakes surrounding the Flin Flon base metal smelter in Manitoba, Canada, were used to evaluate the effectiveness of recent reductions in metal emissions. Fish in lakes close to the smelter had higher hepatic concentrations of metallothioneins and were more resistant to Cd toxicity than fish in lakes distant from the smelter (Klaverkamp and Duncan, 1987). Animals may also store metals in intracellular granules, which are generally less labile than metal-protein associations. For instance, in insects, such granules have been implicated in the storage of essential trace metals (reserves) but also in the sequestration and excretion of toxic trace metals such as Pb, mercury (Hg), and Cd (Hare, 1992).

Phytochelatins are the floral equivalent of metallothioneins, and have similar physical (low molecular weight, sulphur-rich) and physiological properties, although their expression is not genetically mediated. Phytochelatin levels in plant tissues have been used to indicate exposure to metals in soil and ambient air. For instance Gawel et al. (1996) suggested that phytochelatins in foliage of declining red spruce in the northeastern United States provided evidence that metal stress may be contributing to forest decline in high-elevation stands.

STRESS PROTEINS

Stress proteins, also known as heat shock proteins (hsp), were first investigated in 1962 by Ritossa, who discovered puffs in the polytene chromosomes of *Drosophila buskii*, which were part of the cells' response to heat shock. The puffs were later associated with the synthesis of newly induced proteins, heat shock proteins. Hsp synthesis can be induced by a variety of stressors, including, but not limited to, salinity and osmotic changes, trace metals, anoxia, heat (and cold) shock, and xenobiotics. Stress proteins have been found in all organisms examined to date, from the smallest prokaryote to humans. However, there appears to be a great deal of homology of stress protein amino acid sequence among most organisms. The general function of stress proteins appears to be protective in nature. Stress proteins are grouped in families according to their apparent molecular weight and measured by comparison to molecular weight standards on electrophoretic gels. These weights cluster around 90, 70, 60, and 20–30 kilodaltons (kDa), and low-molecular-weight, approximately 7 kDa, proteins called ubiquitin.

Table 4.6 summarises information on heat shock/stress proteins in the context of biochemical markers. Stress proteins meet two important criteria for a biochemical marker: They are ubiquitous in the plant and animal kingdom and respond to a wide range of stressors. Additionally there is some expectation that they may have some potential for isolating specific chemical effects from complex mixtures of stressors. Blom et al. (1992) investigated the effects of nine different pollutants on *Escherichia coli* and found that 50% of the proteins induced by any of the nine chemicals were unique to any given chemical.

Several studies have been published showing a correlation between stress protein synthesis and the effect of chemicals at the cellular level. For example, Goering et al. (1993) correlated stress protein synthesis with hepatotoxicity in

Table 4.6. Heat shock proteins

Heat shock protein (hsp)	Reponse	Function	Potential as biochemical indicator	Reference
hsp90	Abundant in normal cells, synthesis increases three- to fivefold upon exposure to stress	May redirect cellular metabolism	Poor when used alone – due to natural abundance and limited response to stressors	Welch (1992)
hsp70	Marked increase in synthesis upon exposure to stress	May stabilise or solubilise and remove denatured proteins	Excellent – one of the most highly conserved proteins	Welch (1992); Stegeman et al. (1992)
hsp60/Chaperonin-60	Increased rate of synthesis in stressed cells	Facilitates translocation and assembly of oligomeric proteins	Good – highly conserved, easily detected response	Stegeman et al. (1992)
hsp20–30	Highly specific to species, found in nuclei of stressed cells	Unknown function	Unknown at present – poorly understood function	Stegeman et al. (1992)
Ubiquitin	Increased synthesis in response to heat and contaminants	May complement function of hsp70	Good – easily detected response	Stegeman et al. (1992)

rat liver following exposure to cadmium. There is a current need to clarify the relationship between cellular effects and functional impairment at the organismal level and to study the response under field conditions at environmentally realistic chemical concentrations.

IMMUNE FUNCTION

Although disease-causing pathogens are biological, not chemical, stressors it is useful when assessing sublethal effects of toxicants to investigate their possible role in predisposing an organism to debilitating disease.

Macrophages are a critical part of the immune system because they protect an organism from disease-causing agents by engulfing, or phagocytising, foreign material. Phagocytic activity in macrophages and leukocytes is an important indicator of immune capacity, or the ability of an organism to combat infection, and may offer evidence of the general health of the organism. In a healthy organism, the capacity for phagocytic activity is likely to be high. Any agent, such as a toxic chemical, that causes a decline in such activity, is likely to contribute at least indirectly to the deterioration of the health of the organism.

Several studies of the suppression of the immune response have been conducted in the field. For example, Weeks and Warriner (1984) established that the macrophage phagocytic activity in the kidneys of two bottom-dwelling fish species – spot (*Leiostomus xanthurus*) and hogchoker (*Trinectes maculatus*) – from a tributary of the Chesapeake Bay, the Elizabeth River, was reduced, compared with that found in "control" fish from the nearby Ware River. The Elizabeth River is highly industrialised, whereas the Ware River has no significant industrial development and has a low human population density. The response was found to be reversible. When the same Elizabeth River fish were held in relatively clean water, their phagocytic activity returned to normal.

Other related end-points shown to elicit significant responses in the same series of investigations were chemotactic efficiency (the ability of the macrophage to migrate toward a stimulus) and chemiluminescence (CL), which is a measure of ROI production during phagocytosis. Both were suppressed in Elizabeth River fish. A variety of chemicals have been tested in the laboratory for their ability to suppress the CL response in both fish and oysters. Among them are metals (aluminium, cadmium, copper and zinc), pesticides (dieldren and chlordane), and cyclic organic compounds such as naphthalene and 2,4-dinitrophenol. However, for most of these chemicals, CL suppression occurs only at concentrations considerably higher than those seen in the natural environment. This raises questions as to whether immune suppression occurs in ambient waters where concentrations are much lower than those causing suppression in the laboratory.

TISSUE PATHOLOGY

Environmental toxicologists have long sought a connection between chemical contamination and the development of neoplasms in fish and mammals. Cause and effect is not easy to establish with respect to specific chemicals or groups of chemicals, although convincing evidence exists that hepatic neoplasms in flatfish from coastal sites are related to sediment contaminants such as PAHs and PCBs (see Case Study 4.2). Multiple neoplasms in several species of fish from the Fox River, Illinois, USA, seem to be related to high levels of PCBs in the river sediments. These cases were reviewed by Mix (1986), and a comprehensive review of histopathological effects of environmental contaminants has since appeared (see also Case Study 4.2).

In addition to hepatic neoplasms, there is reasonable circumstantial evidence for a linkage between epithelial and renal lesions in benthic fish species and organic chemical (principally PAH and PCB) levels in their native sediments.

SCOPE FOR GROWTH

Growth is a sensitive yet integrated means of assessing toxic stress. The ability of an organism to combat the potentially debilitating effects of toxic agents is realised at a price. Regulation, detoxification, and excretion of toxic products and repair of cellular and tissue damage are all processes requiring energy that would otherwise be directed to useful productivity.

Scope for growth (SFG) describes the excess energy available to an organism, after its basic metabolic needs have been met. As such, it represents the energy budget available for somatic and/or germinal growth (i.e., body size and reproductive effort). It is reasonable to assume that an organism under stress, for instance due to chemical exposure, will utilise energy differently from an organism that is not stressed. This difference in energy utilisation, or partitioning, is reflected in its SFG. It must be noted, however, that changes in energy partitioning do not necessarily reflect changes in environmental quality. SFG in organisms may be affected by physical disturbances to their environment, temperature changes, or variations in food supply. A good account of the theory, methodology, and application of SFG is given by Bayne et al. (1985).

In many cases, the measurement of growth is a protracted process. The assessment of growth in most multicellular organisms requires at least a few days, and reproductive performance usually takes even longer to evaluate. SFG measurements in contrast can be taken over a few hours, under controlled laboratory conditions so that sources of stress remain constant.

Organisms of choice in most scope for growth studies in the aquatic environment or in the laboratory have characteristically been bivalve molluscs (usually *Mytilus*), although the approach has also been adopted using copepod crustaceans and fish. The sessile behaviour of molluscs makes them a popular choice for SFG studies because calculations of their energy balance are not affected by energy expended through movement. Study organisms are generally taken from their natural habitat, and rates of respiration, feeding, and excretion are measured over a few hours under laboratory conditions. SFG can then be calculated using the balanced energy equation, which describes the relationship between physiological processes such as food intake, respiration and excretion, and growth potential.

$$C = P + R + E \tag{4.1}$$

where C = total consumption of food energy, P = total production of somatic tissue and gametes, R = respiratory energy expenditure, and E = energy lost through excretion. If A represents the energy absorbed from the food (i.e., $A = C \times$ [Efficiency of absorption]), then

$$P = A - (R + E) \tag{4.2}$$

In this form, the term P, which represents energy available for production of tissues and gametes, has been designated SFG and can be expressed in energy terms ($J g^{-1} hr^{-1}$). Scope for growth values may range from a positive to a negative integer. Several field investigations using natural and transplanted mussel populations have demonstrated an inverse relationship between SFG and measured pollution gradients involving a mixture of chemical contaminants and habitats as diverse as a Scottish oil terminal; a municipal outfall in Bermuda; Narragansett Bay, Rhode Island; San Diego Bay, California; Venice Lagoon, Italy; and Langesundfjord, Norway (see Bayne et al., 1985).

A similar but more qualitative approach utilising a different set of end-points is illustrated in a case study by Rowe and co-workers (Case Study 4.3). This investigation of amphibians in a coal-ash-polluted environment involved a variety of field and laboratory observations culminating in an energy-based scheme for describing animal fitness.

4.4.3 Individual species as indicators or monitors

INDICATORS

Definitions of the concepts of indicators and monitors as used in the present text have been provided in Section 4.2. As shown in the selection of examples in Table 4.7, many types of organism, from unicellular to complex, can be utilised as indicators of environmental conditions at the "whole organism" level.

Plants are particularly useful as indicators of air pollution in that rather characteristic and visible damage is exhibited, directly attributable to particular substances. Furthermore, certain plant species are known to be particularly sensitive to certain pollutants, and these can be set out in test plots, in effect providing integrated dose-response measurements. Perhaps the most well-known indicators of air pollution are lichens, which are particularly sensitive to sulphur dioxide (SO_2). According to Seaward (1993), this was first reported in the scientific literature by Skye in 1958. The best indicator may be the lichen community, rather than a single species, although since lichens are notoriously difficult to identify, an autecological approach may be preferable for practical reasons if lichens are to be used in routine monitoring. Natural communities of lichens have been shown to exhibit zones related to distance from industrial centres, which are sources of this air pol-

Table 4.7. Examples of aquatic and terrestrial indicator species

Indicator	Species	Perturbation	Response	Application	Reference
Terrestrial					
Terrestrial herb	*Epilobium angustifolium*	Radioactivity in substrate	Variation in flower colour due to radiation-induced genetic mutations	Surveys for detection of radioactive minerals	Shacklette (1964)
Lichens	Epiphytic lichen communities	Sulphur dioxide	Modification of lichen community: presence/absence of species	Scale devised for zones of SO$_2$ pollution based on lichen community indices	Hawksworth and Rose (1970)
Lichens	Epiphytic lichen communities, involving 14 species, transplanted on bark discs of trees	Sulphur dioxide	External and internal changes in the condition of the lichen thallus	Correlation between degree of damage to lichen thallus and average SO$_2$ concentration	LeBlanc and Rao (1973)
Lichens	Epiphyic lichens, 12 species	Sulphur dioxide	Distribution of epiphytic lichens with reference to known or expected SO$_2$ concentrations in air, using indices of atmospheric purity; close to the source of SO$_2$, only four species of lichen were found; increased numbers of species occurred with increasing distance from the source	Specifically, the study compares ground-level pollution from Sudbury, Ontario, (pre-1970) with the distribution of epiphytic lichens; more generally, authors conclude that the lichen method provides a quick way of assessing and mapping long-range pollution from a point source	LeBlanc et al. (1972)
Lichen flora	*Hypnogymnia physoides*, *Evernia prunasri*, *Parmelia* spp and *Usnea subfloridana* (these species show varying ranges of tolerance to sulphur dioxide)	Sulphur dioxide, multiple sources, in Greater London, UK	Species composition of lichens on trees in Greater London, UK, compared with similar studies over previous 15 years	Improvement in air quality as measured by decreased levels of SO$_2$ between the mid 1960s and 1979 can be tracked through the recolonisation of trees by some pollution-sensitive lichens	Rose and Hawksworth (1981)
Lichens that colonise man-made substrates	*Lecanora muralis*	Sulphur dioxide	Reinvasion of the lichen in West Yorkshire, UK, following abatement of air pollution	The lichen has reinvaded, but its response is slow and cannot be modelled in terms of a simple SO$_2$ threshold: possibly other parameters affect lichen as pollution levels fall	Henderson-Sellers and Seaward (1979)

Predatory birds	Spanish imperial eagle (*Aquila heliaca adalberti*), California condor (*Gymnogyps californianus*)	Organochlorine (PCBs, DDE) pollution	Eggshell thinning; egg infertility	Assessment of food chain effects; assessment of recovery after source removal.	Hernandez et al. (1988)
Aquatic Mayfly larvae	*Baetis rhodani*	Lake acidification	Species not present in low pH systems	Surveys of low alkalinity lakes: early warning of acidification	Raddum and Fjellheim (1984)
Diatoms	Community of many species	Lake acidification	Species composition and abundance determined for living and/or fossil diatoms	Preferences for acid conditions (acidophilic) or alkaline (alkalophilic) conditions and varying conditions in between enable the calculation of "diatom-inferred pH", for current or historical lake chemistry reconstruction	Dixit et al. (1992)
Diatoms	Range of diatom species from a dataset of numerous lake sediments	Acid sensitivity based on acid neutralizing capacity, as related to sulphur deposition	Species composition and abundance	Development of a diatom-based paleolimnological model based on a dose-response function that can be used to set critical SO_2 load values for a site	Battarbee et al. (1996)
Mayfly larvae	*Hexagenia limbata*	Lake nutrient status	Dominant species under mesotrophic conditions	Great Lakes water and sediment quality assessment	Edwards and Ryder (1990)
Tubificid oligochaetes	*Limnodrilus* spp.	Nutrient enrichment	Dominant species under mesotrophic conditions	Great Lakes water and sediment quality assessment	Edwards and Ryder (1990)
Walleye	*Stizostedion vitreum*	Lake nutrient status	Dominant species under mesotrophic conditions	Great Lakes water quality assessment	Edwards and Ryder (1990)
Amphipod	*Pontoporeia hoyi*	Lake nutrient status	*P. hoyi* in the dominant species under oligotrophic conditions	Great Lakes water and sediment quality assessment	Edwards and Ryder (1990)
Lake trout	*Salvelinus namaycush*	Lake nutrient status	*S. namaycush* is the dominant fish species under oligotrophic conditions	Great Lakes water quality assessment	Edwards and Ryder (1990)
Algae	*Cladophora glomerata*	Lake nutrient status	Dominant species under eutrophic conditions		McHardy and George (1985)

lutant, and the term *lichen desert* has been used to describe the loss of lichen species in areas close to pollution sources. In 1970 Hawksworth and Rose published a system of zonal mapping using epiphytic lichens for the estimation of SO_2 exposure. With increasing distance, the numbers of lichen species increases until "background" or nontoxic levels of the pollutant result in all the normal species being present. Furthermore, if the process is reversed and air pollution decreases, as it has done as a result of emission controls or to closure of an operation, the recovery is tracked quite reliably by the lichen flora. Transplanted lichens have also been used in air pollution monitoring. A review by Seaward (1993) deals in more detail with lichens and sulphur dioxide.

Other plant species, not shown in the Table 4.6, are known to be sensitive to specific air pollutants and can be planted in test plots in areas of potential impact such as coal-fired power plants and smelters. Thus seasonal plots in which *Gladiolus*, "Snow Princess", is planted show sensitivity to hydrogen fluoride (HF) while certain varieties of tobacco are used in test plots because of their known sensitivity to ozone. This type of application is slightly different from the use of native flora and faunal species as indicators but is, nevertheless, based on the same principle of substance-specific sensitivity of a species or variety to some perturbation in the environment. Case Study 4.4 elaborates on this approach.

Algae have also been advocated as biological indicators. One particular group of algae deserves special mention. The diatoms, members of the Bacillariophyceae, include freshwater and marine species of unicellular and colonial algae. Members of this group frequently show very specific preferences for chemical conditions in their respective aquatic environments. Diatoms can be identified, with suitable expertise, by the markings on their silicified cell walls. Furthermore, these cell walls persist with the markings faithfully preserved, after death of the cells, providing the potential for tracking historical changes in the chemistry of water. A core of profundal sediment is taken and dated by isotopic or pollen analysis, and the history (chemical or climatic or both) is reconstructed from the community composition at different times (e.g., Baron, 1986). The best example of this is probably derivation of the so-called diatom-inferred pH, a technique that gained wide use with the rise of concern for acidification, which began in the mid-1970s and continues to the present time.

The method is not without problems, not the least because of the difficulty of accurate identification of the diatom species and varieties. However, in the hands of experts and with careful attention to quality control and interpretation, studies of diatoms have provided historical evidence about past environments, where no direct information is available. Further readings are provided on this topic, which has a large specialised literature of its own.

Invertebrate and vertebrate animals, especially in the aquatic system, provide examples of indicators. Probably the most notable for freshwaters are trophic level indicators. Concern for excess nutrients in surface waters, especially in the Great Lakes, has led to studies of suitable indicators of water quality. Some of these are shown in Table 4.7. Note that both the loss of a species (as of lake trout) or the increase in a species' abundance (as of tubificid worms) can be applied as indicators.

Fish are frequently advocated as indicators of water quality, since fish are often, although not invariably, more sensitive to environmental pollution than are members of lower trophic levels. Furthermore, fish are frequently the most conspicuous resource that the public are concerned with. The lake trout (*Salvelinus namaycush*), which has rather rigorous requirements for high concentrations of dissolved oxygen and low water temperature, is considered a good indicator of oligotrophic conditions. Its disappearance from freshwaters typically accompanies the early stages of eutrophication (see also Section 6.10). The lake trout is also a highly valued species and will be "lost" rather early in the series of events known as cultural eutrophication. So while it would qualify as an early warning indicator, in terms of integrity of the ecosystem, one might question its value as an indicator. Perhaps the most "practical" use of the lake trout as an indicator would be in the context of recovery. Cultural eutrophication is potentially reversible, so the return of the lake trout would be an indicator of a return to oligotrophic conditions.

Disruption of a community may result in a subsequent repopulation by a small number of opportunistic species having high reproductive rates. Following chemical pollution, the species composition of the community may be biased toward tolerant species, one or more of which may function as an indicator species of the altered chemical environment. The term *indicator species* is not confined to chemically tolerant species, however. For example, the clam *Parvilucina tenuisculpta*, the polychaeta *Capitella capitata*, and the ostracod crustacean *Euphilomedes* spp. have all been associated with organically rich discharges such as sewage outfalls. *C. capitata* is an opportunistic species having good tolerance for low oxygen and an attraction to organically rich sediments as a food source. *Euphilomedes* uses discharged organic particles as food, and *P. tenuisculpta* is capable of utilising a variety of energy sources including hydrogen sulphide in low oxygen environments. If a particular outfall or series of outfalls under investigation is also the source of other contaminants such as metals, these, too, may co-correlate with abundance of such indicator species even though the relationship may be an indirect one. Within this context, the definition of ecological indicator becomes somewhat broader than that portrayed in Table 4.3, which is strictly based on chemical sensitivity, and foreshadows some of the more community-based concepts dealt with later in this chapter.

Some attempts have been made to develop computer-based programs for identifying indicator species based on field habitats. For example, Twinspan is a two-way indicator species analysis that first constructs a site typology and then uses it to obtain a classification of the species according to ecological preferences (Hill, 1979). A variant of this method developed by Dufrene and Legendre (1997) differs from Twinspan in the way it combines the relative abundance of a species with its relative frequency of occurrences at the various groups of sites. It incorporates a randomised procedure, which, unlike Twinspan, uses an indicator index for a given species that is independent of the other species' relative abundance.

The current section addressed indicator organisms. Even though the scope is quite large, practical applications of indicator organisms for assessment and in regulatory use are, to date, quite limited. As discussed in Chapter 12, most standards for toxic substances, although based on biological response, use chemical parameters rather than biological ones. As shown in Case Study 4.1, organisms do not always "conform" to the chemical standards that have been developed. In the final analysis, perhaps a combination of both should be advocated and used in an integrated manner, as for example in Ontario's Sediment Quality Guidelines (Jaagumagi, 1992).

MONITORS

Vectors for uptake and accumulation of chemicals from the environment include direct uptake from the medium – air, water, soil, or sediment – as well as intake through the various normal mechanisms of inhalation, digestion, and so on. Plants and some simple microorganisms can, of course, only reflect exposure from the medium, be it water, soil, sediment, or a combination of these, whereas animals and many microorganisms can be exposed directly from the external medium and also through uptake from ingested material. When selecting organisms for biological monitoring, these potential routes of exposure need to be taken into account. In many instances, the relative significance of the respective routes is not well understood. For example, the respective importance of the uptake of mercury by fish, directly from the medium (water) or from ingestion (of food items and other particles), is still not clear. Although the uptake of mercury by fish has been studied in some detail, the underlying mechanism is still incompletely understood. Considerable debate still exists over the degree to which some epibenthic animals reflect the chemical constitution of the sediment, interstitial water, or overlying water (see Warren et al., 1998). In this regard, much may depend on feeding strategy.

For plants, because the situation concerning routes of uptake is somewhat more clear-cut, in that, with few exceptions, plants take their nutrients and their contaminants directly from the external medium, the use of plants as biomonitors for chemicals therefore appears less complicated than the use of animals. Yet plant monitors obviously tell only part of the story in terms of toxic substances in the environment. In the context of indicator organisms, plants were seen as having their major application for assessment of air pollution, and arguably the same is true in the context of monitors. Case Study 4.4 develops this concept.

The theoretical basis for biological monitoring or biomonitoring rests on the assumption that the biological monitor takes up the contaminant into tissues, resulting in concentrations that reflect the exposure of the contaminant in the environment. As discussed in the introductory section of this chapter, organisms may be superior to a chemical determination of a contaminant in air, soil, water, or sediment in that the organism "sees" only the biologically available component of the contaminant. This may be quite different from the so-called total amount that is measured by most analytical instruments with routine sample preparation.

In a simple setting, uptake can be shown to be related to exposure, at least over a certain range of concentrations. This is illustrated in Figure 4.1A, which is a graph of the uptake of cadmium by caddis fly larvae under controlled laboratory conditions. As expected, the higher the exposure, the higher the tissue concentration. Another set of data from the same laboratory is shown in Figure 4.1B. Zinc uptake has been measured, for two field populations of the same species. The striking message in Figure 4.1B is that, even within a species, the actual amount of Zn in the tissue can vary a great deal for a given exposure, depending upon the population. The possible explanations for these differences will be discussed later. At this point, they need to be considered in the context of biological monitoring.

The attractive and apparently simple idea of using biological monitors to reflect environmental conditions turns out to be very complex. Ideally, any biological monitor should be calibrated in the laboratory prior to using it in the field. At the very least, some statistical relationship should be established from ambient environmental conditions concerning the relationship between a contaminant and the substrate.

But the mechanism(s) of uptake are often incompletely understood, as are the relative contributions of the contaminants from different compartments of the environments (water, air, food, etc.). Furthermore, as referred to earlier and discussed in more detail in Section 6.2.3, biological availability is an essential component of the concept of biological monitoring. Yet rarely is the availability known in a complex, "real world" situation. Indeed, the biological monitor is advocated to circumvent the incomplete understanding, in a chemical sense, of the availability of a contaminant. So calibration becomes problematic, once one moves beyond a controlled situation in the laboratory.

If the availability of a contaminant in water is considered to be problematic, it is arguably more so for the soils and sediments, which have the additional complicating factors of solid phases and frequently more complex geochemistry than the liquid phase.

Considerable effort has been put into examining the relationship between metal uptake into aquatic biota and concentrations in sediments or water. A review by Outridge and Noller (1991) of studies on aquatic macrophytes concluded that the total metal in sediment was unsatisfactory in terms of finding a relationship with the metal in tissue. They quoted a 1985 review by Campbell and others, which showed that fewer than 30% of the studies found significant correlations between total metal in sediment and metal in rooted aquatic plants. The latter authors suggested that the poor relationship could be explained in part by the fact that a large proportion of metals in sediments is unavailable to plants. The unavailable portion is, however, unlikely to be consistent, either among elements or among sediments.

The former reviewers observed that, in the real world, the water column may also be a source of contaminant for uptake by rooted aquatic plants. A great deal of work, including a large literature on marine and freshwater invertebrates, has been published on metal accumulation, and this subject is revisited from other perspectives in Section 6.2.6.

It would appear that, in general, patterns of uptake can be predicted more reliably for organic substances than for metals. This is related to our superior understanding of the mechanisms by which the organisms react with organic molecules (see Section 7.8). Models have been constructed for the purpose of predicting bioconcentration factors for hydrophobic organic molecules (see Section 4.6.2).

In practice, a quantitative relationship between contaminant concentration in the environment and contaminant in the tissue of the monitor is rarely established. This means that for most monitoring studies, data can realistically be used for qualitative assessment only. But this does not diminish the value of the approach. Spatial and temporal comparisons of the tissue concentrations of contaminants in selected organisms have provided excellent means of tracking changes in exposure, with relative rather than absolute values being assessed.

Table 4.8 provides a number of examples of biological monitors, selected not as a comprehensive coverage of the broad topic, but to give a general impression of the range of monitors that have been studied and, in some cases, put into practical use. In the following text, some more detail is provided for some of the examples in the table.

Lichens and mosses have been used rather extensively as monitors of atmospheric deposition. These lower plants have neither cuticle nor transport systems and obtain their nutrients from entrained and dissolved materials that fall upon their surfaces, as well as gaseous substances in the air. Accordingly, contaminants are also encountered in this rather direct manner and may be strongly concentrated in lichen and moss tissue relative to their generally low levels in air and precipitation. Monitoring systems using moss and lichen have been developed quite successfully for a number of airborne contaminants. In Scandinavia, for instance, the forest mosses *Hylocomium splendens* and *Pleurozium schreberi* and the peat moss *Sphagnum fuscum* have been used to monitor atmospheric deposition of trace metals [e.g., Cd, Cu, nickel (Ni), Pb]. Researchers in Sweden and Norway have set up and implemented an impressive system for monitoring long-range transport of metals and more recently trace organics in feather mosses (Ruhling et al., 1997).

The feather moss *Hylocomium splendens* is a particularly good monitor because its typical habitat in the Nordic countries is open sites in ombrotrophic bogs, where it grows at ground level in large, readily recognised stands. Furthermore, its growth habit is such that annual growth increments can be identified and recent growth can be separated from the rest of the plant. Figure 4.5 illustrates some of the results for lead in Norway. Concentrations in the moss immediately illustrate two points: (a) the deposition is much greater in the south of Norway than in the northern parts and (b) concentrations have decreased between 1977 and 1985. The former point

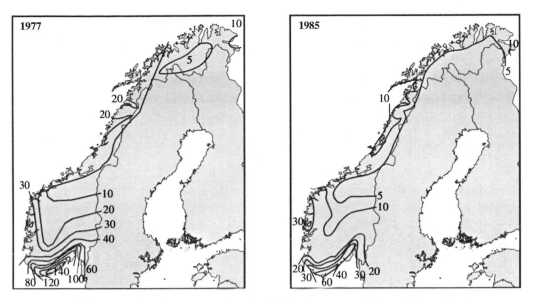

Figure 4.5 Lead concentrations in the feather moss *Hylocomium splendens*, a monitor for atmospheric deposition in Norway. These maps show some of the results from the Norwegian national deposition surveys of a number of elements, carried out in 1977 and 1985, respectively. Lead concentrations in the moss (ppm, dry weight) decreased from southern to northern Norway in both surveys, and decreased between 1977 and 1985. Modified from Steinnes (1995).

has been used to show that long-range atmospheric transport of lead (as well as other elements and compounds) from mainland Europe is a major source of lead for Norway, since the prevailing winds are from Europe. The second point relates to the control of lead in gasoline, introduced in many countries early in the 1980s, resulting in a great decrease in atmospheric lead. Automobile emissions were the greatest single contributor to airborne lead, at least in the northern hemisphere.

Lower plants as well as tree foliage have been used on a more local scale to demonstrate the relationship between point sources of pollution such as smelters and deposition from the atmosphere. Lichens, in a general sense, resemble mosses in their mode of accumulation of material from the atmosphere, although the mechanisms are probably not identical. The subject of the mechanisms by which lichens accumulate substances from deposition is beyond the scope of this chapter. Further reading is provided in Section 4.13. The lichen *Cladina* has been widely used to monitor deposition of contaminants from local sources and with the support of sophisticated analytical and statistical techniques; the accumulation of material in the lichens can be used for source identification. One such study was made on *C. rangiferina*, providing a comparison of sulphur and lead in lichens from Eastern Canada and the Northwest Territories of Canada, respectively. Lichens have also been used in source identification on a local scale as shown in this and some other examples in Table 4.8.

Table 4.8. Examples of terrestrial and aquatic species used for environmental monitoring

Monitor	Species	Perturbation	Parameter(s) measured	Application	Reference
Terrestrial					
Moss	Feather moss *Hylocomium splendens*	Atmospheric contamination	Concentrations of Pb, Cd, V, and Zn in moss tissue	Assessment of the source(s) of atmospheric contaminants in the arctic: Long-range transport or local point sources?	Ford et al. (1995)
Moss	Feather moss *Hylocomium splendens*	Inorganic contaminant deposition from atmosphere, wet or dry, local source or long-range transport	Concentrations of Na, Mg, V, Cr, Mn, Fe, Ni, Cu, Zn, As, Rb, Sr, Mo, Cd, Sb, Ba, La, Pb, and Bi determined in moss tissue	Spatial and temporal monitoring for long-range transported airborne contaminants in Norway; long-range transported elements (V, Cu, Zn, As, Mo, Cd, Sb, Pb, and Bi) generally correlated with wet deposition	Berg et al. (1995)
Moss	Feather moss *Hylocomium splendens*	Atmospheric deposition of PCBs	Concentrations of a range of PCB congeners in 1977, 1985 and 1990 respectively, in mosses from remote sites in Norway	Over the time period, total PCB concentrations have declined in all samples from all sites	Lead et al. (1996)
Mosses	*Hylocomium splendens*, *Pleurozium schreberi* (feather mosses) *Sphagnum* spp. (bog mosses) *Hypnum cupressiforme* (epiphytic moss)	Evaluation of naturally growing moss as a monitor of metal deposition	Review of studies on metals in mosses in Norway with specific concern for polar regions	Evaluation of the use of mosses as monitors of atmospheric deposition of heavy metals recommends (a) calibration of moss data against precipitation data and (b) assess contribution from sources other than air pollution	Steinnes (1995)

	Species	Topic	Findings	Reference
Lichens	*Umbilicaria muhlenbergii, Cladonia rangiferina, Lecanora* spp., etc.	Contaminants in atmospheric deposition	Review of processes by which lichens take up atmospheric deposition includes particle trapping, ion exchange, passive and active uptake	Lichens have value as monitors of atmospheric deposition, from local to global scale; caution concerning interspecies differences; understanding of mechanisms of uptake will improve the value of this approach	Nieboer and Richardson (1981)
Lichens	*Cladina rangiferina*	Deposition of sulphur and lead in Eastern Canada	Measured S and Pb in lichens from Eastern Canada and from the Northwest Territories, Canada	Marked differences in lichen S and Pb reflect differing emission/deposition rates in the two regions	Zakshek et al. (1986)
Lichens	Epiphytic lichens	Lead contamination of the boreal forest of Quebec between 47° and 55°N and along St. Lawrence valley between 45° and 48°N	Lead isotopic composition for epiphytic lichens as well as for higher vegetation and lacustrine sediments	Source identification is possible by using isotope ratios in lichens; in this study, lichens along the St. Lawrence valley show dominant inputs of lead from U.S. sources, and between 48° and 53°N a significant amount of the lead is from Noranda's smelting activities	Carignan (1995)
Lichens	*Cladina mitis*	Trace element patterns following closure of uranium mines	Concentration of 20 elements including uranium were determined in *C. mitis*, sampled in July 1992 (i.e., 9 years from closure of the Agnew mine and 2 years after Quirke mine at Eliot Lake ceased production)	Results showed the presence of 11 elements in the lichen, with concentrations of U much lower than in *C. rangiferina* 10 years earlier; in spite of mine closure, U and Th in lichen thallus remained higher near sources of ore dust	Fahselt et al. (1995)

Table 4.8. *(Continued)*

Monitor	Species	Perturbation	Parameter(s) measured	Application	
Annual growth rings of trees	Sugar maple *Acer saccharum*, sycamore *Acer pseudoplatanus*	Atmospheric metal deposition, trace element pollution in soil	Metals are determined in individual tree rings, representing the environmental exposure to the metal(s) at the time when the ring was laid down	Used for temporal monitoring for point source or long-range transported airborne contaminants, including historical reconstruction of pollution scenarios	Watmough (1997)
Gull eggs	Herring gull	Xenobiotic organic contaminants in the general environment, from various sources, likely to be in the human food chain, including organochlorine pesticides (e.g., DDT and derivatives) and PCBs	Contaminant concentrations in eggs	Tracking the time-course of contamination and checking for recovery in terms of decreased concentrations in the Great Lakes basin	Ryckman et al. (1997)
Aquatic Marine bivalves	Mussel *Mytilus edulis*	Contamination of marine sediments from industrial and municipal sources, including local sources and long-distance transport	Tissue concentrations of various metals in various tissues, which have been accumulated from water, sediment, and food sources	The so-called Mussel Watch; used for tracking local and generalised contamination of marine sediments; also can provide an indication of levels of contaminants in edible seafood	Goldberg et al. (1978)

138

Organism	Species	Pollution	Application	Comments	Reference
Marine bivalve	Filter feeding marine bivalve *Corbicula* sp.	Trace metals entering estuaries from various sources including upstream sites	Concentrations of Ag, Cd, Cr, Cu, Pb, and Zn within a broad salinity range in tissues of the bivalve as well as in sediments	Demonstration of value of intensive sampling in assessing fate and effects of trace metals in estuaries; case study concerns characterisation of regional distribution metals within a broad salinity range where Sacramento River enters San Francisco Bay	Luoma et al. (1990)
Freshwater bivalves	Clams *Andonta* sp.	Water and sediment contamination by inorganic and organic contaminants	Metal concentrations in transplanted and native animals determined and compared with those in reference or "clean" sites	Comparison of different sites, including those connected with historical or current industrial activity along the Niagara River	Niagara River Toxics Committee (1984)
Freshwater bivalves	Zebra mussel *Dreissenia polymorpha*	Point source or diffuse contamination of freshwater sediments by metals	Spatial and temporal comparisons of metal concentrations in tissues		Kauss (personal communication)
Algae	*Cladophora glomerata*		Bioaccumulation of metals and some organics from water		Niagara River Toxics Committee (1984)
Aquatic floating plant	Water hyacinth *Eichornia crassipes*	Pollution by nutrients and metals in tropical water systems	Concentrations of metals in roots and floating shoots	Successful as a monitor of metals in tropical systems	Gonzalez et al. (1989)
Aquatic macrophytes	Pipewort *Eriocaulon septangulare*	Metal pollution in lakes (water and/or sediment)	Bioaccumulation of Zn, Cr	Evaluation of various freshwater plants revealed that *E. septangulare* accumulated the highest Zn and Cr; sediment Zn was the best predictor of plant Zn	Reimer and Duthie (1993)

The leaves of trees, particularly nondeciduous trees, have been used to monitor deposition of toxic substances, but in general it appears that lower plants are more amenable to large-scale monitoring. The wood of trees incorporates material in the course of its growth, and, for some elements, the annual growth rings of trees appear to record faithfully exposure at the dates when they were laid down. Thus, tree rings can be used to monitor historical exposure. The technique benefits from analytical procedures that can detect accurately the concentration of elements in relatively small samples of material. The use of tree rings as monitors for reconstructing the history of contaminant deposition is a relatively new approach, and its practitioners advise caution in its use. The use of tree rings for monitoring metals is revisited in Section 6.10.4.

A rather extensive monitoring system for organochlorine contaminants in the Great Lakes has been developed by measuring concentrations of various persistent organics (DDT, dieldrin and other insecticides, PCBs, and dioxins) in the eggs of the herring gull *Larus argentatus*. These types of contaminants are discussed in more detail in Chapter 7. They are persistent, and are known to biomagnify and to have harmful effects on organisms at high trophic levels, such as fish-eating birds. In the early 1970s, waterbirds in the Great Lakes were among the most heavily contaminated in the world (Ryckman et al., 1997). Putting aside for the moment the possible adverse effects of these contaminants on the birds, which have also been studied, and considering them in the context of biomonitoring organisms, it is notable that the Canadian Wildlife Service (CWS), part of the Canadian Federal Department of the Environment, has been tracking contaminants in gull eggs for several decades. The herring gull meets most of the criteria for a biomonitor. The species breeds in all five of the Great Lakes. It stays in the lakes year-round, it is a predator, and its colonial nesting habits mean that its eggs are easily sampled. By using eggs (one per nest per year is collected), it is not necessary to kill adult or young birds. Furthermore, the species breeds in other regions of Canada and other parts of the world, which means that these results from the Great Lakes can potentially be compared with those from other areas.

Figure 4.6 shows a sample of the types of data that can be obtained, in this case for PCBs. Spatial and temporal trends are identifiable. Since the start of the programme in the 1970s, concentrations of almost all the contaminants have decreased, which almost certainly reflects more stringent regulations and decreased use of these persistent compounds. Yet even those substances that are apparently no longer being released de novo into the environment are still detectable in the gull eggs. More details are provided as additional reading in Section 4.13.

Monitors have also been developed for contaminants in the aquatic system. The filamentous alga *Cladophora*, which grows prolifically in nutrient-rich waters (Whitton, 1970), offers promise as an aquatic monitor for metals and for trace organics. Rooted aquatic plants may accumulate contaminants from both the bottom sediments and/or the water column. Most studies, however, show poor relationships between sediment-metal concentrations and resultant tissue burdens.

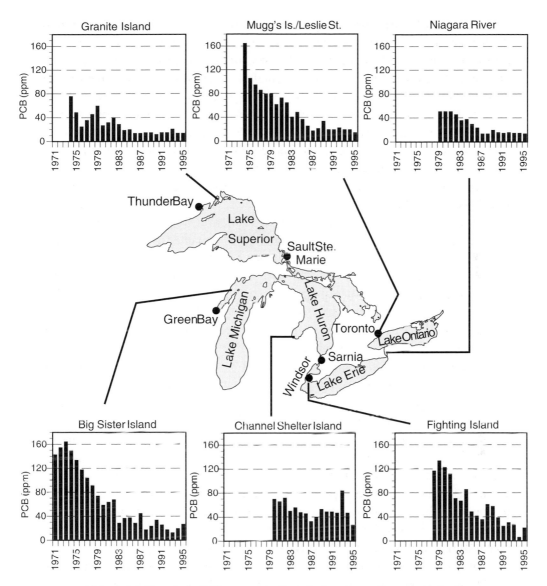

Figure 4.6 Trends in PCB concentrations in herring gull eggs at six sites on the Great Lakes. The maps show the location of six of the monitoring sites at which gull eggs have been collected in the biological monitoring programme for persistent organic contaminants in the Great Lakes of North America. Concentrations of PCBs vary considerably among sites, mostly related to the proximity to urban and/or industrial centres. Long-range atmospheric transport also contributes to the PCB load in the various sites. There has been a general decline in concentration of PCBs and other persistent organics over time. Modified from Ryckman et al. (1997).

This may be due to the generally poor availability of sediment-associated contaminants as discussed earlier, or it may indicate the greater importance of the water vector for contaminant uptake in aquatic plants. Although fewer studies have been conducted with free-floating plants, relationships between contaminant levels in water and their concentrations in tissue of free-floating plants are generally more predictable.

Invertebrates such as clams *Anodonta* spp. (freshwater bivalve); *Mytilus edulis*, *M. californiarius*, and *Ostrea sandvicensis* (marine clams); and *Crassostrea virginica* (marine oyster) offer an attractive means of monitoring pollutants. Recently the invading zebra mussel (*Dreissena polymorpha*) has been promoted as a monitor for metals (e.g., Camusso, 1994). It is prolific and occurs across a range of salinities from fresh to slightly saline. The adult stages of these organisms have a number of attributes that are valuable as monitors: they are sessile, they are relatively long-lived, they are common and easy to recognise in the field. Many are filter feeders, processing large volumes of water, thereby acting as integrators of aquatic contamination. They also make good experimental animals, a feature that may be important in the context of "calibrating" the monitor to relate exposure and body burden. Marine bivalves are used in the now classical (and worldwide) study called the Mussel Watch.

In 1984, the Mussel Watch programme was initiated as part of the National Status and Trends Program of NOAA (*http://www.gsf.de/UNEP/mussel.html*). Tissue concentrations of both inorganic (trace metals) and organic (DDT, PCBs, PAHs) contaminants have been measured in the marine mussels *Mytilus edulis*, *M. californianus*, and *Ostrea sandvicensis*, as well as the oyster *Crassostrea virginica* from approximately 200 sites along the U.S. coastline. In addition, concentrations of chemicals in sediment and water have been measured to estimate the effects of contaminants on estuarine and coastal areas and to indicate any relationships between environmental levels and bioaccumulation. Since its inception, the programme has been extended to many other parts of the world.

4.4.4   Surrogates for ecosystem indicators

Surrogate organisms are species that effectively integrate the physical, chemical, and biological properties of an ecosystem and thereby provide an indication of the "health" of the system. Because stresses in nature rarely occur in isolation, or at a single hierarchical level, it is often more practical to adopt a community approach when attempting to measure ecosystem response to perturbation, and, properly chosen, this type of response can identify early signs of environmental damage.

At least three criteria must be satisfied for an organism to qualify as a surrogate species:

1. It must be a strong integrator of the food web of which it is part.
2. It must be abundant and widely distributed within the system.
3. It must be easily recognised and ecologically relevant.

Community-level responses that might indicate environmental stress include:

- Decrease in mean size of organisms
- Reduction in biomass

- Specialist organisms replaced by generalists
- Reduction in species richness (number of species)
- Change in community composition (relative abundance)
- Reproductive impairment in some species
- Change in food web structure and length
- Decrease in predator numbers and increase in prey

A fairly detailed study of the use of surrogates for assessing ecosystem health in the Great Lakes has been provided in a report to the Great Lakes Science Advisory Board (Edwards and Ryder, 1990).

4.5 Community and higher level indicators: The ecological approach to toxicology

Some of the problems involved with extrapolating from whole organism responses to population and community effects were addressed in Chapter 2 (see Figure 2.1). Most toxicity data for individual organisms come from laboratory investigation, whereas community studies largely involve field investigation. After we make the transition from the laboratory to the field, we are confronted by a host of complicating factors, including species-species interactions, chemical mixtures, and a variety of extrinsic environmental variables. In the face of numerous factors beyond our control, we are forced to take an ecological approach to determining the effect of chemical and other stressors on ecosystems.

4.5.1 Interspecies effects of toxic substances

Each organism within an ecosystem is uniquely adapted to a particular range of environmental factors. Within each species, the collective population response to a particular environmental parameter might be expected to follow a normal distribution. Where two species overlap in range, the presence of a chemical stress may favour the more tolerant species. A hypothetical example is shown in Figure 4.7 where an increase in a particular toxicant favours species A over species B. Subtle variants of this may exist, whereby toxicant stress might affect the relative response to a second stress such as temperature or oxygen availability.

Several things can happen that can change this relationship. In Figure 4.7B, species B becomes more resistant to the toxicant and can compete more successfully with A. The increasing encroachment of B on A (Figure 4.7B) may involve an increase in the population of B, through access to more resources. The size of the equilibrium zone might also increase, although this is not necessarily the case and much would depend on the relative sizes of the shaded areas where competition results in the extinction of one or other of the species. The boundary lines separating the two species will, themselves, depend on the genetic diversity of each species and may be better represented by broad bands representing 95% confidence limits rather than sharp lines. The equilibrium point itself is also subject to such variability and may best be described as an area of probability. In Figure 4.7C, the

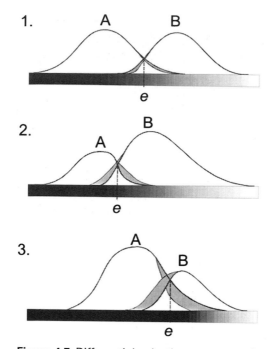

Figure 4.7 Differential selection pressures down a pollution gradient. Situation 1 indicates a preexisting condition wherein two competing populations are in equilibrium around an equilibrium point (e) within a pollution gradient indicated by the differentially shaded bar at the bottom. Shaded portions of overlapping curves represent areas of probability of competing populations reflecting differential pollution tolerance within each population. Situation 2 reflects an increase in pollution tolerance in population B (relative to A) within the same pollution gradient. Situation 3 shows the effect of increasing pollution as indicated by a shift of the gradient to the right (assuming same pollution tolerance as in 1).

degree of pollution may spread, resulting in the encroachment of A on B and forcing the equilibrium to the right. It should be emphasised that the more resistant species may themselves be severely compromised by the toxicant. Although A may hold a selective advantage over B regarding its tolerance of the toxicant, it may pay a significant price in terms of growth, reproduction, and mortality.

An example of how toxicant stress may affect interspecies relationships is shown in a series of experiments by Sanders and co-workers who investigated the effect of low levels of selected toxicants on natural phytoplankton assemblages in estuarine water (Sanders and Riedel, 1998). Filtered river water was entrained in a laboratory and split between control chambers and chambers receiving sublethal levels of toxicants [arsenic (As), Cd, Cu]. Species succession was followed over the next 3 weeks, indicating dominance by centric diatoms and, later, neonate diatoms in control chambers, whereas cyanophytes became dominant in metal-dosed water (Figure 4.8).

Experiments such as this illustrate effects at the primary producer level. Effects at higher trophic levels are complicated by a variety of interactive factors includ-

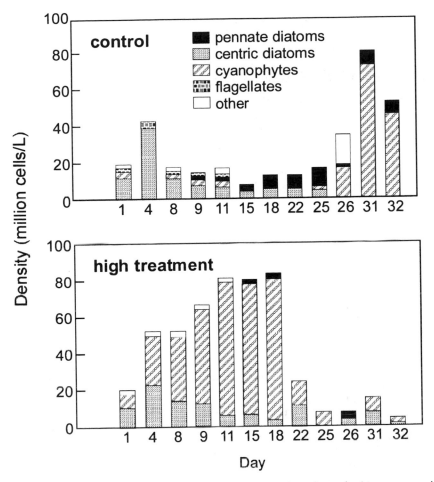

Figure 4.8 Change in species succession in estuarine phytoplankton exposed to sublethal concentrations of trace metals. After Sanders and Riedel (1998).

ing competition for food resources. Therefore, changes at the primary producer level may result in fundamental changes further up the food chain depending on whether surviving species are more or less acceptable as food for the next trophic level. For example, the shift from centric diatoms to cyanophytes seen in Figure 4.8 represented a sharp decline in food quality when presented to a copepod consumer, *Acartia tonsa*. Despite the much higher density of cyanophytes, their extremely small size resulted in a similar biomass in treated and untreated media. However, Acartia fed the cyanophyte-dominated assemblage showed markedly reduced fecundity and survival relative to controls (Sanders and Riedel, 1998). Based on results of experiments such as these, Sanders and Riedel (1998) produced a conceptual model for ecosystem-wide effects of trace metals (Figure 4.9). In Figure 4.9, the size of circles denotes organism size/abundance, and arrows denote feeding relationships.

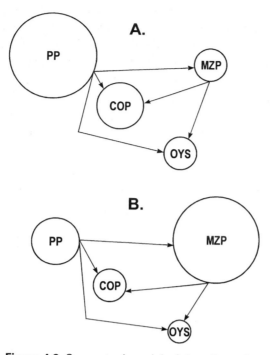

Figure 4.9 Conceptual model of the effect of trace metals on an estuarine ecosystem. Size of circles signify size/abundance of organisms. Arrows denote feeding relationships. PP = phytoplankton, MZP = microzooplankton, COP = copepods, OYS = benthic filter feeders. (A) normal community containing larger phytoplankton species, dominant copepods and benthic filter feeders. (B) Trace element-stressed community containing smaller phytoplankton species (small diatoms, flagellates, cyanophytes), reduced feeding by copepods and benthic filter feeders and dominance by microzooplankton and microbial heterotrophs. After Sanders and Riedel (1998).

4.5.2   Interaction between and among trophic levels as affected by toxic substances

Another view of the food web takes a "top-down" approach wherein the removal of a top carnivore releases a secondary carnivore to crop down a herbivore population, thereby causing an increase in algal biomass. In demonstrating this "cascade" effect in a stream community, Power (1990) reported both qualitative and quantitative changes in the algal community following the elimination of the top predator. Such cascade effects would not be clear-cut in cases where omnivory is prevalent in a food chain (i.e., both secondary carnivore and herbivore would be available as food for the primary predator). Both species interaction and reproductive rate will affect the rate of population or community recovery following perturbation by a toxicant. The time for a community to return to equilibrium following disruption is often called resilience or elasticity. In the absence of ecological considerations, it might be assumed that larger organisms would be more resistant to toxic effects. However, in a community adversely affected by toxic chemicals, food-chain dependency and long reproductive cycles often make the larger, longer-lived species more vulnerable. It follows that increased resilience would therefore result from a faster reproductive rate. Even though there are counterarguments to this strategy under certain circumstances (Stearns, 1976), unstable environments following chemical or physical disruption tend to favour "r-selection" or rapid population growth. One would expect such environments to be dominated by a relatively few opportunistic species. More stable environments enabling competition and avoidance of predation will favour K-selection characterised by delayed reproduction, relatively small reproductive effort, a few offspring, and a high assimilation efficiency.

In addition to direct toxic effects of chemicals on organisms, pollution may cause disruption of the food chain (e.g., through the process of eutrophication). Excess nitrogen and phosphorus, through both point sources and nonpoint sources, may result in both qualitative and quantitative changes at the primary producer level. Lake Erie and the Chesapeake Bay have both experienced eutrophication through anthropogenic inputs of nitrogen (N) and phosphorus (P) from surrounding watersheds. Case Study 4.5 illustrates some of the complexities of food web interactions resulting from anthropogenic activity in the Chesapeake Bay.

A major by-product of eutrophication is increased oxygen utilisation resulting in an increase in hypoxic and anoxic conditions in the warm summer months. During the 1950s, a fivefold increase in oxygen demand was seen in the sediments of Lake Erie and in the central basin increasingly anoxic conditions denied refuge to several cold water fish species such as lake trout, blue pike, and lake whitefish. Accelerating land clearance in the catchment area of this and other Great Lakes contributed to the deposition of clay, which compacted at the sediment surface to form a hard impervious layer that had poor oxygen exchange and was unsuitable for fish egg survival.

4.5.3   Population and community end-points

Populations and communities can be characterised in a variety of ways, which may be qualitative or quantitative. Indices may be based on changes in structure of a particular population or community, or they may measure changes in function. Others may be a mixture of structure and function. Population measurements tend to be largely structural (e.g., the relative abundance of different size/age classes). However, timed components such as speed of sexual maturity and rate of hatch/development represent "bottom-up" determinations, which may contribute functional data on reproductive efficiency, particularly when combined with information on nutrient intake. The ratio of respiration to photosynthesis in a macrophytic or algal population or community represents a "top-down" approach to ecosystem function. Other functional indices include the production/biomass ratio and decomposition rates.

POPULATION-BASED END-POINTS

Relatively simple populations, such as algae, may be quantified in terms of cell numbers or overall biomass. Surrogates (see Section 4.4.4), such as chlorophyll determination, may also be used as a measure of biomass. More dynamic, functional measurements include doubling times, photosynthesis, and respiration rates. Most population models are based on measurements of reproductive performance (fecundity) and on the relative size/age structure of the population determined by a census at appropriate time intervals.

Reproductive performance continues to be a commonly used sublethal end-point in toxicity bioassays. Percentage hatch and survival of the young are integral components of life-cycle tests, although they are rarely examined further in acute and chronic bioassays. Initially developed as a tool for the study of population dynamics, the life table provides a more detailed evaluation of reproductive success through the determination of the intrinsic rate of natural population increase, r. This parameter is derived from the equation:

$$\sum l_x m_x e^{-rx} = 1 \tag{4.3}$$

where l_x = number of living females on day x/number of females at the start of the live table, x = time (days), and m_x = number of female offspring produced on day x/number of living females on day x.

Note that the equation specifically focuses on the survival of females as being most important to reproduction. In species where sexual differentiation is impossible in very young stages, a 1:1 male/female ratio is assumed, and the daily offspring production is simply divided by two.

Life tables as determinants of toxic chemical effect have been confined to a few species of copepod, ostracod, and amphipod crustaceans and have been summarised most recently by Bechmann (1994).

Stage-based demographic models, often called Leslie matrix models, also use empirical data from different life stages, although their critical functional components are estimates of probabilities of transition from one developmental stage to the next. Using a fish as an example, the life-cycle may be represented as being composed of four stages: egg, larva, juvenile, and adult. Connecting each stage are the transition paths (Figure 4.10). Thus, for the egg stage, there are two possible transition paths. Each path is associated with a probability. F_{11} is the probability that on a specific day the egg will remain in the egg stage. F_{12} is the probability that the egg will hatch to yield a viable larva. There are no other transitions defined for this stage [i.e., an egg cannot hatch into a juvenile or an adult ($F_{13} = F_{14} = 0$)]. Failure to hatch ($F_{12} = 1$) represents complete mortality. For the proportion of larvae that hatch, they face a daily probability of remaining a larva (F_{22}) and a daily probability of becoming a juvenile (F_{23}). No other transitions are defined (i.e., a larva cannot regress to an egg nor may it become an adult and spawn). In addition to the projection matrix, F (Figure 4.10),

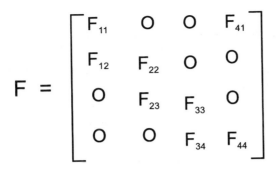

$$F = \begin{bmatrix} F_{11} & O & O & F_{41} \\ F_{12} & F_{22} & O & O \\ O & F_{23} & F_{33} & O \\ O & O & F_{34} & F_{44} \end{bmatrix}$$

Figure 4.10 Projection matrix for a stage-based demographic (Leslie matrix) model illustrating possible transition paths.

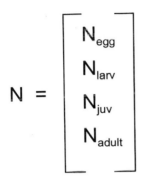

$$N = \begin{bmatrix} N_{egg} \\ N_{larv} \\ N_{juv} \\ N_{adult} \end{bmatrix}$$

Figure 4.11 Four-stage population distribution vector used in coordination with projection matrix to predict future population composition.

the life-cycle diagram can also be represented as a population distribution vector N (Figure 4.11), which defines the number of animals in each stage.

When multiplied together, these two matrices predict the distribution of animals in each stage on the next day. Similarly, when multiplied together t times, the resulting matrix $N(t)$ defines the distribution of animals on the tth day as

$$N(t) = F^t \cdot N(0) \tag{4.4}$$

This equation forms the basis of all population projection models. Similar population vectors and projection matrices can be designed for a variety of species, although it is of particular importance to understand the rate of development so that a census of a particular population is timed appropriately to avoid missing transitional steps. For example, increased nutrient availability may result in a speeding up of the life-cycle of a particular species, in which case more frequent sampling would be required to take into account the more rapid transition from one developmental stage to the next. Changes in population dynamics such as these create practical difficulties when using structural characteristics. From the theoretical point of view, a common modification to matrix models such as these has been the introduction of a density-dependent component in recognition of the fact that, irrespective of influences from toxic chemicals and other extraneous stresses, populations have been described as having intrinsic controls. With or without refinements, such as the introduction of density dependence, matrix models have been applied to a variety of contaminant situations. For example, the U.S. Environmental Protection Agency, Office of Pesticide Programs, used a matrix approach based on age-specific mortality to model declines in raptors exposed directly or indirectly to granular carbofuran. The model predicted declines in population numbers based on age-related toxicity data, particularly from mature birds, and contributed to the suspension of registration of carbofuran in the United States.

Demographic models can be fairly straightforward extensions of the reproductive potential described in Equation 4.4 or more sophisticated projection models, which include estimates of the probability of advancing from one developmental stage to the next. Barnthouse and co-workers (1989) applied both approaches in the development of models designed to quantify the impact of chemical contaminations on fish populations.

As population models become more sophisticated, it is important that they integrate anthropogenic disturbances, either chemical or physical, with a variety of other intrinsic or extrinsic factors. They must include such considerations as the effect of macro- and microclimate on mortality, reproductive success, and dispersal and timing of chemical exposure. For example, if the chemical contaminant is a short-lived pesticide, its effect will greatly depend on the degree to which exposure coincides with the reproductive cycle: whether it occurs at the beginning of the cycle, the end of the cycle, or outside the reproductive period. The impact may depend on the demographic parameters themselves. For example, Emlen and Pikitch (1989) demonstrated that, in large vertebrates such as hawks and deer, the most important demographic characteristic influencing temporal mean population density was adult mortality, with fecundity and juvenile mortality of secondary importance. In small rodent and bird species, however, juvenile mortality was the primary demographic factor. Both timing and longevity of chemical exposure may, therefore, be critical in determining its effect, which is dictated by its application time and half-life in the environment. Additionally, the relative toxicity of the chemical to different life stages must be taken into account.

COMMUNITY INDICES

From a purely quantitative perspective, many indices of community diversity may be used for biotic assessment in polluted areas. Several of these are summarised in Table 4.9. Diversity indices are essentially measures that combine evenness and richness with a particular weighting for each. Species diversity indices have often been used as measures of community perturbation by chemical pollutants and other agents. Pollution usually results in the decline of those native species unable to adapt to toxic stress, although this may be partially compensated by an increase

Table 4.9. Examples of community diversity indices used in pollution assessment

Margalef $d = \dfrac{s-1}{\ln N}$

Pielou $E = \dfrac{H}{H_{max}}$

Brillouin $H = \dfrac{1}{N} \ln \dfrac{N!}{n_1! n_2! \dots n_3!}$

Shannon-Weaver $H' = -\sum \dfrac{n_i}{N} \log_2 \left(\dfrac{n_i}{N} \right)$

where s = number of species
N = number of individuals
n_i = number of individuals of the ith species
d = community diversity
E = evenness
H, H' = diversity per individual

in opportunistic species capable of exploiting the "vacant" niche(s). Increases seldom offset species depletion, although there are rare instances of the creation of increased species richness through the proliferation of habitat as a result of a change in land-use practice (Crawford and Titterington, 1979).

Figure 4.12 (Courtemanch and Davies, 1987) depicts a range of scenarios which may occur in a community downstream from a polluted outfall. Changes in community structure are defined in terms of the total number of taxa (richness) in the polluted section (a) compared with the number of taxa in the upstream (unpolluted) community (b). These two parameters, together with the number of taxa common to both the unpolluted and polluted environments (c), have been incorporated into a coefficient of community loss (I), which is the ratio of taxa eliminated by pollution ($a - c$) to the taxa remaining in the polluted community:

$$I = (a - c)/b$$

b may include some replacement species not found in the unpolluted community. Of the different variants pictured in Figure 4.12, five are unequivocal from the management point of view, where two represent acceptable changes and three denote unacceptable changes. However, the three most likely scenarios are seen as lacking important functional data required to judge the acceptability or otherwise of the observed changes. In some respects the overlapping circles seen in Figure 4.12, scenarios 3 and 4, are analogous to the overlapping curves in Figure 4.7 illustrating interspecific interactions, except that species richness replaces population size. Even with a simple community index such as the coefficient of community loss, information on the size of different species populations is lost and bias may be

Changes in Community Composition	Schematic Relationship	Management Decision
No change	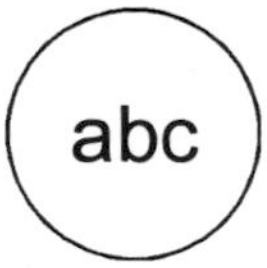 abc	Acceptable
No loss of taxa Increased richness below	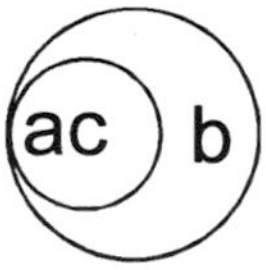 (ac) b	Acceptable
Partial loss of taxa Increased richness below	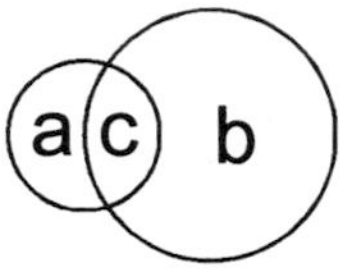 a(c) b	Criteria needed
Partial loss of taxa Partial replacement below	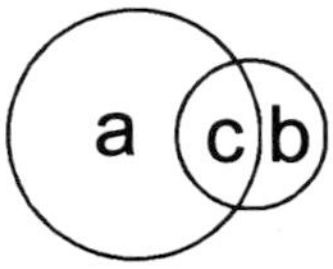 a (c)b	Criteria needed
Partial loss of taxa No replacement	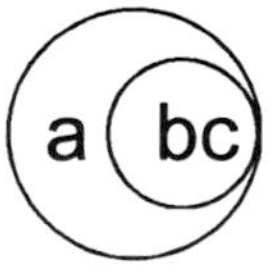 a (bc)	Criteria needed
Total loss of taxa Increased richness below	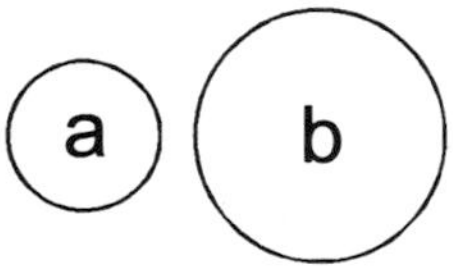 a b	Unacceptable Improbable
Total loss of taxa Partial replacement below	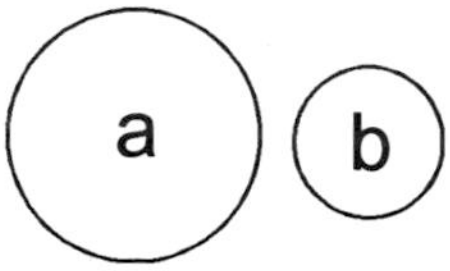 a b	Unacceptable
Total loss of taxa No replacement	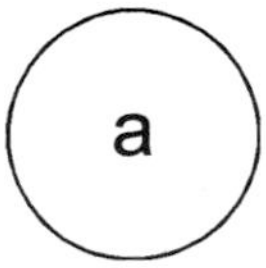 a	Unacceptable

Figure 4.12 Possible scenarios for changes in stream communities downstream from a polluted outfall (Courtemanch and Davies, 1987). See text for explanation of symbols.

Table 4.10. Illustration of calculation of data for plotting log-normal distribution

Number of individuals per species		Number of species	%	Cumulative %
Geometric class	Arithmetic class			
I	1	8	19.5	19.5
II	2–3	13	31.7	51.2
III	4–7	3	7.3	58.5
IV	8–15	4	9.7	68.2
V	16–31	5	12.3	80.5
VI	32–63	2	4.9	85.4
VII	64–127	2	4.9	90.3
VIII	128–255	2	4.9	95.2
IX	256–511	2	4.9	100

Data from Oslofjord, June 1978 (Gray and Mizra, 1979).

introduced through the inclusion/exclusion of a few "rare" species as a result of sampling error.

A community index that focuses on both population size and species richness is the log-normal distribution wherein the geometric distribution of individuals per species may be plotted against cumulative percentage of species in a community. Table 4.10 illustrates the basis of the log-normal plot, which has been shown for a large number of different kinds of communities to yield a straight line. Although there is no unifying biological theory explaining this distribution (May 1975), Gray and Mizra (1979) postulated that the straight-line plot is indicative of an "equilibrium" population, defined as one in which species' immigration and emigration are stable and "the proportions of individuals per species remains fairly constant." They further suggested that a change in the slope of the log-normal plot may indicate pollution stress. Figure 4.13 shows log-normal plots of benthic community data from Loch Eib, Scotland, over a 10-year period encompassing the pollution of this water body in the late 1960s. At the onset of pollution, a transition phase is identified wherein the log-normal relationship breaks down (Figure 4.13B). Following an extended period of accommodation to the polluted situation, the log-normal relationship is reestablished with a characteristically more shallow slope to the line that extends over a broader geometric class range than the unstressed community. Other structural variables that describe communities are associated with stability and recovery. These are described in Table 4.11 together with some measures of complexity.

A shortfall of several of these community structural measurements is their blindness to the differing roles that species may play in different communities. Furthermore, many fail to recognise that differences in characteristics between communities may result from a complex of factors that may have nothing to do with environmental pollution. This fact is really more of a deficiency in the design

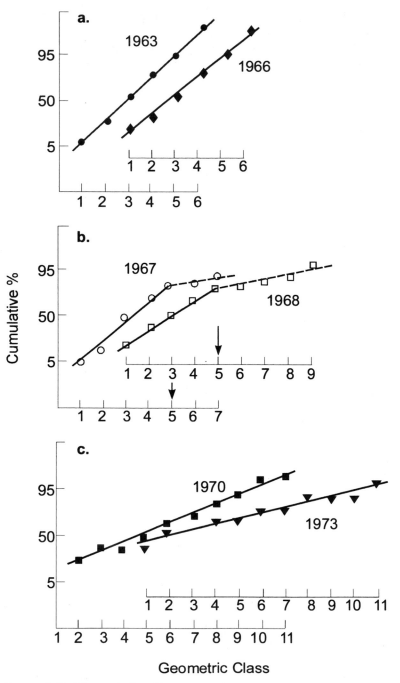

Figure 4.13 Change in benthic community structure in a Scottish loch over a 10-year period before and after pollution episode. After Gray and Mizra (1979).

Table 4.11. Structural characteristics of communities and their definitions

Variable	Definition	Units
Stability		
Homeostasis	Maintenance of a steady state in living organisms through feedback control processes.	
Stable	A system is stable if variables all return to the initial equilibrium following perturbation. A system is locally stable if this return applies to small perturbations and globally stable if it applies to all possible perturbations.	Nondimensional and binary (unstable, stable)
Sustainable	A system that can maintain its structure and function indefinitely. All nonsuccessional (i.e., climax) ecosystems are sustainable, but they may not be stable.	Binary
Resilience	How fast the variables return to equilibrium following a perturbation. Not defined for unstable systems.	Time
Resistance	The degree to which a variable is changed following a perturbation.	Nondimensional and continuous
Persistence	The time taken for a variable to change.	Time
Variability	The variance of population densities over time, or allied measures such as the standard deviation or coefficient of variation.	
Complexity		
Species richness	The number of species in the ecosystem.	Integer
Connectance	The number of actual interspecific interactions divided by the possible interspecific interactions.	Dimensionless
Interaction strength	The magnitude of interspecific interaction: the size of the effect of one species' density on the growth rate of another species'.	Dimensionless
Evenness	The variance of the species abundance distribution.	
Diversity indices	Measures that combine evenness and richness with a particular weighting for each.	Bits
Change		
Perturbation	A change to a system's inputs or environment beyond the normal range of variation.	
Stress	A perturbation with a negative effect on a system.	
Subsidy	A perturbation with a positive effect on a system.	

Table 4.12. Parameters for index of biological integrity of fish from running waters (Midwest United States)[a]

Parameter	Rating		
	5[b]	3	1
Species richness and composition[c]			
Total number of native fish species			
Number and identity of benthic species			
Number and identity of pelagic species			
Number and identity of long-lived species			
Number and identity of intolerant species			
Percentage of individuals as tolerant species	<5	5–20	>20
Trophic composition			
Percentage of individuals as omnivores	<20	20–45	>45
Percentage of individuals as insectivores	>45	45–20	<20
Percentage of individuals as top carnivores	<5	5–1	<1
Fish abundance and condition[d]			
Number of individuals in sample			
Percentage of individuals as hybrids or exotics	0	>0–1	>1
Percentage of individuals with disease, tumours, or skeletal abnormalities	0–2	>2–5	>5

[a] After La Point (1995).
[b] Scoring of each metric as 5, 3, or 1 depends on whether a site deviates slightly from the reference site. Threshold values are usually established at the 5th and 50th percentile [i.e., values below the 5th percentile (strong deviation) are scored as 1, values between the 5th and 50th percentiles (moderate deviation) are scored as 2, values above the 50th percentile (slight deviation) are scored as 5] (Karr et al. 1986).
[c] Expected value for individuals varies with stream order and region.
[d] Expected value for numbers of individuals varies with stream order and physical habitat.

of many monitoring programs and insufficient data collection rather than a problem with the indices themselves. A more appropriate criticism of indices is that they commonly include mean values for components, thereby losing information on the variability associated with these parameters. Notwithstanding these problems, there has been renewed interest in recent years in measures of community structure and function as means of determining ecosystem damage, and indices of biotic integrity have been incorporated into ambient water standards by several state agencies in the United States.

The index of biotic integrity (IBI) was initially constructed for use with fish communities, and, as such, a summary of its principal characteristics in relation to aquatic systems in the midwest United States is shown in Table 4.12. IBIs have also been formulated for invertebrate fauna, and, even though they deal with different groups, they essentially have a similar construction with provision for ephemeral species, insect/noninsect taxa, tolerant species, and so on. IBIs have expanded to analysing community changes along pollution gradients in coastal water, freshwater, and estuarine environments (Weisberg et al., 1997). Many IBIs are based on the idea that species respond to improvements in habitat quality in three progressive stages: the abundance of the organisms increases, species diversity increases, and the dominant species change from pollution-tolerant to pollution-sensitive species. The comparison with reference communities incorporated into the IBI approach represents one of its strengths yet illustrates a fundamental difficulty of evaluating pollution effects on a complex system involving so many potential interactive effects: the recurring question of how to establish a baseline or control. The standard approach to measuring changes in community structure is to compare polluted sites to sites free of anthropogenic stress (i.e., habitat-specific reference sites). This procedure defines conditions at the reference sites and then assigns categorical values for various metrics at sites exposed to anthropogenic stress. In many IBI-based studies, there is increasing focus on pivotal species that may be economically important or endangered or that have some other strategic significance.

An inexpensive multispecies approach adopted by Cairns and co-workers (1990) involved the measurement of protozoan periphytic communities, which were allowed to develop on artificial substrates before being introduced into through-flow dosing chambers. In a typical experiment, periphytic communities were obtained by suspending polyurethane foam artificial substrates in pond water for a 2-week period and then transferring them to a dosing system. Within each treatment receptacle, duplicate colonised substrates acted as sources of colonists for barren substrates suspended in the same box. Cairns et al. (1990) employed this technique to measure the joint toxicity of chlorine and ammonia to periphytic communities. In addition to protozoan species richness, in vivo fluorescence was used as an index of algal biomass, and dissolved oxygen was used to measure community metabolism. Differences in taxonomic composition of newly colonised (initially barren) substrates were measured after 7 days of toxicant exposure. Comparisons were made using Hendrickson's M statistic applied in a stepwise manner by sequentially eliminating treatment groups until heterogenicity in number of positive matches in taxa was no longer significant. Biologically significant levels of toxicant were defined as IC_{20}s (i.e., concentrations causing a 20% inhibition of response relative to the controls). Multiple regression analysis was used to construct a surface response model. With the inclusion of both linear and interactive terms for ammonia and chlorine in the model, Cairns et al. (1990) were able to account for 73% of the variation in species richness.

Experiments such as this illustrate effects at the primary producer level. Effects at higher trophic levels are complicated by a variety of interactive factors including competition for food resources. Therefore, changes at the primary producer level may result in fundamental changes farther up the food chain depending on whether surviving species are more or less acceptable as food for the next trophic level (Section 4.5.1, last paragraph).

4.5.4   Ecosystem equilibrium. Fact or fiction?

An increasing number of ecologists have questioned the notion of an equilibrium state as applied to a natural or experimental ecosystem. Treatises on chaos theory and complexity have given rise to arguments that ecosystems may behave in a far more complex manner than was previously imagined.

As recently as the 1970s, the dominant view of population development portrayed a system that, starting from some initial growth point, increased to an asymptotic equilibrium. The population was seen to oscillate about the equilibrium on a regular cycle. Deviations from such a cycle were ascribed to stochastic events or external influences and the underlying equilibrium was always assumed to exist. This way of thinking was changed by mathematical experiments of May (1974) and May and Oster (1976). They showed, by taking the logistic equation for population growth,

$$N_{t+1} = N[1 + r(1 - N/K)] \tag{4.5}$$

where N = population size at time t, N_{t+1} = population size at the next time interval, K = carrying capacity of the environment, and r = intrinsic rate of increase over the time interval; the population equilibrium can be altered dramatically by setting different values for r. At an r value of 2, the system converged to an equilibrium state through a series of regular oscillations of decreasing magnitude over time, much like the historically accepted paradigm. As r was raised to 3, however, the system initially appeared to approach equilibrium but then began a series of oscillations wherein the timescale doubled and redoubled and then appeared to disintegrate into an apparently random progression exhibiting no discernible regularity. Following periods of apparently completely erratic behaviour, regular or semiregular fluctuations could be reestablished from time to time.

Obviously, this has important implications for the study of ecosystems and their degradation by toxic chemicals. If it transpires that ecosystems are fundamentally nonlinear, then they could not be expected to return to a putative equilibrium condition following their alteration by physical disturbance or chemical contamination. Consequently, terms such as *resilience*, *stability*, and *recovery* would have no meaning, at least within the context of several of the structural descriptors of ecosystems mentioned thus far. At a time when demands for evaluating ecosystem stress have never been higher, the scientific community is therefore presented with a dilemma. We continue to base many of our notions of ecosystem organisation on metaphorical concepts implying homeostasis, yet we suspect that some core components may be flawed.

In reevaluating the current status of measures of ecosystem stress, encouragement must be taken from studies where we have documented large changes in an environment associated with pollution. We have recorded extreme cases of environmental degradation, we have sampled what our experience tells us are healthy communities, and our observation of these unequivocal extremes lead us to conclude that more subtle intermediate stages should be quantifiable. Within the context of a circumscribed pollution "event", descriptors of change require some spatial definition for them to be useful. Changes due to birth and death rates will be modified by immigration and emigration to and from the affected zone. The latter are edge effects and will be much more important in a smaller zone of impact where the boundary is large relative to the total area.

One of the problems concerning the need to define ecosystem stress resulting from human activity is that many of the fundamental ideas on ecosystem function have changed quite dramatically over the last 30 years. In a critical review of theoretical and field ecosystem studies, Pimm (1984) reexamined the widely held concept of the inherent stability of more complex ecosystems and concluded that in communities with larger numbers of species:

1. Connectance is inversely related to stability (see Table 4.11 for definitions).
2. Populations are less resilient (see Table 4.11).
3. Species removal results in a higher chance of change in composition and biomass.
4. There is longer persistence of species composition in the absence of species removal.

Additionally, Pimm (1984) concluded that the more connected a community:

1. The fewer species it must have for it to be stable.
2. The greater the likelihood that loss of one species will lead to the loss of further species.
3. The more resilient will be its populations.
4. The more persistent will be its composition.
5. The more resistant will be its biomass if a species is removed.

Not all disturbances are necessarily "bad" for an ecosystem. It has only been recognised relatively recently that periodic fires and insect attack may actually have a reinvigorating effect on forest ecosystems. Obviously, the resilience of the system will depend on the scale of the disturbance. However, it seems no longer valid to rigidly equate stability with health. A certain degree of turbulence is not only tolerable but may actually be desirable, and it is not unusual for populations of organisms within a community to vary by as much as 1,000-fold within acceptable bounds. Absolute population sizes within a community must therefore be viewed carefully, against a background of seasonal and stochastic changes in resource (habitat) and within the context of *relative* species numbers.

Odum (1969) summarised characteristics of ecosystem development in primarily functional terms:

1. Increasing species richness,
2. Greater trophic efficiency,
3. Richer structure for recycling materials,

4. More intense overall system activity,

5. Greater specialisation in trophic interactions.

When viewed in this way, trophic position may assume a greater importance than a particular species, and the system becomes defined in much more "anonymous" terms relating to the efficiency of energy or carbon flow. Degraded systems tend to be more leaky than healthy ones in terms of carbon flow and/or nutrient transfer from lower to higher trophic levels. A simple way of quantifying this is the bottom-top trophic efficiency, which is that fraction of the carbon fixed by autotrophs that eventually reaches top carnivores. Ulanowicz (1995) analysed trophic networks in two tidal salt marsh communities in Florida, one of which suffered chronic thermal pollution from a nearby power plant; the other plant served as a control. He calculated a 20% reduction in trophic flow (defined as $mg\,cm^{-2}$ day^{-1}) in the heat-affected community. While the focus of Ulanowicz' (1995) study was on energetic transfer between components, he reported that two top predators either disappeared (Gulf flounder) or fell in trophic position (sting ray) in the stressed system. He noted that accommodations were made in the system through hierarchical shifts and altered linkages. The underlying assumption is that species deletions will be compensated for by the addition of other species at the same trophic level or through alteration(s) of the network, which maintains energy flow albeit at possibly altered efficiency. Also present in the trophic web may be energy loops, which are often difficult to identify and quantify but which may have a significant effect on bottom-top trophic efficiency.

From a risk assessment perspective (see Chapter 10), the species-neutral trophic flow network as an indicator of ecosystem stress may present a dilemma. Although a particular species A may be effectively replaced by its trophic equivalent B, the fact remains that A may be a threatened or endangered species or a cultural species (e.g., crop). In such cases, the concept of functionality in terms of energy flow may be superseded by the need to protect species valued by humans.

In addition to direct toxic effects of chemicals on organisms, pollution may cause disruption of the food chain through the process of eutrophication. Excess nitrogen and phosphorus, through both point sources and nonpoint sources, may result in both qualitative and quantitative changes at the primary producer level. Lake Erie and the Chesapeake Bay (Case Study 4.5) have both experienced eutrophication through anthropogenic inputs of nitrogen and phosphorus from surrounding watersheds. Eutrophication is discussed in more detail in Section 6.10.

4.6   Modelling

4.6.1   The concepts of modelling

Models are used in most branches of science. Conceptual modelling involves a series of processes such as organising and focusing on the major ideas, thereby forcing the producer and user of the model to identify the major ideas and put them into some kind of structure or order. The process of constructing a model reduces some of the complexity and simplifies the picture, which in turn facilitates developing hypotheses or consolidating ideas. Building a conceptual model may be formal or informal, deliberate or intuitive, and is normally an iterative process.

Mathematical modelling is essentially a quantitative formalisation of conceptual modelling. In similar manner, mathematical models provide organised and simplified versions of the real situation. A model can be considered to be a synthesis of "knowledge elements" about a system (Jorgensen, 1990). The use of models is quite varied. Uses, or more correctly applications, of modelling in environmental toxicology include:

a. The identification of critical processes in the pathways and behaviour of toxic substances in the environment, which aids in identifying research needs resulting from gaps in understanding;
b. Prediction of the outcome of perturbations to a system;
c. Risk assessment of existing as well as yet-to-be registered chemicals;
d. Aiding in the management of toxic substances.

Ecological models such as the Lotka-Volterra model of predator-prey interactions and the Streeter-Phelps model of oxygen balance in a stream were developed quite early in the twentieth century (Jorgensen et al., 1996). These models have been characterised as the first generation of ecological models. The second generation of ecological models can be seen as those developed for physical, chemical, and population processes in the 1950s and 1960s. Jorgensen et al. (1996) refer to the third generation of models as beginning in the 1970s, when very complex models for eutrophication and very complex river models were produced. These were clearly influenced by the rapid evolution and availability of computers, which enabled the users to handle large amounts of complex data.

In the process of the development of this third generation of models, modellers became more critical and recognised the need to understand as much as possible about the ecosystem; concurrently, the science of ecology was becoming more quantitative. The evolution of ecology from a mostly descriptive mode toward a more quantitative mode, which most people would consider as beginning in the 1960s, is normal in the maturation of the science. In addition, certainly the computer played a role in this development, as did the modellers' need for data and the need for sound environmental management.

The resulting fourth generation of ecological models are "characterised by having a relatively sound ecological basis and with emphasis on realism and simplicity" (Jorgensen et al., 1996). Ecotoxicological models began to emerge in the late 1970s. Jorgensen et al. (1996) now identify the fifth generation of ecological models. These include approaches that attempt to account for the complex feedback mechanisms that are characteristic of ecosystems.

The handbook by Jorgensen et al., from which we have been quoting, refers to 1,000 environmental models that have appeared from the mid-1980s to the mid-1990s, of which some 400 are included in the book. These numbers are provided

Table 4.13. An historical view of the development of ecological and environmental models

Generation[a]	Approximate date of introduction	Examples of model and type
First	1920	Streeter-Phelps: Oxygen balance in stream Lotka-Volterra: Predator-prey interaction
Second	1950 1960	Population dynamics: More complex models River models: More complex models
Third[b]	1970	First models of eutrophication: Applications to environmental management River models: Extremely complex models
Fourth	mid 1970s to 1980s	Ecological models: Fixed procedure, balanced in complexity Ecotoxicological models: Processes validated
Fifth	1990	Ecological models that reflect better understanding of ecosystem processes, including complex feedback mechanisms

[a] Based on Jorgensen et al. (1996).
[b] The availability of computer technology to some extent was responsible for the third generation of models, which were exceedingly complex.

to give the student some indication of the scale of the field. Clearly, in the present text, only a very limited treatment of models in environmental toxicology can be included. The five generations are summarised in Table 4.13.

To introduce the concepts of modelling in environmental or ecotoxicology and to point out some of the potential technical and scientific problems, as well as the "need" for such models in environmental science, it is instructive to consider the differences between environmental chemistry and classical chemistry. Figure 4.14, based on Jorgensen (1990), summarises these differences. In a sense, they should be obvious, but as summarised by this author, the sheer weight of complexity and challenge to the environmental chemist comes into clear focus. In essence the differences involve:

a. Far more chemical compounds in the environment as compared to those in the laboratory or chemical plants;

b. The simultaneous occurrence of many processes in the environment, in contrast to the relatively small number of controlled processes in "pure" chemical experiments;

c. Functions in the environment that vary and that are, to some extent, random but that are kept constant by design in the laboratory;

d. Living organisms that interact with chemical compounds in the environment but that are identified and controlled in the laboratory if organisms are involved (e.g., in fermentation).

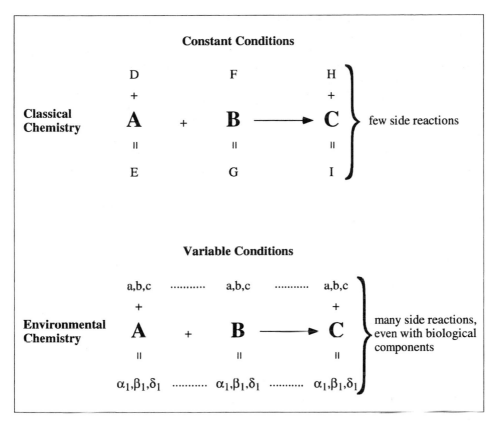

Figure 4.14 The difference between classical and environmental chemistry. Modified from Jorgensen (1990).

A fundamental concept in environmental toxicology is that of dose-response. The exposure to a substance in water, in air, or in food results in a dose that is normally expressed as intake as mass per mass per time (e.g., $mg\,kg^{-1}\,day^{-1}$), or as a concentration ($mg\,L^{-1}$, $\mu g\,g^{-1}$). Thus, it is important to know, or to be able to predict, the quantity or dose resulting from an exposure to a chemical substance in the environment that produces an adverse effect. As indicated in previous chapters, this can be done by laboratory studies with known and controlled concentrations of chemicals or from field observations where concentrations are measured (see Table 2.1).

Such measurements of concentration can be fraught with difficulties in that the substance of concern may be very challenging analytically; it may be unstable or volatile. Of the almost 100,000 chemical substances of commerce, only about 1,000 can be determined satisfactorily by current methods. Furthermore, often the active or bioavailable form may be either unknown or difficult to determine chemically.

An alternative to measurement is to *calculate* the concentrations of various chemicals in various media. This type of calculation is often referred to as mathematical modelling or mass balance modelling. The concept can be applied at various scales to biotic and abiotic systems. Although the most commonly modelled substances are organic compounds, models can be made for a variety of chemicals, including inorganic substances such as arsenic. But since their behaviour is more clearly related to chemical structure, particularly their hydro-phobic/hydrophilic properties, modelling has been most successfully used to predict the behaviour and fate of organic substances such as DDT, PCBs, and PAHs (see Chapter 7) – with more general success than modelling for inorganic substances. This is not to say that the behaviour of inorganic substances is not related to their structure, but simply that, at present, our understanding of "the rules" is less complete.

Models can involve simple, "back-of-the-envelope" calculations, or they can be very complex, requiring computers. The increasing availability and ease of use of computers and the increased power of personal computers over the past two decades have facilitated the development and application of models that describe the behaviour of chemicals in the environment. Particularly attractive now are models that can be downloaded from the Internet and used on personal computers. The reader is referred to *www.trentu.ca/envmodel*; *www.cwo.com/~herd1/*; *ww.epa.gov/epahome/models.htm*; and *www.ei.jrc.it/report/exp.html*.

Essentially, models attempt to quantify the phenomena about which there is already some knowledge; models can be used to test or confirm that our understanding of mechanisms is consistent with observations. Models also provide a predictive capability and can identify errors in hypotheses and highlight gaps in information. They are also used in regulation and environmental management.

Of particular interest in environmental toxicology are models of uptake, clearance, and metabolism of chemicals by organisms, particularly fish. For a given chemical substance, the external concentration in water and in food can be used to deduce internal concentrations or body burdens in the organism, which in turn can be related to effects. Pharmacokinetic models can even yield concentrations in specific tissues.

Mass balance models can be extended to treat the movement and transformations of chemicals within communities and between trophic levels. Other types of models express variation in concentrations with space (e.g., air dispersion models using Gaussian Plume equations). Further types of models can also be used to estimate the response over time of populations or communities to perturbations, taking into account reproduction, life stages, competition, and predation. In the present text, only mass balance models are treated in detail.

4.6.2   Mass balance models

Most models in current use are "box" models in which the mass balance is applied to a defined volume, or envelope, of the environment. Such an envelope must be defined by the modeller or the user and may be large or small, coarse or refined, in short chosen according to the questions being asked and the information that is available. The envelope may be a section of a river, a block of soil, or even a single organism. Conditions within the envelope are assumed to be constant (i.e., the contents are "well-mixed" and only one concentration applies). This greatly simplifies the mathematics. The principle of conservation of mass (as defined by Lavoisier) is applied. The number of "boxes" or components of the model in principle is unlimited and can include air, water, soil, sediment, and organisms (with defined properties), but in a practical sense, the model should not be too complex. Complex models tend to demand excessive amounts of data, they are often too site-specific, and they are difficult to grasp mentally and understand. The mass balance concept involves two principles.

1. Whatever quantity of chemical is added to a system, that quantity must all be accounted for in the model. For example, one adds 1,000 g of pesticide to a 1-ha field. After 10 days there may be 400 g left in the field, 100 g has evaporated, 300 g has degraded microbially, 50 g has reacted photolytically, 120 g has been removed in runoff, and 30 g has been leached to groundwater. The aim of the model is to estimate these quantities and confirm or validate them against monitoring data. Eventually, the model can be used to predict the fate of pesticide added to other systems.

2. If there is an input of chemical to a system at a particular rate the routes, compartments and resulting concentrations must be shown. Input rate minus output rate = inventory change. In a lake receiving effluent, or fish being exposed to contaminated water and food, the basic equation has units such as grams per hour. For example, if the fish is exposed to PCB in water at $1\,ng\,L^{-1}$ and consumes food containing $1\,\mu g\,g^{-1}$ of PCB, the following questions must be addressed: At what rate will the concentration of PCB in the fish increase? Will the concentration reach a steady state when input is equal to output? If so, what will be the concentration at steady state?

The dynamic solution involves solving differential equations. The steady-state solution is algebraic and simple. The primary challenge is to write the appropriate equations describing these processes and to obtain the correct parameter values for the chemicals, the media in which they move, and the biological systems. This process involves obtaining reliable information on the following.

Equilibrium partitioning [e.g., partitioning between air and water (K_{aw}) and partitioning between octanol and water (K_{ow})]. The octanol-water partition coefficient is a good indicator of the tendency of a substance to move from water to lipid. Essentially octanol is a convenient surrogate chemical for lipid or fat. The more lipophilic the substance, the greater its tendency to move from water into living material.

Degrading reactions, including metabolism.

Transport processes (e.g., evaporation, deposition from air to soil and water, deposition from water to sediment, feeding and respiration rates).

These concepts are illustrated by Figure 4.15 and in the following three examples.

1. THE NATURE OF ENVIRONMENTAL MEDIA:
   EVALUATIVE ENVIRONMENTS

Figure 4.16 illustrates a model for eight compartments of the environment into which a pollutant might be distributed. It provides what Mackay (1991) has called an "evaluative" environment, which is, in essence, a qualitative description of environmental media considering, for example, an area 1 km × 1 km, including air, water, soil, and sediment. Mackay emphasises that the quantities here are "purely illustrative" and that "site-specific values may be quite different". Because this type of approximation, including the manner in which estimates have to be made, is fundamental to the modelling process, some of the steps that went into Mackay's construction of this model are now treated in some detail. In particular, the assumptions that led to the relative volumes of the eight respective compartments are considered.

a. *Air.* The troposphere is that layer of the atmosphere that is in most intimate contact with the surface of the Earth, which if assumed to be at uniform density at atmospheric pressure would be compressed to about 6 km. In the 1-km square world, the volume becomes $1,000 \times 1,000 \times 6,000$, or $6 \times 10^9 \, m^3$.

b. *Aerosols.* Particulate matter in the atmosphere includes material from natural and anthropogenic sources, such as condensed water vapour and dust particles, the amounts of which vary from site to site. Based on real data for aerosols, we estimate that particles have an average density of $1.5 \, g \, cm^{-3}$ at a concentration of $30 \, \mu g \, m^{-3}$, which corresponds to a volume fraction of particles of 2×10^{-11}. In an evaluative air volume of $6 \times 10^9 \, m^3$, $0.12 \, m^3$ of solid material is suspended in the atmosphere.

c. *Water.* Water covers approximately 70% of the Earth's surface. The depth to which this water occurs obviously varies from oceans to surface water on land surfaces. In this evaluation environment, the depth is selected to be 10 m, yielding a water volume of $7 \times 10^6 \, m^3$.

d. *Suspended sediment.* The measured range of particulate matter in water varies greatly, from very clear lakes to muddy streams. But many data are available on which to base our estimate. Assuming a density of $1.5 \, g \, cm^{-2}$ and a concentration of $7.5 \, g \, m^{-3}$, the volume fraction of particles is 5×10^{-6}, so in the $7 \times 10^6 \, m^3$ of water, there are $35 \, m^3$ of particles, identified as suspended sediment.

e. *Aquatic biota.* Estimates of the volume fraction of fish biomass in a lake are varied and, according to Mackay, tend to be unreliable. For the purposes of modelling organic contaminants, which tend to biomagnify, or metals, which bioconcentrate in fish, in this evaluative environment all the living material in a lake is considered to be fish. The total concentration

(A) EQUILIBRIUM PARTITIONING

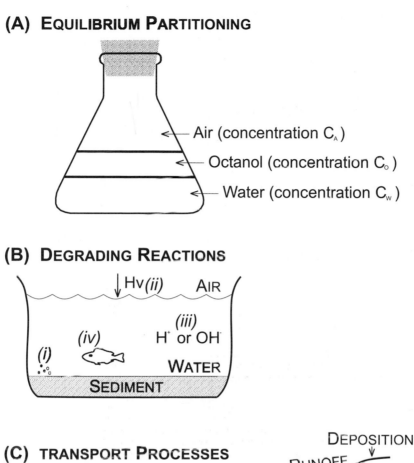

Air (concentration C_A)

Octanol (concentration C_O)

Water (concentration C_W)

(B) DEGRADING REACTIONS

Hv *(ii)*   AIR

(iii)
H⁺ or OH⁻

(iv)

(i)

WATER

SEDIMENT

(C) TRANSPORT PROCESSES

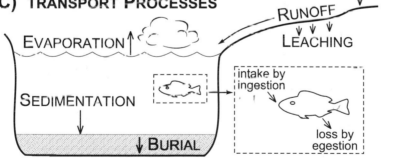

DEPOSITION

RUNOFF

LEACHING

EVAPORATION

intake by
ingestion

SEDIMENTATION

BURIAL

loss by
egestion

Figure 4.15 Concepts used in mass balance models. (A) Equilibrium partitioning. With the three phases air, octanol, and water in the container, once equilibrium is reached the concentrations C_A, C_O, and C_W all remain constant, all other things being equal. The ratios are the equilibrium partition coefficients, K (e.g., $K_{AW} = C_A/C_W$, and $K_{OW} = C_O/C_W$). (B) Degrading reactions. These processes alter the chemical structure of the substance, usually irreversibly. In water in a lake or pond, chemicals may be degraded by (i) biodegradation, (ii) photolysis, (iii) acid or base hydrolysis, or (iv) metabolism in fish liver. (C) Transport processes. These processes move the substance from one phase to another, but the structure of the substance remains unchanged. Processes in a typical watershed as illustrated include evaporation, sedimentation, burial, deposition, leaching, runoff, intake through ingestion, and loss by egestion.

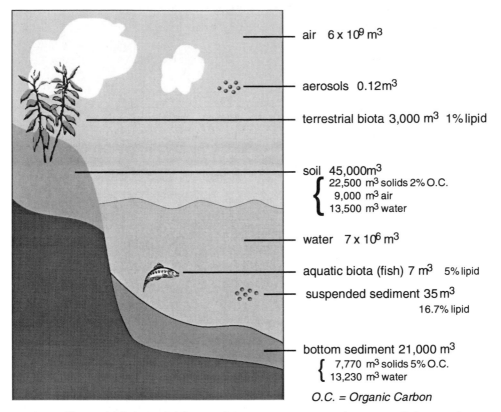

air $6 \times 10^9 \, m^3$

aerosols $0.12 m^3$

terrestrial biota 3,000 m^3 1% lipid

soil 45,000m^3
{ 22,500 m^3 solids 2% O.C.
  9,000 m^3 air
  13,500 m^3 water

water $7 \times 10^6 \, m^3$

aquatic biota (fish) 7 m^3 5% lipid

suspended sediment 35 m^3
                    16.7% lipid

bottom sediment 21,000 m^3
{ 7,770 m^3 solids 5% O.C.
  13,230 m^3 water

O.C. = Organic Carbon

Figure 4.16 A model for an eight-compartment environment. O.C. organic carbon. Based on Mackay (1991).

is estimated as 1 ppm by volume, which produces $7 m^3$ of biota in the 1-km area. It is recognised that there may be relatively large volumes of biomass of other trophic levels (e.g., aquatic macrophytes in the littoral zone), but their roles in terms of mass balances are too poorly quantified at present for them to be usefully included in the model.

f. *Bottom sediments.* Bottom sediments are heterogeneous, even in a given ecosystem, in terms of composition with depth and with space. It is crucially important that sediments be included in mass balance models. From an ecological standpoint, bottom sediments provide important habitat for biota, particularly for invertebrates and microorganisms, which are important links in the food chain to fish. From a mass balance modelling standpoint, sediments serve as a repository for contaminants and can be viewed as sinks or reservoirs of chemicals. For the evaluative environment, a sediment depth of 3 cm with a density of 1.3 g/cm^3 is assumed. The total volume of the bottom sediment is $2.1 \times 10^4 m^3$ or $21,000 m^3$ of which water is 63% by volume (13,230 m^3) and 37% by volume (7,770 m^3) are solids. The solids contain 10% organic matter or 5% organic carbon.

g. *Soils.* For illustrative purposes, the soil is treated as an area covering 1,000 m × 300 m × 15 cm, the depth to which agricultural soils are ploughed. Thus, the soil compartment is 45,000 m³. Further refinement of this compartment, based on a density of soil solids of 2,400 kg m⁻³ and 1,000 km m⁻³ for water, provides for the relative volumes of solids, air, and water as 22,500, 9,000, and 13,500 m³, respectively.

h. *Terrestrial biota.* Even if one limits a consideration of terrestrial biota to the standing stock of vegetation, there is tremendous variation and uncertainty. For the model, a typical "depth" of plant biomass is 1 cm so that terrestrial plant biomass would occupy a volume of 3,000 m³.

Deposition processes. The model needs to take into account the dynamic nature of the ecosystem. Even disregarding the biological processes, which are quantitatively small in comparison with other components, deposition processes of material from air to water and land, from water to sediment, and from terrestrial to aquatic compartment are taken into account. Details of these processes and their respective treatments in the model are beyond the scope of this text, but further reading is provided in Section 4.13.

Fugacity. Many models, such as the one illustrated in Figure 4.16, are written in terms of concentration of a chemical in the various environmental compartments or phases. Mass balance equations are then written and solved in terms of concentrations, process rate parameters, partition coefficients, phase volumes, and flow rates.

An alternative is to use fugacity as a descriptor of chemical quantity. As described by Mackay (1991), fugacity is related to chemical potential; essentially it is the chemical's partial pressure and can be seen as the chemical's tendency to escape. As an example, the higher the K_{OW} of a compound, the lower its fugacity in lipids of living material; thus, it tends to migrate from water to living matter (in other words, to bioconcentrate). Fugacity depends not only on the nature of the chemical, but also on the nature of the medium in which it is dissolved and on temperature. When two phases are in equilibrium with respect to chemical transfer, equal fugacities prevail. A more complete definition is provided in the steady-state mass balance model.

2. THE STEADY-STATE MASS BALANCE MODEL

Figure 4.17 illustrates the results of a one-compartment model for the distribution of a chemical in a lake or pond with the following input values:

a. The volume of water;
b. The rate of inflow and outflow of water, from which, given the volume of water, the residence time of water can be calculated;
c. The rate of inflow and outflow of suspended sediment;

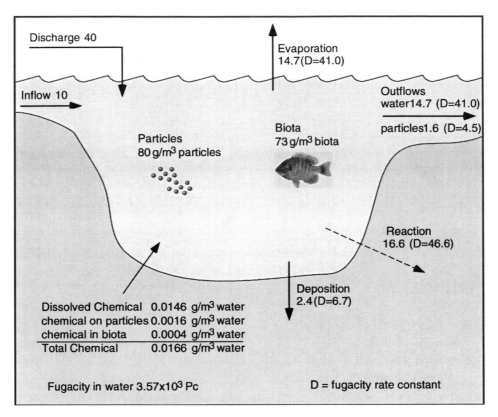

Figure 4.17 Steady-state mass balance of a chemical in a lake. D is the fugacity rate constant; see the definition of fugacity in Section 4.6.2, "3. Uptake/clearance of a chemical by a fish". Based on Mackay and Paterson (1993).

d. The deposition of suspended sediment;
e. The reaction rate of the chemical in the water column;
f. The rate of evaporation;
g. The air-water partition coefficient of the chemical;
h. The biota-water partition coefficient of the chemical;
i. The concentration of particles in water;
j. The concentration of biota in water;
k. The rate of discharge of chemical;
l. The concentration of chemical in the inflowing water.

The aim of the model is to calculate (1) how the chemical partitions between the water, particles, and biota; (2) the steady state or constant concentration in the system of water, particles, and fish; and (3) all the loss rates.

The calculation can be undertaken in two ways: as a conventional concentration calculation or as a fugacity calculation. Details of the calculations are provided in Mackay and Paterson (1993) and in Mackay (1994).

3. UPTAKE/CLEARANCE OF A CHEMICAL BY A FISH

The third example is a fish bioconcentration model. The biological/ecological phenomenon of bioconcentration is one of crucial importance in aquatic systems, mainly for persistent organic chemicals, such as certain organochlorines (see Chapter 7). The potential for harm to humans and other biota resulting from the release of even small amounts of chemical into the environment is increased many times by the concentrations of the chemical that are attained in higher trophic levels. The ecological basis for this is explained in Chapter 7. The task of the modeller is to determine the steps in the bioaccumulation process that influence the concentration in the tissue and to make appropriate quantitative estimates of these. An understanding of the chemical structure of organic compounds and the way in which the structure is related to the behaviour of the chemical is central to this type of modelling.

The model illustrated in Figure 4.18 illustrates the various processes in which the chemical is involved and which are incorporated into this particular model. They include

- a. Uptake from water via the gills;
- b. Uptake from food;
- c. Loss via the gills;
- d. Loss by egestion;
- e. Loss by metabolism;
- f. Dilution of tissue concentration through growth of the organism.

The expressions for input and output processes in Figure 4.18 are expressed in two alternative ways: as a rate constant (K) or in fugacity (D) form. The explanations of the terms and equations are provided in Mackay (1991). Fugacity (from *fugare*, to flee or escape) is a thermodynamic criterion of equilibrium of a substance in solution in two phases. It is closely related to chemical potential and can be regarded as an idealised partial pressure. When a substance such as benzene achieves equilibrium among phases such as air, water, and sediment, it has a common fugacity, in all these media, but differing concentrations. Fugacity plays the same role for distribution of mass in solution as temperature plays for distribution of heat. D values are fugacity rate constants.

Such a model, because it quantifies the exposure routes for the fish, may elucidate the primary route of exposure or correct a prior erroneous assumption about the major route of exposure.

EFFECTS MODELLING

Following the estimation of the accumulation of a chemical in an organism, further modelling, also based on chemical structure, assesses effects and may utilise the so-called quantitative structure-activity relationship (QSAR) approach (see Section 5.5). It is obvious that this approach not only provides a means of assessing the

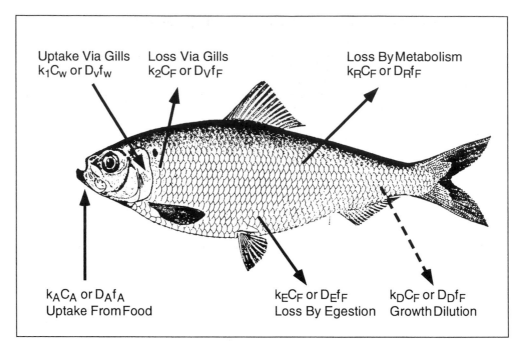

Uptake Via Gills
k_1C_W or D_Vf_W

Loss Via Gills
k_2C_F or D_Vf_F

Loss By Metabolism
k_RC_F or D_Rf_F

k_AC_A or D_Af_A
Uptake From Food

k_EC_F or D_Ef_F
Loss By Egestion

k_DC_F or D_Df_F
Growth Dilution

Figure 4.18 A model for bioaccumulation by fish. The terms for uptake and loss include the following: K values are the rate constants: K_1 = the uptake rate constant; K_2 = the depuration rate constant; K_R = the degradation rate constant; K_A = the food uptake rate constant; K_E = the egestion rate constant; K_D = the growth dilution rate constant. C values are concentrations: C_W = the dissolved concentration in fish; C_F = the concentration in whole fish; C_A = the concentration in food. f_W, f_F, and f_A are the fugacity transport rate constants corresponding to the concentrations C_W, C_F and C_A. D values (e.g., D_V) are fugacity rate constants corresponding to the rate constants such as K_1. The groups K_1C_W and D_Vf_W, respectively, are essentially equivalent methods of expressing the same process rate. Based on Mackay (1991).

behaviour and fate of chemicals already in the environment but also provides one way of assessing the risks of new chemicals. Increasingly, regulatory agencies are accepting model predictions and risk assessments for ranking the environmental significance of potentially toxic substances.

4.6.3   Some other models for use in environmental toxicology

In addition to the preceding examples, a number of different models that also address the fate of chemicals are available. In addition to the mass balance type, models include chemical transformation models such as Henrikson's acidification model (Henrikson et al., 1988), which was used to calculate the permissible emissions of acid gases in the context of determining a target loading of acid for sensitive lakes. Table 4.14 lists just a few models in an attempt to illustrate the range of those that have application to environmental toxicology. Not all these models directly involve the living organism. For example, WHAM (Windermere Humic Acid Model) is a chemical equilibrium speciation model, which can predict the

Table 4.14. Examples of mathematical modelling applied to environmental toxicology

Perturbation/toxic substance	Type of model/model characteristics	Reference
Organic chemical in a water body	Environmental fate: fugacity	Mackay (1991)
LAS	Environmental fate: exposure analysis modelling system	Burns et al. (1981)
LAS	Simplified lake/stream analysis	DiToro et al. (1981)
Metals in combination with organic matter	WHAM designed to calculate equilibrium chemical speciation of metals in surface and ground waters, sediments and soils	Tipping (1994)
PAHs	Environmental fate: transport, degradation, and bioaccumulation	Bartell et al. (1984)
Mercury	Distribution among compartments in a reservoir	Harris (1991)
Sulphuric acid (acidic deposition) in fresh waters	Steady-state chemistry, prediction of critical loading with respect to lake acidification	Henrikson et al. (1988)

chemical forms of metals, with particular reference to the effect of natural organic substances (humic acids). Toxicologists know that organic substances have a major influence on the toxicity of metals, through their ability to affect the bioavailability. Thus, WHAM, although it deals exclusively with chemistry, has real value to the environmental toxicologist. Each of the models was designed with a specific purpose in mind, a point that should be emphasised when considering the advantages and disadvantages of modelling in environmental toxicology.

Comparisons have been made among different models that address the same or similar problems. Games (1983) compared three types of models for estimating exposure to linear alkylbenzene sulphonate (LAS) and concluded that "each model is applicable at a particular stage in hazard assessment". This appears to reinforce the maxim that in applying any method or approach in science, it is critically important to start with a clear understanding of the question that one is to address.

4.6.4 Advantages, limitations, and pitfalls in the modelling for environmental toxicology

The advantages of the modelling approach were described in Section 4.6.3. Briefly, these advantages include the capacity of models to

    a. Synthesise knowledge;
    b. Analyse the properties of entire systems;

    c. Through sensitivity analysis, identify the rate-limiting or key steps within a system;

    d. Identify deficiencies in the quality or quantity of information about a system or a substance;

    e. Provide recommendations for critical loadings (e.g., emissions) of a substance to the environment that will produce no adverse consequences to an ecosystem;

    f. Provide estimates of the future response of the system.

The latter two properties in particular lead to the application of models in risk assessment and environmental management more generally.

In 1994 the Society for Environmental Chemistry and Toxicology (SETAC) carried out an exercise in which four fate models were compared by running them for a similar environment and a selection of chemicals with similar properties. The results are reported in detail by Cowan et al. (1995), but the point to be made here is that initially there were very large differences in the results from the four models. Upon analysis, these were related to (1) differences in interpretation of input data; (2) use of different concentration units (e.g., simple things like wet versus dry weight as a basis for concentration); and (3) differences in the selection of transport coefficients among media. When these differences were resolved, results from all four models were in good agreement, despite differences in structure of the models.

The biggest limitations to the modelling approach would appear to be the often inadequate quality of the input data and questions concerning the validity of the mathematical expressions used for describing the various processes. Overconfidence in computed results, either by the modeller or by the user, can result if these limitations are not recognised. Mackay (1994) emphasises this by pointing out that the models "should be viewed as merely tools, not as an end in themselves. They must therefore be designed to satisfy a stated need."

4.7   Examples of methods and approaches for community or higher level responses

One of the dilemmas facing environmental toxicologists is the oft-repeated concern that laboratory experiments, while invaluable for the elucidation of mechanisms at the organismal or lower level, do not represent the "real world". The present text has pointed this out in several places, and indeed much of the current chapter addresses those approaches that attempt to overcome this problem. In Section 4.5, we discussed some of the community and higher level indicators of the effects of potentially toxic substances. The present section addresses some of the experimental approaches that have been used for assessment of these higher level responses. For these approaches, either field studies or complex built laboratory ecosystems seem to be essential. The most widely used approaches involve experiments on "synthetic" ecosystems, usually termed microcosms; the enclosure of a representative segment of a real ecosystem, usually termed mesocosms; or the manipulation of an entire ecosystem such as a lake. The enclosure may be extremely large; nevertheless, the "whole ecosystem" has to have boundaries, which are normally defined by the user. Consequently, there is in reality a continuum rather than distinct difference between these types of experimental methods. Nevertheless, because there are some rather consistent differences between the typical application of enclosures compared with whole system manipulation, it is convenient to treat them separately.

4.7.1   Enclosures: Microcosms and mesocosms

Microcosms and mesocosms are experimental systems set up to re-create part or all of a particular ecosystem. These systems range from constructed systems such as simple aquaria, with perhaps only one representative species at each trophic level, to large enclosures (e.g., limnocorrals, test plots) in the external (i.e., real-world) environment, which may be tens of cubic meters in volume.

MESOCOSMS

A typical mesocosm for an aquatic system is a tube constructed of some inert plastic material, set in a lake, with its base open, but immersed in the sediments and its upper extremity made as a collar that floats above the lake surface, limiting exchange of water with the bulk of the lake but allowing atmospheric inputs. Several earlier designs were large plastic bags, termed limnocorrals, suspended in the water column. Different scale versions of these were deployed in enclosed or semienclosed water bodies such as Loch Ewe in Scotland or Sanisch Inlet in British Columbia, Canada. More recent designs have incorporated a benthic or even shoreline component, and the recently constructed SERF (Shoreline Environmental Research Facility) facility at Corpus Christi, Texas, USA, includes a wave maker to simulate shoreline marine conditions. Some mesocosms take the form of artificial streams that simulate flowing freshwater systems. It should immediately be obvious that even though these types of experimental system are more complex and realistic than a flask or a plant pot in a laboratory; nevertheless, there are limitations (e.g., in terms of water and nutrient flow) between these types of enclosure and the real-world situation.

According to Shaw and Kennedy (1996), the use of mesocosms evolved from farm pond and monitoring studies in the 1970s and 1980s that were designed to test ecological risk assessment for pesticide registration decisions. A variety of mesocosms (e.g., lotic/lentic mesocosms) has since been used to study the fate and effects of anthropogenic chemicals in aquatic environments. In this context, the terms *micro-* and *meso-* refer only to scale and have no ecological relevance per se.

Generally experiments with these simplified ecosystems involve setting up replicate treatment and control systems where treatments involve dosing with potentially toxic or perturbing chemicals over an environmentally realistic range of concentrations. Measurements of predetermined parameters or end-points are made on all replicates prior to and after treatment. Such parameters can include physical measurements (e.g., light and temperature), geochemical characteristics (e.g., dissolved and suspended chemicals, pH measurements), and a variety of biological indices ranging from the biochemical to the "community" level. The pretreatment data should establish the baseline state of the system as well as indicate interreplicate variability in these same parameters. After treatment, the parameters are typically measured at intervals, providing a time-course of responses. Thus, in common with standard toxicity bioassays, replicated systems that are dosed over a concentration gradient are created, and treatments are compared with undosed control systems.

Experiments with enclosures can yield information on community and higher level responses to contaminants that cannot be obtained in the laboratory with single-species tests; furthermore, biological responses can be integrated with measured physical and chemical changes. Over all, these types of experiments, carried out as they frequently are under field conditions, should be more realistic than laboratory experiments.

However, there are potential drawbacks associated with these potential advantages. Although mesocosms may reach considerable complexity, their capacity to represent a truly natural ecosystem is open to question. Typical problems involve making a decision on the time required for the system to respond to an applied perturbation, limitations of nutrients if inflow and outflow are prevented, and the implications of including or excluding certain communities, especially larger predators, when the mesocosm is set up. Further problems are related to the degree of boundary or edge effects, wherein the surface of the enclosure itself creates an artificial environment such as a substrate for bacterial or algal growth. Filters designed to control the import and export of material to and from a mesocosm may cause similar problems, although many would argue that such a control is necessary to minimise the influence of predators. Nevertheless, an appropriate degree of flushing, of food input, and of nutrient balance are all important considerations, and differences in these parameters may result in conflicting conclusions. Some mesocosm designs actually allow species migration from the ambient environment as a deliberate feature, with rate and scope of recolonisation as a specific end-point. Variants on this include measurement of recolonisation/recovery following termination of chemical exposure.

Terrestrial enclosures suffer from comparable drawbacks. Although nutrient exchange may be less of a problem on land than for aquatic enclosures, other factors such as the exclusion or atypical representation of predators and the possibility of the defining structure (the "edge") becoming itself part of the system and becoming colonised are problematic.

A major concern with mesocosms is the establishment of an equilibrium situation within the experimental context. Typical questions include

1. The choice of sites for replication of treatments and control containers;
2. The length of time that replicated systems should be allowed to run before dosing begins;
3. The degree to which colonisation from the ambient environment should contribute to the establishment of the test communities;
4. The length of time the system should be allowed to run after dosing begins (or ends);
5. The frequency of sampling, particularly since some types of sampling will be destructive or will modify the system. Usually sampling regimes will be dictated by the life-cycles of the component organism. Many systems start with a quantified "seed" population consisting of one or more species from different trophic levels.

Considerations such as boundary or edge effects (mentioned earlier), the size and feeding strategies of organisms, and competition for space will also influence decisions concerning the size of the systems to be used. The question of scale is addressed by considering some recently published studies.

Table 4.15 summarises the advantages and limitations of mesocosms according to Graney et al. (1995).

MICROCOSMS

Microcosms, literally little worlds, such as aquaria and terraria are in a sense intermediate between the single species laboratory bioassay and the field enclosure. These will not be discussed in as much detail as mesocosms. Briefly, the advantages of microcosms as compared with mesocosms follow.

1. Microcosm experiments can be performed more quickly than can mesocosm studies.
2. Replication is more convenient and less costly for microcosms than for mesocosm studies.
3. Microcosms can address a wider range of ecological issues than can laboratory tests.
4. Microcosms allow studies to be conducted using radioactive tracers; this would be either impractical or perhaps illegal in field enclosures.

Inevitably, there are limitations associated with the use of microcosms. These limitations follow.

1. Microcosms cannot normally accommodate large organisms such as fish and mammals, at least in the adult forms.
2. The confined physical system is likely to distort chemical exposure regimes.
3. If destructive sampling is used, the small size of the microcosm means that only a limited number of samples can be taken in a time-course.
4. With most microcosm designs, recolonisation is not possible.

Table 4.15. Advantages and limitations of mesocosm (enclosure) studies in environmental toxicology

Advantages	Limitations
Test system contains a functioning ecosystem.	Communities of large organisms, particularly predators, are rarely included, for practical reasons (size).
The system can be sustained over a period of time.	Scaling factors and edge effects must be considered.
Conditions within the system can be monitored before and after addition of the chemical of interest.	Enclosures are expensive to construct and maintain.
Replicates and untreated controls can be set up in an appropriate statistical design.	With time, changes that are related to the effect of the enclosure occur.
Components of the ecosystem can be manipulated (i.e., isolated, removed, or augmented) to test hypotheses concerning the effects of contaminants.	Systems tend to show variability among replicates even at time zero, and replicates tend to diverge with time.
Environmental conditions and thus exposure are more realistic than those in laboratory studies.	Environmental conditions are uncontrolled, although they can be monitored. Unplanned events such as the introduction of a foreign object (e.g., bird excrement) to one of the replicates can destroy an entire experiment.
The effects of chemicals on a number of species with different sensitivities can be investigated simultaneously.	
Ecosystem level effects and interactions among species can be investigated.	

[a] Based on Graney et al. (1995).

5. The normally short time period of experiments means that recovery will not be detected.

Table 4.16 provides a few examples of studies in environmental toxicology for which micro- and mesocosms have been used and further reading is also provided. Case Study 4.6 gives greater detail about the application of this approach.

4.7.2   Whole system manipulations

The present text has already pointed out in several places that laboratory experiments, while invaluable in the elucidation of mechanisms at the organismal or lower level, poorly represent the real world. Even though enclosures may go some way toward addressing these concerns, perhaps the ultimate experimental approach to environmental toxicology is exemplified by the so-called whole system manipulation. The application of potentially toxic substances in field plots or sites, or to enclosures in the field, leads logically to the application of chemicals to whole lakes, forests or streams. In a sense, the so-called field trials (not addressed as such in this chapter) that have been used for many decades to test pesticides and

Table 4.16. Examples of studies in environmental toxicology utilising microcosms and mesocosms

Type of system	Organism/community	Substance(s)	Duration of test	End-point(s)	Reference
Sediment/water, compared with water alone	Midge *Chironomus riparius*	11 (eleven) water soluble neutral lipophilic compounds	24 hours	Toxicity and accumulation	Lydy et al. (1990)
Static laboratory assay	Natural phytoplankton assemblages reared in the laboratory	Zinc (Zn)	24 days	Changes in density of algal populations over time; Shannon Weaver diversity index; equitability and richness calculated	Loez et al. (1995)
Enclosure (limnocorrals) in the Bay of Quinte, Lake Ontario	Natural assemblages	Nutrients	3 years	Chlorophyll-*a*, phosphorus concentration in water, soluble reactive phosphorus concentration in water, oxidation-reduction potential of sediments, algal biomass, zooplankton grazing rates	Lean and Charlton (1976)
Farm ponds	Natural assemblages	Endosulfan	25 weeks	Water chemistry, including pesticide, pytoplankton zooplankton, periphyton, macroinvertebrates, macrophytes, fish, ecosystem metabolism	Fischer (1994)
Artificial ponds	Existing communities plus stocked fish	Various pesticides, singly and in combination, part of the requirement for registration	Usually at least a full season, with dosing every 2 weeks during summer	Water chemistry including the pesticides, primary production, macrophytes, zooplankton density and species composition, community structure at various trophic levels	La Point and Fairchild (1994)
Coastal ocean	Benthic invertebrates and fish	Chemically dispersed and undispersed oil	1–2 weeks	Toxicity and bioaccumulation	Coelho et al. (1999)
Prairie grassland enclosures	Small mammal communities	Anticholinesterase pesticide, Diazinon	2–30 days	Reproductive performance	Sheffield and Lochmillen (2001)

fertiliser effects are themselves whole system manipulations and foreshadowed some of the research work discussed later. A major difference between the more traditional type of field trial and the more recent large-scale manipulations lies in the parameters of response that are monitored. It is fair to say that for most, although not all, field trials the end-points of interest are, understandably, the efficacy or other effects of the application of interest on the target organism, usually a crop. Wider ranging responses, which one might term ecosystem effects, are not normally assessed in field trials.

One of the most notable early examples of a whole system manipulation approach was a series of studies of eutrophication in the Experimental Lakes Area (ELA) of northwestern Ontario (Schindler et al., 1978). The ELA is an area of the Precambrian shield with a large number of small lakes available for scientists to study and use for experiments. Application of various combinations of nutrients – carbon (C), N and P, all of which were at the time candidates for the primary cause of cultural eutrophication – were made to sections of whole lakes. This early work provided conclusive evidence of the pivotal role of phosphorus as a nutrient in cultural eutrophication in a manner that other approaches could not do nearly as effectively or dramatically. And as far as convincing any sceptic (and there were plenty) of the role of phosphorus, it has been claimed with considerable justification that the aerial photograph of the dense green coloration resulting from the algal bloom on the treated lake, compared with the deep blue clarity of the untreated portion of the lake, was far more dramatic than any data or graphs from laboratory tests could ever be. Plate 4.1 shows the result of the experiment.

Whole lake manipulations also made possible considerable progress in our understanding of the effects of acidification on fresh water ecosystems (Schindler, 1990). A wide spectrum of biological and geochemical parameters can be monitored in such whole system experiments, providing information on population to ecosystem level effects. Reversibility of acidification can also be assessed in whole lake studies, as was done in the ELA following the cessation of additions of acid. Our understanding of the fundamental structure and function of whole systems, aside from ecotoxicological effects, can also benefit from these types of study.

In the ELA cadmium spike experiment (Malley, 1996; Case Study 4.7), low concentrations of cadmium labelled as ^{109}Cd were added over a 5-year period to a whole lake. The pathways, transformations, and effects of the added metal could be monitored. The study represented a unique research opportunity, which was well exploited. However, this type of research is precariously sensitive to funding because maintaining facilities and conducting experiments are extremely expensive. Indeed, the cadmium spike experiment is not only an example of usefulness of the approach but also of the impact of truncated funding (Case Study 4.7).

In the 1980s, the RAIN (Reversing Acidification in Norway) project began. It consisted of experiments on catchments in which the effects of "drastic changes in precipitation chemistry" on soil and surface water deacidification and acidification, respectively, were monitored over a period of more than 5 years (Wright et

Plate 4.1 Whole-system manipulation illustrating the role of phosphorus in eutrophication. Legend: Lake 226, Experimental Lakes area, Ontario, Canada, was manipulated to address the role of phosphorus, nitrogen, and carbon respectively, in eutrophication. At the time, there was still considerable controversy about the effectiveness of controlling phosphorus entering freshwater lakes (Schindler 1974). In 1973, the far basin in the photograph was fertilised with phosphorus, nitrogen, and carbon and within two months was covered by an algal bloom (pale, solid area in the phtograph). The near basin, which received similar quantities of nitrogen and carbon but no phosphorus, did not develop a bloom (dark, transparent area in the photograph). Photograph supplied by and reproduced by permission of David Schindler.

al., 1988). For an acidified catchment in southern Norway, acidic precipitation was excluded by use of a transparent canopy and "clean" precipitation was added beneath the canopy. For the opposite process, two pristine catchments were acidified by additions of sulphuric acid or sulphuric and nitric acids, respectively. This process was also an expensive venture with financial support from a number of agencies. No biological data have been reported from this project, but the geochemical changes in soil and runoff provided information that contributed to a better understanding of the long-term effects of acidification, as well as a better appreciation of the prospects for recovery should the acidity of precipitation decrease.

Even whole lake manipulations may have problems of design as well as interpretation and applicability. Arguably, the biggest problems concern replication and the choice of an appropriate control or reference site, respectively. In the Little Rock Lake acidification experiment in Wisconsin, the lake was divided into two halves, one of which was treated and the other left as a control. It was later apparent that the fish communities on the two "halves" were not strictly comparable, so although a great deal of information was gained, and is still being gained, from this experiment, nevertheless, there was strictly speaking no conventional control for the treatment.

The ELA lake manipulations have also utilised the "split" lake concept, typically with a barrier in the form of a curtain dividing a lake into two halves; ELA researchers have also applied the concept of a "matching" reference system which remains untreated. Large-scale manipulations of stream or river systems have generally used sites upstream of the manipulation as a control or reference. Manipulations of terrestrial ecosystems have generally sought nearby sites comparable in all respects to the test system for use as reference. In all these approaches, unanticipated events or conditions sometimes limit the interpretation of the results.

Land-based whole system manipulations have also been reported. Between 1961 and 1976 at various sites in the United States, scientists conducted a series of experiments which examined the effects of gradients of ionising radiation on forest and field ecosystems. Woodwell and co-workers reported on the results of some of these experiments for the Brookhaven Irradiated Forest Experiment. Some of the results from these ecosystem-scale experiments could not have been anticipated and were "rich with surprises" (Woodwell and Houghton 1990). This work is discussed in more detail in Section 8.6.

The Hubbard Brook Forest ecosystem in New Hampshire has also been the subject of manipulations. A large-scale experiment involved forest disturbance in the flora of clear-cutting over an entire watershed, 15.6 ha, with the cut trees being left on the ground. In addition to the cutting, an herbicide treatment was applied to prevent regeneration for the next 3 years. Data on nutrient cycling and hydrologic changes in the watershed and streams (S. Findlay et al., 1997) provided the basis of nutrient budget calculations as well as modelling. The Hubbard Brook watershed was also studied intensively in the context of acidic deposition, although the deposition per se was not experimental. The existing database on nutrient budgets provided an excellent opportunity to assess the impact of acidic precipitation on a forested watershed and the related streams (Likens, 1985).

Other studies have treated forest ecosystems with contaminants and have looked at various community and higher level responses. Hutchinson et al. (1998) treated a section of sugar maple (*Acer saccharum*) forest in Ontario, Canada, with acidified precipitation, with and without fertiliser additions to soils, and monitored a number of biological indicators. The authors concluded that the types of results indicate "a complex interaction of events which are not reproduced in pot trials and must be fully understood before the impact of acid rain on sugar maple forests can be evaluated".

Table 4.17 lists examples of whole system manipulations.

Radioactive tracers have the potential to provide useful information concerning the pathways of contaminants in ecosystems. For obvious reasons, radioactive materials will not normally be permitted for use in open experiments in the field and, therefore, cannot usually be utilised in whole system manipulations. Under some conditions (e.g., the cadmium spike experiment, see Case Study 4.7), however, short-lived isotopes can be used. Stable isotopes (see Section 6.2.4) offer all the advantages of radioactive isotopes with few of the concerns for safety. It is anticipated that in the near future, for many elements or substances of concern, stable isotopes will provide valuable tools for improving our understanding of environmental pathways and transformations of contaminants. Stable isotopes therefore have promise in large-scale field manipulations, as well as in simpler systems.

There may be legal barriers to carrying out large-scale manipulations on field sites. Furthermore, in a practical sense, the major limitation for whole system manipulations is not scientific but rather pragmatic, namely that of cost.

4.8 The role of technical advances in methods for environmental toxicology

All branches of science benefit from technical advances, even though no amount of gadgetry will compensate for lack of original ideas and sound design and interpretation. Environmental toxicology has been the beneficiary of new technology and refinement of older technology. This connection was referred to in principle in Chapter 1 and will arise in subsequent chapters. For the sake of completeness, it is also included in the present chapter because "approaches" frequently depend on or are limited by technology.

For convenience, categories of advances that appear to be of particular relevance to environmental toxicology follow.

a. Analytical chemistry with accompanying electronic interfaces;
b. Electron microscopy and associated analytical techniques;
c. The availability of computers, particularly personal computers.

ANALYTICAL CHEMISTRY
Chemical determination of substances in the abiotic and biotic components of ecosystems has always been an integral component of environmental toxicology. Recent decades have seen improvements in analytical methods and, perhaps of even more significance to the toxicologist, improvements in instrumentation. Detection limits for determining inorganic and organic substances have been lowered by

Table 4.17. Examples of studies in environmental toxicology which involved whole-system manipulation

Type of system	Treatment, substance(s)	Duration	End-point(s)	Reference
Stream	Sulphuric acid		Invertebrate drift	Hall et al. (1980)
Lake	Phosphate and nitrate, alone and in combination	1 year (or less)	Water chemistry, primary production, community structure at various trophic levels	Schindler et al. (1978)
Lake	Sulphuric acid added to water column until target pH was reached, then additions ceased and recovery was tracked	Several years	Major chemical parameters, species changes, and community indices for all trophic levels	Schindler (1990)
Lake	Additions of labelled cadmium to the whole lake; associated enclosure and laboratory experiments for the same parameter	Several years	Geochemistry of added cadmium, water chemistry, zooplankton, and macroinvertebrate community structure; food chain effects of cadmium measured in parallel laboratory experiments	Malley (1996)
Forest catchment	Reversibility of acidification by acid exclusion, protecting whole catchments from acidic deposition and increasing deposition of strong acids by addition to natural precipitation	5 years	Chemistry of runoff, some soil chemistry	Wright et al. (1988)
Sugar maple and jack pine forest stands	Precipitation treated with sulphuric and/or nitric acid, with and without soil fertiliser amendments	2–5 years	Soil chemistry, mycorrhizae, various physiological indicators of tree and lichen condition, lichen morphology, feather moss and lichen community structure	Series of papers; see Scott et al. (1989), Hutchinson et al. (1998)
Oak-pine forest	Chronic gamma radiation	15 years	Plant communities monitored for diversity, coefficient of community, percentage similarity, effects on early successional communities, at different distances from the source and over time	Woodwell and Rebuck (1967), Woodwell and Houghton (1990)

orders of magnitude, and this too has aided the toxicologist because many of the effects of concern can be evoked at extremely low concentrations of a toxicant. In addition to conferring obvious advantages to the quality of research and monitoring, these technical improvements have also had other effects. The problem of inadvertent contamination by ambient chemicals, as well as simple lack of accuracy for some older methods, have the effect of placing in doubt the reliability of many data published in earlier decades. Prior to the publication and acceptance of analytical results, particular care is required for quality assurance and quality control of sampling and determination, and most protocols for regulatory or other prescribed methods of assessment define this quite rigorously. Access to expertise in analytical chemistry is still essential for the proper determination of environmental contaminants, and a "black box" approach to an instrument, even if it is apparently simple, is not acceptable.

ELECTRON MICROSCOPY

The relationship between structure and function is of interest to biologists, and microscopy has always played a significant role in the biological sciences. Perhaps less well-used than some other approaches in environmental toxicology, as evidenced by the relatively few references in the present text to microscopy, light and electron microscopy continue to evolve with more sophisticated instrumentation. As for any specialised field, technical expertise is essential. Perhaps the relative lack of application of microscopy to environmental toxicology results from a limited collaboration among the required disciplines. A notable example of the potential for electron microscopy to be applied to toxicology as a chemical analytical method is the capacity of the energy dispersive X-ray and other techniques, which permit the detection of extremely low concentrations of elements in specific organelles or other ultrastructures of cells.

THE COMPUTER

Many volumes and weeks and months have been devoted to analysing the impact of the computer on scientific endeavours. Space does not permit more than a brief reference to the influence of computers on the particular branch of science that is our subject. In ecology alone, the collection of data has been revolutionised by the knowledge that data can be stored and processed in ways that were not practical before. Reference has already been made to the way in which instrumentation in analytical chemistry has been aided by electronic interfaces. Modelling, referred to in Section 4.6, has also benefited from the increased availability and convenience of operation of personal computers. In all these contexts, the computer has aided studies in environmental toxicology. A relatively recent development in the use of computers for this science is the use of the Internet for sharing and communicating information.

In common with any other advance in technology, for most users the computer has to be recognised as a tool, not as an end in itself. Therefore, the caution that has been expressed already in terms of needing expertise for appropriate training and use of the technology, even if it appears to be operationally simple, applies to computers.

4.9   Choice of approaches

CHOICE OF BIOLOGICAL SCALE

Great advances have been made in the scientific considerations relating to the study and understanding of biological indicators over the past decade. The preceding sections covered a range of approaches that can be applied at various levels of complexity across the biological spectrum, from the biochemical, subcellular level to the whole ecosystem. When choosing methods and approaches for practical purposes, it is important to consider the scale of the indicator. Table 4.18 summarises the wide range of scale for approaches and methods in which the effects of toxic substances have been assessed.

Table 4.18 also indicates the limitations and the best use(s) of each approach. It is provided not only as a summary but as a practical guide for a student who is planning an assessment. Through these examples and some theoretical considerations of biological indicators, the user can consider the techniques as well as the advantages and the disadvantages of the various approaches on a problem-specific basis. When utilising biological indicators or monitors for assessment, careful consideration should be given to the most appropriate methods for the particular question or issue that is being addressed.

CHOICE OF METHODS

As an adjunct to Table 4.18, Figure 4.19 attempts to match or link laboratory and field approaches, again with emphasis on the biological or ecological scale of the method. Concerning laboratory tests, discussions have been published concerning the question of how representative laboratory tests or laboratory organisms are of conditions in nature. A common-sense response is that of Chapman (1983) who points out that if the goal of the laboratory tests is made clear, "when appropriate test parameters are chosen, the response of laboratory organisms is a reasonable index of the response of naturally occurring organisms". The issue of extrapolation from laboratory toxicity results to the field continues to evoke heated discussion.

CHOICE OF TEST ORGANISMS OR SYSTEMS

Once the general method has been selected, the choice of the appropriate biological monitor organism or system will also depend on the environmental media of interest (i.e., sediment, soil, water, air). As illustrated in the preceding section on indicator and monitoring organisms and systems, different organisms are exposed to and accumulate contaminants through different pathways. Common sense will normally dictate the type of organism or system that is most appropriate for the medium, but the choice is also limited by the availability of test organisms in suit-

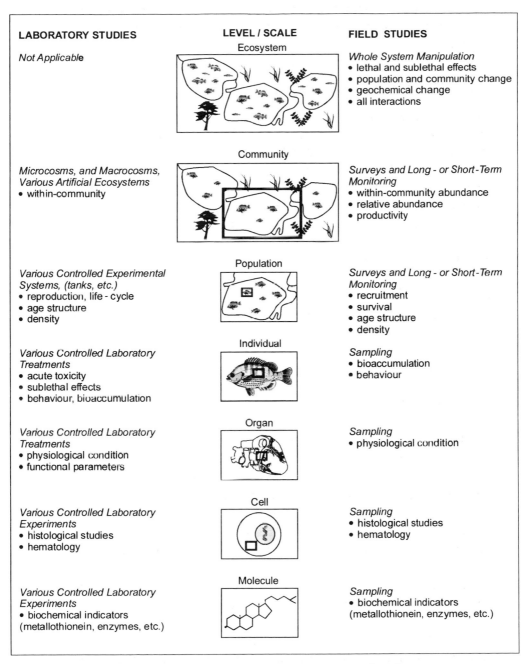

LABORATORY STUDIES   **LEVEL / SCALE**   **FIELD STUDIES**

Not Applicable

Ecosystem

Whole System Manipulation
- lethal and sublethal effects
- population and community change
- geochemical change
- all interactions

Community

Microcosms, and Macrocosms, Various Artificial Ecosystems
- within-community

Surveys and Long - or Short - Term Monitoring
- within-community abundance
- relative abundance
- productivity

Population

Various Controlled Experimental Systems, (tanks, etc.)
- reproduction, life - cycle
- age structure
- density

Surveys and Long - or Short - Term Monitoring
- recruitment
- survival
- age structure
- density

Individual

Various Controlled Laboratory Treatments
- acute toxicity
- sublethal effects
- behaviour, bioaccumulation

Sampling
- bioaccumulation
- behaviour

Organ

Various Controlled Laboratory Treatments
- physiological condition
- functional parameters

Sampling
- physiological condition

Cell

Various Controlled Laboratory Experiments
- histological studies
- hematology

Sampling
- histological studies
- hematology

Molecule

Various Controlled Laboratory Experiments
- biochemical indicators (metallothionein, enzymes, etc.)

Sampling
- biochemical indicators (metallothionein, enzymes, etc.)

Figure 4.19 Approaches to assessment over a range of scales.

able quantity and genetic uniformity. This explains why relatively few species of organisms appear in the literature on toxicity testing, and why biochemical level tests are often preferred. Relevance to the real world continues to challenge the assessor of toxic substances.

Table 4.18 Summary of approaches for the assessment of biological and ecological effects of toxic substances

Type of study	Advantages	Limitations	Best use(s)
Whole system manipulation	Cause and effect generally defined. Incorporates all community-level and within-system biogeochemical processes. Can follow for long periods of time, including impact and recovery as needed.	Experiments large and costly. For reasons of cost and practicality, experiments rarely replicated.	To identify community interactions and biogeochemical processes, direct and indirect. Useful for comparison with field surveys.
Field surveys, including the use of indicators	Real-world information, potentially incorporating all community interactions, biogeochemical processes, and natural variability.	Cause and effect only inferred, since a number of factors covary, thus confounding interpretation of results. If surveys attempt to include impacted and unimpacted sites, difficult to find representative "reference" or control sites.	To identify potential problems as well as real current problems, to develop and evaluate hypotheses of effect of contaminants at community or population level, to compare with experimental results.
Long-term monitoring, chemical and/or biological	Fewer problems with confounding factors than those related to field surveys. Real-world responses to change in the environment can be captured.	Cause and effect can only be inferred. Very long time-courses may be needed to distinguish real-time trends from natural variability.	For comparison with experiments and field surveys, to confirm the validity of experimental results and see if spatial patterns determined from surveys reflect expected trends in time.
Mesocosm/microcosm experiments	Cause and effect clearly defined. Some community-level interaction can be incorporated, depending on the scale and complexity of the experimental system. Manageable size allows for replication and repetition of experiments.	Large variability among replicates tends to be the rule, probably because true replication is difficult to achieve. Experimental units generally too small to incorporate larger organisms such as predatory fish and larger mammals, thus some important community processes are often excluded.	Useful for study of community-level responses and interactions among the smaller-bodies communities. Useful for making preliminary assessment of food chain transfer of contaminants (e.g., pesticides).

	Advantages	Disadvantages	Comments
Field bioassays, whole organisms	Test conditions simulate the real world more effectively than do laboratory bioassays.	Many factors vary simultaneously; therefore, specific cause and effect relationships can only be inferred. There are significant uncertainties in extrapolating from population responses in simple field bioassays to population and community-level responses.	Useful to confirm under field conditions the occurrence of toxic responses to a contaminant. Useful to compare with laboratory bioassays, to check that all factors have been accounted for (e.g., light/dark or temperature variations that affect toxic response).
Laboratory bioassays, whole organisms	Cause and effect readily defined. Experiments are relatively cheap and quick to perform and are amenable to quality assurance and control.	Test conditions poorly represent the real world. There are uncertainties in extrapolating from laboratory responses to expected changes at the population or community level. Typically, there is poor agreement between laboratory and field tests, even for the same species.	Controlled experiments can be used to test the effects of toxic substances under specific sets of conditions and can be used to screen "new" chemicals for acute or chronic toxicity.
Field bioassays, biochemical indicators	Experiments are rapid and relatively inexpensive to perform, amenable to comparisons with identical tests on organisms exposed under controlled laboratory conditions.	Response may be influenced by other factors beyond the contaminant of concern (i.e., may not be entirely specific), thus potentially confounding interpretation.	May provide a link, if other parameters are also measured concurrently, between biochemical and higher level responses. If the biochemical indicator has been shown to be reliable and substance- or group-specific, field measurements may be useful as integrators of exposure to contaminants.
Laboratory tests, biochemical indicators	Experiments are rapid and relatively inexpensive to perform. Cause and effect clearly defined for the biochemical response, but, due to the short-term nature of response, it is rarely possible to link it to a functional response.	Extrapolation to whole organism or community function not easy to determine.	Mechanisms of toxicity can be determined through controlled experiments. Substance-specific responses can be determined.

RESOURCES

In most instances, the availability of resources such as time and funding will have considerable influence on a programme of assessment. This is not to say that scientific considerations are less important. On the contrary, when resources are limited, or when there is urgency to come to a management decision, the choice of scale and method based on scientific understanding and reasoning becomes even more important.

4.10 Case studies

Case Study 4.1. Benthic invertebrate communities in metal-contaminated sites exceeding criteria for acceptable sediment quality

The port of Mont-Louis, Quebec, Canada, is located on the Gaspé Peninsula, in the St. Lawrence Gulf. The site had been used by Mines Gaspé for transporting copper concentrate over 2 or 3 years in the latter part of the 1980s as well as for other commercial activities. Previous studies, dating back to 1984, indicated that copper concentrations in sediments were elevated, and in the summer of 1997 this was confirmed. The present study was initiated when Transport Canada stopped using the port.

In 1997, a joint investigation was initiated with Mines Gaspé 9, Transport Canada, and the Canadian Department of Public Works. The objectives were to carry out:

> A chemical characterisation of the sediments, including sequential extraction of metals and acid-volatile sulphides;
> A biological survey of the benthic communities in the vicinity of the docks (i.e., observed effects); and
> Toxicity testing of sediments (i.e., potential effects)

(Beak International, Inc., 1998; Prairie and Lavergne, 1998).

Two sources of criteria for copper in sediments were considered:

> The St. Lawrence Centre (le Centre Saint-Laurent, Montreal, Quebec)
> > Interim value for the threshold for minor effects on indigenous organisms (Seuil d'effets mineurs, SEM), 55 ppm
> > Interim value for the threshold for major effects on indigenous organisms (Seuil d'effets nefastes, SEN), 86 ppm
> The Canadian Council of Ministers of the Environment (CCME)
> > The threshold effects level (TEL), 18.4 ppm
> > The probable effects level (PEL), 108 ppm

> > > > > > (Prairie and Lavergne, 1998).

These two different sets of criteria illustrate one of the dilemmas of attempting to conform to criteria or standards, namely that the bases of the criteria and the recommended values differ among agencies. In this case, the sediment criteria from the CCME appear to be conceptually different from those of the Centre Saint-Laurent in

Quebec. The apparent discrepancy may also reflect in part the uncertainty surrounding the relationship between the toxicity of chemicals in sediments.

Chemical analyses

Sediments for chemical analysis were sampled within 100 m of one side of the dock, at 18 sites at 4 depth intervals: 0–10, 10–50, 50–100, and 100–150 cm. Metal enrichment (mainly copper) was often observed at/near the surface layer of the sediments. Nickel, arsenic, and chromium were also detected but generally at lower concentrations. Copper in sediments ranged from 11,000 to 12,000 ppm, with a zone of contamination extending approximately 40–50 m to the south of the dock. The site selected as a reference had all values below the SEM for most of the substances of concern.

Sequential extraction procedures (adapted from Tessier et al., 1979) indicated generally that 90% of the copper was present in the operationally defined fractions F4 (organically bound) and F5 (in the crystalline matrix).

Benthos

Four sites were sampled for identification and enumeration of benthic organisms. In terms of number of species, crustaceans and polychaetes were the most numerous, but molluscs comprised more than 45% of total numbers of organisms at test sites and 85% at the control site. Indices included determination of the number of taxa (richness) and the Shannon-Wiener index of diversity. There was no correlation between either of these indices and the total copper concentration in sediments. Indeed, the lowest score for the index of diversity was obtained for the site with the lowest concentration of copper. The sandy nature of the substrate at the reference site makes comparison with test sites quite difficult (Beak International, Inc., 1998).

Toxicity tests

Toxicity tests were made using two different standard organisms, according to the recommended Environment Canada protocols: the marine amphipod *Eohaustorius estaurius* (mortality) and the sea urchin *Lytechinus pictus* (fecundity).

The highest mortality in the amphipod test was observed for the sediment from the reference site (Figure 4.20A). The sea urchin test showed very high loss of fecundity for all the samples, with the greatest loss for the sample from the reference site (Figure 4.20B).

Conclusions

For all the correlations that were examined, the study showed none of any significance. This implies no significant relationships between copper concentration and mortality of the amphipods, nor between copper concentrations and fecundity of the urchins. Similarly for chromium and nickel, no significant correlations were found between the chemical and biological parameters. Furthermore, there were no significant correlations between any of the following: amphipod mortality and richness, amphipod mortality and diversity, urchin loss of fecundity and richness, and urchin loss of fecundity and diversity.

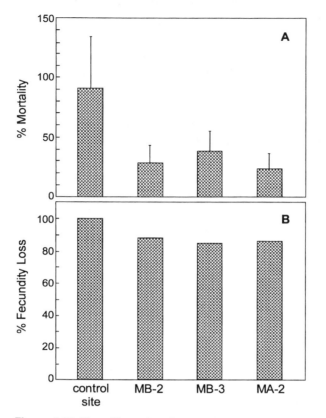

Figure 4.20 The effect of sediments from stations at varying distances from the Mont-Louis dock, on (A) a marine amphipod (*Eohaustoris estuaris*) and (B) sea urchin (*Lytechinus pictus*). The amphipod test showed highest mortality at the control site, MF-1. The fecundity test for the sea urchin showed high loss of fecundity in all sites, but the highest loss was for the control site MF-1. Based on data in Beak International, Inc. (1998).

Regardless of other contaminants in the sediments, many of the samples of sediment had concentrations of copper that exceeded thresholds for minor and severe effects. Yet when indices of community structure and composition were compared with copper concentration, there was no significant relationship.

Several possible explanations for these findings follow.

1. Much of the measured metal was not biologically available.
2. The organisms had become tolerant to metal contaminant, suggesting that the criteria were not appropriate.
3. The criteria are excessively stringent for normal sites.

The authors of the study (Beak International, Inc., 1998) concluded that the major contaminants in the sediments, mainly copper, were for the most part not biologically available. This conclusion was supported by the results of the sequential extraction procedure. Thus the measured concentrations, although exceeding the criteria, did not result in severe environmental effects.

Case Study 4.2. Biomarkers of organic chemical contamination in fish from Puget Sound

Over the last 20 years, intensive studies have been made of the relationship between PAH and PCB exposure and the development of lesions and other toxic symptoms in bottom-dwelling fish, notably English sole (*Pleuronectes vetulus*) from Puget Sound, Washington, and other marine sites on the west coast of the United States.

Much of the concern has focused on hepatic neoplasms, although Myers et al. (1998) emphasise consideration of a broad spectrum of neoplastic, nonneoplastic, and other types of lesions that have been positively related to toxic exposure. Many such lesions are involved with the stepwise histogenesis of hepatic neoplasms and are associated with changes in serum chemistry indicative of liver dysfunction.

Specific questions addressed by this ongoing series of investigations are: (a) How can actual toxic chemical exposure be determined? (b) Can hierarchical effects be determined at different scales of cellular organisation?

Chemical exposure

For conservative pollutants, exposure is often inferred from measurement of chemicals in ambient media (e.g., water, sediment) and body tissues. Although this may be appropriate for conservative pollutants readily accumulated by biota, tissue analysis may not be appropriate for chemicals such as PAHs, which are readily metabolised and may not have a long half-life in the body. Investigators at the U.S. National Oceanographic and Atmospheric Administration Laboratory at Seattle, Washington, therefore, adopted a mixed strategy to quantify chemical exposure. Liver PCB exposure may be satisfactorily determined from liver tissue concentrations, and PAH exposure is estimated by screening levels of metabolites, fluorescent aromatic compounds (FACs) in the bile. Like other vertebrates, fish are capable of extensively metabolising PAHs and excreting them into the gallbladder.

Chemical effects

Concomitantly with tissue PCB and biliary FAC measurements, several other biomarkers are routinely measured to understand the relationship between dose (exposure) and potentially adverse effects at different levels of cellular organisation.

Because DNA modification is an essential, if not sufficient, criterion for chemically induced carcinogenesis, an important component of this suite of biomarkers is a ^{32}P-postlabelling assay (i.e., products are labelled following chromatographic separation) for detecting DNA-xenobiotic adducts in liver tissue. When correlated with benzo[a]pyrene (B[a]P) dose, B[a]P adducts showed a linear dose-response relationship over a 50-fold dose range. Estimates of the persistence of B[a]P-DNA adducts vary, although several reports indicate that such compounds may remain for several weeks after B[a]P exposure.

In laboratory experiments, arylhydrocarbon hydroxylase was rapidly induced after organic chemical exposure. Typically, investigators injected fish with sediment extract from differentially contaminated sites and found a relationship between AHH activity and PAH levels in sediments. Although it might be argued that such a procedure represents an unnatural means of exposure, the approach is justified as a means of establishing a dose-response relationship that may then be extrapolated to a field situation. In field-caught specimens of English sole, Stein et al. (1992) established significant relationships between several different biomarkers and the degree of contamination at respective collection sites. In Stein et al. (1992, Figure 4.21), six biomarkers in fish collected from variously contaminated sites in Puget Sound are compared. Biomarkers measured included biliary FACs, hepatic PCB concentration, hepatic DNA-Xenobiotic adducts, total hepatic glutathione (TH-GSH), hepatic AHH activity, and hepatic ethoxyresorufin-o-deethylase activity. The two most contaminated sites, Duwamish Waterway and Hylebos Waterway (Commencement Bay) in the southeast part of Puget Sound, were heavily contaminated with PCBs and selected PAHs relative to other sites sampled. Common underline in Figure 4.21 indicates sites that were not significantly different.

Stein et al. (1992) constructed a cumulative bioindicator response (CBR) using normalised values (maximal response = 100) for each of these biomarkers and giving weighting factors of 0.5 each for AHH and EROD activity in view of the fact that they were highly correlated and appeared to reflect the activity of the same cytochrome P450 enzyme. Thus,

$$CBR = PCBs_{norm} + FACs_{norm} + 0.5(AHH_{norm} + EROD_{norm}) \\ + DNA - Xenobiotic\ adducts_{norm} + TH\text{-}GSH_{norm} \qquad (4.6)$$

The CBR has since been used as an effective index of response with good potential for application to other areas of contamination.

In a parallel study with extended components in other contaminated coastal areas, English sole from Commencement Bay and other contaminated sites proved to have a higher prevalence of neoplasms than fish from reference sites. Using a stepwise logistic regression analysis, which is a common epidemiological tool for examining multiple risk factors, it was determined that significant overall risk factors for hepatic lesions included total low-molecular-weight aromatic hydrocarbons, total high-molecular-weight aromatic hydrocarbons, and total PCBs in sediments. Low-molecular-weight and high-molecular-weight biliary FACs also proved to be significant risk factors for neoplasms in this analysis.

Although enzyme biomarkers were not included in the risk analysis for neoplasms, there is now good correlative evidence from these studies of hierarchical effects at different scales of biological organisation. A coherent rationale now emerges. It can be followed in a logical manner from induced MFO enzymes capable of producing reactive intermediate metabolites, which, in turn, react with DNA (as adducts). These, presumably, affect DNA function and correlate with tissue-level consequences of carcinogen exposure, neoplastic lesions. Overall, these investigations offer some of the most convincing evidence to date of the progression of toxicological end-points over the scale: subcellular-cellular-tissue-individual organism.

In a more recent extension of these investigations, decreased levels of the hormone estradiol and associated depression of spawning activity are seen in English sole from contaminated areas of Puget Sound such as Duwamish Waterway relative to fish from

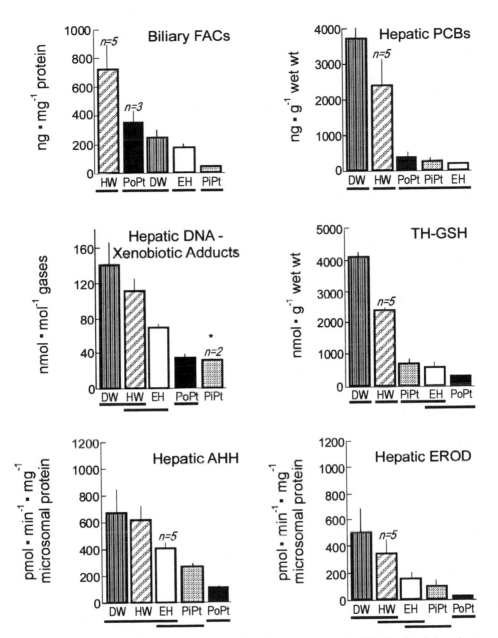

Figure 4.21 Six biomarkers of xenobiotic exposure and effect in English sole (*Pleuronectes vetulus*) from five variously contaminated sites in Puget Sound, Washington. Sites: Hylabos Water (HW), Duwamish Water (DW), Eagle Harbor (EH), Pilot Point (PiPt), Poinell Point (PoPt). After Stein et al. (1992).

more pristine sites. This implies that the scale of end-points discussed previously may yet be extended to the population level. A comprehensive suite of biomarkers currently used to indicate toxic chemical exposure and effect in Puget Sound fish is shown in Figure 4.22.

EXPOSURE ⟶ BIOCHEMICAL RESPONSE ⟶ EFFECTS at TISSUE/
INDIVIDUAL LEVEL

Biliary fluorescent aromatic compounds (PAHs) Tissue (liver) chemical concentrations (PCBs)	Hepatic cytochrome P450 Hepatic GSH Hepatic DNA xenobiotic adducts Hepatic DNA damage Plasma estradiol	Hepatic lesions Ovarian maturation Spawning success

Figure 4.22 Biomarkers of exposure and effect used to characterise impact of sediment-borne organics on fish from Puget Sound. After Varanasi et al. (1992).

Case Study 4.3. The effect of coal-ash pollution on bullfrogs: An energy budget approach

For several years the bullfrog *Rana catesbeiana* has been used as a model species in determining the adverse effects of coal-ash pollution at the U.S. Department of Energy's Savannah River Site. Sediments at the study site have been chronically enriched with several trace metals characteristic of coal-ash including As, Cd, Cr, Cu, Hg, and Se.

Studies have been focused on the early premetamorphic life stages of frogs: embryos and larvae. These life stages represent particularly convenient study organisms in view of the unusually long larval period (1–3 years). Larvae are, therefore, exposed to water and sediment-bound contaminants for long periods of time prior to their transition to the terrestrial environment.

In addition to observations of larval mortality at polluted sites relative to specimens taken from reference sites, there were three principal areas of study: morphology, behaviour, and physiology.

Morphology

Two categories of deformity were recorded in specimens from polluted environments. These affected (a) the oral region and (b) the tail region.

a. The incidence of larvae with missing labial teeth was as high as 96% and 85% in larvae from the deposition and drainage swamp (both polluted). In pristine reference areas, only 3% of collected specimens showed such abnormalities. Rowe et al. (1996) provided correlative evidence that oral abnormalities in bullfrogs were related to conditions at polluted sites. Evidence for a causal relationship was supplied by a reciprocal transplant study involving embryos from polluted and unpolluted sites. Such a study was designed to differentiate between the effects of environment versus maternal/genetic history. Regardless of their source, 97–100% of larvae raised in the ash-polluted site exhibited oral abnormalities, whereas less than 1% of larvae reared at the reference site were abnormal.

b. Tail abnormalities were found in 37% of larvae from the settling basin and 18% of larvae from the drainage swamp. Larvae from two reference sites had 0 and 4% deformities. Abnormalities took the form of lateral tail flexures and were significantly correlated with whole body concentrations of As, Cd, Cr, and Se.

Behaviour

Two behavioural effects were observed in larvae from polluted environments, and both were apparently related to morphological deformities:

> *Feeding behaviour.* Larvae from polluted environments showed a significant difference in feeding abilities relative to controls when presented with particulate food, and their ability to use periphyton as a dietary resource was nearly eliminated. This trait was almost certainly related to their poorly developed labial teeth.

> *Swimming behaviour.* Swimming behaviour was adversely affected in larvae from polluted environments in two respects. First, these larvae swam at less than half the speed of control larvae. Second, their response to a physical stimulus was much slower than control larvae, requiring twice the number of stimuli to cover the same distance. Both behavioural aspects clearly diminished their ability to evade predators. Swimming speed was highly correlated with axial abnormalities such as tail flexure, and the escape response probably had an additional neurophysiological component.

Physiology

An examination of maintenance energy expenditure (recorded as standard metabolic rate, SMR) revealed that field-collected tadpoles from polluted sites had SMR, 40–97% higher than larvae from a reference site. Results were collected at three temperatures and during spring and autumn. This trend was reinforced in a reciprocal transplant experiment similar to that used to investigate labial teeth abnormalities (Rowe et al., 1998).

In a scheme that integrates morphological, behavioural, and physiological impacts of (principally trace metal) pollutants on bullfrogs, Rowe (personal communication) represents the relationship between net energy assimilated and energy allocation with the equation:

$$N = M + S + G + R$$

where N = net energy assimilated, M = maintenance, S = storage, G = growth, and R = reproduction. Thus, if N decreases and/or M increases, then the remaining energy available for S, G, and R must be diminished and may even become negative. The equation is essentially the same as the scope for growth equation (Equation 4.1 in Section 4.4.2).

A scheme summarising interactions among measured responses and their potential ramifications for larval bullfrogs from coal-ash polluted sites is shown in Figure 4.23.

Case Study 4.4. Phytotoxicology assessment for Nanticoke Generating Station: Biological indicators and monitors of air pollution

Nanticoke is a village in a predominantly rural area of southwestern Ontario, Canada, on the north shore of Lake Erie. Historically, agriculture was the major land use and economic activity in the area. An industrial complex consisting of a thermal generating station, a steel mill, and an oil refinery was planned. In 1971, the Ontario

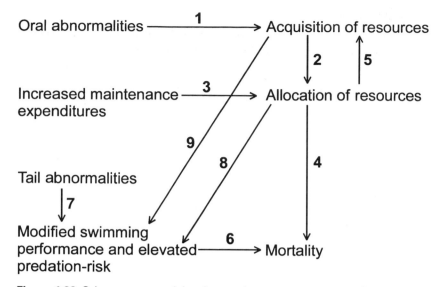

Figure 4.23 Scheme summarising interactions among measured responses of bullfrogs (*Rana catesbelana*) to fly-ash pollutants. Numbered arrows refer to the following measured or hypothesised relationships: 1 = limited ability of orally deformed tadpoles to consume periphyton; 2 = decreased energy for allocation to various processes due to limited intake of resources; 3 = modified energy allocation patterns due to high expenditures of energy on maintenance; 4 = mortality resulting from failure to meet maintenance costs; 5 = limited ability to find and procure resources due to limited energy available for foraging activity; 6 = mortality resulting from decreased predator recognition and escape ability; 7 = decreased swimming speed in tadpoles and tail abnormalities; 8 = decreased swimming performance due to limited energy available; 9 = modified foraging behavior in orally deformed tadpoles resulting from necessity to locate consumable resources (Rowe et al., unpublished).

Ministry of the Environment (MOE), Phytotoxicology Section, began prepollution studies for air quality and vegetation and soil conditions. Construction for the various industries took place over the period from 1973 to 1980. The generating station, one of the largest coal-fired thermal power plants in the world, was completed in 1978. MOE's studies were continued after the industrialisation had taken place.

The combustion of coal, the refining of oil, and the manufacture of steel produce emissions including oxides of sulphur and nitrogen, fluoride, and a large number of metals. There is particular concern for precursors of acidic deposition, for ozone formation, and for the contamination of forage for local beef and dairy herds with fluoride and metals.

The study provides a model of a planned pre- and postindustrialisation study, incorporating both chemical and biological indicators and monitors of potential contaminants related to the development. The possibility of preexisting (regional) ozone problems in the area, as well as the complication of three "new" potential air pollution sources, required careful sample site design and premonitoring studies. A total of 32 study locations were established on five radial transects from the Nanticoke Generating Station (GS) location.

The objectives were

1. The determination of concentrations of fluoride, sulphur, and other elements in foliage before and after the industries became operational;
2. The determination of the endemic occurrence of insect, disease, oxidant, and physiological injury on vegetation;
3. The establishment of permanent sample plots of selected seedling trees and clonal ramets for annual observation and measurement;
4. The establishment of permanent sample plots comprised of existing wood-lots and/or plantation trees for the purpose of annual observation and measurement;
5. The establishment of sample plots on which annual indicator plant species are grown for observation, measurement, and sampling;
6. Annual vegetation surveillance and sampling to be conducted at all permanent sample plot locations;
7. The establishment of a network of sulphation and fluoridation candle stations to determine background ambient air quality levels;
8. The correlation of air sampling data recorded by Ontario Hydro and the Ministry of the Environment with vegetation and soil data (MOE, 1976).

For the purposes of the present case, only objectives 1, 3, and 5 are addressed.

In 1978, the GS, with a capacity of 1,000 MW of electricity, was complete. The refinery, capable of producing 160,000 barrels of oil products per day, also came on stream in 1978. The steel plant was not completed until 1980.

Contaminants in foliage

Between preoperating measurements and the 1979 operating year, no increases in foliar concentrations of iron, arsenic, selenium, chlorine, or fluoride were found in native vegetation (MOE, 1981).

Forage plants

McLaughlin (1986) reported on chemicals in forage in the vicinity of the Nanticoke industrial complex, from 1978 to 1985. He compared the values for the various sample sites to those from a database known as the Phytotoxicology Upper Limits of Normal. This value is derived by taking the average value of all background or control forage samples (i.e., taking into account the normal variability) and adding three standard deviations. "Statistically this represents 99% of all natural variation within the sampled population. Concentrations above these can usually be associated with an extraneous contamination source" (McLaughlin, 1986).

From 1978 to 1985, the MOE Phytotoxicology sampling showed that "the vast majority of forage consumed by local cattle has chemical concentrations well within normal background levels" (McLaughlin, 1986). Chlorine and sulphur slightly exceeded the Phytotoxicology upper limit of background, but the sulphur excesses tended to occur most frequently at the sampling locations most distant from the industrial sources. In terms of toxicity to cattle, no appropriate references were found for Cl and S. Iron was the only element that showed a consistent gradient of concentration with distance, and its source was clearly the steel mill. However, in terms of acceptability for cattle forage, the provincial agriculture and food ministry concluded that the iron concentrations presented no problem.

Permanent plots

Indicator species of trees were transplanted for perennial (permanent) plots. A fluoride-sensitive type of white pine (*Pinus strobus* var.) was brought from Cornwall Island, Ontario, a sulphur-dioxide-sensitive white birch (*Betula papyrifera*) was brought from Sudbury, Ontario, and for an ozone indicator, a local common ash (*Fraxinus* sp.) was planted.

These plots were set up in 1971, and the first comprehensive examination was made of randomly tagged trees in 1977. Visual examinations were repeated in 1978 and 1979. There was no evidence of fluoride or sulphur dioxide damage. Ozone damage was detected, but the occurrence was scattered and not related to distance from the source.

The annual indicator plot programme

Five locations, aligned along a line northeast (downwind of the GS) to southwest (upwind of the GS), were selected for indicator plots. At each of these five plots, the following pollutant-specific plants were established: Bel W 3 tobacco for ozone, alfalfa (both sensitive and resistant varieties) and blackberry (Lowden) for sulphur dioxide, and gladiolus (Snow Princess) for fluoride.

Results were provided for 1979, covering at least one full season of operation of the GS and the oil refinery, although these may not have been running at full capacity for all the time.

During 1979, ozone damage was greatest in the plots upwind of the GS, which was in agreement with the monthly chemical measurement of ozone. The ozone concentrations generally declined with distance from the north shore of Lake Erie, suggesting that ozone precursors carried from the southwest were most abundant at the lakeshore.

The SO_2 indicator plants were examined weekly and showed no evidence of damage.

The gladiolus plants showed no visible fluoride injury, and this portion of the indicator study was terminated in September 1979. Leaves were collected for fluoride determination and results showed no significant difference between the preoperation dates in 1978 and during operation in 1979. All values were very low (MOE, 1981).

In conclusion, the results presented in the 1981 and 1985 reports, as well as the other components of the Nanticoke study, which are not discussed here but are shown in the list of objectives, demonstrated that to date, for the new industrial complex at Nanticoke, very little measurable impact has resulted from emissions. Ozone is the only air pollutant that reached phytotoxic levels, and the ozone was not related to the new industries. Studies continued until 1985, at which time there were still no demonstrable impacts, and the MOE programme was scaled down. Chemical monitoring is still required for the major air pollutants to comply with air quality standards.

Table 4.19. Annual nitrogen and phosphorus inputs to two tributaries of the Chesapeake Bay[a]

Location	Time period	Annual nutrient loading	
		Total nitrogen load ($kg\,N \times 10^6$/yr)	Total phosphorus load ($kg\,P \times 10^6$/yr)
Patuxent River	Pre-European	0.37	0.01
	1963	0.91	0.17
	1969 71	1.11	0.25
	1978	1.55	0.42
	1985–86	1.73	0.21
Potomac River	Pre-European	4.6	0.12
	1913	18.6	0.91
	1954	22.6	2.04
	1969 71	25.2	5.38
	1977–78	32.8	251
	1985–86	32.1	3.35
	1985–86	35.5	2.93

[a] Loading rates for the pre-European period (prior to 1600) were made by using nitrogen and phosphorus release rates from mature forests not exposed to significant atmospheric deposition of nitrogen and phosphorus; estimates for other periods were based primarily on direct measurements (Boynton, 1997).

The utility of plants, ideally in conjunction with other measures, as biological indicators and monitors of air pollution was also upheld by this programme.

Case Study 4.5. Chesapeake Bay: A study of eutrophication and complex trophic interactions

Formed from the flooded river valley of the Susquehanna River, Chesapeake Bay on the east coast of the United States is the largest and most productive estuary in the country. Increasing fertilisation of the bay from sewage and agriculture began soon after European settlement in the early 1600s and accelerated during the 1950s. Table 4.19 shows estimates of annual nitrogen and phosphorus inputs to two Chesapeake Bay tributaries from pre-1600 to the present. The ratios of diffuse (nonpoint) sources to point sources for these nutrients are approximately 1:1 for the Patuxent River and 3:2 for the Potomac River. Atmospheric wet deposition accounts for approximately 12–15% of N input to most of Chesapeake Bay.

Associated changes in plant communities accelerated during the late 1950s and early 1960s. These changes were characterised by a decline in submerged aquatic vegetation (SAV) in shallow water. This decline was attributed to an increase in the turbidity of water resulting from excess phytoplankton accompanied by inorganic particulate loading from land-clearing activity. Increased phytoplankton production was characterised by both algal blooms in the water body and the promotion of algal growth on the leaves of submerged aquatic macrophytes. During the same period, hypoxic ($<2\,mg\,O_2\,L^{-1}$) and anoxic conditions in the central channel of the bay expanded both spatially and temporally during the summer months, driven by the bacterial decomposition of excess phytoplankton. Anoxia was increased during years of high freshwater runoff, which sharpened the stratification of the bay and increased the stagnation of bottom water. Anoxic conditions in sediments produced reducing conditions during the summer, which accelerated P release. Maximum N input relative to P occurred during the spring runoff when phytoplankton were P-limited. Increased P release from sediment later in the summer coincided with a period when phytoplankton were N-limited.

In addition to the influence of increased nutrient input, excess phytoplankton production has been exacerbated by a steady decline in Chesapeake Bay of the eastern oyster *Crassostrea virginica*. Formerly a dominant species (and fishery) in the bay, it has been estimated that standing stocks of 100 years ago were capable of filtering the whole volume of the bay within a few days. In accelerated decline since 1981, the current population is now a remnant of that which existed a few decades ago. Remedial measures for the fishery are currently a matter of vigorous debate.

Decline in submerged aquatic vegetation has been attributed to an increase in the turbidity of water resulting from excess phytoplankton accompanied by inorganic particulate loading from land-clearing activity. Nutrient enrichment also promotes the growth of algae on the leaves of submerged vegetation. During the 1970s, at least ten species of submerged aquatic plant were almost completely eliminated from bay estuaries along with refuge habitat for a variety of fish, invertebrate, and waterfowl populations.

The role of deteriorating water quality on declining Chesapeake Bay fisheries is unknown but is likely to be secondary to loss of habitat and fishing pressure. Sharp declines in stocks of anadromous species such as the American shad (*Alosa sapidissima*) and the striped bass (*Morone saxatilis*) have raised questions about possible chemical or acid rain contamination of upstream spawning habitats. The recent apparent rebound of the Chesapeake Bay striped bass (rockfish) catch, following a 1984 ban and then severe restrictions on the fishery, suggests that fishing pressure was a major contributing factor in this case. For shad, loss of habitat through the building of dams on the Susquehanna River has been heavily implicated, although both species have demonstrated considerable sensitivity to acid rainfall, which is a feature of precipitation in the Chesapeake Bay watershed.

Atmospheric studies have indicated significant aerial input of PAHs and nitrogen from NO_x to the Chesapeake Bay watershed. The latter implies that, unlike phosphate, nitrogen input to the system may be only marginally controllable through sewage treatment. States adjacent to the Chesapeake Bay watershed have set goals for a 40% reduction in nutrient loading to the Bay by 2005, although these goals appear difficult to meet. Localised outbreaks of *Pfiesteria* (toxic dinoflagellate) infestation in the bay have refocused attention on nonpoint source nutrient (mainly phosphate) enrichment from agricultural practices such as chicken farming.

Case Study 4.6. The use of lentic mesocosms in toxicity testing

Under the Federal Insecticide, Fungicide, and Rodenticide Act (FIFRA) the U.S. Environmental Protection Agency has conducted tests using lentic freshwater mesocosms to determine the effects of pesticides on zooplankton, macroinvertebrate, or fish populations during relatively short periods of time, or on whole ecosystems over longer periods of time. These mesocosms are artificially constructed clay-lined or concrete ponds ranging in size from 0.01 to 0.1 ha, in volume from 100 to 1,000 m³, and in depth from 1 to 2 m. The majority of these studies have been conducted for pesticide registration and have mainly focused on either pyrethroid or organophosphate pesticides, although noninsecticide chemicals such as fluorine, cadmium, and phenols have also been studied (Graney et al., 1995).

Esfenvalerate (EV), one of the most toxic pyrethroid insecticides, had significant adverse effects on bluegill sunfish (*Lepomis macrochirus*), zooplankton, and macroinvertebrates during a 2-week exposure in clay-lined mesocosms (Fairchild et al., 1992). In this study, bluegill sunfish were sensitive to EV at concentrations between $0.25 < x \leqslant 0.67 \,\mu g\,L^{-1}$. Zooplankton and macroinvertebrates were sensitive to EV at concentrations $\leqslant 0.25 \,\mu g\,L^{-1}$. Depression of cladoceran and copepod populations, however, led to relaxation of predation and competitive pressure on rotifers, which increased in numbers. Studies on the carbamate insecticide carbaryl indicated an initial reduction in mesocosm zooplankton populations followed by a recovery; also the extent of recovery was dependent on the timing of the carbaryl application. Addition of azinphos-methyl organophosphate pesticide to lentic mesocosms caused an increase in copepods, which was speculated to result from decreased predation due to insecticide-induced mortality of bluegill sunfish in the test system (Giddings et al., 1993). Indirect effects such as this demonstrate that outdoor mesocosms are useful tools for ecological assessment of pesticides, even though multiple exposure regimes are critical to the interpretation of results. However, the large and expensive effort required to monitor such systems effectively has prompted many to question how big mesocosms have to be to provide a realistic simulation of the natural environment (Crossland and LaPoint, 1992). Johnson et al. (1993) and Morris et al. (1993) evaluated the ecological impact and fate of Baythroid (cyfluthrin) insecticide using mesocosms (635 m³) and microcosms (2 m³). Johnson et al. (1993) found that microcosms yielded higher densities of some taxonomic groups, whereas other groups were more abundant in mesocosms (Table 4.19). Members of the Trichoptera and Tanypodinae were equally represented in both types of system.

Case Study 4.7. The cadmium spike experiment, Experimental Lakes Area

Long-range transport of atmospheric pollutants (LRTAP) can result in contamination of ecosystems at considerable distances from the original sources of the chemicals. Early observations and research on LRTAP emphasised acid precursors, but it soon became clear that other substances including volatile organic substances and metals were also implicated. Cadmium is a relatively volatile metal, which is transported in the atmosphere. It is mobilised by anthropogenic activities such as coal burning and smelting, indeed the ratio of anthropogenic to natural emissions is in the order of 7 : 1. Cadmium is toxic to aquatic life at low ($ng\,L^{-1}$) concentrations, while its toxicity to humans is much less, although it is suspected of being a carcinogen. In any case, drinking water guidelines are in the microgram-per-litre levels, which are unlikely to protect aquatic life.

This metal was chosen for a whole lake addition study at the Experimental Lakes Area of northwestern Ontario. The objectives of the study were

1. To assess the adequacy of the Canadian Water Quality Guideline for the protection of aquatic life of $0.2\,\mu g\,L^{-1}$;
2. To assess the effects of cadmium at water concentrations less than $0.2\,\mu g\,L^{-1}$;
3. To follow the fate, distribution, and routes of uptake of cadmium into the aquatic system;
4. To observe the importance of the sediments as a sink for cadmium;
5. To determine and integrate the biochemical, cellular, and whole organism responses to cadmium;
6. To identify some responses that might serve as early warning indicators of impending ecological damage; and
7. To assess the predictive value of results from small-scale, laboratory, micro-cosm, or mesocosm experiments against the response of individuals and populations exposed in natural ecosystems.

The lake selected for the experiment, known as Lake 382, is a small soft water lake, of hardness $7.0\,mg\,L^{-1}$ as $CaCO_3$ and calcium $2.2\,mg\,L^{-1}$, with a relatively large population of lake trout for lakes in this area.

Beginning in June 1987, cadmium as chloride labelled with ^{109}Cd was added to the epilimnion of the lake. Target concentrations for water were 100, 100, 120 and $160\,ng\,L^{-1}$, respectively, for 1987, 1988, 1989, and 1990 and for at least four further years was $200\,ng\,L^{-1}$, this being the Canadian Water Quality Guideline for the protection of life in soft waters. Dosing was initially done manually twice per week with increasing attempts to avoid great fluctuations of cadmium in water through space and time. From 1989 to 1991, an electronically timed injection pump was used for safe, virtually continuous addition (Holoka and Hunt, 1996).

Annual loadings of cadmium ranged from 641 to 1,546 g cadmium per year (Lawrence et al., 1996). In addition to the cadmium added for the experiment, there was some input from atmospheric deposition and runoff to the lake during the course of the experiment, calculated to total $46\,g\,yr^{-1}$. Compared with the approximately 1 $kg\,yr^{-1}$ of cadmium added from the experiment, these other sources were deemed to be negligible. Values for cadmium in water close to the target values were reached in 9 to 51 days after addition was started in each year and were maintained for 85–153 days each year from 1987 to 1992.

Prior to making additions of cadmium to the lake, a finite numerical difference model was used to estimate the concentrations of cadmium in the epilimnion that would result from various additions (Figure 4.24). The model was based on previous observations of the fate of radiotracers in ELA lakes. It assumed that dissolved cadmium partitions quickly between dissolved and particulate phases. Dissolved cadmium can be transferred to the sediment via the boundary layer where it can be absorbed and distributed between pore water and solid phase. Particulate cadmium can settle to the bottom sediment. Sediment cadmium can mix with deeper layers. Both particulate and dissolved forms of cadmium can leave via the lake outflow (Lawrence et al., 1996).

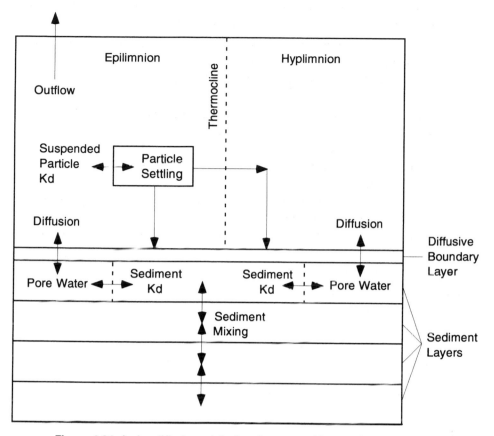

Figure 4.24 A simplified model of pathways and interactions of a substance introduced into a lake.

Use of the radioactive [109]Cd label facilitated tracing the fate and accumulation of the added metal because many analyses could be performed on biotic and environmental samples economically. This isotope has a short half-life. Over the course of the experiment, cadmium concentrations were measured in water, in sediment cores, and in biota, including tissue distribution, and the effects of cadmium on algal production, phytoplankton diversity and community structure, zooplankton community structure, mussel physiology, and fish (lake trout and white sucker) structure and function were determined.

The results for the physical/chemical behaviour of cadmium agreed with the model in that cadmium moved quickly from the water column to the bottom sediments and so continuous addition of cadmium was necessary to maintain the target concentrations in the epilimnion. In addition to sedimentation of Cd-laden particles, other factors that influenced the epilimnetic concentrations of cadmium included seasonal changes in the depth of the epilimnion – rainfall and outflow. However, for this particular lake, 99% of the water column cadmium was "lost" within a year of the addition; binding to suspended particulates and incorporation into the sediments were the major mechanisms for this loss from the water column. Even though cadmium tends to associate with organic and sulphide fractions of sediment, which would immobilise the metal, sediment diagenesis is expected to result in the release of the more mobile forms of cadmium, which means that cadmium associated with the sediments could be recycled into the water column (Stephenson et al., 1996).

Cadmium accumulated in organisms from all the trophic level that were examined, up to 30 to 40 times background level for some organisms. For smaller organisms, the cadmium concentrations tended to stabilise quite early in the course of the experiment, whereas long-lived organisms such as large fish and mussels continued to accumulate in 1992, the last year of addition (Malley, 1996).

Zooplankton represent the freshwater community most sensitive to cadmium. Effects had been anticipated at $200\,ngL^{-1}$, based on the results of in situ experiments that were carried out by the research group prior to the whole lake additions. However, in the treated lake, there were no obvious changes in species composition or declines in populations. Nor was there any evidence of adverse effects on the phytoplankton community of the lake (Findlay et al., 1996). Lake trout and white sucker showed no adverse effects either, but the cadmium concentrations in the posterior kidney of the lake trout had reached $9.5\,\mu g\,g^{-1}$ (wet weight) by 1992. This is approaching the value of $15\,\mu g\,g^{-1}$, which is associated with histopathological lesions that may lead to decline of kidney function.

Metallothionein can be used as a biochemical indicator of exposure to cadmium (see Section 4.4.2). White suckers from the treated lake had significantly higher MT than did fish from a nearby reference lake, and after 3 years of treatment, mussels in Lake 382 had two to four times more MT in their body parts than those from a reference lake. These kinds of subtle chronic effects are unlikely to show up in any other type of experiment. They may show up in field surveys, but cause-effect relationships are notoriously difficult to establish under uncontrolled conditions.

Scientific and technical components are only part of most environmental issues. The inclusion of socioeconomic and political components is quite familiar to toxicologists, although one normally associates these with real-world problems where there are multiple stakeholders, rather than with experimental research.

As described in the introductory paragraph of this case study, additions of cadmium were to have continued for at least 8 years. It should be evident from the account to date that this cadmium spike experiment was a major, complex undertaking, involving strong senior scientific management and large numbers of highly trained personnel and requiring substantial funding. Not only funding, but continuity of funding needs to be in place in order to meet the planned objectives of assessing the long-term effects of small but continuous contamination by the selected metal.

The experiment was begun under an agreement between the Canadian Federal Department of Fisheries and Oceans (DFO) and the Ontario provincial Ministry of Natural Resources (OMNR) who owns the ELA lands. The completion of the plan was thwarted through a series of events. In April 1992, prior to the start of the sixth year of additions, the Ontario Environment Ministry (then the Ministry of Environment and Energy) released a report "Candidate Substance List for Ban or Phase-outs" as part of

Ontario's commitment to the Great Lakes Water Quality Agreement to reduce substances that are persistent, subject to bioaccumulation, and toxic. Cadmium at this point was on a secondary list of candidates. In 1992, it was moved to the primary list (Malley, 1996). The 1992 additions of cadmium to Lake 382 were interrupted a day after they began. One might consider that this would add a complication to the already difficult task of bringing the cadmium to the planned level and maintaining it there. Work was allowed to proceed about 3 weeks later, after three favourable peer reviews of the experimental design and approach. But the Ontario government allowed the experiment to proceed for 1 year only.

Cadmium was now on a list of substances that could not be approved for addition to whole lake ecosystems. Yet the experiment, still incomplete, was designed to elucidate the behaviour, fate, and effects of low concentrations of cadmium on an aquatic ecosystem, goals which one would consider highly relevant to regulatory agencies. Even with the "consensus among the researchers that further Cd additions to Lake 283 were scientifically desirable to meet the goals of the experiment" (Malley, 1996), federal and provincial (Ontario) ministries would not approve continuation of the experiment.

The long-term goal of assessing the accumulation of cadmium by long-lived organisms was no longer feasible. The last year of funding for the experiment was 1996. After the last additions of cadmium, which were made in 1992, the researchers had to revise their plan and emphasise the dynamics of the existing cadmium, including the possible release and redistribution of cadmium from the sediments.

Despite its untimely termination, the whole lake experiment contributed to the important result that Water Quality Guidelines are not the appropriate regulatory vehicle for controlling persistent, sediment-seeking contaminants, such as cadmium, that continuously leave the water column and accumulate in the sediments. However, porewater metal concentrations may be compared to Water Quality Guidelines, as is the current U.S. EPA approach.

4.11 Questions

1. Identify the main approaches that are used in classical toxicology. Indicate reasons why these approaches do not meet all the requirements for environmental toxicology.

2. List the approaches that have been, or are being, developed to investigate the effects of contaminants on ecosystems or components of ecosystems.

3. For each of the approaches that were listed in Question 2, summarize the advantages and disadvantages. For each approach, indicate the situations where it would be most appropriate.

4. Critically evaluate top-down versus bottom-up methods that have been used to determine the effect of contaminants at the population and community level. Give an outline of a Leslie Matrix model. Describe how you

would collect input data for such a model in a contaminated field environment.

5. Provide definitions of *biological indicators* and *biological monitors* as these terms are used in the present text. Compare and contrast the respective properties of indicators and monitors as used in the present text.

6. "The biochemical indicator (biomarker) is the ideal indicator for providing an inexpensive, rapid and sensitive response of a living system to a contaminant". Discuss this statement critically.

7. An estuarine environment has been contaminated by a mixture of PAHs and PCBs. What kind of chemical and biological measurements would you make to determine the threat posed by these contaminants on (a) the estuarine ecosystem and (b) any food resources that may be related to that ecosystem?

8. Certain species or populations of a species of organisms are said to show tolerance to certain contaminants. Explain (a) the scientific definition(s) of tolerance; (b) the possible mechanism(s) by which tolerance arises; (c) the possible or demonstrated mechanisms of tolerance for named contaminants or groups of contaminants; (d) the significance of tolerance in the context of biological indicators and monitors.

9. List the potential applications of modelling in environmental toxicology.

10. Prepare a conceptual model for the fate and effect(s) of a hydrophobic organic substance released into the water for an aquatic ecosystem. Include at least six labelled compartments.

11. Illustrate a model for the various processes in which the chemical is involved in the course of the exposure of a fish in water to a persistent organic contaminant. You should include uptake from water via the gills, uptake from food, loss via the gills, loss by egestion, loss by metabolism, and dilution of tissue concentration through growth of the organism.

12. As a research project, quantify this model using fugacity equations.

13. Tabulate the potential advantages and disadvantages of the application of models in environmental toxicology.

14. Describe a case study of a study of an investigation of the effect(s) of one or more toxic substances in a named type of environment that used a whole system manipulation approach. Critically appraise the approach and suggest additional or alternative approaches for study of the same problem.

4.12 References

Antonovics, J. (1975) Metal Tolerance in Plants: Perfecting an Evolutionary Paradigm. In *International Conference on Heavy Metals in the Environment*, vol. II, ed. Hutchinson, T. C. pp. 169–86, Toronto, Ontario.

Aoki, Y., Hatakyeama, S., Kobayashi, N., Sumi, Y., Suzuki, T., and Suzuki, K. T. (1989) Comparison of cadmium-binding protein induction among mayfly larvae of heavy metal resistant (*Baetis thermicus*) and susceptible species (*B. yoshinensis* and *B. sahoensis*), *Comparative Biochemistry and Physiology*, **93C**, 435–57.

Barnthouse, L. W., Suter, III, G. W., and Rosen, A. E. (1989) Inferring population-level significance from individual-level effects: An extrapolation from fisheries science to ecotoxicology. In *Aquatic Toxicology and Environmental Fate*, ASTM STP 1007, eds. Suter, III, G. W., and Lewis, M. A., pp. 289–300, American Society for Testing and Materials, Philadelphia.

Baron, J. (1986) Sediment diatom and metal stratigraphy from Rocky Mountain lakes with special reference to atmospheric deposition, *Canadian Journal of Fisheries and Aquatic Sciences*, **43**, 1350–62.

Battarbee, R. W., Allott, T. E. H., Juggins, S., Kreiser, A. M., Curtis, C., and Harriman, R. (1996) Critical loads of acidity to surface waters: an empirical diatom-based palaeolimnological model, *Ambio*, **25**, 366–9.

Bayne, B. L., Brown, D. A., Burns, D. A., Burns, K., Dixon, D. R., Ivanovici, A., Livingstone, D. R., Lowe, D. M., Moore, M. N., Stebbing, A. R. D., and Widdows, J. (1985) *The Effects of Stress and Pollution on Marine Animals*, Praeger Scientific, New York.

Beak International, Inc. (1998) Caracterisation des sediments au site de Mont-Louis. Rapport d'analyse et d'interpretation. Version Preliminaire Vol. I.

Bechmann, R. K. (1994) Use of life tables and LC_{50} tests to evaluate chronic and acute toxicity effects of copper on the marine copepod *Tisbe furcata* (Baird), *Environmental Toxicology and Chemistry*, **13**, 1509–17.

Berg, T., Royset, O., and Steinnes, E. (1995) Moss (*Hylocomium splendens*) used as biomonitor of atmospheric trace element deposition: Estimation of uptake efficiencies, *Atmospheric Environment*, **29**, 353–60.

Blanck, H., Wängberg, S.-Å., and Molander, S. (1988) Pollution Induced Community Tolerance (PICT) – A New Ecotoxicological Tool. In *Functional Testing of Aquatic Biota for Estimating Hazards of Chemicals*, eds. Carins, J. J., and Pratt, J., pp. 219–30, American Society for Testing Materials, Philadelphia.

Blom, A., Harder, W., and Matin, A. (1992) Unique and overlapping pollutant stress proteins of *Escherichia coli*, *Applied and Environmental Microbiology*, **58**, 331–4.

Boynton, W. R. (1997) Estuarine Ecosystem Issues on the Chesapeake Bay. In *Ecosystem Function and Human Activities*, eds. Simpson, R. D., and Christense, J. N. L., pp. 71–93, Chapman and Hall, New York.

Brown, B. E. (1977) Uptake of copper and lead by a metal-tolerant isopod *Asellus meridianus* Rac, *Freshwater Biology*, **7**, 235–44.

Burns, L. A., Cline, D. M., and Lassiter, R. R. (1981) *Exposure Analysis and Modelling System (EXAMS): User Manual and System Documentation*. U.S. Environmental Protection Agency, Athens, GA.

Butler, M., Haaskew, A. E. J., and Young, M. M. (1980) Copper tolerance in the green alga, *Chlorella vulgaris*, *Plant Cell and Environment*, **3**, 119–26.

Cairns, J. J., Niederlehner, B. R., and Pratt, J. R. (1990) Evaluation of joint toxicity of chlorine and ammonia to aquatic communities, *Aquatic Toxicology*, **16**, 87–100.

Campbell, P. G. C., Tessier, A., Bisson, M., and Bougie, R. (1985) Accumulation of copper and zinc in the yellow water lily *Nuphar variegatum*: Relationships to metal partitioning in the adjacent lake sediments, *Canadian Journal of Fisheries and Aquatic Sciences*, **29**, 729–45.

Camusso, M. (1994) Use of freshwater mussel *Dreissena polymorpha* to access trace metal pollution in the lower River Po (Italy), *Chemosphere*, **29**, 729–45.

Carignan, J. (1995) Isotopic composition of epiphytic lichens as a tracer of the sources of atmospheric lead emissions in southern Quebec, Canada, *Geochimica et Cosmochimica Acta*, **59**, 4427–33.

Chapman, G. A. (1983) Do Organisms in Laboratory Toxicity Tests Respond Like Organisms in Nature? In *Aquatic Toxicology and Hazard Assessment: Sixth Symposium*, ASTM STP 802, eds. Bishop, W. E., Cardwell, R. D., and Heidolph, B. B., pp. 315–27, American Society for Testing and Materials, Philadelphia.

Coelho, G. M., Aurand, D. V., and Wright, D. A. (1999) Biological uptake analysis of organisms exposed to oil and chemically dispersed oil. Proceedings of the 22nd Arctic and Marine Oil Spill Programs Technical Seminar, Environment Canada, Ottawa, pp. 685–94.

Collier, T. K., and Varanasi, U. (1991) Hepatic activities of xenobiotic metabolizing enzymes and biliary levels of xenobiotics in English sole (*Parophrys vetulus*) exposed to environmental contaminants, *Archives of Environmental Contamination and Toxicology*, **20**, 462–73.

Courtemanch, D. L., and Davies, S. P. (1987) A coefficient of community loss to assess detrimental change in aquatic communities, *Water Research*, **21**, 217–22.

Cowan, C. E., Mackay, D., Feijtel, T. C. J., van der Meent, D., DiGuardo, A., and Mckay, N. (1995) *The Multimedia Fate Model: A Vital Tool for Predicting the Fate of Chemicals*, SETAC Press, Pensacola, FL.

Crawford, H. S., and Titterington, R. W. (1979) Effects of silvicultural practices on bird communities in upland spruce fir stands. In *Management of Northeastern and North Central Forests for Non-Gambirds*, U.S. Forest Service Technical Report No. NC-51, Washington, DC.

Crossland, N. O., and LaPoint, T. W. (1992) The design of mesocosm experiments, *Environmental Toxicology and Chemistry*, **11**, 1–4.

Dillon, P. J., and Rigler, F. H. (1974) The phosphorus-chlorophyll relationship in lakes, *Limnology and Oceanography*, **19**, 763–73.

DiToro, D. M., O'Connor, D. J., Thomas, R. V., and St. John, J. P. (1981) Simplified analysis for fate of chemicals in receiving waters. Water Pollution Control Federation 54th Annual Conference, Detroit, October 1981.

Dixit, S. S., Smol, J. P., Kingston, J. P., and Charles, D. F. (1992) Diatoms: Powerful indicators of environmental change, *Environmental Science and Technology*, **26**, 22–33.

Dufrene, M., and Legendre, P. (1997) Species assemblages and indicator species: The need for a flexible asymmetrical approach, *Ecological Monographs*, **67**, 354–66.

Edwards, C. J., and Ryder, R. A. (eds.) (1990) Biological surrogates of mesotrophic ecosystem health in the Laurentian Great Lakes. Report to the Great Lakes Science Advisory Board. Windsor, Ontario.

Elton, C. (1927) *Animal Ecology*, Sidgwick and Jackson, London.

Emlen, J. M., and Pikitch, E. K. (1989) Animal population dynamics: Identification of critical components, *Ecological Modelling*, **44**, 253–73.

Fahselt, D., Wu, T.-W., and Mott, B. (1995) Trace element patterns in lichens following uranium mine closures, *The Bryologist*, **98**, 228–34.

Fairchild, J. R., La Point, T. W., Zajicek, J. L., Nelson, M. K., Dwyer, F. J., and Lovely, P. A. (1992) Population-, community- and ecosystem-level responses of aquatic mesocosms to pulsed doses of a pyrethroid insecticide, *Environmental Toxicology and Chemistry*, **11**, 115–29.

Findlay, D. L., Kasian, S. E. M., and Schindler, E. U. (1996) Long-term effects of low cadmium concentrations on a natural phytoplankton community, *Canadian Journal of Fisheries and Aquatic Sciences*, **53**, 1903–12.

Findlay, S., Likens, G. E., Hedin, L., Fisher, S. G., and McDowell, W. H. (1997) Organic matter dynamics in Bear Brook, Hubbard Brook Experimental Forest, New Hampshire, USA, *Journal of the North American Benthological Society*, **16** (1), 43–6.

Fischer, R. (1994) Simulated or Actual Field Testing: A Comparison. In *Aquatic Mesocosm Studies in Ecological Risk Assessment*, SETAC Special Publication Series, eds. Graney, R. L., Kennedy, J. H., and Rodgers, J. H., pp. 35–46, Lewis, Boca Raton, Ann Arbor, London, Tokyo.

Ford, J., Landers, D., Kugler, D., Lasorsa, B., Allen-Gil, S., Crecelius, E., and Martinson, J. (1995) Inorganic contaminants in Arctic Alaskan ecosystems: Long-range atmospheric transport or local point sources?, *Science of the Total Environment*, **160/161**, 323–35.

Foster, P. L. (1982) Metal resistances of Chlorophyta from rivers polluted by heavy metals, *Freshwater Biology*, **12**, 41–61.

Games, L. M. (1983) Practical applications and comparisons of environmental exposure assessment models. In *Aquatic Toxicology and Hazard Assessment: Sixth Symposium*, ASTM STP 802, eds. Bishop, W. E., Cardwell, R. D., and Heidolph, B. B., pp. 282–99, American Society for Testing and Materials, Philadelphia.

Gawel, J. E., Ahner, B. A., Friedlan, A. J., and Morel, F. M, M (1996) Role for heavy metals in forest decline indicated by phytochelatin measurements, *Nature*, **381**, 64–5.

George, S. G. (1983) Heavy metal detoxication in the mussel *Mytilus edulis*: Composition of Cd-containing kidney granules (tertiary lysozomes), *Comparative Biochemistry and Physiology*, **76C**, 53–7.

Giddings, J. M., Biever, R. C., Helm, R. L., Howick, G. L., and deNoyelles, J. F. J. (1993) The Fate and Effects of Guthion (Azinphos Methyl) in Mesocosms. In *Aquatic Mesocosm Studies in Ecological Risk Assessment*, eds. Graney, R. L., Kennedy, J. H., and Rodgers, J. H., pp. 369–496, Lewis, Boca Raton, FL.

Goering, P. L., Fisher, B. R., and Kish, C. L. (1993) Stress protein synthesis induced in rat liver by cadmium precedes hepatotoxicity, *Toxicology and Applied Pharmacology*, **122**, 139–48.

Goldberg, E. D., Bowen, V. T., Farrington, J. W., Harvey, G., Martin, J. H., Parker, P. L., Risebrough, R. W., Robertson, W., Schneider, E., and Gamble, E. (1978) The mussel watch, *Environmental Conservation*, **5**, 101–25.

Gonzalez, H., Lodenius, M., and Otero, M. (1989) Water hyacinth as indicator of heavy metal pollution in the tropics, *Bulletin of Environmental Contamination and Toxicology*, **43**, 910–14.

Graney, R. L., Giesy, J. P., and Clark, J. R. (1995) Field Studies. In *Fundamentals of Aquatic Toxicology*, ed. Rand, G. M., pp. 257–305, Taylor and Francis, Washington, DC.

Gray, J. S., and Mizra, F. B. (1979) A possible method for the detection of pollution-induced disturbance on marine benthic communities, *Marine Pollution Bulletin*, **10**, 142–6.

Grue, C. E., Hart, A. D. M., and Mineau, P. M. (1991) Biological consequences of depressed brain cholinesterase activity in wildlife. In *Cholinesterase-Inhibiting Insecticides*, ed. Mineau, P., pp. 152–209, Elsevier, New York.

Hall, R. J., Likens, G. E., Fiance, S. B., and Hendrey, G. R. (1980) Experimental acidification of a stream in the Hubbard Brook experimental forest, New Hampshire, *Ecology*, **61**, 976–89.

Hare, L. (1992) Aquatic insects and trace metals: Bioavailability, bioaccumulation and toxicity, *Critical Reviews in Toxicology*, **22**, 327–69.

Harris, R. C. (1991) A mechanistic model to examine mercury trends in aquatic systems. Master of Civil Engineering Thesis, McMaster University, Hamilton, Ontario.

Hawksworth, D. L., and Rose, F. (1970) Qualitative scale for estimating sulphur dioxide pollution in England and Wales using epiphytic lichens, *Nature (London)*, **227**, 145–8.

Henderson-Sellers, A., and Seaward, M. R. D. (1979) Monitoring lichen reinvasion of ameliorating environments, *Environmental Pollution*, **19**, 207–13.

Henrikson, A., Dickson, W., and Brakke, D. F. (1988) Estimates of Critical Loads to Surface Waters. In *Critical Loads for Sulphur and Nitrogen*, ed. Nilsson, J., pp. 87–120, Nordic Council of Ministers, Copenhagen.

Hernandez, L. M., Gonzalez, M. J., and Fernandez, M. A. (1988) Organochlorines and metals in Spanish Imperial Eagle eggs, 1986–87, *Environmental Conservation*, **15**, 363–4.

Hill, M. O. (1979) *TWINSPAN: A Fortran Program for Arranging Multivariate Data in an Ordered Two-Way Table by Classification of the Individuals and Attributes*, Cornell University, New York.

Holoka, M. H., and Hunt, R. V. (1996) Automated addition of toxicant to a whole lake, *Canadian Journal of Fisheries and Aquatic Sciences*, **53**, 1871–5.

Hutchinson, T. C., Watmough, S. A., Sager, E. P. S., and Karagatzides, J. D. (1998) The impact of simulated acid rain and fertilizer application on a mature sugar maple (*Acer saccharum* Marsh) forest in central Ontario Canada, *Water, Air, and Soil Pollution*, **109**, 17–39.

Jaagumagi, R. (1992) *Development of the Ontario Provincial Sediment Quality Guidelines for Arsenic, Cadmium, Chromium, Copper, Iron, Lead Manganese, Mercury, Nickel and Zinc*, Ontario Ministry of the Environment, Toronto.

James, M. O., and Little, P. J. (1981) Polyhalogenated biphenyls and phenobarbital: evaluation as inducers of drug metabolizing enymes in the sheepshead *Archosargus probatocephalus*, *Chemico-Biological Interactions*, **36**, 229–48.

Johnson, P. C., Kennedy, J. H., Morris, R. G., Hambleton, F. E., and Graney, R. L. (1993) Fate and Effects of Cyfluthrin (Pyrethroid Insecticide) in Pond Mesocosms and Concrete Microcosms. In *Aquatic Mesocosm Studies in Ecological Risk Assessment*, eds. Graney, R. L., Kennedy, J. H., and Rodgers, J. H., pp. 337–71, Lewis, Boca Raton, FL.

Jorgensen, S. E. (1990) Modelling Concepts, Chapter 2. In *Modelling in Ecotoxicology*, ed. Jorgensen, S. E., pp. 15–35, Elsevier, New York.

Jorgensen, S. E., Halling-Sorensen, B., and Nielsen, S. N. (1996) *Handbook of Environmental and Ecological Modelling*, Lewis, CRC Press, Boca Raton, FL.

Karr, J. R., Fausch, K. D., Angermeier, P. L., Yant, P. R., and Schlosser, I. J. (1986) *Assessing Biological Integrity in Running Waters: A Method and Its Rationale*, Special Publication No. 5, Illinois National History Survey, Champaign, IL.

Klaverkamp, J. F., and Duncan, D. A. (1987) Acclimation to cadmium toxicity by white suckers: Cadmium binding capacity and metal distribution in gill and liver cytosol, *Environmental Toxicology and Chemistry*, **6**, 275–89.

Klerks, P. L., and Weis, J. S. (1987) Genetic adaptation to heavy metals in aquatic organisms: A review, *Environmental Pollution*, **45**, 173–205.

Landis, W. G., and Yu, M.-H. (1999) *Introduction to Environmental Toxicology: Impacts of Chemicals upon Ecological Systems*, Lewis, Boca Raton, FL.

La Point, T. W. (1995) Signs and measurements of ecotoxicity in the aquatic environment. In *Handbook of Ecotoxicology*, eds. Hoffman, D. J., Rattner, B. A., Burtan, G. A., Jr., and Cairrs, J., Jr., Lewis, CRC Press, Boca Raton, FL.

La Point, T. W., and Fairchild, J. F. (1994) Use of Aquatic Mesocosm Data to Predict Effects in Aquatic Mesocosms: Limits to Interpretation. In *Aquatic Mesocosm Studies in Ecological Risk Assessment*, eds. Graney, R. L., Kennedy, J. H., and Rodgers, J. H., pp. 241–55, Lewis, Boca Raton, Ann Arbor, London, Tokyo.

Lawrence, S. G., Homolka, M. H., Hunt, R. V., and Hesslein, R. H. (1996) Multi-year experimental additions of cadmium to a lake epilimnion and resulting water column cadmium concentrations, *Canadian Journal of Fisheries and Aquatic Sciences*, **53**, 1876–87.

Lead, W. A., Steinnes, E., and Jones, K. C. (1996) Atmoshperic deposition of PCBs to moss (*Hylocomium splendens*) in Norway between 1977 and 1990, *Environmental Science and Technology*, **30**, 524–30.

Lean, D. R. S., and Charlton, M. N. (1976) A Study of Phosphorus Kinetics in a Lake Ecosystem. In *Environmental Biogeochemistry: Carbon, Nitrogen, Phosphorus, Sulphur and Selenium Cycles*, vol. 1, ed. Nriagu, J. O., pp. 283–94, Ann Arbor Science Publishers, Ann Arbor, MI.

LeBlanc, F., and Rao, D. N. (1973) Effects of sulphur dioxide on lichen and moss transplants, *Ecology*, **54**, 612–17.

LeBlanc, F., Rao, D. N., and Comeau, G. (1972) The epiphytic vegetation of *Populus balsamifera* and its significance as an air pollution indicator in Sudbury, Ontario, *Canadian Journal of Botany*, **50**, 519–28.

Likens, G. E. (1985) An experimental approach for the study of ecosystems, *Journal of Ecology*, **73**, 381–96.

Loez, C. R., Topalian, M. L., and Salibian, A. (1995) Effects of zinc on the structure and growth of a natural freshwater phytoplankton assemblage reared in the laboratory, *Environmental Pollution*, **88**, 275–81.

Luoma, S. N., Dagovitz, R., and Axtmann, E. (1990) Temporally intensive study of trace metals in sediments and bivalves from a large river-estuarine system: Suisun Bay/Delta in San Francisco Bay, *Science of the Total Environment*, **97/98**, 685–712.

Lydy, M. J., Bruner, K A., Fry, D. M., and Fisher, S. W. (1990) Effects of Sediment and the Route of Exposure on the Toxicity and Accumulation of Neutral Lipophilic and Moderately Water-Soluble Metabolizable Compounds in the Midge, *Chironomus riparius*. In *Aquatic Toxicology and Risk Assessment*, vol. 13, ASTM STP 1096, eds. Landis, W. G., and van der Schalie, W. H., pp. 140–64, American Society for Testing and Materials, Philadelphia.

Macfie, S. M., Tarmohamed, Y., and Welbourn, P. M. (1994) Effects of cadmium, cobalt, copper and nickel on growth of the green alga *Chlamydomonas reinhardtii*: The influence of the cell wall and pH, *Archives of Environmental Contamination and Toxicology*, **27**, 454–8.

Mackay, D. (1991) *Multimedia Environmental Models: The Fugacity Approach*, Lewis, Boca Raton, FL.

(1994) Fate Models. In *Handbook of Ecotoxicology*, ed. Calow, P., pp. 348-67, Blackwell Scientific Publications, Oxford.

Mackay, D., and Paterson, S. (1993) Mathematical Models of Transport and Fate. In *Ecological Risk Assessment*, ed. Suter, G. W., pp. 129–51, Lewis Chelsea, MI.

Malley, D. F. (1996) Cadmium whole-lake experiment at the Experimental Lakes Area: An anachronism?, *Canadian Journal of Fisheries and Aquatic Sciences*, **53**, 1862-70.

Mathys, W. (1980) Zinc Tolerance by Plants. In *Zinc in the Environment*, ed. Nriagu, J. O., John Wiley and Sons, New York.

May, R. M. (1974) *Stability and Complexity in Model Ecosystems*, Princeton University Press, Princeton, NJ.

(1975) Patterns of species abundance and diversity. In *Ecology and Evolution of Communities*, eds. Cody, M. L., and Diamond, J. M., pp. 81–120, Belknap Press, Cambridge, MA.

May, R. M., and Oster, G. F. (1976) Bifurcations and dynamical complexity in simple ecological models, *American Naturalist*, **110**, 573–99.

McCarthy, J. F., and Shugart, L. R. (1990) *Biomarkers of Environmental Contamination*, Lewis, Boca Raton, FL.

McHardy, B. M., and George, J. J. (1985) The uptake of selected metals by the green alga *Cladophora glomerata*, *Symposia Biologica Hungarica*, **29**, 3–19.

McLaughlin, D. L. (1986) Seasonal and annual variation of some chemicals in forage in the vicinity of the Nanticoke industrial complex 1978 to 1985. ARB-091-86-Phyto, Ontario Ministry of the Environment, Air Resources Branch, Phytotoxicology Section.

Mix, M. C. (1986) Cancerous diseases in aquatic animals and their association with environmental pollutants: A critical review of the literature, *Marine Environmental Research*, **20**, 1–141.

MOE. (1976) Phytotoxicology assessment studies conducted in the Nanticoke area – 1974. Air Resources Branch, Phytotoxicology Section, Vasilof, G., Plant Physiologist.

MOE. (1981) Results of Phytotoxicology studies conducted in the Nanticoke Area – 1979. Air Resources Branch, Phytotoxicology Section, Vasilof, G., Plant Physiologist.

Moriya, K., and Horikoshi, K. (1993) Isolation of a benzene degradation, *Journal of Fermentation and Bioengineering*, **76**, 168–93.

Morris, R. G., Kennedy, J. H., Johnson, P. C., and Hambleton, F. E. (1993) Pyrethroid Insecticide Effects on Bluegill Sunfish in Microcosms and Mesocosms and Bluegill Impact on Microcosm Fauna. In *Aquatic Mesocosm Studies in Ecological Risk Assessment*, eds. Graney, R. L., Kennedy, J. H., and Rodgers, J. H., pp. 373–95, Lewis, Boca Raton, FL.

Mort, S. L., and Dean-Ross, D. (1994) Biodegradation of phenolic compounds by sulfate-reducing bacteria, *Microbial Ecology*, **28**, 67–77.

Mulvey, M., and Diamond, S. A. (1991) Genetic Factors and Tolerance Acquisition in Populations Exposed to Metals and Metalloids. In *Metal Ecotoxicology: Concepts and Applications*, eds. Newman, M. C., and McIntosh, A. W., pp. 301–21, Lewis, Chelsea, MI.

Myers, M. S., Johnson, L. L., Hom, T., Collier, T. K., Stain, J. E., and Varanasi, U. (1998) Toxicopathic hepatic lesions in subadult English sole (*Pleuronectes vetulus*) from Puget Sound, Washington, U.S.A.; Relationships with other biomarkers of contaminant exposure, *Marine Environmental Research*, **45**, 47–67.

National Research Council of Canada (NRCC). (1985) The role of biochemical indicators in the assessment of ecosystem health – Their development and validation. NRCC No. 24371, Associate Committee on Scientific Criteria for Environmental Quality, National Research Council of Canada.

Niagra River Toxics Committee. (1984) Report of the Niagra River Toxics Committee. Environment Canada, Ontario Ministry of the Environment, New York State Department of Environmental Conservation, and U.S. EPA.

Nieboer, E., and Richardson, D. H. S. (1980) The replacement of the nondescript term "heavy metals" by a biologically and chemically significant classification of metal ions, *Environmental Pollution*, **1**, 3–26.

   (1981) Lichens as monitors of atmospheric deposition. In *Atmospheric Pollutants in Natural Waters*, ed. Eisenreich, S., pp. 339–99, Ann Arbor Science, Ann Arbor, MI.

Odum, E. P. (1969) The strategy of ecosystem development, *Science*, **164**, 262–70.

Outridge, P. M., and Noller, B. N. (1991) Accumulation of toxic trace elements by freshwater vascular plants, *Reviews of Environmental Contamination and Toxicology*, **121**, 1–63.

Petersen, J. R. C. (1986) Population and guild analysis for interpretation of heavy metal pollution in streams. In *Community Toxicity Testing*, ASTM STP 920, ed. Cairns, J. J., pp. 180–98, American Society for Testing and Materials, Philadelphia.

Pimm, S. L. (1984) The complexity and stability of ecosystems, *Nature*, **307**, 321–6.

Posthuma, L., and Van Straalen, N. M. (1993) Heavy metal adaptation in terrestrial invertebrates: A review of occurrence, genetics, physiology and ecological consequences, *Comparative Biochemistry and Physiology*, **106C**, 11–38.

Poulter, A., Collin, H. A., Thurman, D. A., and Hardwick, K. (1985) The role of the cell wall in the mechanism of lead and zinc tolerance in *Anthoxanthum oderatum* L, *Plant Science*, **42**, 61–6.

Power, M. E. (1990) Effects of fish in river food webs, *Science*, **250**, 811–15.

Prairie, R., and Lavergne, Y. (1998) Caracterisation des sediments autour des quais de Mont-Louis et Sandy Beach. Presentation, August 1998.

Price, N. M., and Morel, F. M. M. (1990) Cadmium and cobalt substitution for zinc in a marine diatom, *Nature*, **344**, 658–60.

Raddum, G. G., and Fjellheim, A. (1984) Acidification and early warning organisms in freshwater in western Norway, *Internationale Vereinigung für Theoretische und Angewante Linnologie – Verhandlungen*, **22**, 1973–80.

Rand, G. M. (1995) *Fundamentals of Aquatic Toxicology: Effects, Environmental Fate and Risk Assessment*, Taylor and Francis, Washington, DC.

Reimer, P., and Duthrie, H. C. (1993) Concentrations of zinc and chromium in aquatic macrophytes from the Sudbury and Muskoka regions of Ontario, Canada, *Environmental Pollution*, **79**, 261–5.

Ritossa, F. M. (1962) A new puffing pattern induced by a temperature shock and DNP in *Drosophila*, *Experientia*, **18**, 571–3.

Rose, C. I., and Hawksworth, D. L. (1981) Lichen recolonisation in London's cleaner air, *Nature*, **287**, 289–92.

Rowe, C. L., Kinney, O. M., Fiori, A. P., and Congdon, J. D. (1996) Oral deformities in tadpoles (*Rana catesbeiana*) associated with coal ash deposition: Effects of grazing ability and growth, *Freshwater Biology*, **36**, 723–30.

Rowe, C. L., Kinney, O. M., Nagle, R. D., and Congdon, J. D. (1998) Elevated maintenance cools in an anuran (*Rana catesbeiana*) exposed to a mixture of trace elements during the embryonic and early larval periods, *Physiological Zoology*, **71**, 27–35.

Ruhling, A., Rasmussen, L., Pilegaard, K., Makinen, A., and Steinnes, E. (1987) Survey of atmospheric heavy metal deposition in the Nordic countries in 1985. Nordidisk Ministerrad 1987.

Russel, G., and Morris, O. P. (1970) Copper tolerance in the marine fouling alga *Ectocarpus siliculous*, *Nature* (*London*), **228**, 288–9.

Ryckman, D. P., Weseloh, D. V. C., and Bishop, C. A. (1997) Contaminants in Herring Gull Eggs from the Great Lakes: 25 Years of Monitoring Levels and Effects, #EN 222/6-1997E. Environment Canada, Downsview, Ontário.

Sanders, J. G., and Riedel, G. F. (1998) Metal accumulation and impacts in phytoplankton. In *Metal Metabolism in Aquatic Environments*, eds. Langston, W. J., and Bebianno, M. J., Chapman and Hall, London.

Schindler, D. W. (1974) Eutrophication and recovery in experimental lakes: Implications for lake management. *Science*, **184**, 897–9.

Schindler, D. W. (1990) Natural and anthropogenically imposed limitations to biotic richness in fresh waters. In *The Earth in Transition: Patterns and Processes of Biotic Impoverishment*, ed. Woodwell, G. M., pp. 425–62, Cambridge University Press, Cambridge.

Schindler, D. W., Fee, E. J., and Ruszcynski, T. (1978) Phosphorus input and its consequences for phytoplankton standing crop production in the Experimental Lakes Area and in similar lakes, *Journal of the Fisheries Research Board of Canada*, **35**, 190–6.

Scott, M. G., Hutchinson, T. C., and Feth, M. J. (1989) A comparison of the effects on Canadian boreal forest lichens of nitric and sulphuric acids as sources of acidity, *New Phytologist*, **111**, 663–71.

Seaward, M. R. D. (1993) Lichens and sulphur dioxide air pollution: field studies, *Environmental Research*, **1**, 73–91.

Severne, B. C. (1974) Nickel accumulation by *Hybanthus floribundus*, *Nature*, **248**, 807–8.

Shacklette, H. T. (1964) Flower variation of *Epilobium angustofolium* L., *Canadian Field-Naturalist*, **78**, 32.

Shaw, J. L., and Kennedy, J. H. (1996) The use of aquatic field mecocosm studies in risk assessment, *Environmental Toxicology and Chemistry*, **15**, 605–7.

Sheffield, S. R., and Lochmiller, R. L. (2001) Effects of field exposure to Diazinon on small mammals inhabiting a semienclosed prairie grassland ecosystem. I. Ecological and reproductive effects, *Environmental Toxicology and Chemistry*, **20**, 284–96.

Shirley, M. D. F., and Sibley, R. M. (1999) Genetic basis of a between-environment trade-off involving resistance to cadmium in *Drosophila melanogaster*, *Evolution*, **53**, 826–36.

Skye, E. (1958) Luftföroreningars inverkan pa busk–och Lladlavfloran kring skifferoljeverket I Narkes Kvantorp, *Sven. Bot. Tidskr.*, **52**, 133–90.

Stearns, S. C. (1976) Life-history tactics: A review of the ideas, *Quarterly Review of Biology*, **51**, 3–47.

Stegeman, J. J., Brouwer, M., DiGiulio, R. T., Farlin, L., Fowler, B. A., Sanders, B. M., and Van Veld, P. A. (1992) Molecular Response to Environmental Contamination: Enzyme and Protein Systems as Indicators of Chemical Exposure and Effect. In *Biochemical, Physiological, and Histopathological Markers of Anthropogenic Stress*, eds. Huggett, R. J., Kimerle, R. A., Mehrle, J. P. M., and Bergman, H. L., pp. 235–335, Lewis, Boca Raton, FL.

Stein, J. E., Collier, T. K., Reichert, W. L., Casillas, E., Ham, T., and Varnasi, U. (1992) Bioindicators of contaminant exposure and sub-lethal effects: Studies with benthic fish in Puget Sound, WA, *Environmental Toxicology and Chemistry*, **11**, 701–14.

Steinnes, E. (1995) A critical evaluation of the use of naturally growing moss to monitor the deposition of atmospheric metals, *Science of the Total Environment*, **160/161**, 243–9.

Stephenson, M., Bendell-Young, L., Bird, G. A., Brunskill, G. J., Curtis, P. J., Fairchild, W. L., Homolka, M. H., Hunt, R. V., Lawrence, S. G., Motycka, M. F., Shwartz, W. J., Turner, M. A., and Wilkinson, P. (1996) Sedimentation of experimentally added cadmium and ^{109}Cd in Lake 382, Experimental Lakes Area, Canada, *Canadian Journal of Fisheries and Aquatic Sciences*, **53**, 1888–902.

Sullivan, J. T., Cheng, T. C., and Chen, C. C. (1984) Genetic selection for tolerance to nicosamide and copper in *Biomphalaria glabrata* (Mollusca:Pulmonata), *Tropenmedizin und Parasitologie*, **35**, 189–92.

Terhivuo, J., Pankakoski, E., Hyvarinen, H., and Koivisto, I. (1994) Pb uptake by ecologically dissimilar earthworm (Limbricidae) species near a lead smelter in south Finland, *Environmental Pollution*, **85**, 87–96.

Tessier, A., Campbell, P. G. C., and Bisson, M. (1979) Sequential extraction procedure for the speciation of particulate trace metals, *Analytical Chemistry*, **51**, 844–51.

Tipping, E. (1994) WHAM – A chemical equilibrium model and computer code for waters, sediments, and soils incorporating a discrete site/electrostatic model of ion-binding by humic substances, *Computers and Geosciences*, **20**, 973–1073.

Tranvik, L., and Eijsackers, H. (1989) On the advantage of *Folsomia fimetaroides* over *Isotomiella minor* (Collembola) in a metal polluted soil, *Oecologia (Heidelberg)*, **80**, 195–200.

Trelease, S. F., di Somma, A. A., and Jacobs, A. L. (1960) Seleno-amino acid found in *Astragalus bisulcatus*, *Science*, **132**, 530–5.

Turner, R. G. (1969) Heavy Metal Tolerance in Plants. In *Ecological Aspects of the Mineral Nutrition of Plants*, ed. Rorison, I. H., pp. 399–410, British Ecological Society Symposium 9, Blackwell Scientific Publications, Oxford.

Turner, R. G., and Marshall, C. (1972) The accumulation of zinc by subcellular fractions of roots of *Agrostis tenuis* Sinth: In relation to zinc tolerance, *New Phytologist*, **71**, 671–5.

Ulanowicz, R. E. (1995) Trophic Flow Networks as Indicators of Ecosystem Stress. In *Food Webs: Integration of Patterns and Dynamics*, eds. Polis, G. A., and Winemiller, K. O., Chapman and Hall, New York.

Varanasi, U., Stein, J. E., Johnson, L. L., Collier, T. K., Casillas, E., and Myers, M. S. (1992) Evaluation of Contaminant Exposure and Effects in Coastal Ecosystems. In *Ecological Indicators, Proceedings of an International Symposium*, ISBM# 1851667113, eds. McKenzie, D. H., Hyatt, D. E., and McDonald, V. J., Fort Lauderdale, FL, October 16–19, 1990.

Warren, L. A., Tessier, A., and Hare, L. (1998) Modelling cadmium accumulation by benthic invertebrates *in situ*: The relative contributions of sediment and overlying water reservoirs to organism cadmium concentrations, *Limnology and Oceanography*, **43**, 1442–54.

Watmough, S. A. (1997) An evaluation of the use of dendrochemical analyses in environmental monitoring, *Environmental Reviews*, **5**, 181–210.

Weeks, B. A., and Warriner, J. E. (1984) Effects of toxic chemicals on macrophage phagocytosis in two estuarine fishes, *Marine Environmental Research*, **14**, 327–35.

Weisberg, S. B., Ranasinghe, J. A., Schaffner, L. C., Diaz, R. J., Dauer, D. M., and Frithsen, J. B. (1997) An estuarine benthic index of biotic integrity (B-1B1) for the Chesapeake Bay, *Estuaries*, **20**, 149–58.

Welch, W. J. (1992) The mammalian stress response – Cell physiology, structure-function of stress proteins, and implications for medicine and disease, *Physiological Reviews*, **72**, 1063–81.

Whitton, B. A. (1970) The biology of *Cladophora* in fresh waters: Review paper, *Water Research*, **4**, 457–76.

Wikfors, G. H., Neeman, A., and Jackson, P. J. (1991) Cadmium-binding polypeptides in microalgal strains with laboratory-induced cadmium tolerance, *Marine Ecology Progress Series*, **79**, 163–70.

Woodwell, G. M., and Houghton, R. A. (1990) The Experimental Impoverishment of Natural Communities: Effects of Ionizing Radiation on Plant Communities, 1961–1976. In *The Earth in Transition: Patterns and Processes of Biotic Impoverishment*, ed. Woodwell, G. M., pp. 3–24, Cambridge University Press, Cambridge.

Woodwell, G. M., and Rebuck, A. L. (1967) Effects of chronic gamma radiation on the structure and diversity of an oak-pine forest, *Ecological Monographs*, **37**, 53–69.

Wright, R. F., Lotse, E., and Semb, A. (1988) Reversibility of acidification shown by whole-catchment experiments, *Nature (London)*, **334**, 670–5.

Zakshek, E. M., Puckett, K. K., and Percy, K. E. (1986) Lichen sulphur and lead levels in relation to deposition patterns in eastern Canada, *Water, Air, and Soil Pollution*, **30**, 161–9.

4.13 Further reading

Diatoms as indicators: Dixit, S. S., J. P. Smol, J. P. Kingston, and D. F. Charles. 1992. Diatoms: Powerful indicators of environmental change. *Environmental Science and Technology*, **26**, 22–33.

Plants as estuarine indicators: Lytle, J. S., and Lytle, T. F. 2001. Use of plants for toxicity assessment of estuarine ecosystems. *Environmental Toxicology and Chemistry*, **20**, 68–83.

Great Lakes biomonitoring: Ryckman, D. P., D. V. C. Weseloh, and C. A. Bishop. 1997. Contaminants in Herring Gull Eggs from the Great Lakes: 25 Years of Monitoring Levels and Effects, #EN 222/6-1997E. Environment Canada, Downsview, Ontario.

Effect of xenobiotics on cellular and subcellular systems: Malins, D. C., and G. K. Ostrander. 1994. *Aquatic Toxicology. Molecular, Biochemical and Cellular Perspectives*. Lewis, Boca Raton, FL.

Lichens as indicators: Richardson, D. H. S. 1988. Understanding the pollution sensitivity of lichens, *Botanical Journal of the Linnaean Society*, **96**, 31–43.

Surrogates and ecosystem health: Edwards, C. J., and R. A. Ryder (eds.). 1990. Biological Surrogates of Mesotrophic Ecosystem Health in the Laurentian Great Lakes. Report to the Great Lakes Science Advisory Board, Windsor, Ontario.

Environmental models: Mackay, D. 1991. *Multimedia Environmental Models: The Fugacity Approach*. Lewis, Boca Raton, FL.

Environmental models: Mackay, D., L. A. Burns, and G. M. Rand. 1995. Fate Modeling. In *Fundamentals of Aquatic Toxicology: Effects, Environmental Fate and Risk Assessment*, ed. Rand, G. M., 2nd ed., pp. 563–86, Taylor and Francis, Washington, DC.

Whole system manipulations: Gorrie, P. 1992. Lakes as laboratories, *Canadian Geographic*, **112** (2), 68–78.

5

○ ○

Factors affecting toxicity

5.1   Introduction

In Chapter 2, we saw that the exposure of organisms to toxic agents in the environment may result in adverse effects if the exposure concentration and time exceed certain thresholds. A primary goal of toxicologists is to establish a quantitative relationship between toxic exposure and degree of effect. Usually this takes the form of a dose-response curve, although in environmental toxicology we often use the concentration of the toxicant in the exposure medium as a substitute for dose. In doing so, we are assuming that the ambient toxicant concentration has a direct relationship to the dose accumulated by the organism, even though we may choose not to quantify the dose during a bioassay. Although such an assumption seems reasonable, a variety of extraneous factors, individually and collectively, may significantly affect this relationship by varying the bioavailability of a toxic chemical or by altering its metabolism.

Chemical bioavailability may be influenced by a number of abiotic parameters such as temperature and water chemistry, as well as a variety of biological factors relating to the physical condition of the organism. Morphological and biochemical differences between or among organisms of different size and/or taxonomic group may have an enormous influence on how they react to different types of toxic agents, and, as described in Chapter 4, prolonged exposure to sublethal toxicant concentrations may actually lead to changes in chemical tolerance. There are, then, a multitude of factors that are likely to affect the expression of toxicity, either within, between, or among species. Therefore, in making comparisons between different data relating to the same toxic agent, it is important to have as much information as possible on the taxonomic relationship between exposed organisms, their developmental state, the exposure conditions, and the form in which the toxic agent is presented to the biota.

Usually, toxic chemicals appear in the environment as complex mixtures wherein the toxicity of one or more chemicals is affected by others in the mixture. As studies of the toxicological action of different chemicals have progressed, it has become possible to draw some inferences concerning the way specific groups of chemicals are likely to interact. These investigations have also provided a large body of information on the relationship between certain structural properties of chemicals and the way they behave toxicologically. Toxicologists have used these structure activity relationships or quantitative structure-activity relationships to make assumptions about the likely toxicity of chemicals based on properties such as size, polarity, lipid solubility and the presence of certain functional groups. Such information has proven useful in the registration and regulation of chemicals for which there may be little or no specific toxicological data.

This chapter discusses some of the principal biotic and abiotic factors affecting the toxicity of chemicals and other agents either by altering their availability to the organism or by affecting their mode of toxic action.

5.2 Biotic factors affecting toxicity

5.2.1 Taxonomic group

Toxicity testing is designed to be protective. In selecting organisms from a broad range of ecosystems, our intent is to define the boundaries of damage to these ecosystems caused by man's activities. In doing so, we need to determine the maximum acceptable concentration for toxicants of concern. To achieve this goal, toxicologists require information from the most sensitive end-points and the most sensitive species. However, environmental toxicity data are notoriously variable from species to species and from chemical to chemical. Therefore, compromises have been made in terms of species selection. In Table 2.2, we listed several species in common usage in toxicity bioassays in Europe and North America. An advantage of using relatively few species is that comparative chemical toxicity can be determined intraspecifically for large numbers of toxicants, thereby enhancing the reliability of the information. A disadvantage is that, in a regional sense, some habitats are poorly covered (e.g., standard test species may not be indigenous to a particular area and, therefore, not appropriate for that environment). In such a situation, an indigenous species, having local economic and/or ecological importance, may be selected for testing in parallel with standard test species. A typical case in point is the striped bass (*Morone saxatilis*). Larvae from this species are only available from hatcheries for a few weeks of the year between late April and late May, making them unsuitable for standardised toxicity testing. Nevertheless, concern over the striped bass fishery in the 1970s and 1980s in estuaries such as the Chesapeake Bay and San Francisco Bay led to intensive use of *M. saxatilis* larvae for toxicity bioassays. One result of these assays was the demonstration of a high degree of acid sensitivity in the larvae of this species, an outcome that could not have been foretold through the use of surrogate standard test species, many of which are acid-tolerant.

Reviews of several large data sets from aquatic bioassays indicate that arthropods are generally more sensitive than fish, and fish are more sensitive to chemical contaminants than larval amphibians (Mayer and Ellersieck, 1986). Algae and aquatic macrophytes are usually less sensitive than aquatic animals. In freshwater tests, daphnids (e.g., *Daphnia magna*) are usually among the most sensitive crustaceans, and salmonids are frequently the most sensitive fish. In the marine environment, mysid and panaeid shrimp are frequently the most sensitive crustaceans (Suter and Rosen, 1988), although larval bivalves are probably the most sensitive organisms used in saltwater testing. In estuarine environments it is not always possible to employ standard freshwater and saltwater test organisms and so the use of regional species is often more appropriate. The euryhaline copepod (*Eurytemora affinis*) has proven to be a sensitive estuarine test species and, as mentioned earlier, larvae of the anadromous striped bass (*Morone saxatilis*) have shown sensitivity similar to that of salmonids in response to toxicants such as trace metals and acidity.

Among 82 toxic chemicals tested using six or more species of freshwater animals, Mayer and Ellersieck (1986) reported that the highest toxicity values for a particular toxicant averaged 256 times the lowest values for that toxicant and ranged from 2.6X to 166,000X. In cases where toxic action is related to life stage, very large intraspecific differences in sensitivity may occur according to life stage. Wilson and Costlow (1987) reported nearly four orders of magnitude difference in diflubenzuron toxicity between larval and adult ovigerous female grass shrimp (*Palaemonetes pugio*).

Toxicity assays conducted on *P. pugio* and the killifish (*Fundulus heteroclitus*) at approximately 3-week intervals over a period of >6 years gave 48-hr LC_{50}s ranging from 0.26 to 2.39 for *P. pugio* and from 22.4 to 60.5 for *F. heteroclitus*. The mean LC_{50} of 80 tests for *P. pugio* was $1.11 \pm 0.57\,mg\,L^{-1}$ (SD) with a coefficient of variability of 52%. The mean LC_{50} of 63 tests using *F. heteroclitus* was $37.3 \pm 11.7\,mg\,L^{-1}$ (SD) with a coefficient of variability of 31%.

Studies of aquatic organisms and terrestrial animals such as birds indicate that, despite differences in species sensitivity, no single species stands out as being consistently more or less sensitive than the others. Toxicologists have, therefore, sought to identify combinations of a few test species that have strong predictive capability for determining toxicity to most or all species within any particular system. Mayer and Ellersieck (1986) concluded that LC_{50} data from a daphnid/rainbow trout combination could be divided by a factor of 25 to cover the "most sensitive" species. Giesy and Graney (1989) used a similar method to estimate 80% species coverage based on a plot of distribution of species sensitivity (Figure 2.9). The idea of using one or two representatives from a few select taxa is reflected in most regulatory testing today.

Premanufacturing registration (PMR) for new products generally involves testing with an aquatic crustacean (usually *Daphnia*) and at least one species of fish (usually fathead minnow, *Pimephales promelas*, or rainbow trout, *Salmo gairdneri*). Preliminary (first-tier) testing typically involves an algal species or macrophyte depending on the projected use of the chemical product. First-tier tests also include LC_{50} and LD_{50} assays involving at least one mammalian species (e.g., mice or rats) and a bird (mallard, *Anas platyrhynchos*, or bobwhite quail, *Colinus virginianus*).

Several studies have focussed on interspecific comparisons of chemical toxicity, particularly between standard aquatic test species. Generally speaking, sensitivity to toxic chemicals can be equated with taxonomic relationship, which means that reasonable extrapolation may be made within species, genera, and families.

5.2.2 Age/body size

Life stage and body size usually have a significant influence on toxic response, a fact recognised in pharmacology, where drug dosage is almost always calibrated to age or some measure of body size (usually weight). However, in environmental toxicology, it is not easy to find good comparative information from different life stages of the same species obtained under similar experimental conditions. Larval fish are generally more sensitive than adults, and eggs are usually the most resistant life stage because of the protection provided by the egg membranes.

Small animals have a larger surface area:volume ratio, which leads to fast chemical uptake per unit weight. This is further enhanced by higher ventilatory and metabolic rates in smaller animals. Many of these parameters may be age-dependent within a species, but age-independent from species to species (i.e., smaller species are likely to be more sensitive than larger species). Age-dependent sensitivity in some cases may result from the incomplete development in young organisms of enzyme systems capable of detoxifying a variety of chemicals. For example, the mixed function oxidase system in the human new-born infant is very poorly developed and only reaches its maximum capacity at puberty. Although data are sparse, it is likely that a comparable situation exists in other vertebrates. In some instances, a particular pesticide may be designed to affect a specific developmental life stage. For example, a moult inhibitor such as diflubenzuron (Dimilin) is very toxic to arthropods as they progress through several moults to the adult stage. However, after the final moult is reached, the toxicity of the pesticide declines dramatically.

5.3 Abiotic factors affecting toxicity

5.3.1 Temperature

Ectothermic metabolism increases approximately twofold for every 10°C change in temperature (i.e., the $Q_{10} = 2$). Changes in metabolism may be reflected by changes in respiratory rate, chemical absorption, detoxification, and excretory rates, all of which may affect chemical toxicity. Kyono-Hamaguchi (1984) reported that medaka (*Oryzias latipes*) exposed to diethylnitrosamine at cold temperatures developed fewer liver tumours than at higher temperatures. Gill and Walsh (1990) demonstrated marked temperature dependence of the metabolism of benzo(a) pyrene to phase II metabolites in toadfish (*Opsanus beta*) but noted that substrate affinity of the phase II enzyme system may diminish with an increase in temperature. Both basal activity and induction of cytochrome P450 (CYP1A1) are influenced by environmental temperature.

Most toxicants exhibit a two- to fourfold change in toxicity for every 10°C change in temperature. Usually the correlation between temperature and toxicity is positive although the toxicity of some organochloride and pyrethroid pesticides may decrease at higher temperatures. Such instances probably relate to the increased metabolism of the compound. A negative metabolic effect of elevated temperature may be increased oxygen usage. Cairns et al. (1975) suggested that thermal death may be the result of tissue anoxia. Consequently, the toxicity of any chemical that increases metabolic demand or inhibits oxygen uptake and utilisation may be enhanced at higher temperatures. In aquatic organisms this may be further exacerbated by lower oxygen solubility in water at higher temperatures. Additionally, the solubility of many toxic chemicals may increase at elevated temperatures.

At the cellular level, a rise in temperature causes an increase in lipid fluidity resulting in increased membrane permeability, and, even though enzymes may exhibit a generalised increase in activity, some may move beyond their optimal functional range.

Temperature has a marked effect on target enzymes of several xenobiotic chemicals. For example, several enzymes are known to exist in two forms depending on ambient temperature. In the brain of the rainbow trout (*Oncorhynus mykiss*), acetyl cholinesterase may exist as two isozymes, one after acclimation to 2°C and another following acclimation to 17°C; both are present after acclimation to 12°C. Their optimal K_m values reflect their synthesis temperatures (i.e., maximal substrate affinities are 2°C and 17°C, respectively). It therefore follows that, whether or not the substrate is acetylcholine or a competitive inhibitor, temperature changes are likely to have both qualitative and quantitative effects on enzyme activity (or inhibition thereof) according to prevailing temperature conditions.

Temperature probably has a more complicated influence on chemical toxicity than is recognised in the literature. Two parameters are of overriding importance: acclimation temperature and experimental temperature. Acclimation temperature has different meanings in the literature. Within the context of a bioassay it is defined as the temperature at which organisms are held prior to toxicant exposure. It may or may not differ from the experimental temperature, although usually organisms are acclimated to the experimental temperature for several hours before the initiation of the assay. In view of the fact that so many reported environmental toxicity data are collected under differing conditions with respect to these parameters, the relationship between acclimation temperature, experimental temperature, and chemical toxicity has probably not been given the attention it deserves.

In a classic series of experiments, Brett (1960) was able to demonstrate that both high- and low-temperature acclimation shifted the tolerance zone for young sockeye salmon (*Oncorhynchus nerka*) as determined for survival, growth, and

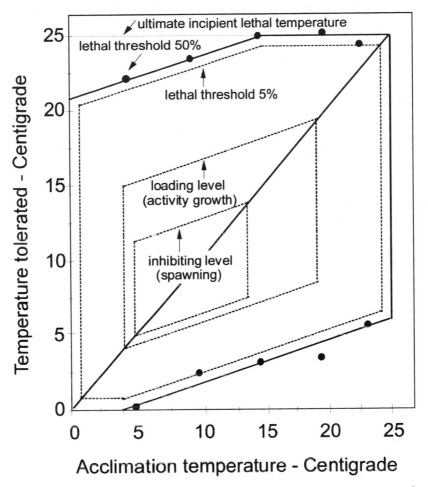

Figure 5.1 Plot of acclimation temperature versus tolerance temperature for three toxic end-points (mortality, growth, spawning). Areas inside respective dotted lines indicate zones of tolerance. After Brett (1960).

reproduction. Figure 5.1 shows a plot of acclimation temperature against incipient high and low lethal temperatures for this species. The data indicate, for example, a rise in the upper lethal temperature of about 1°C for every 4°C change in acclimation temperature. Within the polygon describing lethal temperatures, smaller polygons have been constructed for reproductive and locomotor activity. Acclimation occurs at different rates for different species with heat acclimation usually several days faster than cold acclimation.

Temperature tolerance has a host of interactive effects with other environmental factors such as photoperiod, oxygen, and salinity. In some cases, genetically "thermal tolerant" races within a species survive at temperatures that would be lethal to less tolerant races of the same species. It will be clear from this and other sections in this chapter that a variety of biotic and abiotic factors influence both temperature and toxic chemical tolerance and that exposure time plays an important part in the type of interactive effects that may result.

Hodson and Sprague (1975) demonstrated lower Zn toxicity in Atlantic salmon (*Salmo salar*) at 3°C compared with 19°C when determined as a 24-hr LC_{50}. Paradoxically, when the assay was extended to 2 weeks, the 19°C LC_{50} remained unchanged but the 3°C LC_{50} decreased approximately sixfold thereby reversing the temperature effect. Such a study indicates that caution must be exercised when interpreting temperature effects from acute toxicity tests.

A major problem in interpreting temperature/toxic chemical interactions is the fact that, even within tests conducted on a single organism, both chemical and temperature tolerance may show substantial differences at different levels of cellular organisation, and finding a mechanistic linkage between specific chemical and temperature effects is therefore very difficult.

One area of research where some progress has been made is in the study of heat shock proteins. As the name suggests, heat shock or stress proteins are synthesised in response to a variety of stresses including elevated temperature and several toxic chemicals. These stresses are discussed in greater detail in Chapter 4 (Figure 4.6). Most of this research has dealt separately with the induction of hsps by chemicals or by temperature. Nevertheless, it seems reasonable to assume that a response to one stress will affect an organism's reaction to another source of stress, and this is borne out by the results of at least one recent paper (Müller et al., 1995), which demonstrated that sublethal preexposure to heat resulted in higher tolerance to subsequent chemical stressors.

5.3.2   pH and alkalinity

pH may affect chemical toxicity in a variety of ways. In very acid conditions (<pH 5), the hydrogen ion itself may be detrimental to aquatic life, usually by increasing the permeability of the gill epithelium thereby causing the loss of important electrolytes. An integrated model of the effect of acid on the physiology of fish is shown in Figure 5.2 and is briefly described as follows:

> Hydrogen ions pass into the fish through gills down a concentration gradient.
>
> High concentrations of hydrogen ions in the external medium inhibit active Na^+ versus H^+ (or the ammonium ion (NH_4^+)) exchange through the gill.
>
> Passive permeability of the branchial epithelium to the sodium (Na^+), potassium (K^+), chloride (Cl^-), and perhaps other ions is increased. These electrolytes diffuse out of the fish.

Acid entering the extracellular fluid is initially buffered by protein (mainly haemoglobin) and bicarbonate ions. The resultant CO_2 is easily washed out of the gills, and metabolic acidosis results. Later, acid penetrates intracellular fluid in exchange for K^+ resulting in intracellular acidosis.

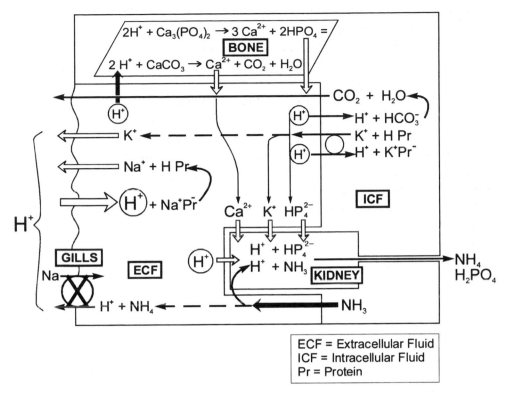

Figure 5.2 Integrated model of the effect of excess acidity on the physiology of a fish. Modified from Wood and McDonald (1982).

Acid may also enter bone tissue where it can be buffered by calcium carbonate ($CaCO_3$) and calcium phosphate [$Ca_3(PO_4)_2$]. Elevation of plasma K^+, net branchial K^+ efflux, and renal excretion of K^+, Ca^{2+}, and inorganic phosphate (PO_4) are the consequence. As ions are lost from the extracellular fluid through the gills and kidney, they are replaced by electrolytes from the intracellular fluid. Salt depletion of the entire body results.

Hydrogen ions also influence the toxicity of trace metals, either by affecting their chemical speciation in water or by competing with metals for sites on biological membranes. In acid conditions, aluminium toxicity often acts synergistically with hydrogen ion toxicity (Section 6.2.2). Between pHs 4.8 and 6.0, a variety of Al hydrolysis products damage the gill epithelium by causing apoptosis and necrosis of ion transporting cells (see Section 2.3.1) thereby inhibiting O_2 and CO_2 diffusion through an increase in mucus production. Deposition of Al on gills associated with polymerisation of low-molecular-weight Al complexes at pHs as high as 6.4 may also inhibit gaseous exchange.

Chemical speciation of copper in water is also highly dependent on pH and the concentration of carbonate (CO_3). In a freshwater system of pH 7.6 open to the atmosphere, the cupric ion (Cu^{2+}) concentration represents 24% of the total copper present with 46% as cupric carbonate ($CuCO_3$). In the same system but closed to the atmosphere with the carbonate concentration fixed at $10^{-4}\,M$ and the precipitation of tenorite or malachite (oxides) suppressed the calculated distribution is 31% Cu^{2+}, 37% cupric hydroxide ($CuOH^+$), 3% cuprous hydroxide [$Cu(OH)_2$], and 29% $CuCO_3$. For the same closed system but with a fixed carbonate concentration of $10^{-3}\,M$, the calculated distribution is 9% Cu^{2+}, 10% $CuOH^+$, and 79% $CuCO_3$. At a pH of 6, the Cu^{2+} component may exceed 90% of total copper present, but this may again be modified by the carbonate present. In highly acidified systems, H^+ is more likely to be associated with the sulphate ion (SO_4^{2-}) rather than the carbonate ion (CO_3^-), and the Cu^{2+} will be the dominant copper species. Several studies have indicated that the free copper ion is largely responsible for copper toxicity in the freshwater environment, athough there is not universal agreement on this. For example, Erickson et al. (1996) showed that acute copper toxicity apparently increases with increase in pH when expressed in terms of measured cupric ion activity, although the 96-hr LC_{50} expressed as total copper actually increases at higher pHs. This disparity between these curves probably results from decreased competition for binding sites at the biological surface between the free cupric ion (Cu^{2+}) and the proton (H^+) as the pH increases. Alternatively, the apparent increase in the toxicity of the free Cu^{2+} ion at higher pH may reflect the formation of a ternary HO-Cu-gill surface complex at higher pH values, and the contribution of this surface complex to the toxic response.

pH may also affect the bioavailability of lipophilic metal species present in solution, which may traverse the plasma membrane without forming a surface complex. For example, acid conditions tend to favour the predominance of mercuric chloride ($HgCl_2$) for inorganic mercury and monomethyl mercury chloride (CH_3HgCl) for methylmercury. Such neutral chloride complexes exhibit greater lipid solubility and cross membranes faster than the hydroxo complexes [e.g., $Hg(OH)_2$], which predominate at high pHs. The K_{OW} for $HgCl_2$ is 70X that of $Hg(OH)_2$. Un-ionised organic molecules are also more lipid-soluble and, therefore, more able to cross cell membranes than ionised forms. Weak acids such as nitrophenols become increasingly ionised at higher pHs where their LC_{50}s are 20 to 50 times those seen at lower pHs. Conversely, the toxicity of ammonia is associated principally with the un-ionised form, which predominates at higher pHs.

In Section 3.2.4, we showed how the dissociation of weak acids and bases could be explained in terms of pH and the acid dissociation constant according to the Henderson-Hasselbach equations (Figure 3.3). For acids the un-ionised, protonated form I (HA) is the most lipid soluble and easily absorbed; whereas, for bases, the protonated form is charged and, therefore, less soluble in membranes.

pH also affects the aqueous solubility of weak acids and bases. The total solubility S_T of a weak acid [the sum of the ionised (A^-) and un-ionised (HA) forms] increases at higher pHs according to the equation (Notari, 1987)

$$S_T = S_0\left(1 + 10^{pH - pK_a}\right)$$

where S_0 is the intrinsic solubility of the undissociated form.

Conversely, the total solubility of a weak base [the sum of the ionised (R_3NH^+) and ionised (R_3N) forms] increases at low pHs according to the equation (Notari, 1987)

$$S_T = S_0(1 + 10^{pK_a - pH})$$

The development of acid conditions in the aquatic environment has most often been associated with acid mine drainage (see Section 9.2.4) and atmospheric fallout of sulphate from fossil fuel burning, in the form of sulphuric acid (Section 9.6.2). The vulnerability of a particular system to acid input may be greatly affected by the buffering capacity of the water, which may, in turn, have much to do with the underlying geology. This buffering capacity is referred to as alkalinity and is chiefly a function of the carbonate (CO_3^{2-}), bicarbonate (HCO^{3-}), and hydroxide (OH^-). Other compounds such as phosphates, borates, silicates, and some organics may also contribute to the alkalinity of water.

5.3.3 Salinity

In contrast to many other water chemistry parameters, the study of the effect of salinity on chemical uptake and toxicity is characterised by a large number of field studies conducted in estuaries. As focal points of commerce and industry, estuaries may become contaminated by a variety of chemicals that interact with salinity to varying degrees with respect to their bioaccumulation and their toxicity to aquatic organisms.

In a field situation, care must be taken to differentiate between salinity effects per se and other factors that may correlate with this. For example, several studies have indicated that, in estuaries, trace metal concentrations in biota tend to increase in less saline media. Irrespective of other influences, such a situation could arise if metal input were associated with an upstream freshwater source. Depending on the source of input, another factor that could correlate with salinity is organic carbon, which might significantly affect the bioavailability of a broad range of chemicals.

In a review of laboratory studies, Hall and Anderson (1995) examined the relationship between chemical toxicity and salinity for 173 data sets covering a broad range of metals, petroleum hydrocarbons, and several industrial and agricultural compounds. In 55% of the cases, toxicity was negatively correlated with salinity, in 18% toxicity was positively correlated with salinity, and in 27% no significant correlation was seen. The toxicity of most metals increased at low salinities, particularly for crustaceans. Few organic chemicals showed consistent correlation between toxicity and salinity, although in most cases the toxicity of organophosphates increased with increasing salinity.

Studies of the effect of salinity on trace metal accumulation by biota have indicated the influence of both biological and chemical factors. The physiology of an organism may affect trace metal uptake and toxicity in a number of ways.

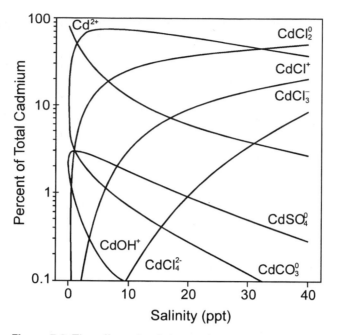

Figure 5.3 The effect of salinity on the speciation of dissolved cadmium.

Euryhaline species and anadromous species clearly have a wide range of salinity tolerance, although optimal salinity may vary from species to species. Outside its optimal salinity, an organism will devote a disproportionate amount of energy to ionic or cell volume regulation and may be more susceptible to toxic stress. Inter-action between trace metal uptake and ionic regulation sometimes occurs, and, for some aquatic species, a relationship apparently exists between cadmium uptake/toxicity and the calcium concentration of the medium, whether or not this is related to salinity.

For several metals such as cadmium (Cd), copper (Cu), silver (Ag), and zinc (Zn), the effect of salinity on their bioavailability is related to their chemical spe-ciation. For example, the most bioavailable form of cadmium is the free cadmium ion (Cd^{2+}), which predominates in freshwater. In increasingly saline media, cadmium forms chloride complexes such as $CdCl^+$ and $CdCl_2$, which are less available and, therefore, less toxic (Sunda et al., 1978). The effect of salinity on the chemical speciation of cadmium is shown in Figure 5.3.

Mason et al. (1996) demonstrated that the growth rate of *Thalassiosira weis-floqii* exposed to mercury was inversely related to salinity, although unlike the Cd model described earlier, salinity-related mercury toxicity was most closely related to the uncharged form $HgCl_2$. Figure 5.4 shows that the proportion of total mercury represented by $HgCl_2$ fell from 23% at 10 ppt to 3% at 35 ppt. Higher salinities caused the formation of complexes such as $HgCl_3^-$, which were apparently less bioavailable than the nonpolar form.

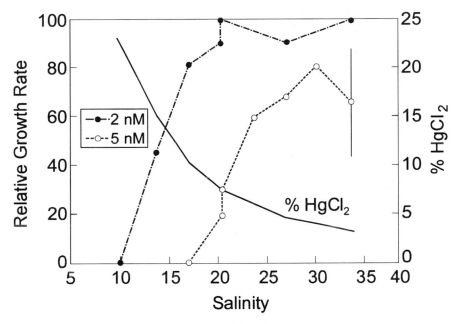

Figure 5.4 Effect of salinity on the toxicity of mercury to *Thalassiosira weissflogii* (Wright and Mason, 2000).

5.3.4   Hardness

The principal components of water hardness are the divalent ions calcium and magnesium. Hardness is defined by the U.S. Environmental Protection Agency in terms of $CaCO_3$ equivalents: $0-75\,mg\,L^{-1}$ for soft water, $75-150\,mg\,L^{-1}$ for moderately hard water, $150-300\,mg\,L^{-1}$ for hard water, and $>300\,mg\,L^{-1}$ for very hard water. Hardness alone has no effect on the speciation of other cations such as metals, although it is usually correlated with pH and alkalinity, which do. There is some evidence that hardness affects the toxicity of organic surfactants, although the data are variable and often confounded by pH differences (Henderson et al., 1959).

   The correlation among hardness, alkalinity, and pH often makes it difficult to differentiate among the effects of these parameters, although some studies (e.g., Miller and Mackay, 1980) have demonstrated that, at a fixed combination of pH and alkalinity, increasing water hardness is associated with a reduction in metal toxicity. Winner (1985) reported increased bioaccumulation of copper by *Daphnia pulex* in soft water but no effect of hardness on copper toxicity. Notwithstanding disputes concerning the relative influence of correlates, the large body of information indicating the ameliorative effect of hardness on metal toxicity has led to adoption by the U.S. Environmental Protection Agency and other regulatory bodies in Canada and Europe of separate water quality criteria for metals such as Cd, Cu, lead (Pb), nickel (Ni), and Zn in hard and soft waters.

5.3.5   Chemical mixtures

Most studies of chemical toxicity have been conducted using individual chemicals, yet in the environment, organisms are usually exposed to complex chemical mixtures. Effluents, leachates, and many surface waters contain numerous toxic chemicals that may interact in varying degrees to produce joint toxicity. The concept of joint toxicity is inherent in much of our thinking about chronic toxicity-related conditions such as carcinogenicity where the action of one chemical may promote the toxic action of another. However, standards for regulating chemicals in the environment are almost invariably based on individual chemicals. Generally speaking, chemicals within the same chemical class tend to have similar toxic action, and, to make toxicological predictions, chemists are interested in major chemical groups within a mixture (e.g., PAHs, nitrosamines, halogenated hydrocarbons, metals). Within a particular chemical group, single chemicals are often selected for toxicological study as being "representative" of such a group. For example benzo(a)pyrene is often selected as a representative PAH and TCDD as a representative chlorinated hydrocarbon. Use of QSAR analysis and toxic equivalents provides means of characterising the toxicity of multiple chemicals within a mixture in terms of the representative chemical. The basis for such an approach is that chemicals with similar toxic action show additive toxicity when acting in concert. This is dealt with in more detail in Section 5.5.

In fact, there are several reasons why chemical mixtures may not act in an additive fashion, particularly where different chemical classes are involved. For example, one chemical or group of chemicals may affect the chemical speciation of another, thereby altering its bioavailability. A particular chemical may act by altering membrane permeability to one (or more) other chemicals, or chemical competition for receptor sites may occur. Any combination of these effects may occur, resulting in more-than-additive or less-than-additive toxicity. Such complexities are often best studied using an empirical, experimental approach. Some differences in terminology are associated with assays of mixtures, although, for the most part, they appear to be semantic rather than interpretive differences. Probably the simplest model describing the toxicity of mixtures employs the toxic unit in which the concentration of each individual toxicant is expressed as a fraction of its LC_{50}. In the case of simple additive toxicity, the 50% response of a mixture of chemicals is arbitrarily set at unity. If the toxicity of a chemical component of the mixture is diminished, its LC_{50} will be higher, and the sum of toxic units will exceed 1. If the toxicity of a component is expressed more highly in the mixture, its LC_{50} falls, and the sum of toxic units becomes less than 1 (Table 5.1).

Könemann (1981) introduced a mixture toxicity index based on five categories of joint toxic action: antagonism, no addition (independent action), partial addition, concentration addition (simple additive toxicity as described previously) and supra addition [potentiation of the toxic action(s) of one or more of the components of the mixture]. These actions are described in terms of an isobologram in

Table 5.1. Sum of toxic action

$$\frac{A_m}{A_i} + \frac{B_m}{B_i} = \text{sum of toxic actions(s)}$$

A_i and A_m are LC_{50}s of chemical A expressed individually and in the mixture. B_i and B_m are LC_{50}s of chemical B expressed individually and in the mixture.

When S is additive,

$$S = 1: \qquad \frac{A_m}{A_i} + \frac{B_m}{B_i} = \frac{1}{2} + \frac{1}{2} = 1$$

When S is additive, when S is less than additive,

$$S > 1: \qquad \frac{A_m}{A_i} + \frac{B_m}{B_i} = \frac{1}{2} + \frac{2}{2} = 1.5$$

and the toxicity of B underexpressed in the mixture.

When S is more than additive,

$$S < 1: \qquad \frac{A_m}{A_i} + \frac{B_m}{B_i} = \frac{1}{2} + \frac{0.5}{2} = 0.75$$

and the toxicity of B overexpressed in the mixture.

Figure 5.5. The terms *potentiation* and *synergism* are often used interchangeably in the literature, and although some authors ascribe subtle differences of meaning to these terms, for most purposes they describe similar phenomena.

A simple way of understanding the concepts illustrated in Figure 5.5 is to consider two chemicals A and B having 96-hr LC_{50}s of $1\,mg\,L^{-1}$ and $10\,mg\,L^{-1}$, respectively. These two concentrations represent unity (one toxic unit) on the *y*-axis and *x*-axis, respectively. Assuming that the mixture contains $1\,mg\,A\,L^{-1}$ and $10\,mg\,B\,L^{-1}$, a 50% mortality after 96 hr would indicate no interaction between the chemicals. Less than 50% mortality at 96 hr would indicate an antagonistic relationship between A and B. If the interaction between A and B were strictly (concentration) additive, the percentage mortality would be equivalent to two toxic units of A or B (i.e., $2\,mg\,A\,L^{-1}$, $20\,mg\,B\,L^{-1}$). In cases of partial addition the mortality at 96 hr would be greater than 50% but less than that seen in the strictly additive situation. Supra addition causes 96-hr mortality greater than that resulting from simple concentration addition. This scheme is summarised in Table 5.2.

An example of how this approach is used to characterise the toxicity of mixtures is illustrated in Figure 5.6, which is a composite isobologram showing joint toxic action of seven organic chemicals versus 1-octanol. Octanol was used as one of the two chemicals in each of eight binary tests (two hexanol data sets are included). Data indicate simple additive toxicity, as might be expected from a suite of compounds having similar modes of toxic action. Noninteractive effects would

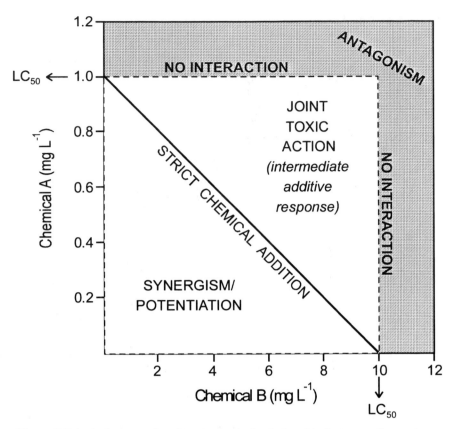

Figure 5.5 Isobologram showing theoretical relationship between the toxic action(s) of two toxicants in a mixture. Axes show the relative contribution of toxicants *A* and *B* to the mixture.

Table 5.2. Toxic mixture model – definition of terms

> 96 hr LC_{50} for chemical $A = 1\,mg\,L^{-1}$
> 96 hr LC_{50} for chemical $B = 10\,mg\,L^{-1}$

Assuming A and B added in mixture of $1\,mg\,L^{-1}$ and $10\,mg\,L^{-1}$, respectively.

Description of mortality at 96 hr	Definition
(a) Exactly 50% of organisms die	No interaction (organism appears to respond to one chemical only)
(b) <50% of organisms die	Antagonism (together, A and B are less toxic than one chemical alone)
(c) >50% of organisms die	Addition
(i) Percentage dead is equivalent to $2\,mg\,L^{-1}$ A or $20\,mg\,L^{-1}$ B	Strict addition/concentration addition
(ii) Percentage dead >50% but less than (c) (i)	Infraaddition
(iii) Percentage dead > (c)(i)	Potentiation

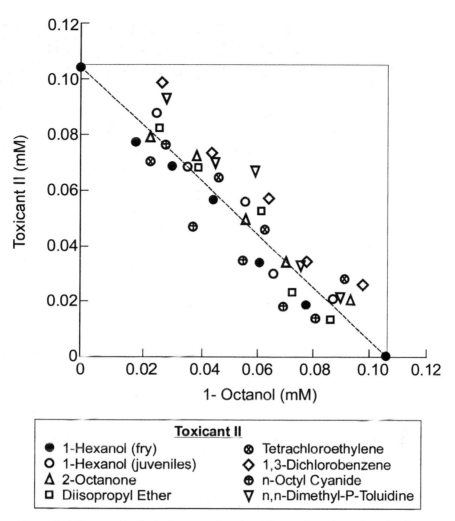

Figure 5.6 Composite isobologram showing joint toxic action of each of seven organic compounds versus 1-octanol on fathead minnows. Modified from Broderius and Kahl (1985).

be expected more from mixtures of different classes of chemicals. Clearly the experiments required to generate the information necessary to establish these relationships are very time consuming and have only limited application to environmental situations involving complex chemical mixtures.

An extension of the linear isobole approach was developed by Haas and Stirling (1994). This method has the advantage of not being limited to LC_{50} concentrations and makes use of Margueles equations, which are commonly used to describe excess energy in thermodynamics.

For a mixture of two chemicals X and Y, a_1 and a_2 are the weight fractions of X_1 and Y_2, respectively. By fitting all the data (from experiments with single components as well as experiments with mixtures) to ideal and excess function models, the model with the best fit can be determined statistically. If the model with the best fit has an excess function of zero, then additive toxicity is assumed. If the best fit model has an excess function that differs from zero, then there is some degree of interaction between the chemicals in the mixture. Synergism and antagonism are indicated by, respectively, negative and positive values for the excess function.

A more comprehensive integrated approach to chemical mixtures in a field situation is the PAH model (Swartz et al., 1995), designed to predict the probability of acute toxicity of sediments contaminated by a mixture of several PAHs. The model assumes that the chemicals are accumulated principally from the interstitial water and uses an equilibrium partitioning model to predict the interstitial water concentration from total sediment analysis. Ten-day LC_{50} (interstitial water) values obtained from acenaphthene, phenanthrene, and fluoranthene were extrapolated to other compounds using a QSAR regression analysis based on octanol water partition coefficients; that is,

$$\log 10\text{-day } LC_{50} \text{ (interstitial water)} = 5.92 - 1.33 \log K_{OW}$$
$$(r = -0.98, \, p < .001) \tag{5.1}$$

The organisms used for toxicity bioassay were sensitive marine and estuarine amphipod crustaceans. Toxic units for the model were defined as $PAH_{iw}/10\text{-day}$ $LC_{50(iw)}$ and were summed for the whole PAH mixture assuming concentration additivity. A summary of the complete model (Swartz et al., 1995) is shown in Figure 5.7.

5.3.6   Dissolved organic carbon

Metals such as copper can form soluble complexes with a wide range of compounds from simple ligands such as Cl—, OH—, and amino acids to colloids and large complex molecules such as clays, hydrous metal oxides, humic substances, and polysaccharides. In seawater, organic complexes may be dominant. Advances in the use of selective ion electrodes (see Section 6.3.1) have allowed better measurement of ionic copper in saline water (salt interfered with such measurements in older models). These, and other studies of ligand competition, have confirmed that organic complexes may account for more than 95% of the total copper present in seawater. In soft water lakes, copper toxicity may be related to both pH and dissolved organic carbon (DOC), with a tenfold reduction in DOC resulting in an approximately tenfold reduction in LC_{50} (Welsh et al., 1993).

Natural dissolved organic matter is found in a variety of forms and concentrations in the aquatic environment. In freshwater, recalcitrant complexes such as humic acid and fulvic acid comprise 50–80% of dissolved organic matter. Dissolved organic molecules may range from amino acids having a molecular weight (MW) of less than 1,000 to very large macromolecules of MW > 100,000. Dissolved organic compounds may possess a wide variety of oxygen-containing functional groups, although phenolic and carboxylic groups are dominant. Other functional groups associated with dissolved organic matter include amines, ketones, quinones, and sulphydryl ligands.

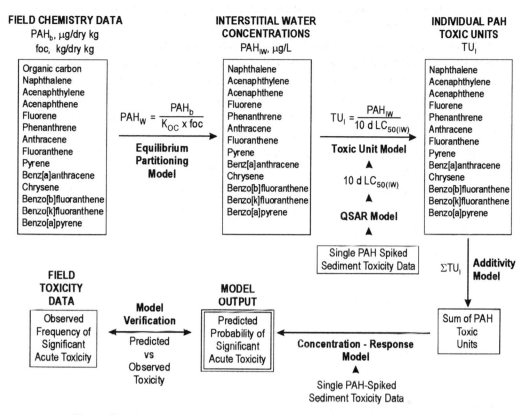

Figure 5.7 Summary of field model used to determine joint toxicity of polyaromatic hydrocarbon mixture (ΣPAH) in sediments (after Swartz et al., 1995).

Many published studies demonstrate the amelioration of metal toxicity through the formation of metal-organic complexes at the expense of the more bioavailable free metal ion, yet a small but significant number of studies indicate that metal bioavailability may actually be increased through the formation of organic complexes. In considering this, it is important to remember that DOC cannot be viewed simply as an amorphous entity even though it is often measured and reported as such. Different organic ligands may be differentially bioavailable by virtue of the relative strength of the organic bonds. The concept of different bond strengths is embodied in the "onion" theory (Mackey and Zirino, 1994) which conceptualises metal-organic ligands as layered with weaker bonds towards the surface of the complex. Further complications may arise through interactions between metal-organic ligands and other aspects of water chemistry such as pH and components of water hardness. In other words, the same elements that compete with potentially toxic trace metals for receptor sites on an organism, may also compete with organic ligands that would otherwise sequester metals in an unavailable form.

To make sense of the sometimes contradictory information concerning the effect of DOC on chemical bioavailability, we need to better understand events happening at or close to the surface of the receptor membrane. It is known that most dissolved organic molecules such as humic acid accumulate at both organic and inorganic surfaces and may be adsorbed to the surface: a process that, for inorganic surfaces, may involve a ligand exchange reaction. The degree and strength of the surface interaction is determined by the nature of the available surface ligands, in particular their charge, as well as those associated with the organic solute. Surface adsorption of organic matter to membrane may also be affected by pH. Campbell et al. (1997) demonstrated that adsorption of fulvic acid to whole *Chlorella* cells was inversely related to pH. How such phenomena affect the bioavailability of other compounds remains open to question and may depend on how the organic matter (DOC)-chemical complex is oriented relative to the receptor site on the membrane itself. Questions yet to be answered include:

- Does the presence of DOC alter the charge on the membrane in a way that alters its receptivity to other chemicals?
- Does the presence of the DOC change the nature or configuration of the surface ligands on the membrane in a way that alters chemical receptivity?
- Does the DOC-chemical complex present the chemical to the membrane in such a way as to enhance its passage across the membrane?
- Does the DOC-chemical complex remain in high concentration at or close to the membrane surface in a configuration that inhibits the interaction between the chemical and the receptor site on the membrane (i.e., does the DOC "compete" with the membrane)?

These conflicting influences sometimes create differences between field observations and results from controlled laboratory experiments. A corollary of this is that greater attention needs to be paid to the role of DOC in laboratory experiments involving natural water. For example, in a laboratory bioassay in which natural water, say from an estuary, might be diluted with tap water or deionised water to create a salinity gradient, the natural DOC in the estuarine water will also change, despite the fact that this is not the water quality parameter of primary interest. Unless this is taken into account, the role of DOC may be overlooked.

5.4   Role of particulates

In addition to respiratory inhalation and stomatal or cuticular entry of aerial pollutants into plants, there are two basic routes of exposure by which organisms accumulate chemicals: (1) direct transport across biological membranes in dissolved form and (2) ingestion of contaminated particulate material. The latter applies only to heterotrophs; except for certain air pollutants, plants can only be affected by chemicals in solution.

Following digestion, the latter may end up in largely soluble form prior to uptake into the organism, yet toxicologists tend to make a distinction between the initial form(s) in which toxicants are presented to organisms because this form often has an important effect on their bioavailability. In the aquatic environment, sorption to organic or inorganic particulate material may play an important role in controlling the availability of a broad range of chemicals. At the same time, many so-called particulates are, in fact, food for the organism; therefore, it is not possible to conveniently place particulate material into either of the foregoing biotic or abiotic categories. In aquatic toxicology, this subject has often been dealt with quite differently by geologists, chemists, and biologists, and to an extent, this section reflects such different approaches. However, one of the recent advances in aquatic toxicology has probably been a realisation on the part of both natural and biological scientists that neither approach is entirely satisfactory in explaining toxic chemical bioavailability. There is an increasing recognition that models explaining chemical bioavailability (and toxicity) need to incorporate both physical chemistry and physiology.

In the sedimentary environment, where particulate material is the dominant phase, a major consideration in determining chemical bioavailability is the degree to which a particular chemical will partition between the particulate and the aqueous phase. This relationship is described by the partition coefficient, which is largely dictated by the hydrophobicity of the chemical. Chemicals with high hydrophobicity (i.e., low water solubility) will tend to be driven out of solution and adsorb to the particulate phase. Such is the case with nonpolar organic compounds such as PAHs and PCBs. The organic carbon content of the sediment has proven to be a good predictor of its affinity for such compounds because of their tendency to bind to humic material. Partitioning organic chemicals between the sediment phase and interstitial water follows the relationship

$$\frac{C_{w}}{C_{s}} = K_{oc} \cdot f_{oc}$$

where C_{w} is chemical concentration in interstitial water, C_{s} is the chemical concentration in sediment, K_{oc} is the partition coefficient for the organic phase, and f_{oc} is the fraction of organic carbon in sediment. It has been shown that the octanol-to-water partition coefficient of a compound may approximate K_{oc}, thereby providing a convenient means of assessing the binding potential of a nonpolar organic compound in a sediment (see also Section 5.5). Thus, sediment-to-water or soil-to-water partition coefficients can be predicted from the following general equation:

$$\log K_{oc} = a \cdot K_{ow} + b$$

Several studies have indicated that bioaccumulation of organic chemicals in bulk sediments may depend on their concentration in interstitial water (Ankley et al., 1991), which may be considerably higher than in overlying waters. However, organisms that ingest sediment by burrowing activity or through filter feeding may accumulate chemicals from both particulate and aqueous phases, and it is important to understand the feeding behaviour of an organism when considering bioaccumulation of chemicals (see the following section).

Concentration of ionic organic compounds, metals, and metalloids in sediments is less affected by the organic carbon content of the sediment and more so by ionic influences such as pH, Eh, and cation exchange capacity. Nitrobenzenes and nitrophenols, for example, may form strong ionic bonds with mineral surfaces. Clays and amorphic oxides of silicon (Si) and aluminium (Al) may adsorb oxyanions metal and metalloids. Even with ionic organic compounds, the organic carbon content of the sediment may still influence adsorption. For example, both pH and organic carbon content are dominant factors in the partitioning behaviour of pentachlorophenol and the phenoxyacid pesticide Silvex.

Trace metals in sediments may be complexed as carbonates and aluminosilicates, adsorbed to oxides of iron and manganese, or incorporated into both soluble (low-molecular-weight humic) and insoluble organic complexes. Association with metal oxides may be altered by such influences as pH and oxidation state. Lower pH may result in the dissolution of hydrous oxides and release of trace metals. Metals may also be displaced from oxides and minerals at low pHs by hydrogen ions. Increasing salinity may cause trace metals to be displaced from particulate material by the major ionic components of seawater, sodium (Na), potassium (K), calcium (Ca), and magnesium (Mg). In estuarine environments, this may result in localised concentrations of trace metals in the water column at the turbidity maximum, where freshwater first encounters salt. The association of trace metals with metal oxides has been negatively correlated with their bioavailability. Luoma and Bryan (1978) found good correlation between Pb concentration in bivalve tissue and Pb : iron (Fe) ratios in associated sediments. They hypothesised that Pb availability may be controlled by iron oxide in the sediments.

Many important processes affecting chemical bioavailability in the sedimentary environment are controlled directly or indirectly by the oxygen status of sediments. Iron and manganese are more soluble in their reduced state and tend to precipitate as oxyhydroxides in oxygenated surface sediments. Cadmium, copper, nickel, and zinc are more soluble in the presence of oxygen, but may precipitate as sulphides under anoxic conditions. It has been postulated that the bioavailability and toxicity of cationic metals in sediments may be controlled by the formation of sulphides under anoxic conditions, which limit free metal ion concentrations to very low levels under equilibrium conditions (di Toro et al., 1992). Weak acid extraction of sulphides, termed acid volatile sulfides (AVS), and metals extracted during the same process have been used as a measure of metal bioavailability. Molar ratios of AVS to simultaneously extracted metals (SEM) less than one are taken to indicate that those metals would be available for bioaccumulation and toxicity; in other words,

$$AVS/SEM < 1 = \text{Bioavailable metal}$$

Some reports of AVS : SEM ratios less than 1, which do not apparently result in increased metal toxicity, have been attributed to the possible influence of colloidal carbon, although its role in determining metal bioavailability has yet to be clarified. Total percent sediment carbon has been shown to decrease metal bioavailability in the sedimentary environment. Analysis of clams and sediment from a dredge-spoil facility near Baltimore Harbor in the Chesapeake Bay indicated that even though sediment metal concentrations were typically highest in sediments with high total organic carbon (TOC), the biota-sediment concentration factor (BSCF) *decreased* with increasing TOC (Mason and Lawrence, 1999).

In studies of chemical bioaccumulation by benthic animals, it is important to consider the feeding habitat and behaviour of the organisms under investigation. Even though trace metal concentrations in some benthic species may reflect those of associated sediments, other sediment dwellers may rely on the overlying water as a source of food and, therefore, metals (Warren et al., 1998). A principal drawback of the AVS-based approach is its reliance on porewater as a controlling factor in metal bioavailability. Although this may be appropriate for some benthic organisms under highly contaminated conditions, many acquire trace metals through particulate ingestion. Under these circumstances, the AVS approach cannot be applied and, therefore, should not be regarded as a general paradigm (Lee et al., 2000).

Processes such as sedimentation, resuspension, and bioturbation may be critical in determining chemical distribution in the sediment, and the overall chemical mass will be heavily influenced by the size characteristics of the sediment. Fine silty sediments have a comparatively very large surface area compared with coarse grain (sandy) sediments, and differences in total metal concentration between two sites may simply reflect differences in sediment size profile rather than differences in chemical loading. Percent organic carbon and percentage silt:sand represent basic sediment characteristics for such comparisons. Refinements include an enrichment factor (EF) that is used to normalise the concentration of a contaminant trace metal to a sediment component unlikely to be substantially altered by man's activities (e.g., iron). Turekian and Wedepohl (1961) published trace metal levels for a "standard shale", assumed to be relatively free from contamination. Using this information, the EF is defined as the ratio of metal (M):Fe at site x compared with M:Fe for the standard shale.

$$EF = \frac{M_x/Fe_x}{M_{ss}/Fe_{ss}}$$

The EF is assumed to be independent of sediment size characteristics. Because the M:Fe ratio for the standard shale is assumed to be free from anthropogenic influence, an EF > 1 signifies chemical enrichment relative to the standard shale, thereby indicating trace metal input to the system resulting from human influence.

5.4.1 The importance of food
Much of the foregoing discussion of factors affecting chemical bioavailability in the sedimentary environment takes the view that the primary chemical source is the interstitial water and assumes that sedimentary particles and refractory

soluble complexes essentially "compete" with the organism in terms of chemical bioavailability.

In fact, the situation is complicated by the fact that many bottom-dwelling animals, particularly burrowing members of the benthic infauna, may process large volumes of sediment as food, thereby stripping off and digesting associated chemical contaminants. Depending on the nature of their respective digestive juices, such organisms may be capable of processing a proportion of "bound" chemicals from sediments. For example, Langston (1990) showed that the burrowing clam *Scrobicularia plana* was able to accumulate tributyltin (TBT) efficiently from sediment particles, and that this represented the most important source of TBT in this species. Several attempts have been made to simulate this biologically labile fraction by "digesting" sediment using a variety of dilute mineral acids, organic acids, bases, and oxidising agents. Weston and Mayer (1998) demonstrated that the bioavailability of benzo(a)pyrene from sandy sediments could be reasonably approximated with an in vitro digestion process using the digestive juice from the gastrointestinal tract of the large burrowing polychaete *Arenicola braziliensis*.

Where trophic relationships have been well established between different organisms, it has been possible to design experiments that follow the food-chain transfer of different chemicals. In an aquatic food chain, a typical study might measure the relative efficiency of chemical transfer between two or more trophic levels. For example, a study conducted by Lawson and Mason (1998) investigated the transfer of inorganic mercury ($HgCl_2$) and monomethyl mercury between a primary producer, the diatom *Thalassiosira weissflogii*; a primary consumer, the estuarine copepod *Eurytemora affinis*; and a secondary consumer, represented by larval sheepshead minnows (*Cyprinodon variegatus*). For copepods feeding on algae,

$$\text{Assimilation efficiency} = \frac{(A - B) \times C}{[(A - B) \times C] + D}$$

where A = Hg in copepods after feeding; B = Hg in copepods before feeding; C = number of copepods; D = Hg in fecal pellets.

Assimilation efficiency of mercury (Hg) in fish feeding on copepods was determined using preexposed copepods. In this case,

$$\text{Assimilation efficiency} = \frac{(F - E) \times G}{H \times I}$$

where E = average Hg concentration in control fish; F = Hg concentration in fish following exposure; G = final fish weight; I = number of contaminated copepods eaten. The authors found that the assimilation efficiency of monomethyl mercury by sheepshead minnows (*Cyprinodon variegatus*) was much higher than that of inorganic mercury (76% versus 37%). They postulated that higher assimilation efficiencies may be associated with toxicants found principally in the cytoplasm of prey organisms as opposed to those bound to cell membranes.

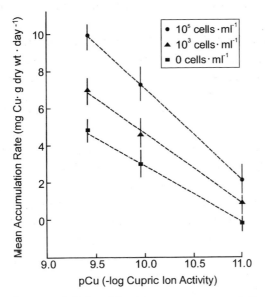

Figure 5.8 Role of food in copper accumulation by the American eastern oyster (*Crassostrea virginica*). Oysters were fed different daily concentrations of algae, *Thalassiosira pseudonana* cultured at three different cupric ion activities (i.e., pCu of 9.5, 10.0, and 11.0; pCu = −log$_{10}$ molar cupric ion concentration). The pCu of exposure media were 9.5, 10.0, and 11.0, respectively (Wright and Mason, 2000).

Variants on this experimental design are commonly used to determine relative assimilation efficiencies between trophic levels. An important experimental consideration is that animals should be allowed to purge their guts prior to analysis, to eliminate artefacts caused by residual (and, therefore, unassimilated) material in the digestive tract.

In many aquatic organisms, chemical accumulation from solution across the outer integument or through specialised epithelia such as gills may be as important as uptake through the digestive tract. Differential chemical uptake from food and water may be investigated by providing test organisms with controlled food rations of known chemical composition and comparing chemical bioaccumulation with unfed controls having access to the chemical only in soluble form. Figure 5.8 shows results from such an experiment, which was designed to investigate the importance of food as a source of copper for the oyster *Crassostrea virginica*. In this case, the food source was the diatom *Thalassiosira pseudonana*, which was grown at three different copper levels (shown in Figure 5.8 as cupric ion activity; pCu, $10 = 10^{-10} M Cu^{2+} L^{-1}$), and supplied at two different densities. The results from the unfed controls clearly indicate the overriding importance of water as a source of copper in this instance, although food contributes significantly to copper uptake. Obviously, the precise proportion contributed by food will vary according to relative food availability and the relative chemical concentration in food versus water.

In a field situation involving chemical analysis of mixed assemblages of organisms of different and, perhaps uncertain, trophic status, a different approach is required. The simplest means of establishing relative trophic position is by examining gut contents of different species to establish prey or food preferences. A more holistic approach is to measure isotopic ratios of light elements such as carbon and nitrogen in different species. During their passage through the food chain, lighter isotopes tend to be differentially excreted relative to heavier isotopes (i.e., the heavier isotope ^{15}N tends to be retained in preference to ^{14}N). The $^{15}N : ^{14}N$ ratio may, therefore, be used as an indicator of relative trophic level, and, when combined with an inventory of concentrations of chemical contaminants in different species, estimates may be made of their relative trophic transfer efficiencies.

5.5   Quantitative structure-activity relationships

Quantitative structure-activity relationships have been developed to help researchers predict toxic properties of chemicals from such physicochemical properties as molecular volume, molecular connectivity, electronic charge distribution, and octanol-to-water partition coefficient.

The idea behind the creation of QSARs is quite straightforward, and the method used to develop the relationships is easy to understand. It is assumed that each substructure of a compound makes a specific contribution to the properties of the compound. Thus, similar compounds (e.g., having the same functional groups) should have similar modes of action on target organisms or receptors. Simple or multiple regression is usually used to relate the properties of the chemicals of interest to one or more physicochemical parameters, which represent the properties of the substructures. Mathematical expressions used to describe the relationships are

$$Y = b + aX \qquad \text{for simple linear regression}$$

or

$$Y = b + a_1X_1 + a_2X_2 + a_3X_3 + \dots \qquad \text{for multiple regression}$$

where Y is the estimate of the property of interest, usually some form of toxic endpoint, X (or X_1) is the known physicochemical parameter(s) or descriptor(s), and a and b represent the slope and interval of the regression equation, respectively. If the value of R^2 (coefficient of determination) is high (e.g., 0.99), as much as 99% of the variability associated with dependent variable Y can be explained by the independent variables being considered. In Table 5.3 a range of different QSARs are summarised for a variety of species.

5.6   Implications for future environmental regulation

Most scientific endeavours undergo distinct phases beginning with the intense, and often somewhat indiscriminate, data gathering, leading to the characterisation of some general rules or patterns. In environmental toxicology, from the 1960s

Table 5.3. Examples of quantitative structure activity relationships

Species	Chemicals	Toxic end-point(s)	Independent variables in QSAR	R^2	Reference
Fathead minnow (*Pimephales promelas*)	Polychlorinated dibenzodioxins	Bioconcentration factor	K_{ow}, molecular volume, Molecular redundancy index, dipole moment, oK index	0.94	de Voogt et al. (1990)
Carp, guppies, golden orfe, fathead minnow, brine shrimp, *Daphnia magna*	77 organic chemicals include benzenes, anilines, phenols	LC_{50}	K_{ow}	0.74 (golden orfe), 0.95 (brine shrimp, guppies)	Zhao et al. (1993)
8 fish species (*Daphnia magna*)	Phenylurea herbicides	LC_{50}	K_{ow}	0.82–0.99	Nendza (1991)
3 algae (*Scenedesmus; Microcystis; Nostoc*)	Phenylurea herbicides	IC_{50} for Hill reaction, photosynthesis	K_{ow}, electronic properties	0.93	Takemoto et al. (1985)
Sole (*Solea solea*), flounder (*Platichthys flesus*), copepod (*Tisbe battaglia*)	Chlorophenols	24-hr, 96-hr LC_{50}	K_{ow}, dossociation constant, molecular connectance (regressions calculated on 2 out of 3)	0.84–0.99	Smith et al. (1994)
Zebra fish (*Brachydanio rerio*)	Chloro- and nitroanilines	Bioconcentration factor	K_{ow}	0.93	Kalsch et al. (1991)
Photobacterium (= *Vibrio*) *phosphoreum*	Metal ions	Bioluminescence inhibition (EC_{50})	Various variables including electronegativity, ionic radius, softness index	0.73–0.84	McCloskey et al. (1996)
Tetrahymena pyriformis	Alkly, nitro, halogenated, alkoxy, and aldehyde substituted phenols	50% population growth inhibition	K_{ow}, molecular connectivity, electronic energy descriptors	0.75	Cronin and Schultz (1996)

onward, we developed an appreciation of the principal classes of toxicant and their relative toxicity to different biological taxa. This information has served as the basis for many of the guidelines governing the worldwide regulation of environmental contaminants.

Having established these basic principles governing environmental release of toxic chemicals, we have entered the more difficult phase of incorporating an enormous variety of modifying influences into our understanding of how toxic chemicals affect the environment. Chapter 4 dealt with numerous subtle effects of toxic agents on different levels of biological organisation. In this chapter, we outlined several parameters affecting the toxicity and bioavailability of the chemical contaminants themselves, acting both singly and in combination.

Already the QSAR approach has been well developed for the registration of new industrial, pharmaceutical, and agricultural chemicals. However, the incorporation of concepts such as differential bioavailability into legislation governing environmental contaminants is a much more difficult proposition. Current guidelines are based almost exclusively on total concentrations of single chemicals. Yet, there has long been a recognition that physicochemical modifiers of toxicity and the speciation of elements and compounds must eventually be taken into account in providing effective and realistic environmental legislation. A typical dilemma may be posed as follows. If, in a particular system, >90% of a toxic contaminant is complexed in a form that is unavailable to biota, it would be unrealistic to regulate it on the basis of its total concentration. In a different system, >90% of the toxicant may be bioavailable and would require different criteria.

This subject represents one of the more active and challenging research areas in environmental toxicology and has led to some rewarding collaborations between biologists and chemists, particularly in the field of membrane physiology and biophysics. As a useful text on several aspects of this subject, the student is directed to a collection of papers edited by Hamelink et al. (1994).

Treatment of toxic chemical mixtures remains an extremely difficult subject, principally because of the complexity of appropriate experimental design. This aspect will, no doubt, be simplified by the use of increasingly sophisticated (chemical) structural and statistical models coupled with toxic biological end-points, which are manageable in large matrix experiments. In field situations, more empirical approaches, such as toxicity reduction evaluations (see Chapter 2), are likely to play an increasingly important part in determining specific toxic components of chemical mixtures.

5.7   Questions

1. Give a summary of test organisms commonly used for laboratory toxicity testing. What factors influence inter- and intraspecific variability? Give an example of at least one quantitative strategy that has been adopted to provide protection from toxic chemicals for the majority of aquatic species.

2. Describe the different ways in which temperature affects chemical toxicity. What experimental precautions need to be observed in laboratory bioassays to account for the effect of temperature?

3. Define pH and alkalinity. How does pH affect the chemical form of a toxicant in water? Describe the different ways in which chemical speciation affects the bioavailability of toxic chemicals.

4. Give an account of the factors influencing the concentration and bioavailability of (a) trace metals and (b) organic compounds in estuarine sediments.

5. Define the terms *synergism*, *antagonism*, and *additive toxicity*. Describe and compare the different strategies that have been used to quantify the toxicity of chemical mixtures in laboratory assays. Give an example of how one or more of these approaches have been modified to determine the toxicity of a mixture of chemicals in the ambient environment.

6. *Dissolved organic carbon* is a somewhat amorphous term used to describe a variety of naturally occurring organic solutes in the aquatic environment. What are the principal components of DOC in fresh and estuarine water? What role does DOC play in determining the bioavailability of trace metals in the aquatic environment?

7. How would you conduct experiments to investigate the importance of food as a source of toxicant to an aquatic organism? How would you determine (a) assimilation efficiency and (b) trophic transfer efficiency?

8. How have quantitative structure activity relationships been employed in environmental toxicology? Give a specific account of the use of the octanol : water partition coefficient as a QSAR.

5.8   References

Ankley, G. T., Phipps, G. L., Leonard, E. N., Benoit, D. A., Mettson, V. R., Kosian, P. A., Cotter, A. M., Dierkes, J. R., Hansen, D. J., and Mahony, J. D. (1991) Acid-volatile sulfide as a factor mediating cadmium and nickel bioavailability in contaminated sediments, *Environmental Toxicology and Chemistry*, **10**, 1299–307.

Brett, J. R. (1960) Thermal Requirements of Fish – Three Decades of Study. In *Biological Problems of Water Pollution*, ed. Tarzwell, C. M., pp. 110–17, U.S. Department of Health, Education & Welfare, Robert A. Taft Sanitary Engineering Center, Cincinatti, OH.

Broderius, S. J., and Kahl, M. (1985) Acute toxicity of organic chemical mixtures to the fathead minnow, *Aquatic Toxicology*, **6**, 307–22.

Cairns, J., Jr., Heath, A. G., and Parker, B. G. (1975) Temperature influence on chemical toxicity to aquatic organisms, *Journal Water Pollution Control Federation*, **47**, 267–80.

Campbell, P. G. C., Twiss, M. R., and Wilkinson, K. J. (1997) Accumulation of natural organic matter on the surfaces of living cells: Implications for the interaction of toxic solutes with aquatic biota, *Canadian Journal of Fisheries and Aquatic Sciences*, **54**, 2543–54.

Cronin, M. T. D., and Schultz, T. W. (1996) Structure-toxicity relationships for phelols to *Tetrahymena pyriformis*, *Chemosphere*, **32**, 1453–68.

de Voogt, P., Muir, D. C. G., Webster, G. R. B., and Govers, H. (1990) Quantitative structure-activity relationships for the bioconcentration in fish of seven polychlorinated dibenzo-dioxins, *Chemosphere*, **21**, 1385–96.

di Toro, D. M., Mahony, J. D., Hansen, D. J., Scott, K. J., Carlson, A. R., and Ankley, G. T. (1992) Acid volatile sulfide predicts the acute toxicity of cadmium and nickel in sediments, *Environmental Science and Technology*, **26**, 96–101.

Erickson, R. J., Benoit, D. A., Mattson, V. R., and Nelson, J. H. P. (1996) The effects of water chemistry on the toxicity of copper to fathead minnows, *Environmental Toxicology and Chemistry*, **15**, 181–93.

Giesy, J. P., and Graney, R. L. (1989) Recent developments in the intercomparisons of acute and chronic bioassays and bioindicators, *Hydrobiologia*, **188/189**, 21–60.

Gill, K. A., and Walsh, P. J. (1990) Effects of temperature on metabolism of benzo(a)pyrene by toadfish (*Opsanus beta*) hepatocytes, *Canadian Journal of Fisheries and Aquatic Sciences*, **47**, 831–8.

Haas, C. N., and Stirling, B. A. (1994) New quantitative approach for analysis of binary toxic mixtures, *Environmental Toxicology and Chemistry*, **13**, 149–56.

Hall, J. L. W., and Anderson, R. D. (1995) The influence of salinity on the toxicity of various classes of chemicals to aquatic biota, *Critical Reviews in Toxicology*, **25**, 281–346.

Hamelink, J. L., Landrum, P. F., Bergman, H. L., and Benson, W. H. (1994) *Bioavailability: Physical, Chemical, and Biological Interactions*, Special SETAC Publication Series, Lewis, Boca Raton, FL.

Henderson, C., Pickering, Q. H., and Cohen, J. M. (1959) The toxicity of synthetic detergents and soaps to fish, *Sewage and Industrial Wastes*, **31**, 295–306.

Hodson, P. V., and Sprague, J. B. (1975) Temperature-induced changes in acute toxicity of zinc to Atlantic salmon (*Salmo salar*), *Journal of the Fisheries Research Board of Canada*, **32**, 1–10.

Kalsch, W., Nagel, R., and Urich, K. (1991) Uptake, elimination, and bioconcentration of ten anilines in zebrafish (*Brachydanio rerio*), *Chemosphere*, **22**, 351–63.

Könemann, H. (1981) Quantitative structure-activity relationships in fish toxicity studies. Part I. Relationships for 50 industrial pollutants, *Toxicology*, **19**, 209–21.

Kyono-Hamaguchi, Y. (1984) Effects of temperature and partial hepatectomy on the induction of liver tumors in *Oryzias latipes*, *National Cancer Institute Monograph*, **65**, 337–44.

Langston, W. J. (1990) Bioavailability and effects of TBT in deposit-feeding clams, *Scrobicularia plana*. Proceedings of the 3rd International Organotin Symposium, Monaco. pp. 110–13.

Lawson, N. M., and Mason, R. P. (1998) Accumulation of mercury in estuarine food chains, *Biogeochemistry*, **40**, 235–47.

Lee, B.-G., Griscom, S. B., Lee, J.-S., Choi, H. J., Koh, C.-H., Luoma, S. N., and Fisher, N. S. (2000) Influences of dietary uptake and reactive sulfides on metal bioavailability from aquatic sediments, *Science*, **287**, 282–4.

Luoma, S. N., and Bryan, W. G. (1978) Factors controlling the availability of sediment-bound lead to the estuarine bivalve *Scrobicularia plana*, *Journal of Marine Biological Association of the United Kingdom*, **58**, 793–802.

Mackey, D. J., and Zirino, A. (1994) Comments on trace metal speciation in seawater, or do "onions" grow in the sea?, *Analytica Chimica Acta*, **284**, 635–47.

Mason, R. P., and Lawrence, A. L. (1999) Concentration, distribution, and bioavailability of mercury and methylmercury in sediments of Baltimore Harbor and Chesapeake Bay, Maryland, U.S.A., *Environmental Toxicology and Chemistry*, **18**, 2438–47.

Mason, R. P., Reinfelder, J. R., and Morel, F. M. M. (1996) Uptake, toxicity and trophic transfer of mercury in a coastal diatom, *Environmental Science and Technology*, **30**, 1835–45.

Mayer, F. L., Jr., and Ellersieck, M. R. (1986) Manual of acute toxicity: Interpretation and data base for 410 chemicals and 66 species of freshwater animals. Resource Publication 160, U.S. Fish and Wildlife Service, Washington, DC.

McCloskey, J. T., Newman, M. C., and Clark, S. B. (1996) Predicting the relative toxicity of metal ions using ion characteristics: Microtox (R) bioluminescence assay, *Environmental Toxicology and Chemistry*, **15**, 1730–7.

Miller, T. G., and Mackay, W. C. (1980) The effects of hardness, alkalinity, and pH of test water on the toxicity of copper to rainbow trout (*Salmo gairdneri*), *Water Research*, **14**, 129–33.

Müller, W. E. G., Koziol, C., Kurelec, B., Dapper, J., Batel, R., and Rinkevich, B. (1995) Combinatory effects of temperature stress and nonionic organic pollutants on stress protein (hsp70) gene expression in the freshwater sponge *Ephydatia fluviatilis*, *Environmental Toxicology and Chemistry*, **14**, 1203–8.

Nendza, M. (1991) Predictive QSAR estimating ecotoxic hazard of phenylureas: aquatic toxicology, *Chemosphere*, **23**, 497–506.

Notari, R. E. (1987) *Biopharmaceutics and Clinical Pharmacokinetics: An Introduction*, 4th ed., Dekker, New York.

Smith, S., Furay, V. J., Layiwola, P. J., and Menezes-Filho, J. A. (1994) Evaluation of the toxicity and quantitative structure-activity relationships (QSAR) of chlorophenols to the copepodid stage of a marine copepod (*Tisbe battagliai*) and two species of benthic flatfish, the flounder (*Platichthys flesus*) and sole (*Solea solea*), *Chemosphere*, **28**, 825–36.

Sunda, W. G., Engel, D. W., and Thuotte, R. M. (1978) Effects of chemical speciation on the toxicity of cadmium to the grass shrimp *Palaemonetes pugio*: Importance of free cadmium ion, *Environmental Science and Technology*, **12**, 409–13.

Suter, II, G. W., and Rosen, A. E. (1988) Comparative toxicology for risk assessment of marine fishes and crustaceans, *Environmental Science and Technology*, **22**, 548–56.

Swartz, R. C., Schults, D. W., Ozretich, R. J., DeWitt, T. H., Redmond, M. S., and Ferrano, S. P. (1995) ΣPAH: A model to predict the toxicity of polynuclear aromatic hydrocarbon mixtures in field-collected sediments, *Environmental Toxicology and Chemistry*, **14**, 1977–87.

Takemoto, I., Yoshida, R., Sumida, S., and Kamoshita, K. (1985) Quantitative structure-activity relationships of herbicidal N-substituted phenyl-N-methoxy N-methylureas, *Pesticide Biochemistry and Physiology*, **23**, 341–8.

Turekian, K. K., and Wedepohl, K. H. (1961) Distribution of the elements in some major units of the earth's crust, *Geological Society of America Bulletin*, **72**, 175–92.

Warren, L. A., Tessier, A., and Hare, L. (1998) Modelling cadmium accumulation by benthic invertebrates *in situ*: The relative contributions of sediment and overlying water reservoirs to organism cadmium concentrations, *Limnology and Oceanography*, **43**, 1442–54.

Welsh, P. G., Skidmore, J. R., Spry, D. J., Dixon, D. G., Hodson, P. V., Hutchinson, N. J., and Hickie, B. E. (1993) Effect of pH and dissolved organic carbon on the toxicity of copper to larval fathead minnow (*Pimephales promelas*) in natural lake waters of low alkalinity, *Canadian Journal of Fisheries and Aquatic Sciences*, **50**, 1356–62.

Weston, D. P., and Mayer, L. M. (1998) Comparison of *in vitro* digestive fluid extraction and traditional *in vivo* approaches as a measure of PAH bioavailability from sediments, *Environmental Toxicology and Chemistry*, **17**, 830–40.

Wilson, J. E. H., and Costlow, J. D. (1987) Acute toxicity of diflubenzuron (DFB) to various life stages of the grass shrimp, *Palaemonetes pugio*, *Water, Air, and Soil Pollution*, **33**, 411–17.

Winner, R. W. (1985) Bioaccumulation and toxicity of copper as affected by interactions between humic acid and water hardness, *Water Research*, **19**, 449–55.

Wood, C. M., and MacDonald, D. G. (1982) Physiological mechanisms of acid toxicity to fish. In *Acid Rain/Fisheries: Proceedings of an International Symposium on Acidic Precipitation and Fishery Impacts in Northeastern North America*, ed. Johnson, R. E., American Fisheries Society, Bethesda, MD.

Wright, D. A., and Mason, R. P. (2000) Biological and chemical influences on trace metal toxicity and bioaccumulation in the marine and estuarine environment, *International Journal of Environment and Pollution*, **13**, 226–48.

Zhao, Y. H., Wang, L. S., Gao, H., and Zhang, Z. (1993) Quantitative structure-activity relationships – Relationships between toxicity of organic chemicals to fish and to *Photobacterium phosphoreum*, *Chemosphere*, **26**, 1971–9.

5.9   Further reading

Factors affecting bioavailability of toxicants: Hamelink, J. L., P. F. Landrum, H. L. Bergman, and W. H. Benson. 1994. *Bioavailability: Physical, Chemical, and Biological Interactions*, Special SETAC Publication, CRC Press, Boca Raton, FL.

Organism-level effects on toxicity and chemical mixtures: Suter, G. 1993. Organism-Level Effects. In *Ecological Risk Assessment*, ed. Suter, G. W., pp. 175–246, Lewis, Chelsea, MI.

6

○ ○

Metals and other inorganic chemicals

6.1 Introduction

Inorganic contaminants include metals, metalloids, and a number of relatively simple molecules such as phosphate and ammonia. The ways in which inorganic substances can become problematic in the environment, and thus be considered as contaminants, are frequently as a result of their being mobilised or modified chemically by human activities. In contrast to organic contaminants (Chapter 7), many of which are xenobiotic, many inorganic contaminants occur naturally; ecosystems do not distinguish between natural and anthropogenic substances. In terms of their regulation, as well as their management, inorganic contaminants are expected to differ in many respects from xenobiotic substances.

Another point of contrast between inorganic and organic contaminants is that a number of inorganic substances not only are potentially toxic but also are required as nutrients. Such substances exemplify the oft-quoted statement that the dose defines the poison (Paracelsus, c 1493–1541, cited in Rodricks, 1993).

In the context of dose-response, those substances that are nutrients as well as potential toxicants can be put into a simple model, as shown in Figure 6.1, which invokes the concepts of deficiency, sufficiency, and toxicity, successively. For an element such as copper, this succession is theoretically a relatively simple series of transitions. The dose of copper is shown as supply, on the x-axis, and the response as growth rate is shown on the y-axis. Very low doses of copper can result in nutrient deficiency, with below optimum growth; as the supply increases, up to a certain point, there is a positive response. This is the range over which copper is required as a micronutrient. After the optimum concentration is reached, there is no further positive response and normally there is a plateau, as seen in Figure 6.1. At increasingly higher doses, the copper is in excess and begins to have harmful effects, which cause growth rate to decrease (i.e., it has reached concentrations that are toxic). The logical result of further increase in copper dose beyond that shown on the graph would be death.

For other metals, depending upon whether they are required in trace amounts, in large amounts, or are not required at all, the dose-response patterns vary. This

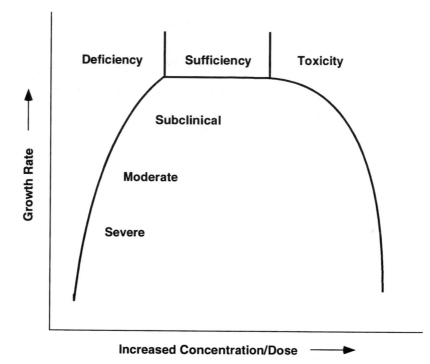

Figure 6.1 A schematic representation of a dose-response for a micronutrient such as copper. Using growth rate as an end-point, the response to increased doses of a micronutrient such as copper includes stimulation from a less-than-optimum dose up to the point of sufficiency. For a certain range of concentration, there is no further change in response, and the growth rate reaches a plateau. As the dose reaches higher concentrations, growth rate decreases because the metal is now exerting toxic effects. Modified from Thornton (1995).

concept has been captured as three basic patterns of dose-response, as illustrated in Figure 6.2. In this figure, stimulation is shown as a positive response, and inhibition is shown as a negative response, both on the y-axis. The type A pattern, exemplified by a macronutrient such as calcium, is not discussed in any further detail in this chapter. The type B pattern, for metals that are required in small amounts but are toxic at higher concentrations, is exemplified by copper and selenium (Sections 6.7 and 6.9, respectively), whereas the type C pattern, where there is no positive response, is exemplified by mercury (Hg), lead (Pb), cadmium (Cd), and nickel (Ni) (Sections 6.4, 6.5, 6.6, and 6.8, respectively).

For those substances required as nutrients, the borderline or transition point between the dose which is beneficial and that which is potentially harmful may be difficult to determine, particularly in the real world. The reasons for this include firstly the often narrow range of concentration in the medium (water, soil, etc.) which represents the optimum for the biological system. Second, and more impor-

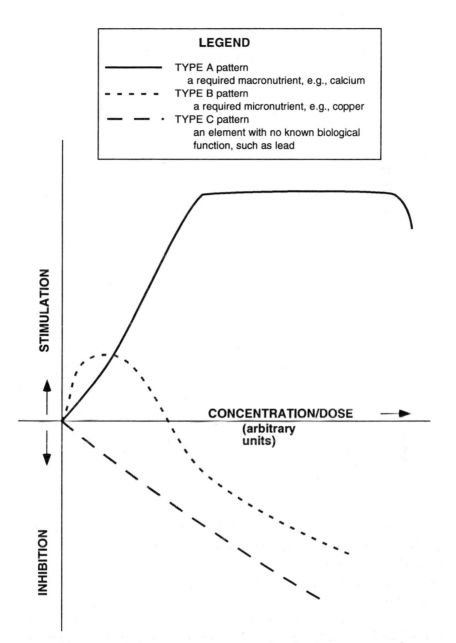

Figure 6.2 A schematic representation comparing the dose-response of a macronutrient element, a micronutrient element, and a nonessential toxic element. The type A pattern is typical of a macronutrient such as calcium. As the dose is increased, there is stimulation up to the point of optimum dose. Increases beyond this elicit no further response for a wide range of concentration. Finally, at extremely high doses, the response may decrease, and theoretically there could be a dose that is high enough to be inhibitory. The type B pattern is typical of a micronutrient such as copper, with stimulation over a low range of low dose, a short plateau, followed by increasing inhibition at higher doses. This is the same response illustrated in Figure 6.1. The type C pattern is typical of a substance such as mercury or cadmium that has no biological function and that is consistently toxic. There is no stimulatory effect, and inhibition increases with dose.

tantly, the response depends on the chemical form of the element, which affects its biological availability. The element selenium (Se) exemplifies the influence of chemical form on availability, as described later. (More details on selenium are provided in Section 6.9.)

Livestock require selenium, which they normally acquire from the pasture plants on which they feed. In agriculture, concentrations of Se below $0.5\,\mu g/g$ in soil are associated with pasture plants that contain inadequate amounts of Se to support livestock. In contrast, soils containing 2–$80\,\mu g/g$ Se may give rise to selenium poisoning in cattle. Selenium exists in a number of chemical forms, which are not equally available to the pasture plants. For example, selenate [Se(VI)] in soil is soluble, mobile, and available to plants, whereas selenite [Se(IV)] is considered to be biologically unavailable. Unfortunately, the relationship between chemical form and biological availability is rarely understood as clearly as in this example of selenium. For substances that have no known biological requirement, the concept of biological availability is equally important. Chemical form and biological availability are discussed more fully in Sections 6.2.2 and 6.2.3.

Some generalisations can be made concerning the toxic effects of metals on living systems. Trace metals may bind to different functional groups on an enzyme and alter enzyme activity. Usually these reactions result in enzyme deactivation, although binding of metallic cations to enzymes can also stimulate their catalytic function. Experiments on fish have demonstrated that copper (Cu) and lead (Pb) in vitro inhibit alkaline phosphatase, but in vivo, exposure to Cu or Pb may stimulate enzyme activity. Variations in activity of enzymes that catalyse essentially irreversible (rate-limiting) reactions may have important implications for certain aspects of intermediary metabolism. For example, cadmium alters glucose metabolism in mammals by affecting the activity of rate-limiting enzymes such as glucose-6-phosphatase.

Within the scope of this chapter, only a small sample of inorganic substances can be dealt with in detail. The metals and metalloids that are generally considered to be of major concern in environmental toxicology have been selected for the appendix. Of necessity, only a small number can be dealt with in detail in the text.

In the present chapter, a selection of these was also made to illustrate some of the common features and differences among metallic toxic substances and metalloids, as well as to highlight some of the metals that are of major concern. In addition, two other inorganic substances that have caused major environmental problems, namely phosphorus (P) and fluorine (F), will also be discussed. Arguably the most important of these is phosphate. Itself a nutrient that is frequently a limiting nutrient in natural systems, excesses or imbalance of phosphate are not of themselves directly toxic. Excess phosphorus results in an ecosystem phenomenon known as eutrophication, a study of which serves to illustrate many principles of ecotoxicology (see also Section 9.7.3 and Case Studies 4.5 and 9.1).

6.2 The properties and environmental behaviour of metals and metalloids

6.2.1 General properties of metals and metalloids

All elements of the periodic table can be classified as metals or nonmetals, with the exception of a small number of so-called metalloids. The main physical and chemical properties that chemically distinguish metals from nonmetals include

- Having a characteristic lustre (in the elemental state);
- Being malleable and ductile;
- Characteristically forming positively charged ions (cations);
- Having higher melting points and boiling points than nonmetals (i.e., solid at normal temperature and pressure);
- Having high density compared with nonmetals;
- Having the ability to form basic oxides;
- Usually being good conductors of heat and electricity;
- Often displacing hydrogen from dilute nonoxidising acids.

Figure 6.3 shows three main subgroups in the context of the periodic table of elements: metals, nonmetals, and metalloids. The metalloids display some of the same properties as metals, and a number of them are of considerable interest for their environmental toxicology. In contrast to true metals, which form cations, the metalloids such as arsenic (As) and selenium can form compounds in which they behave either as cations or anions (see, for example, selenate and selenite, Section 6.1).

Within the metal group of elements, chemists, biochemists, and toxicologists have attempted to further classify or categorise the elements, in terms of their function (e.g., essential versus nonessential, major versus minor nutrients, potentially toxic versus benign), as well as in terms of their electrochemical properties. Furthermore, in the literature, confusion has sometimes resulted from the use of various qualifying terms for metals, including trace metals and heavy metals. Heavy metals are generally characterised as those with specific density greater than $6\,\mathrm{g\,cm^{-1}}$, and this property includes most, but not all, of the most toxic elements. Several authors have suggested better ways of characterising the various metals and have proposed rejecting the nondescript and undefined term *heavy metals*. Nieboer and Richardson in 1980 presented an argument for classifying metals according to their coordination chemistry, which demonstrates their respective tendency to seek oxygen (O) or nitrogen (N)/sulphur (S) groups in bonding. The biochemical roles of metal ions in biological systems, including their toxicity, match well with the proposed classification, based as it is on their bonding preferences.

The various suggestions for classifying metals, of which Nieboer and Richardson's is arguably the most clearly presented in terms of the biological significance of metals, have the merit that they offer some theoretical bases upon which to predict the behaviour of metals in biological systems. If such predictions were

metals			metalloids		non-metals			
Li	Be		B	C	N	O	F	Ne
Na	Mg		Al	Si	P	S	Cl	Ar
K	Ca	Zn	Ga	Ge	As	Se	Br	Kr
Rb	Sr	Cd	In	Sn	Sb	Te	I	Xe
Cs	Ba	Hg	Tl	Pb	Bi	Po	At	Rn

Figure 6.3 Part of the periodic table of the elements, distinguishing among metals, metalloids, and nonmetals. The main group elements are divided into nine groups (corresponding to the nine vertical columns). The columns are respectively (left to right) the alkali metals, the alkaline earths, the zinc-cadmium-mercury group, the boron group, the carbon group, the nitrogen group, the oxygen group, the halogens, and the noble gases. Modified from Harrison and de Mora (1996).

shown to be of practical use, then the toxicology of metals and other inorganic substances could be addressed, for example by modellers and regulators, in a systematic manner comparable to the approach that is used for many organic substances (see Section 5.5). However, it is fair to say that none of the proposed systems has received sufficient acclaim or come into practical use enough for us to advocate their routine application in the study or practice of toxicology.

For the present text, in keeping with Thornton's 1995 suggestion, the terms *metals* and *inorganic chemicals* have been used to avoid ambiguity or unnecessary complexity.

6.2.2 The mobilisation, binding, and chemical forms of metals in the environment

Metals occur naturally. Ever since early civilisations recognised the enormous potential of metals, human endeavours have been extracting various metals from geological material and, in so doing, have mobilised the elements. Mobilisation normally involves changing the chemical and/or physical form of an element, and sometimes its location. When this happens, it is reasonable to expect that its biological effect(s) may also change. To take a simple example, aluminium (Al) is the most common element in the Earth's crust, but in most natural occurrences, it is

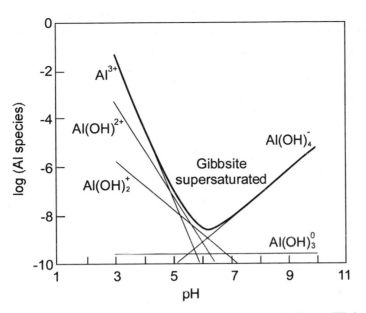

Figure 6.4 Activities of dissolved aluminium species in equilibrium with Al(OH)$_3$ (Gibbsite) at 25°C. The diagram shows the forms of Al that would be present in dilute aqueous solution, from pH 3–10. The heavy line is the sum of individual activities. Modified from Drever (1997) from data by Wesolowski and Palmer (1994).

found as insoluble chemical compounds or complexes such as aluminosilicates. These have little or no effect on biological systems. Weathering releases some Al from rocks and soils, but the elements still remain for the most part in insoluble form. The chemical forms (termed the chemical speciation) in which Al exists are determined in large part by pH, as shown in Figure 6.4 for a simple aqueous solution of aluminium. When acidity increases, aluminium is released from soils and rocks into soil solution by dissolution of aliminium hydroxides [e.g., Al(OH)$_3$ (s)] and by desorption of aluminium bound at the surface of soil organic matter. The dissolved aluminium can then move with the soil solution from watersheds into water bodies. Dissolved aluminium can be toxic to plants and to aquatic biota. Most insoluble forms of Al, although they may have physical effects of aquatic biota through clogging membranous surfaces, are not strictly speaking toxic for the simple reason that they are not biologically available.

The emergence of concern for acidic deposition (Sections 5.3.2 and 9.2) stimulated research not only on the effects of acidification per se but also on the effects of acidification on metals. One of the metals of major interest was Al. As a result, a great deal of information on the geochemistry and environmental toxicology of Al was published. Not only does the chemical speciation of Al show strong pH dependence (Figure 6.4), but the toxicity of Al also shows a strong dependence on hydrogen ion concentration. In the field, the relative influence of hydrogen ion and Al are difficult to factor out because the naturally occurring Al becomes solubilised, and the two potentially harmful ions do not vary entirely independently. It was hypothesised that the adverse effects of acidification on aquatic life, particularly for fish, and on certain forests, was due in part to the increased solubility and consequent mobilisation of Al ions, which accompanied acidification. Note that Al is rarely itself a contaminant in the sense that it is added to ecosystems through human activities; the element in effect becomes a contaminant when the pH decreases below a certain point (see Section 9.6.2).

Metal concentrations in natural waters are normally controlled by sorption reactions, which are inherently pH-sensitive. Because the hydrogen ion has a major influence on many other geochemical and biological processes as well, the hydrogen ion (or pH) is known as a major driving variable in environmental toxicology.

Another factor considered to be a major driving variable is organic matter (OM), particularly dissolved organic carbon. It affects the mobility of metals in a number of complex ways (see also Section 5.3.6). Organic matter, whether natural in origin (simple organic acids such as citrate and acetate, complex organic molecules such as humic and fulvic acids) or synthetically derived [e.g., ethylenediaminetetraacetic acid (EDTA) and nitrilotriacetic acid (NTA)], provides ligands to which cations bind rather readily. The resulting combinations are called complexes.

Equation 6.1 shows the general type of metal-ligand interaction.

$$M^2 + L \overset{K_1}{\rightleftharpoons} ML$$

where K_1 is the conditional equilibrium constant

$$K_1 = \frac{[ML]}{[M^{2+}][L]} \tag{6.1}$$

If the organic ligand has certain properties, being multidentate, with two or more donor groups such that the organic molecule surrounds a central metal cation, the resulting complex is called a chelate and the complexation is termed chelation (Figure 6.5). Students are warned that the term chelation is often used erroneously for any type of complexation. If there is any doubt concerning true chelation, the more general term complexation should be used.

The strength or stability of the metal-ligand binding depends upon the chemical nature of the ligand, the metal itself, and the pH. The strength of binding is expressed as the conditional stability constant (K_1 in Equation 6.1). The significance of the binding strength to biological availability of the metal, and thus the biological effect(s) of the binding, may be very high (see availability and the free ion activity model in Section 6.2.3).

6.2.3   The biological availability of metals in the environment

In Section 6.1, using the example of the relative biological significance of selenite and selenate, the concept of biological availability or bioavailability was introduced.

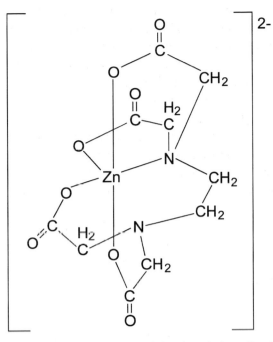

Figure 6.5 The chelation of zinc by ethylenediaminotetraacetic acid. The diagram shows the five rings surrounding the central zinc atom, as an example of chelation, which is a special case of complexation. Modified from Harrison and de Mora (1996).

The concept of bioavailability can be understood by considering the fact that a simple measurement of "total" metal (i.e., of the concentration of the element regardless of its chemical species or the distribution of these) is not a good predictor of its biological effect (stimulatory or inhibitory). To affect a biological system, the metal must be in a form that can be "seen" by the biological system. Put simply, this means that the metal must interact with the surface or the interior of the cell(s). Not all forms of a metal meet the criterion.

The most elementary distinction for forms of elements is between the dissolved and solid phases. Plants can only take up nutrients or contaminants in dissolved form. But animals (e.g., filter-feeders) will take in particulates and, if these have contaminants associated with them, may subsequently assimilate the contaminants. Thus, the elementary distinction between dissolved and solid phases is not a simple one in terms of the biological availability of a contaminant. A sample of water that has been filtered is chemically different from "whole" water. Even the definition of *filtered* is not absolute, but most water samples that are described as filtered have been passed through a filter with 0.45-μm pores. This operational definition of *filtered* is fairly universal in its use. The effect of filtration on the concentration of chemicals in the filtrate will obviously be greater for a water sample that has a large amount of suspended sediment or colloidal material than for a clearer water. Furthermore, different elements will be affected differently by the filtration process. The filter may itself contribute contamination of the sample, which leads to the recommendation, not always addressed, that filters themselves be pretreated to removed trace contaminants. This treatment can be accomplished by purchasing filters that have been pretreated by the manufacturer and have specifications concerning their purity or by treating the filters in the laboratory by processes such as washing with acid. Other concerns related to the act of filtering water include several types of artefacts (Horowitz et al., 1992). These authors warn that the pore size of the filter can change during the process of filtration, as the pores become clogged with solid particles. Such an effect cannot be readily corrected for because its extent will vary with the nature of the solid material, which itself can be a seasonal phenomenon. Thus, apparent seasonal or annual trends in concentrations may not reflect real environmental conditions. These authors point out that the apparently simple designation of concentrations in filtered water may have implications not only for research but also for regulatory aspects of water pollution. Therefore, in any review of water chemistry, it is important to identify whether the water has been filtered. And it is equally important to establish a protocol and rationale for filtration when designing a field-sampling project.

The simplest example of determining availability of a contaminant or a nutrient is provided for plants, particularly simple plants such as algae, which have only one uptake vector. These organisms will not "see" metals if they are not dissolved. Thus, a crude measure of metal availability for plants would be dissolved versus undissolved. It has been known since the mid 1970s that, even if a metal is in dissolved form, it is not necessarily available to the plant. Even before the 1970s, there was qualitative evidence that certain ligands, while maintaining metals in solution, nevertheless changed their availability to biota. Much of the early work on the role of ligands was done with synthetic organic ligands such as EDTA (see Section 6.2.2) and aquatic biota such as algae, planktonic invertebrates, and fish. The effect of EDTA or NTA was to decrease the toxicity or uptake of a metal, as shown in Figures 6.6 and 6.7. Figure 6.6 illustrates for a laboratory experiment that the toxicity of copper to the freshwater shrimp is related to the cupric ion, not to the total amount of copper. Because the ligand NTA has a known binding capacity for copper, the system is in effect chemically defined, and the amount of free ion can be calculated. The explanation for the decrease in metal toxicity in the presence of the complexing ligand is that the metal-ligand complex is not biologically available. Figure 6.7 shows another laboratory experiment with copper and an organic ligand, in this case, EDTA. The parameter of interest here is uptake of copper by a green alga, and copper uptake is limited in the presence of EDTA. Again, the interpretation is that copper-ligand complex is not biologically available. Figure 6.8 shows the binding of various metals with EDTA. The same figure illustrates that different metals have different affinities for EDTA. This fact may be of particular significance in the context of a ligand's effect on a mixture of more than one metal because strongly binding metals may displace those that bind more weakly from sites on ligands.

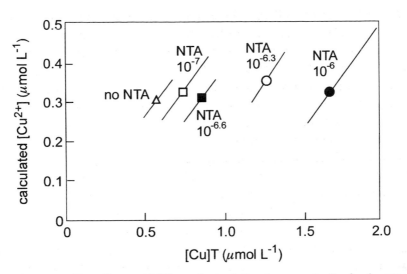

Figure 6.6 The effects of NTA on the toxicity of copper to the freshwater shrimp *Paratyla australiensis*. Results are expressed as LC$_{50}$ values for the freshwater shrimp. Total copper is plotted on the x-axis, and cupric ion (calculated), on the y-axis. As NTA concentrations increase, increasing concentrations of total copper are required to kill 50% of the shrimp population; when the LC$_{50}$ is expressed in terms of the cupric ion, the values remain relatively constant. Modified from Daly et al. (1990).

Naturally occurring organic ligands, such as the humic substances in soils and surface waters, are expected to have similar effects in terms of influencing metal availability, but their composition is variable, and their chemistry is far less well understood than is the chemistry of synthetic organic molecules such as EDTA and NTA. For this reason, much of the early work on the biological effect of organic complexing on metal availability was done on these synthetic molecules.

The most widely held theory concerning the bioavailability of metal is the free ion activity model (FIAM). This model states that the concentration of the free ion, M^{z+}, is the best predictor of a metal's bioavailability. The free ion interacts with either the cell's surface as in Equation 6.2 and/or is transported into the cell. In a sense, the cell with its negatively charged sites can be envisaged as a ligand to which the free metal binds. If the metal outside the cell is not in the free form but is bound relatively weakly to a ligand, the cell surface may compete with the external binding, and ligand exchange can occur, resulting in removal of the metal from the external complex and transfer of the metal to the cell, as shown in Equation 6.3. Campbell (1995) has reviewed the literature on the free ion activity model and its assumptions.

$$M^{z+} + -X\text{-cell} \overset{K_2}{\rightleftharpoons} M\text{-}X\text{-cell} \tag{6.2}$$

where $-X$-cell is a (natural) ligand on the cell surface and K_2 is the conditional equilibrium constant.

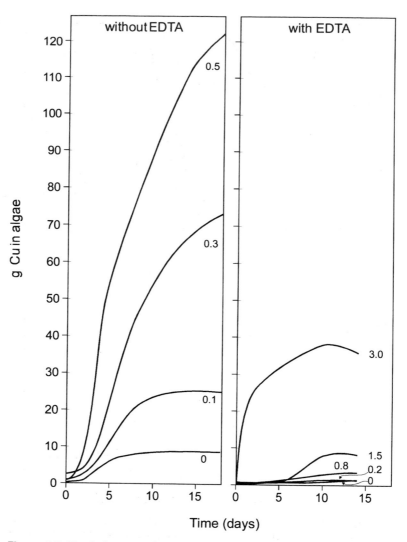

Figure 6.7 The influence of ethylenediaminotetraacetic acid on copper uptake by a green alga. Cultures of the green unicellular alga *Scenedesmus* in defined (modified Bold's basal medium) solution were exposed to copper at varying concentrations of EDTA. Values shown for each curve are the nominal total copper concentrations (mg/L) in the medium (i.e., the exposures). Uptake was related to the EDTA such that very little copper was taken up in the presence of EDTA. Modified from Stokes and Hutchinson (1975).

$$ML + -X\text{-cell} \overset{K_3}{\rightleftharpoons} MM\text{-}X\text{-cell} + L \tag{6.3}$$

where $-X$-cell is a (natural) ligand on the cell surface and K_3 is the conditional equilibrium constant.

Over the past two decades, a great deal of attention has been paid to questions related to metal speciation and metal availability, and a number of technical advances have coincided with scientific interest in the subject. To some extent, this

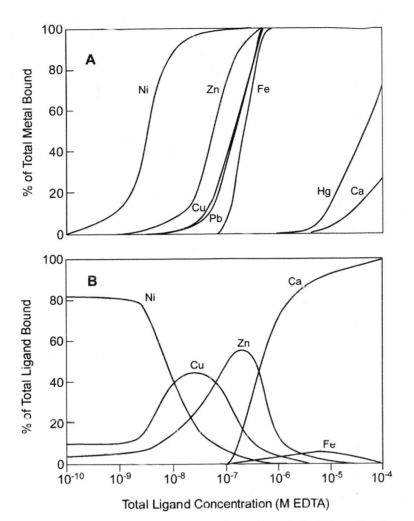

Figure 6.8 The chemical speciation of metals and of ethylenediaminote-traacetic at a fixed pH. The figure shows the calculated titration of a model (defined) freshwater with EDTA. The pH is fixed at 8.10 and the total metal concentrations are as follows: Ca $3.7 \times 10^{-4} M$; Mg $1.6 \times 10^{-4} M$; K $6.0 \times 10^{-5} M$; Na $2.8 \times 10^{-4} M$; Fe(III) $5.0 \times 10^{-7} M$; Cu $5 \times 10^{-8} M$; Hg $10^{-9} M$; Zn $1.5 \times 10^{-7} M$; Ni $5 \times 10^{-9} M$; Pb $10^{-9} M$. (A) The amount of bound metal at varying concentrations of EDTA from 10^{-10} to $10^{-4} M$. (B) The relative amount of ligand that is bound to metals at varying ligand concentrations. EDTA at very low concentrations is present mainly in the nickel complex. When nickel itself becomes titrated as the EDTA concentration exceeds that of nickel, the copper and zinc complexes become prevalent. Finally, when those two metals are titrated, EDTA becomes controlled by its calcium complex. Modified from Morel and Hering (1993).

process has been reciprocal, with the stimulus for the technical advances being provided by the interest in speciation, whereas the technical advances facilitated experimental and field studies on biological availability, which were not previously feasible.

Even though trace metal uptake by eukaryotic organisms has often been regarded as largely a passive process driven by the metal binding capacity of macromolecules, this view is changing. With the impetus of modern molecular techniques, this subject offers an expanding field of study. There is already an extensive literature on metal regulation in bacteria, where a variety of toxic metal transporters are known to confer metal resistance. Specific cation specificities for Ag^+, Cu^+, Cu^{2+}, Cd^{2+}, and Zn^{2+} have been identified in P-type ATPases associated with bacterial membranes (see review by Agranoff and Krishna, 1998).

More recent studies have indicated that the regulation of metal ions may play a broader role in bacterial homeostasis including the generation of electrochemical ionic gradients, which may supply the driving force for secondarily active transport of metabolic precursors. Studies of yeast and human cells have now identified discrete metal transporter proteins, and the study of such proteins in eukaryotes will rapidly transform the science of trace metal physiology.

6.2.4   Approaches for determining the chemical species and availability of metals

It has been pointed out (Section 1.6) that technical advances have facilitated progress in environmental toxicology. This is well illustrated by the methods that have been used for determining metal species.

RECENT TECHNICAL ADVANCES FOR DETERMINING METAL SPECIES

1. Equilibrium modelling of chemical mixtures, with programmes such as MINEQL, MITEQ2A, GEOCHEM, and more recently WHAM, for aqueous systems. These computer programmes calculate the relative amounts of respective chemical species of an element in solution, when the composition of the medium is specified. Equilibrium distribution is assumed (i.e., these are not dynamic models). Table 6.1 shows a sample of the output from such a model for dissolved cadmium. The chemically simple "reconstituted" water has known concentrations of the major ions, including pH. The model output predicts that at pH 7.0 most of the Cd will be in the free ionic form.

The reliability of the model results depends, of course, on the accuracy of the constants for the reactions involved among the inorganic as well as organic ligands that are specified for the system. The assumption of equilibrium may not always be realistic, bearing in mind the dynamic nature of ecosystems. Furthermore, it should be obvious that defined mixtures are more amenable to modelling than are natural waters, soils, and sediments. There are two main reasons for this: (a) the exact composition of natural waters and other substrates is not known; and (b) for a natural ligand such as fulvic acid, the binding constants are not well known. Although models such as WHAM do incorporate natural organic matter, they are not yet as reliable as the models that use only synthetic organic ligands such as

Table 6.1. A sample calculation using MINEQL for the chemical speciation of cadmium in solution

Input	
Na	$1.14 M^{-3}$
CO_3^{2-}	$1.14 M^{-3}$
Ca	$2.94 M^{-4}$
SO_4^{2-}	$7.94 M^{-4}$
Mg	$5.0 M^{-4}$
K	$5.4 M^{-5}$
Cl	$5.4 M^{-5}$
Cd	$1.34 M^{-7}$
pH	7.0
Output	
Cd^{2+} (aqueous)	86%
$CdSO_4$ (aqueous)	11%
$CdHCO_3$ (aqueous)	3.0%

The water chemistry is from a moderately hard reconstituted water.
The system is open to the atmosphere.
All values are in molar units.
All ions are input at the real concentrations; the programme calculates the chemical forms of cadmium at equilibrium.

EDTA. Nevertheless, even given these potential disadvantages, this type of modelling has resulted in respectable progress in providing a chemical basis for understanding the behaviour of metals in a variety of milieus.

2. The development of analytical chemical approaches that determine not total metal but rather distinguish among different chemical species of metal. This can be achieved by pretreatment (e.g., with ion exchange resins), through electrochemical measurements, either voltammetric methods (e.g., anodic stripping voltammetry) or potentiometric, with the use of ion selective electrodes.

In general, these methods have had limited success. Voltammetric methods in general are quite sensitive and can be used to determine environmentally realistic concentrations of free or labile metals, but interpretation is difficult for mixtures of several ligands of different concentrations. Furthermore, there has been some concern that the preparation or analytical process tends to alter the chemical status of the sample. The usefulness of ion selective electrodes (ISEs) has been limited until recently mainly because of detection limits that are much higher than typical environmental or cellular concentrations. This problem may soon be remedied. In addition, ISEs have been developed for only a limited number of elements.

In common with any technological tool, it is most important for the user to be aware of the limitations of the tool. As obvious as this may appear, it is particularly pertinent when researchers and practitioners are crossing the boundaries of traditional disciplines.

3. The development, for solid matrices such as soil and sediment, of sequential extraction methods, using successively stronger leaching agents, to remove various operationally defined fractions such as ion exchangeable, organically bound, and iron and manganese oxide bound. The amount of metal that is removed with each of the fractions can be related to the way in which the metal was bound in the original material.

The main criticism of sequential extraction lies in the fact that the fractions are operationally defined, and the exact chemical nature of each fraction and its corresponding metal binding is not known. Furthermore, since the preparation of each fraction is destructive, one cannot "use" the respective fractions for parallel biological testing. For both of these reasons, interpretation becomes problematic. Nevertheless, this approach has made possible considerable advances in understanding the bioavailability of metals, particularly in sediments, to sediment dwelling animals and rooted aquatic plants.

4. The routine determination of a chemical property of the environment that is apparently related to metal availability. Both pH and DOC have been identified (Section 6.2.2) as having significant effects on metal speciation; however, these variables have too many and varied effects for them to be used routinely as predictors of metal availability in complex systems, even though they may be used as variables in experimental design. The parameter known as acid volatile sulphide has, however, been proposed as a property that, for sediments, may have some predictive value for determining metal availability (Di Toro et al., 1990), although its broad application has been questioned (Lee et al., 2000; see also Section 5.4).

The theoretical basis of the approach is that sulphide-bound metals are not biologically available, and a measurement of the chemical fraction known as AVS can be used to predict metal availability (see also Section 5.4). To date, the practical value of AVS as a routine predictor of metal availability has not been established, but research is continuing

5. The use of stable isotopes. This is one of the most recent developments in the study of metal availability. It relies upon the fact that, for most metals, a number of chemically identical but isotopically distinct forms exist, and that the distinct isotopes can be detected by inductively coupled plasma-mass spectrometry (ICP-MS). The approach is built upon the same principles as the radiotracer approach, but it has advantages over the use of radioactive substances. First, the potential risks related to the handling of radioactive materials are absent. In many examples, several different isotopes of the same element are available, making possible concurrent dis-

tinctive labelling of different compounds of the same element. Thus, for the determination of availability, a number of different chemical forms of an element can be "tagged" with different isotopes of the same element, dosed simultaneously or in parallel, and their fate in an organism or in a whole system can be traced. Very low concentrations of metals can be detected by this method. This means that experimental work can be done with realistic concentrations of elements, such as would occur in the environment, a feature that is often lacking with radioactive labels because of the low specific activity of some isotopes.

At present, this method has enjoyed limited attention because of the relative inaccessibility and complexity of the analytical equipment: At the present stage, it is a research tool. Data are, however, beginning to become available, and the theoretical promise of the method appears to be justified because of the extremely low detection limits and the linkage between the biological system and the chemical speciation.

BIOLOGICAL ASSAYS FOR MEASURING THE AVAILABILITY OF METALS
The advances in biological testing in the context of assaying particular species or forms of metals are limited by the fact that most of the chemical assays require preparation of samples that change the samples, such that the availability of metals in "natural" media is very difficult to determine quantitatively.

Biological tests that have shown promise for determining bioavailability include the following.

6. Laboratory assays for metal toxicity or uptake, controlling the chemical form of a metal by, for example, controlling pH or by using organic complexing agents.

These types of approaches have already been illustrated in Figures 6.6 and 6.7. They work best with known simple or synthetic ligands, such as low-molecular-weight organic acids (e.g., malate, or EDTA and NTA). Until quite recently, the form of the metal was calculated, as for example in Figure 6.8. For natural organic ligands, calculated values for metal binding are less reliable, although they are being developed, for example, in the WHAM model.

Bioassays to determine metal availability in waters with unknown composition have also been carried out. The freshwater green alga *Selenastrum capricornutum* is something of the "white rat" of aquatic plant test systems. When cultured in standard media under standard assay conditions, it can be used by many different workers as a means of determining the presence of toxicity to algae in natural waters. As an example, Chiaudani and Vighi (1978) calculated the 96-h EC_{50} of various metals for this alga and then screened complex waters for toxicity. They concluded that the algae could be used not only as a means of determining the relative toxicities of metals to phytoplankton but also as a test for monitoring the quality of waste waters and natural waters. Such monitoring would not of itself identify the specific pollutant(s) but in many instances would provide more useful information than would chemical analysis of complex waters whose composition was unknown and where even given a comprehensive chemical analysis, the biological availability of the various substances would not be known.

More specific information can be obtained using specific biological response. For example, the cupric ion was shown to inhibit the cellular incorporation of ^{14}C-labelled glucose by a marine bacterial isolate, but organic complexes of copper had no effect (Sunda and Gillespie, 1979). By calibrating this response, a model relationship was established between glucose incorporation and cupric ion activity. After this calibration had been made, the system was used to estimate the binding capacity of natural seawater for the cupric ion. The authors compared the bacterial response to added copper in the natural seawater with that of added copper with a known chelator NTA. They identified the presence of a highly reactive ligand in the natural seawater, its concentration, and its conditional stability constant, as well as estimated the cupric ion activity of the water.

MEASUREMENTS OF METAL CONTENT OR UPTAKE IN LIVING ORGANISMS IN THE FIELD

This approach is based directly on the theory that only the available metal will be taken up by living tissue. It is the basis for biomonitoring as discussed in Section 4.4.3.

7. Multilevel tests, in which several trophic levels are exposed simultaneously to a metal. This type of test is usually carried out in a microcosm or some type of field enclosure (Section 4.7.1).

A major challenge for an understanding of bioavailability is to "match" the respective biological and chemical measurements. The stable isotope technique described earlier has considerable promise for tracing the uptake or toxicity of specific forms of an element, not only in organisms but also in simple systems such as microcosms. Clearly, there is a challenge for the analytical chemist (see Section 6.3) to work in collaboration with the biologist, and vice versa.

Beyond the technical challenges referred to so far in the context of the relationship between chemical form and biological availability, there are other more formidable challenges. It must be understood that ecosystems are dynamic, and that the chemical speciation of any element may be in a constant state of change, at least in the active parts of the ecosystem. This state occurs because ecosystems are subject to fluctuations of physical, chemical, and biological processes, which interact. Such dynamism renders the interpretation of field measurements made at a given point in time quite problematic. On the other hand, if measurements are made in the laboratory, under controlled conditions, with speciation more or less stable, the realism of the results can be called into question, although, from a mechanistic point of view, useful information can be obtained from controlled experiments.

The dilemma of field versus laboratory, or of real world versus controlled experiment, which was discussed in Chapter 4, is nowhere more clearly exemplified than in this context of studies on metal speciation and bioavailability.

6.2.5 The persistence of metals in the environment

The fate of substances that are introduced into ecosystems may be to remain in the original form or to be changed chemically and or biologically into some other compound. The term *persistence* appears to be very useful for organic compounds, which may or may not undergo conversion or degradation after being released into the environment. If the compound remains unchanged for a long period of time, clearly the consequence, if it has any toxic properties, is more serious than if it were degradable into less harmful compounds. In most jurisdictions, at least at the present time, persistence, along with toxicity and the tendency to bioconcentrate, is used in hazard assessment for potentially toxic substances.

Metals by their very nature cannot, however, be degraded, except in the rare case of neutron bombardment, which occurs inside a reactor. Therefore, all metals are deemed persistent in the environment. In general, then, the concept of persistence in the environment may be of little use in the scientific understanding of metal toxicology. But the concept of persistence of a substance in the organism or tissue is a more useful one. For this concept, the term most frequently used is *biological half-life*, which is the time required for elimination from the body of 50% of the dose (see Section 3.3.2). Metals such as lead and mercury have rather long biological half-lives in mammals. Lead accumulates and remains more or less permanently in bone. Lead in bone appears to do no harm, but it may provide an internal source of lead for other organs of the body, even in the absence of any new exposure. Mercury, in the methylated form (see Section 6.4.2), persists in nervous tissue, which is also the target for its most damaging toxic effects. A metal that has a long biological half-life (i.e., is persistent in the living organism) may signal more serious potential long-term harmful effects than a metal that is eliminated rapidly from the body. This concept is referred to in more detail in the sections on specific metals.

6.2.6 Bioconcentration, bioaccumulation, and biomagnification of metals in the environment

When organisms are exposed to metals and metalloids in their respective environments, the elements are taken up into or onto the organism, either actively or passively, depending on the element and on a number of environmental conditions. At equilibrium, a living organism usually contains a higher concentration of metal in its tissues than in its immediate environment (water, sediment, soil, air, etc.). The specific tissue or tissues into which an element concentrates varies, again with the element and with conditions. The term *bioconcentration* applies to the phenomenon whereby a living organism contains higher concentrations of a given substance than the concentration in its immediate source of that substance. Bioconcentration does not imply any specific mechanism. It does not automatically imply that the metal has harmed the organisms. Indeed, a particular metal may reach fairly high concentrations in certain tissues and do no apparent damage to the organism. Bioconcentration may even result from internal detoxification of metals and metalloids (e.g., Section 4.3.2). Bioconcentration may, however, be damaging to the ecosystem, if the consumer of the organism gets a high dose of metal through its food (see Section 4.5).

With time, the process of bioconcentration may continue, such that older or larger organisms will have higher metal concentrations than younger or smaller organisms, and we speak of bioaccumulation over the lifetime of an organism. On the other hand, one can envisage a situation wherein bioconcentration may cease when a steady state is achieved. The ratio of the concentration of a given metal in an organism to the concentration of that metal in its immediate environment (always specified) is called the bioconcentration factor.

The terms *bioconcentration factor* (BCF) and *bioaccumulation factor* (BAF) are used with rather precise meaning in recent literature. BCF is the ratio of the concentration in biota to the concentration in the medium (typically water), measured from experimental exposure of the organism(s) in the laboratory. The BAF is the ratio of the same two measurements from a real or field situation and incorporates the phenomenon of food chain accumulation as well.

Bioconcentration and bioaccumulation of metals occur actively or passively and, in either case, as noted previously, may not necessarily lead to harmful effects. These phenomena have profound implications in many areas of environmental toxicology. For example, they explain why fossil fuels that were laid down millions of years ago may contain high concentrations of potentially toxic elements. Because all living organisms, as well as nonliving parts such as shed carapaces, shells, and hair, tend to bioconcentrate metals to a greater or lesser extent, fossil fuels, formed from the compressed carbonised remains of living material, often contain rather high concentrations of metals. The exact composition varies according to a number of factors, including the type of fuel and the site and geology of formation. When fossil fuels are burned, the gaseous and solid waste products, therefore, may be sources of metals to the environment (Section 9.4.3 and Table 9.3).

The accumulation of a substance through successive trophic levels, which happens if they are not eliminated or excreted or controlled, is called biomagnification. Figure 6.9 illustrates the concept of biomagnification. Biomagnification is a food chain or food web phenomenon and cannot apply to a single organism, trophic level, or community. Although this phenomenon is often cited, it is in fact rare for metals, in that mercury (see Section 6.4) is the only metal that has been unequivocally demonstrated to undergo biomagnification.

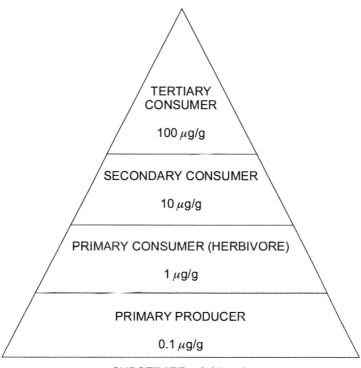

Figure 6.9 A schematic pyramid of biomass for a simple ecosystem, showing successive concentrations of metal. In the conceptual model, the metal is introduced in the substrate. The assumptions are: approximately 10% transfer of biomass from one trophic level to the next; no elimination of metals by the organisms at any trophic level; the only source of metal for subsequent trophic levels is the food. The concentration of metal will increase tenfold at each increasing trophic level. This is termed biomagnification and is a food chain process. Of the metals, biomagnification occurs only with mercury.

6.3 Analytical methods, temporal and spatial distribution of metals and metalloids in the environment

6.3.1 Analytical chemistry

A number of methods of analysis are in common use for determining directly the concentration, and in some cases the chemical form, of metals in environmental samples. Wet chemical methods have, for the most part, been superseded by spectrometry and other physical methods. These changes, along with the development of electronic detectors and recorders, have brought certain analytical tools within the convenient reach of biologists and toxicologists. Previously, these were the exclusive domain of the analytical chemist. The apparent simplification of the analytical methods has its own problems. There is a danger of complacency if the operator is in effect using the machine to produce numbers, rather than having an appreciation of the principles underlying the method. Thus, caution is recommended, and the best approach is a combination of skills and expertise, on a team if not in one person, with analytical chemistry going hand in hand with the biological components.

Working with the environmental toxicology of metals frequently means working at very low concentrations. Aside from the need for realistic detection limits in any method, quality assurance and quality control need to be rigorously applied. Contamination from external or internal sources is a constant threat when working with metals, and artefacts of analysis are also possible. Regrettably, many of the determinations of metals that were made prior to the 1990s may be of limited value, for reasons of inappropriate methodology or because of poor control over contamination. Furthermore, the apparent decline of the concentrations in the environment of a metal such as Hg in recent times may be partially or entirely due to the cleaner sampling and manipulation techniques that are now in use. This question is particularly troublesome in the context of concentrations in air and water, media in which mercury concentrations are in general very low. Unless one has access to historical or archived samples that are free of contamination, questions concerning the true trends for mercury in air and similar issues may remain unanswered.

The following techniques are in regular use for metal determination:

1. Atomic absorption spectrophotometry (AAS), either by flame or with graphite furnace (GFAAS), or with cold vapour hydride generation;
2. Atomic fluorescence;
3. X-ray fluorescence (XRF) and total reflection X-ray fluorescence (TXRF);
4. Polarography;
5. Potentiometry (ion selective electrodes, ISE);
6. Neutron activation (NAA);
7. Inductively coupled atomic emission plasma spectrometry (ICP-AES);
8. Inductively coupled plasma spectrometry with mass spectrometry (ICP-MS).

6.3.2   Historical records

The inherent persistence of metals in the environment means that the legacy of past use of metals may remain in certain sites and provide an historical record of their release and deposition. Lake sediments and soils, if undisturbed, as well as glaciers and other ice sheets, have been sampled and, with their stratigraphy maintained, have provided evidence of the mobilisation and distribution of metals in the past, up to the present day. These techniques require that vertical columns of the substrate material, called cores, are sampled in the field, and that their integrity is maintained by various means, such as freezing. The cores are sliced into fractions and digested or otherwise prepared, and then individual sections of various elements and compounds are chemically analysed. A profile of concentration is obtained; it can, in some conditions, be translated into rates of deposition. Techniques are available using chemical or biological markers for pinpointing the dates of certain events, and, thus, the various strata of the core can be dated.

Some of the results of such investigations can only be considered as qualitatively valid, but in the best examples, excellent quantitative data can be obtained. The assumptions for the technique include the faithful preservation of the deposited material (i.e., that it has been undisturbed since its original deposition), as well as reliable dating techniques. Each of these assumptions has been challenged in the course of the development of the techniques. The approach is comparable to that used by paleoecologists, who use fossil remains, pollen spores, and siliceous remains of plants and animals in sedimented material to reconstruct historical environments. The most useful information concerning the environmental toxicology of past eras combines geochemical with the biological data.

Examples abound for the reconstruction of the historical deposition of metals. Possibly the most striking results for inorganic substances are for lead. The widespread contamination of the biosphere by lead, resulting especially from the use of tetraethyl lead in gasoline (see Sections 6.5.1 and 6.5.2), can be seen in a number of media including ice caps, lake sediments, undisturbed soils, mosses, lichens, and tree rings. An increase in Pb concentration since the 1940s reflects the widespread use of tetraethyl lead, and a recent decrease in concentration in the same media reflects the dramatic decrease in the use of tetraethyl lead since the early 1980s, initially in North America, and more recently in Europe and elsewhere.

The annual growth rings of trees accumulate metal in a manner that is related to the amount in the environment at the time when the ring was laid down. This has been referred to already in Section 4.4.3 and is illustrated in Figure 6.10. If the tree is sufficiently long-lived, then it too can be used to reconstruct the historical record of metals in its environment. Lead in tree rings (Figure 6.10) has declined significantly over the past two decades, and this certainly reflects the recent general improvement in air quality with respect to lead.

6.3.3 Spatial records and source signatures

Several types of media were identified earlier as being of use in historical (i.e., temporal investigations of metal deposition). The same media can be used to investigate the spatial deposition of metals, both of local releases from point sources and from long-range transport. Several of the examples of biological monitors of contaminants, such as lichens, described in Chapter 4 (Section 4.4.3 and Table 4.7), illustrate the same principle of tracing the spatial patterns of contaminants.

Sediments and soils have frequently been used in this way. As an example, the concentrations of nickel in lake sediments were measured at different distances from the Sudbury smelting complex in Ontario. Nickel is not a common contaminant, so it can serve as a "fingerprint" for this particular source. There was a clear relationship between the source of nickel and the concentration in sediments

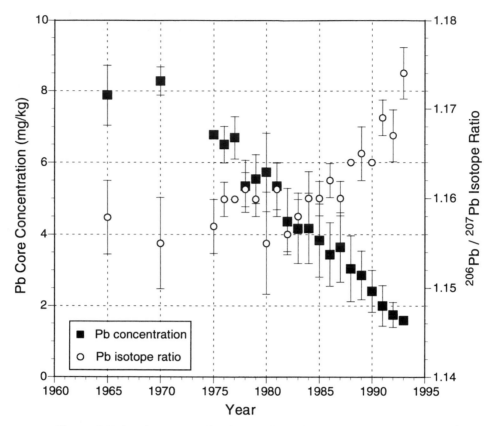

Figure 6.10 Lead concentration in tree rings. Lead concentrations (mg kg^{-1}; closed square symbols) and ratios of $^{206}Pb/^{207}Pb$ (open circles) were measured in annual rings of sycamore (*Acer pseudoplatanus*), sampled from codominant trees at Croxteth, an urban parkland in northwest England. Values are mean ± standard error. The total lead concentrations show a steep decline over the period 1980–95, long before the introduction of unleaded gasoline in the United Kingdom (1986). The lead isotope ratio was high and remained relatively constant between 1965 and 1986, indicating that the majority of the lead in wood derived from urban and industrial sources rather than from motor vehicles. After 1986, lead concentrations continued to decline, but the lead isotope increased. The decrease in lead accompanied by the increase in the ratio reflects the decrease in lead from vehicles. Modified from Watmough et al. (1999).

(Figure 6.11), reflecting the point source of nickel as well as the distance to which the nickel was being distributed through the atmosphere prior to its being deposited.

Patterns of decreasing concentration in a standard medium with distance from an expected source of a given metal can provide evidence of the source of that metal. However, where multiple point sources are involved, such as in urban complexes, or where the sources are diffuse, there may be no such distinct pattern. More refined methods utilise the isotopic pattern of the element from a given source.

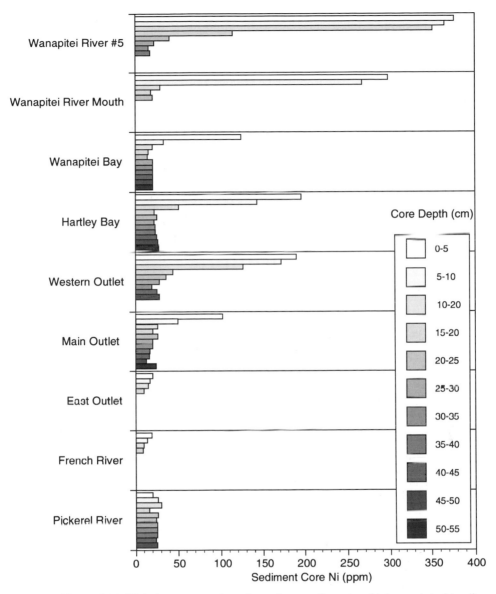

Figure 6.11 Nickel concentrations in surface sediments of lakes, related to distance from the Sudbury smelting complex. Sites were chosen in a river system at different distances from the Sudbury, Ontario, smelting complex. Metals in sediment can originate from the parent geological material or be deposited via water or atmospheric transport from an external source. The pattern of nickel shown in all but the lowest panel in the figure, namely with the highest concentrations at the surface, indicates surface loading. This is typical of an external source, with the highest concentrations during relatively recent times. The pattern of decreasing concentrations of nickel with distance from the industrial complex indicates that the complex is the most likely source of the metal enrichment. Modified from Hutchinson et al. (1975).

For certain elements, the ratio of naturally occurring isotopes is specific to a source, for example, a particular ore body or industrial supply of an element. This pattern is illustrated for lead in Figure 6.10, where the data show a decline of lead in tree rings from 1970 up to the time when the study ended. The value of the isotopic ratio up to 1986, which is relatively constant, is indicative of general urban and industrial sources of lead. The increase in the ratio beginning in 1986 is indicative of a decrease in the contribution of lead from tetraethyl lead, the additive to leaded gasoline. Without the information from the isotopic ratios, the interpretation of the decline in lead could not have been clearly related to the various sources. Once again, the coupling of advances in technology with environmental toxicology has provided a benefit to the toxicologist.

The balance of this chapter deals with the environmental toxicology of individual inorganic contaminants, of which Hg is presented in the greatest detail.

6.4   Mercury

6.4.1   The background to environmental concerns for mercury

Our awareness of the potential for damage to health of humans and other organisms from environmental exposure to mercury, as compared with occupational exposure, which was a much older concern, came from a series of apparently unconnected events. Medical and environmental toxicologists, as well as governments and the public, were alerted by outbreaks of Hg poisoning in Japan, originating from a chemical plant (in the 1950s), and in Iraq, originating from the human consumption of Hg-treated grain seed (in the 1970s). These led to severe and often fatal human health effects (see Section 6.4.3). Furthermore, local massive pollution of the environment by Hg from chlor-alkali plants in Canada and Sweden came to public attention in the 1970s. Prior to this time, the main concern for toxic effects of Hg had been through occupational exposure.

Since the 1970s, largely in response to the recognition of the hazards related to environmental exposure to Hg from industrial point sources, releases of Hg from industrial operations into the environment have been regulated and reasonably well controlled, at least in developed countries. Table 6.2 illustrates this for chlor-alkali plants in Canada. Chlor-alkali plants produced chemicals used in the pulp and paper industry, and Hg, employed as a cathode in the electrolysis of concentrated saline solution to yield chlorine and sodium hydroxide, was released into the environment. Until the mid 1970s, this represented a major category of mercury release into the environment in a number of countries, including Canada and Sweden. The decline in values indicates the effectiveness of regulations. From such types of regulations, which were applied to most known major point sources of mercury, one might anticipate that environmental Hg problems have been largely solved. But Hg is still released into the environment from diffuse sources, including the combustion of fossil fuel. There is also a tendency for "old" deposits of mercury to be

Table 6.2. Mercury discharged in liquid effluents from chlor-alkali manufacturing plants in Canada, 1970–1983

Year	Mercury discharged, kg
1970	67,000
1971	3,000
1972	1,100
1973	500
1974	390
1975	310
1976	300
1977	280
1978	190
1979	150
1980	100
1981	90
1980	50
1983	60

Based on Environment Canada (1985).

reemitted and recycled (see Section 6.4.2), and all emitted Hg can be transported long distances through the atmosphere.

The emphasis of interest in research and in regulation has now shifted from point sources to more widespread Hg contamination, but more significantly, to the factors that influence the processing of Hg in the environment. Mercury is used in this text as a model of an environmental toxicant, and is treated in greater detail than any of the other metals.

6.4.2 The properties, occurrence, and environmental behaviour of mercury

Mercury is an element with chemical symbol Hg (see position in Figure 6.3), atomic number 80, and atomic weight 200.59; in metallic form it volatilises readily at room temperature. Mercury has no known essential biological function. The element takes on different chemical states, three of which are important to our understanding of its environmental behaviour.

- *Elemental or metallic mercury*, which is liquid (hence the name "quicksilver") and volatilises readily at room temperature, is the major form in air and is scarcely soluble in water. It is symbolised as Hg^0.
- *Divalent inorganic mercury*, symbolised as Hg^{2+}, forms salts with various anions and ionises readily. Mercuric salts are sparingly soluble in water, and in the atmosphere Hg^{2+} associates readily with particles and water.
- *Methylmercury* is the most important organic form of mercury. Monomethylmercury, symbolised as CH_3Hg^+, is soluble in water and is

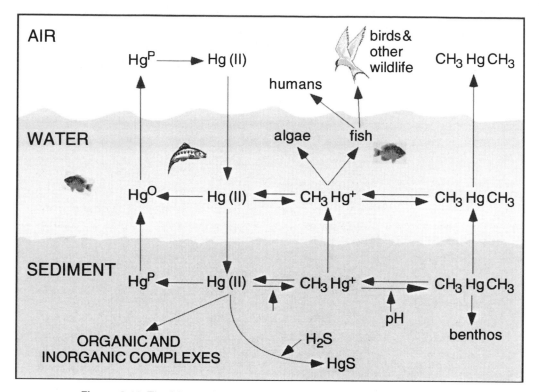

Figure 6.12 The biogeochemical cycle of mercury with emphasis on the aquatic system. This diagram indicates the most important chemical and biological transformations of mercury in an aquatic system such as a lake, river, wetland, or estuary. Sediments and, in some cases, the water column, are the major sites of biological methylation. Simple models like this can be used as the basis for quantifying the fluxes by modelling, by measurement, or by some combination of these. Based on Winfrey and Rudd (1990).

rather stable because of the presence of a covalent carbon–mercury bond. Dimethylmercury, symbolised as $(CH_3)_2Hg$, is less stable than the monomethyl form. It is less water-soluble and is volatile.

Figure 6.12 represents a simplified biogeochemical cycle of Hg, emphasising the aquatic system. It indicates the significance of methylation in the food web accumulation of Hg.

In common with all metals, Hg occurs naturally, and the absolute amount on the planet does not change. Its chemical form and location do, however, change quite readily, and an appreciation of these changes, in conjunction with the chemical forms listed previously, is needed to understand its environmental toxicology. Mercury is released into the atmosphere through a number of human activities and is also released as a result of natural phenomena. Mercury enters the atmosphere through direct and indirect pathways, from soils, rocks, water, and sediment, as well as from living systems. Anthropogenic Hg behaves in exactly the same manner as does the metal that occurs naturally: The behaviour of Hg in the environment depends upon its chemical form and the medium in which it occurs. Furthermore, at present there are no reliable methods by which to distinguish the ultimate source(s) of Hg. Even estimates of natural versus anthropogenic fluxes are quite variable.

In the biosphere, Hg is delivered to terrestrial and aquatic systems from the atmosphere as wet and dry deposition. This is quantitatively by far the most significant route by which ecosystems receive Hg and is a consequence of the volatility of Hg and some of its compounds.

In the natural environment, Hg undergoes chemical transformations, the most significant of which is methylation; methylation is the addition of one or two methyl groups, to divalent Hg, resulting in the organic substance, monomethylmercury (CH_3Hg^+) and dimethylmercury [$(CH_3)_2Hg$]. The process of biological methylation, first described in 1967, was published in English in 1969 (Jensen and Jervclov, 1969). Freshwater and saltwater ecosystems are major sites of methylmercury production.

Until quite recently, the anaerobic methane bacteria were believed to be the main methylators in lake sediments, but this view is no longer supported; in anoxic sediments, sulphate-reducing bacteria are now believed to be the principal agents of methylation. Other microorganisms probably methylate Hg in the water column and in wetlands and soils, but to date none have been identified with certainty.

Abiotic methylation has been also demonstrated, with natural organic humic matter as an agent, but in natural (versus experimental) situations, abiotic methylation probably accounts for less than 10% of methylmercury production.

The methylation of Hg is influenced by a number of environmental factors. These include low pH, which not only stimulates methylmercury production but also increases the proportion of total Hg available as substrate for methylation. At low pH, the availability of Hg appears to be increased because less Hg is lost from water by volatilisation than at neutral or alkaline conditions and because binding of Hg to particulates increases at acidic pH.

Many forms of organic carbon dissolved in water also promote methylation. High DOC is often combined with low pH, and, under natural conditions, assessing the separate respective roles of these two factors in methylation is often difficult. Conditions in wetlands, combining as they do low pH and high organic matter, will usually promote the methylation of Hg. Wetlands then frequently represent a source of methylmercury that is subsequently exported to lakes and rivers. The biochemical reaction most commonly cited for mercury methylation is transmethylation of inorganic divalent Hg from vitamin B_{12} (methylcobalamin, symbolised as L_5Co—CH_3) according to the following reaction:

$$L_5Co—CH_3 + Hg^{2+} \rightarrow L_5Co^+ + CH_3Hg^+ \tag{6.4}$$

Monomethylmercury is not only the form that has the most harmful effects on living organisms but also the form that accumulates in tissues, notably in the

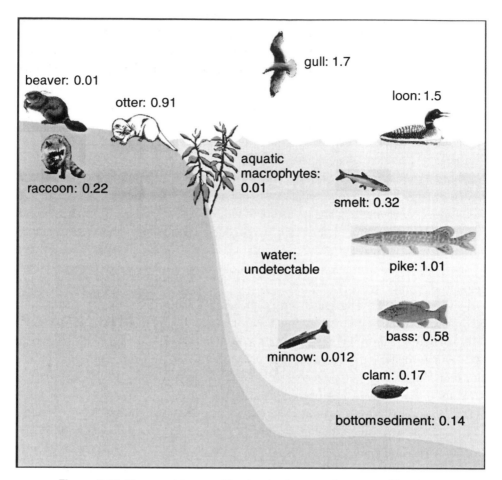

Figure 6.13 Mercury biomagnification in the aquatic system. The data were collected for total mercury in different compartments of a remote lake ecosystem and shows increasing concentrations with higher trophic level (i.e., biomagnification). All values are in micrograms per gram fresh weight, except for sediment concentrations which are expressed in micrograms per gram on a dry weight basis. Modified from Wren et al. (1983).

muscle of fish. Thus, it is not surprising that considerable effort and emphasis have been devoted to the study of the formation and behaviour of this chemical form. As already indicated, there are considerable differences in the physical and chemical properties among the three forms; in its biological behaviour, too, methylmercury has very different properties from elemental or inorganic forms.

Of all the metals that have been studied, Hg is the only one that biomagnifies through the food web (refer to Section 6.2.5). This remarkable amplification of Hg between water and biota is illustrated for a remote Ontario softwater lake in Figure 6.13. This lake had no known anthropogenic sources of Hg, and although there might have been natural sources, the concentration of mercury in water was undetectable by the methods available at the time of the study. Yet the concentrations in carnivores at higher trophic levels were magnified by at least two orders of magnitude compared with the aquatic plants. Of interest is the fact that the bass and pike in this lake had concentrations of Hg that exceeded the advisory for consumption of sport fish, which is 0.5 ppm in most countries. Another striking example of the high concentration factors for Hg in fish is seen in data from a northern Minnesota lake, in which the ratio of Hg in fish tissue to that in water for a 22-inch northern pike is 225,000×. For a long-lived top predator, northern pike, in Lower Trout Lake in Michigan, a BAF of 1.5 million was recorded. It is significant that the form of Hg that accumulates in fish muscle is almost entirely the highly toxic methylmercury.

Biomagnification results in concentrations of Hg in fish muscle that render the fish unfit for human consumption even though the fish show no adverse effects. A notable feature concerning Hg in ecosystems is the relative insensitivity of fish to this element. Fish "are highly resistant to the toxic effects of methylmercury" (Clarkson, 1995). They can tolerate ten times as much methylmercury as can humans and are also more tolerant of Hg than the wildlife that are likely to consume the fish. It has been suggested that storage of methylmercury in fish muscle, a tissue that seems less sensitive to methylmercury than other tissues, functions as the primary detoxification mechanism for methylmercury in fish and at the very least will reduce the exposure of the brain and nervous tissue to methylmercury. The result of this may be positive for the individual fish species but can lead to severe adverse effects at the ecosystem level. Such an example provides an excellent illustration of the complexity of ecotoxicology.

Some historical records of Hg production would lead to human exposure. Mercury was mined at least 2,000 years ago in the Spanish Almaden mines, which are still the largest producers of Hg today. Use of cinnabar as a pigment for red ink in China goes back 3,000 years. Somewhat more recently, Hg has contributed to major scientific discoveries, including the early laws of physics, in the development of which Hg was used in instruments to measure temperature and pressure. Today, uses of Hg reflect its properties, although the bactericidal uses (medications, slimicides, fungicides), which had been significant in terms of products and environmental contamination, are now almost entirely absent in developed countries.

A number of human activities have directly and indirectly altered the cycle of Hg, resulting in exposure of humans and ecosystems to various forms of the element. The major sources of Hg, or processes that mobilise the element, are provided in Table 6.3 for the contemporary situation in the United States. Two different data sets are provided: the disparities between them underline the uncertainty that still surrounds estimates for anthropogenic sources of this element.

For a general picture of the pathways that Hg follows in the environment, a simplified diagram of the global Hg cycle is provided in Figure 6.14. Among experts, there is consensus that the atmospheric pathway for transport is quantitatively by far the most significant. Of the chemical forms in the air, elemental Hg dominates

Table 6.3. Annual anthropogenic emissions of mercury for the United States

	Voldner and Smith (1989) for 1989	U.S. Environmental Protection Agency (1993) for 1992[a]
Combustion sources		
Utilities: coal	113	106
Utilities: oil	1	4
Comm/industries	25	31
Municipal waste incineration	68	58
Sewage sludge incineration	36	2
Medical waste incineration	–	59
Copper smelting	41	1
Lead smelting	5	8
Mercury mining, smelter, recovery	1	7
Chlor-alkali plants	3	7
Battery manufacturing	1	–
Dental amalg preparation	1	1
Dental lab use	2	–
Paint	136	4
Industrial and control instruments	7	1
Electric applications, fluorescent lamps	143	8
Mobile sources	–	5
Portland cement	–	6
Miscellaneous	–	3

[a] All in tonnes per year.

and has a long (1 week or more) residence time in the atmosphere. It is very clear that preindustrial emissions were much lower than those of today: On a global scale, fluxes have increased about five times over the last 100 years, and the atmospheric burden increased by about a factor of three.

As already indicated, sources of Hg include geologic, natural sources. The value for a ratio of natural to anthropogenic emissions of Hg remains controversial, but it is probably of the order of natural:anthropogenic, 1:2. However, this type of calculation, although of concern for regulators, is particularly difficult for Hg and may not even be meaningful. Because of its extreme mobility, Hg deposited from a particular pollution source into an ecosystem may be reemitted later into the atmosphere and in this way contribute to apparently "natural" sources. In the words of Nriagu (1994) in a recent report on the long-range transport of metals, "The ease with which Hg is recycled through the atmosphere obscures its anthropogenic past and industrial Hg probably accounts for a significant fraction of the estimated 'natural flux' of 2,500 t/y".

CURRENT MERCURY BUDGETS AND FLUXES

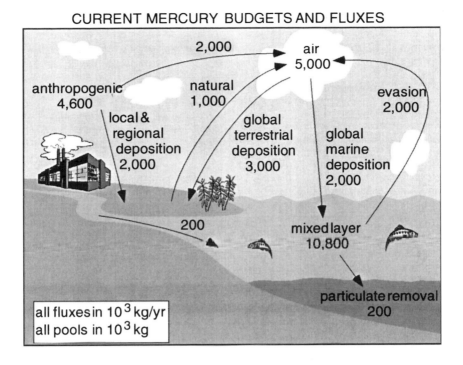

PRE-INDUSTRIAL MERCURY BUDGETS AND FLUXES

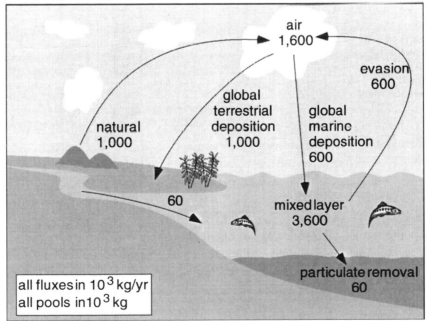

Figure 6.14 Global cycles of mercury. Numbers were obtained from box models, gas exchange models, and particulate flux models based on a variety of environmental analyses. Reactive mercury was operationally defined through $SnCl_2$ reduction of labile Hg to Hg^0. Based on Mason et al. (1994).

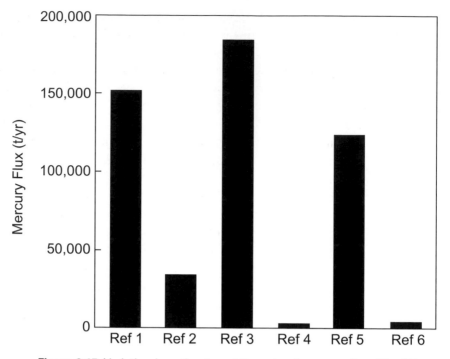

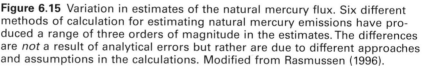

Figure 6.15 Variation in estimates of the natural mercury flux. Six different methods of calculation for estimating natural mercury emissions have produced a range of three orders of magnitude in the estimates. The differences are *not* a result of analytical errors but rather are due to different approaches and assumptions in the calculations. Modified from Rasmussen (1996).

Clearly natural fluxes of Hg cannot be measured directly on a global or even regional scale. The methods used to calculate these "natural" fluxes also influence the values for this parameter, even aside from the problematic definition of "natural flux" of Hg. Estimates of natural global Hg flux vary greatly, as shown in Figure 6.15 in which six different and recent estimates of total natural global Hg emissions are provided: They vary over two to three orders of magnitude.

From the point of view of regulation, there is some reason for attempting to identify the relative importance of natural versus anthropogenic Hg because there is obviously limited value in controlling anthropogenic emissions of an element if these are quantitatively insignificant compared with those of natural origin. However, the increase in global fluxes of Hg over the past century and its toxicity and food chain behaviour are, of themselves, sufficient justification to identify Hg as a substance of major concern. Detailed policy on its regulation will probably continue to be modified as more reliable information emerges through research and monitoring.

6.4.3   The toxicity of mercury and populations at risk

Mercury has a number of adverse effects on all forms of life; specific effects depend upon a number of factors including the form and concentration in which it is administered, the route of uptake, and the type of organism, as well as other contributing factors. Elemental Hg poisoning occurs mainly by inhaling Hg vapour. This type of Hg poisoning is predominately an occupational problem and will not be dealt with in any more detail in this text.

In the context of environmental toxicology, methylmercury is of concern. Mercury, in contrast to most other metals, for which the free ion activity model is at least partially tenable, exists as an organometal (particularly as monomethylmercury), which behaves more as an organic contaminant than as an inorganic one. Monomethylmercury moves readily across biological membranes and thus is the most available and the most toxic form of mercury, with the longest biological half-life (see Section 6.3.4 for definition). In humans, inorganic Hg has a half-life of 6 days, whereas methylmercury's half-life is in the order of 70 days. For a predatory fish such as pike (*Esox lucius*), the half-life of methylmercury is as long as 170 days. Several factors influence the actual concentrations of methylmercury in fish. These factors include the size (age) of the fish, the species and lake water chemistry as well as the incidence of Hg contamination. Figure 6.16 illustrates some fairly typical data for a single lake, showing the effects of species and size. The predatory fish species have higher Hg concentrations than those lower in the trophic chain, and older or larger fish have higher concentrations than younger smaller ones.

An additional observation concerning Hg in fish is the relationship with the ambient pH. Many separate studies have shown that for softwater lakes, even when the Hg in water is at very low concentrations, there is a statistical relationship between pH and Hg in fish, such that all other things being similar, fish from acidic lakes have higher Hg than higher pH lakes. This is seen over a range of pH that supports fish (i.e., there is no harmful effect of acidity per se on the fish). The effect of pH and DOC on methylation described in Section 6.4.2 explains, in part, the possible mechanisms underlying this enhanced bioaccumulation. Several studies have reported on an interaction between Hg and Se, in that Se protects organisms from Hg uptake and toxicity (see Section 6.9). The physiological basis for this interaction remains, however, unclear.

Exposure to methylmercury usually occurs through the ingestion of contaminated food, and the organisms at the greatest risk are high-level consumers, especially humans, domestic animals, and wildlife. For mammals, methylmercury is a neurotoxin, and through its action on the brain, symptoms include aphasia, ataxia, convulsions, and death. At doses higher than those eliciting neurological symptoms, other systems such as kidney, cardiovascular, and digestive can be affected, but these are not considered in the present account. Lower forms of life are less susceptible to Hg poisoning than are higher vertebrates, but the "processing" of Hg through the ecosystem means that the higher trophic levels such as predatory wildlife populations are particularly at risk.

The syndrome in humans resulting from exposure to methylmercury has been called Minamata disease. The first recorded widespread epidemic of environmental methylmercury poisoning, with severe neurological symptoms as described

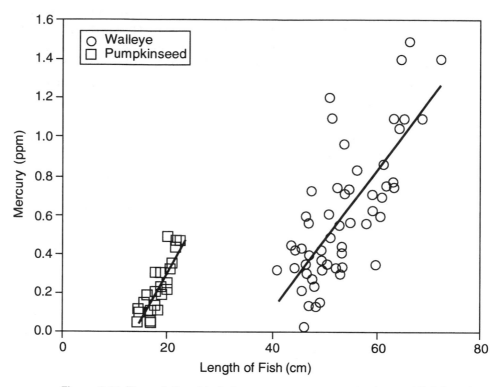

Figure 6.16 The relationship between mercury concentration and fish length for two species of fish from Lake St. Clair, Ontario, Canada. The collections were made from Lake St. Clair in 1990. The graph shows the influence of size, species, and trophic position on mercury in fish tissue. The walleye (*Schizostedion vitreum*) feeds at a higher trophic level than the pumpkinseed (*Leposmis gibbosus*). Mercury concentrations in the larger pumpkinseed and in most of the walleye exceed the 0.5 ppm advisory for sports fish in Ontario. Lake St. Clair has been affected by industrial releases of mercury in the past, and clearly the fish are still accumulating rather high concentrations of mercury. Based on data from the Ontario Sports Fish Monitoring Program, Ontario Ministry of the Environment (1992).

here, occurred in the Japanese fishing village of Minamata, beginning in 1953. There were close to 100 deaths over a period of several years and many more serious illnesses; a particularly tragic fact was that some infants were born with the disease having been exposed in utero through the mother's blood, even though the pregnant woman did not always show symptoms. Thus, one might consider that the mother was actually being detoxified by her foetus. By crossing the placental barrier, Hg was removed from the mother's blood. Although initially the source or agent of the epidemic of neurological disease was not known, it became clear that the cause was Hg, and that the Hg was being released from a chemical plant into the bay where much of the fishing took place. In this particular instance, it is thought that environmental methylation of inorganic Hg was not the major issue, but rather that the Hg was released in the form of methylmercury. Fish and shellfish were major components of the diet of villagers and their domestic animals, and the Hg had accumulated in these food items. In retrospect, nearly 50 years later, with the benefit of the knowledge that has been gained, this explanation may not seem surprising, but at the time it was a complete mystery. Even after the source was suspected, the company apparently attempted to cover up the cause and to disclaim liability, all of which made it harder to establish cause and effect.

The second major epidemic occurred in Iraq in 1971 when a consignment of grain treated with methymercury (a fungicide) and intended for use as seed was diverted to make flour, which was made into bread. Over approximately 6 months after this event, a total of 6,530 cases of poisoning were admitted to hospital and a total of 10,000 people became ill. Hospital records attributed 459 deaths to methylmercury, but it now seems probable that at least 3,723 people died, and many more were permanently damaged.

The means by which the Minamata and the Iraqi people were exposed to methylmercury were quite different. The contaminated fish were a long-term source, with relatively low methylmercury concentrations, whereas the Iraq exposures were to extremely high concentrations but for a very short time. Nevertheless, the resulting manifestation of poisoning was very similar in these as in most other known cases. The overt symptoms are relatively easy to recognise, but there is considerable uncertainty surrounding the possible adverse effects of lower and often chronic exposure. In this context, the determination of a "safe" level in food becomes problematic. Obviously, the concentration of methylmercury is one factor in protecting the receiving population, but because relatively few food sources will accumulate methylmercury, the composition of the diet is also a determining risk factor. In the absence of a truly abnormal event such as the Iraq grain consumption, the consumption of fish is the single most important factor in determining environmental exposure to methylmercury. For humans, diet analysis is difficult but not impossible to accomplish; for wildlife it becomes almost impossible.

The Minamata and Iraqi incidents have been covered in some detail because, tragic though they were, they provide the potential for epidemiologists and toxicologists to make detailed studies of the environmental exposure and to quantify the harmful effects. In fact, only the incident in Iraq provided data that could be applied in this way. For Minamata, the data were less reliable, in part because of the primitive state of analytical methods for Hg at that time, but also because the political climate at the time, particularly surrounding the proof of liability, led to considerable uncertainty in the facts that emerged.

Data from the Iraqi outbreak have formed the basis of consumption advisories for fish. At present, advisories range from 0.5 to 1.0 ppm (fresh weight) in fish. The advisory is based on measurements of total Hg, but nearly 100% of the Hg in fish muscle is, in fact, methylmercury, so the advisory in effect is for methylmercury. The value is based on a complex set of calculations, involving not only dose-response models but also estimates of fish intake for average populations. Because much of the data that have been used to compute this value originated in the human exposure to methylmercury in grain from Iraq, which, as pointed out, was massive in dose and short in duration, there has to be some question as to the relevance of this to the consumption of fish that contains methylmercury. A large margin of safety is built into the advisory, but even the value of the advisory is a subject of controversy. Some recent data have led some scientists to suggest that the advisory may be unnecessarily stringent, whereas others consider that it should be even lower. In a review of epidemiological studies of human exposure to methylmercury in fish, Myers et al. (2000) cast some doubt on the "conventional wisdom" concerning the risk related to methylmercury when the exposure is through fish consumption. The case of Hg illustrates in a number of ways the complexity of regulating environmental toxicants. Further reading is supplied on mercury and human health.

Wildlife is the other major group of communities potentially at risk from consuming methylmercury in fish. Direct consumers of fish include mammals such as the freshwater otter, common mink, seals, walrus, and polar bears and birds such as the common loon, blue heron, gulls, and arctic seabirds. Also at risk are predatory birds that feed on fish consumers. Additionally, although they are not fish consumers, waterfowl, which include benthic invertebrates in their diet, also show elevated Hg.

Studies of Hg in wildlife have examined the tissue distribution of Hg. In a few studies, attempts have been made to relate tissue levels to dietary intake. A second type of study looks at the effects of Hg on wildlife and, in some instances, establishes a dose-response that can be determined through tissue levels. Some studies address both of these aspects.

There are major problems in studying and interpreting the wildlife toxicology of Hg, most of which are common to all wildlife toxicology. Actual consumption patterns for wildlife under natural conditions are almost impossible to establish. Controlled experimental studies can be carried out on mammals, birds, and fish, providing the potential for direct establishment of dose-response relationships, an approach that is clearly not possible with humans. Two controlled studies on the fish-eating mammal, mink (*Mustela bison*) from Ontario and Quebec, showed that prolonged consumption of fish containing 0.9 or 1 ppm methylmercury, respectively, was lethal to at least 50% of the exposed population. If fish in lakes have Hg between 0.1 and 1 ppm Hg, there would appear to be little, if any, margin of safety.

A significant point concerning Hg and wildlife is expressed well by the Minnesota Pollution Control Agency (1994), which points out that, in contrast to humans, wildlife cannot be advised or made to use alternative food sources. Therefore, the implication is that possibly even more caution must be used in making risk assessments for wildlife effects from environmental Hg than for any other populations potentially at risk.

6.4.4 The reservoir problem

The final topic to be covered in this treatment of the environmental toxicology of Hg is the so-called reservoir problem. Over the past two to three decades, attention has been drawn to the increase in Hg concentration of fish after the creation of an impoundment (i.e., after flooding a previously terrestrial ecosystem). Mercury concentrations in fish muscle from sites with no point sources of Hg ranged from 0.08 ppm for small yellow perch to 0.93 for northern pike in a remote site in Quebec; fish in some reservoirs in southern Labrador exceeded 1.5 ppm. In a reservoir in Quebec, La Grande-2, Hg in northern pike reached a maximum of 3.5 ppm, approximately five times higher than that in (unflooded) natural environments (James Bay Mercury Committee, 1993).

This problem arises whenever there is flooding of a terrestrial site and conversion of the site into an aquatic system. For the most part, there seem to be no obvious Hg pollution sources and no relationship with geography (climate) or the natural geology. The reservoir problem only became obvious over the past two to three decades, but since then it has attracted considerable attention. Many of the impoundments that have been studied are related to hydroelectric projects (see also Section 9.4.3). Newly formed lakes have been promoted as potential fish habitat, with the idea of substituting a fishery for previous land-based resource activities such as hunting for food and fur. These resources have been significant, especially for aboriginal people who often rely on them for food and for basic income. Yet if the resulting fish are not acceptable for consumption, then the planned fishery is obviously not going to be of use. There may be health risks from consuming Hg-contaminated fish, and there will certainly be social and economic consequences.

The most widely accepted explanation for the reservoir problem is that DOC from the vegetation in the original terrestrial ecosystem stimulates microbial methylation of existing Hg to such an extent that, combined with biomagnification, predatory fish in recently impounded reservoirs have unacceptably high Hg. The source of the Hg is simply the existing "background" in soils and vegetation. The problem is a result of the factors that are promoting methylation, in combination with the tendency of Hg to biomagnify. There is no simple remedy because there is no controllable source of Hg. Aside from a proposal to strip vegetation from the land prior to flooding, the only resort to date has been restrictions on consumption, using the 0.5 ppm advisory referred to previously.

6.5 Lead

6.5.1 The occurrence, sources, and properties of lead

Lead is a soft grey metal, solid at normal temperatures and melting at 327.5°C, with atomic number 82 and atomic weight 207. The usual valence state of Pb is (II), in which state it forms inorganic compounds. Lead can also exist as Pb(IV), forming covalent compounds, the most important of which, from an environmental viewpoint, are the tetraalkyl leads, especially tetraethyl lead, used for many years as an additive octane enhancer for gasoline.

Lead occurs naturally in trace quantities: The average concentration in the Earth's crust is about 20 ppm. Weathering and volcanic emissions account for most of the natural processes that mobilise Pb, but human activities are far more significant in the mobilisation of Pb than are natural processes. Estimates of natural compared with anthropogenic sources for mobilisation of Pb vary according to the method, the dates, and the media under consideration. Estimates in the 1960s reported that "the amount of lead mined and introduced into our relatively small urban environments each year is more than 100 times greater than the amount of natural lead leached each year from soils by streams and added to the oceans over the entire Earth. There are indications that about nine tenths of lead in the upper mixed zones of the open ocean in the northern hemisphere originates from lead mines, and that the atmosphere of the northern hemisphere contains about 1,000 times more than natural amounts of lead" (Patterson, 1965).

The history of the human use of Pb goes back for about 4,000 years. Records of the use of this soft, easily worked metal date back to the time of the Egyptians and Babylonians. Early uses of Pb included the construction and application of pipes for the collection, transport, and distribution of water. The term *plumbing* originates from the Latin *plumbum*, for lead. The use of Pb in plumbing has continued to the present day, although most jurisdictions no longer use it in new installations. The Romans made extensive use of Pb for supplying water for their baths, as well as in containers for food and drink. There has been speculation concerning the adverse effects of Pb on the Romans, with some suggestions that chronic Pb poisoning among the ruling classes contributed to the decline of the Roman Empire (see Further Reading). Through the Middle Ages and beyond, Pb was used for roofing material, for example, in some of the great cathedrals of Europe. Although its use in plumbing and for construction has now been largely discontinued, older Pb-based systems still remain in use.

The largest single recent use of Pb in a product, and probably the most significant from the point of view of environmental contamination, is the use of tetraethyl lead as an additive to gasoline. Other uses of Pb included lead acetate (sugar of lead) as a sweetener, Pb salts as various pigments in artists' as well as in commercial paints and anticorrosion coatings; in solder, including the seams of canned food and drink; in batteries, as shields for protection against radioactive materials; and as an insecticide; Pb arsenate. Lead has also been used in ceramic glazes and in crystal tableware. Even though many of these practices have been discontinued, their legacy of lead contamination, particularly in soil, remains (Wixson and Davies, 1993).

Lead is now the fifth most commonly used metal in the world. Even though the applications that humans have made of this versatile and useful metal have been very considerable over time, mobilisation of the metal into the environment has varied a great deal among the different uses. It has also varied a great deal over time, and in more recent times some of its releases have been controlled in response to occupational and environmental health concerns. Many of the compounds of Pb are rather insoluble, and most of the metal discharged into water partitions rather rapidly into the suspended and bed sediments. Here it represents a long-term reservoir that may affect sediment-dwelling organisms and may enter the food chain from this route (see Sections 6.5.2 and 6.5.3). Mobilisation of Pb into the aqueous medium (e.g., surface waters) is of concern mainly in connection with soft, acidic (pH < 5.4) waters. Determinations of dissolved Pb in waters for two of the Great Lakes ranged from a minimum of 8.25 pM ($0.0017 \mu g L^{-1}$) for a station on Lake Erie to a maximum of 1,374 pM ($0.275 \mu g L^{-1}$) for a station on Lake Ontario (Coale and Flegal, 1989). Lead concentrations in soft, natural waters are in the order of $0.45 \pm 0.325 \mu g L^{-1}$ (Schut et al., 1986). Typical values for lead in seawater are $1-4 ng L^{-1}$ (Jaworski, 1978). Concerning the concentration of Pb in water supplies, hard or soft water may contain up to $5 \mu g L^{-1}$ of Pb (Jaworski, 1978).

The use of Pb for the distribution of water and in vessels for food and drink has resulted in its mobilisation into drinking water in areas where the geology produces soft, acidic, or poorly buffered water. Lead solder in pipes, water coolers and other equipment also represents a source of Pb through its being mobilised by soft acidic water. Foods and juices that are acidic will similarly solubilise Pb from vessels or from solder. Wine, beer, fruit juices, and canned fruits, indeed anything that has an acid reaction, will tend to contain Pb that has dissolved from the containers in which the products were stored. Certain types of crystal have Pb incorporated into their manufacture, and rather recently there have been warnings that wine stored in crystal decanters may contain Pb that has dissolved out of the glass.

Until recently many paints and anticorrosive materials contained large amounts of Pb, and in fact some such products still do. Aside from the rather special cases where leaded paint is a danger to children who tend to mouth toys and other objects (see Section 6.5.3), Pb in paint is of more widespread environmental concern because it eventually begins to wear or flake; Pb is mobilised into the resulting dust. Lead in dust may represent a direct route of exposure (see Sections 6.5.2 and 6.5.3); furthermore, Pb in particulate form as in dust may also be transported to soils and aquatic systems.

Primary smelting of Pb ore, as well as other metals, notably copper and zinc, whose ores also contain Pb, mobilises Pb. Industrial operations such as secondary Pb smelting, with the crushing and grinding that are involved in recycling Pb from old batteries, pipes, and so on, also mobilise Pb dust into the air and thence potentially to soil and water. Because these secondary smelters tend, for convenience, to be situated in urban centres, often near residential housing, they have attracted considerable attention from the point of view of pollution leading to human exposure. In contrast, mines and primary smelters tend to be located in more remote areas: For these, the concern has been for human occupational exposure and for effects on domesticated animals, wildlife, and other ecosystem components.

Many of the particulates that carry Pb into the air are relatively large and tend to be deposited close to their respective sources. The situation for Pb from gasoline is completely different. Basically, the addition of Pb to gasoline, a practice that was introduced in the 1920s but became more widespread with increased numbers of automobiles in the 1940s, provided an inexpensive method to prevent engine "knock" (i.e., to boost the octane), as well as to provide some lubrication for engine parts. When gasoline containing tetraethyl lead is burned in automobiles or other internal combustion engines, Pb is emitted in very small particles, less than 2 µm in diameter, and these particles travel for greater distance than the previous types of dust. The manifestation of this movement can be seen in lake sediment cores, arctic glaciers, and other historical records, where a distinct increase in Pb concentration is observed, beginning in the 1940s. A corresponding signal, namely a decrease in Pb concentrations in these various monitoring media, followed the control of this source of Pb emissions, which occurred in the mid 1980s in most developed countries. The illustration of Pb deposition into the feather moss *Hylocomium splendens* provided in Chapter 4 (Figure 4.6) also illustrates the decrease in long-range transported lead since the mid 1980s.

Table 6.4 summarises the major sources of Pb for ancient as well as current uses.

6.5.2   The environmental transport and behaviour of lead

The atmosphere is the major vector for the transport of Pb to living receptors as well as to other media. Inorganic Pb in particulate form, as already noted, travels varying distances according to the size of the particles. Particles larger than 2 µm in diameter tend to fall close to the source and contaminate local soil, water, and structures. The smaller particles, as noted, are mainly from the combustion of tetraethyl lead, and these may travel very long distances. Deposition of Pb from the air depends not only on particle size but also on atmospheric conditions such as wind speed, direction, and precipitation. Lead particles, once deposited onto soil or water, do not appear to undergo a complex series of chemical changes, and this condition is in contrast to a more reactive metal such as mercury (Section 6.4). Although there has been some experimental demonstration that Pb can be methylated, there is almost no evidence that methylation or indeed any alkylation is significant in ecosystems and alkyl lead is found only rarely and in very small quantities in the natural environment. Alkyl leads are, as already noted, produced industrially, although this practice has almost ceased since the implementation of the regulations that removed most lead from gasoline.

Therefore, most lead in soil, unless it originates from natural geological sources, has originated from airborne particles. Chemical changes following the deposition of Pb to soil appear to be very limited. Some evidence suggests that, in soil, deposits of lead "age", with some decrease in biological availability over time, but to date this decrease has not been directly related to specific chemical changes in the form of the metal. This trend could be in part a result of physical changes such as increase in particle size. Lead in soil can itself be a source of dust by which Pb is reentrained, but Pb in soil rarely contaminates surface or groundwater through being transported in the dissolved phase. Lead may be directly discharged into water as waste product from a point source, but, in fact, much of the lead that enters fresh- or saltwater bodies originates from atmospheric deposition.

In general, Pb is less reactive in water than many metals. The pH-sensitive nature of the sorption reactions of most inorganic Pb compounds means that, except as already noted, dissolved Pb rarely reaches very high concentrations. The exception is acidified or poorly buffered freshwater systems. These may have an existing lead burden or be receiving Pb from deposition. In either case, the dissolved concentration will increase in comparison with the response of more alkaline, harder waters. The consequence of this increase for exposure and body burden to aquatic biota in acidified systems is discussed in Section 6.5.3. Hard water, alkaline, and salt water systems are, nevertheless, contaminated with Pb; in common with most other contaminants, compounds of this element accumulate to higher concentrations in sediments than in the water column. Some of this may be available to sediment-dwelling organisms and through the food web to those that feed upon them. The food chain behaviour of Pb, however, is quite different from that of mercury. There is no biomagnification. Even though Pb is readily bioconcentrated and bioaccumulated by living organisms, in contrast to mercury and many organic contaminants, concentrations of Pb decrease at successive trophic levels. This latter type of behaviour is, in fact, more typical of metals: Biomagnification is much more typical of persistent organic compounds. For Pb, the tendency is to actually decrease with higher trophic levels, a phenomenon that has been called biominification.

It should be fairly obvious from the foregoing that, once in soils or sediments, Pb is extremely immobile. In contrast to other metals, which may leach from soils into lower strata or into surface water and groundwater, lead and its compounds only dissolve under unusual conditions. Thus, it is not uncommon to find the pattern of surface accumulation in undisturbed soils retained for hundreds of years after deposition. The legacy in the Old World, particularly in the United Kingdom, of lead mines and local smelters from centuries ago is seen in the high Pb concentrations in surface layers of soils.

6.5.3 Environmental exposure and the toxicity of lead

HUMAN HEALTH

Humans have been exposed to lead for many centuries. Because lead accumulates in bones, there are opportunities to compare pre-industrial with modern exposures, as demonstrated by Pb-bone concentrations. However, there is some disparity among various estimates of the Pb burden in bones of preindustrial compared with contemporary vertebrates. Certainly, there have been indications that the bones of humans from Rome contained rather high Pb concentrations. Some analyses of ancient bones reported from the laboratory of Patterson, however, suggested that old skeletons contain only 0.01 to 0.001 as much lead as contemporary skeletons.

Table 6.4. Major anthropogenic sources of lead to the environment

Material/activity/ process	Form of lead released	Immediate receiving medium	Other media contaminated from same source	Main exposure route (receptor)	Comments
Lead plumbing	Dissolved Pb	Distributed (drinking) water	Food cooked or prepared with water	Ingestion (humans, domestic animals)	Mainly of concern for soft, acidic water
Lead solder in plumbing	Dissolved Pb	Distributed (drinking) water	Food cooked or prepared with water	Ingestion (humans, domestic animals)	Mainly of concern for soft, acidic water
Lead in metal vessels for food and drink	Dissolved Pb	Beverages and foods		Ingestion (humans)	Mainly of concern for drinks and foods with acidic reaction
Lead solder for food cans	Dissolved Pb	Beverages and foods		Ingestion (humans)	Mainly of concern for drinks and foods with acidic reaction
Lead in pottery glazes and in crystal glass vessels	Dissolved Pb	Beverages		Ingestion (humans)	Mainly of concern for wine, beer, or fruit juices
Lead pigment in artist's paints	"White lead" [lead carbonate $2PbCO_3.Pb(OH)_2$] and other inorganic compounds	Air		Inhalation (humans); ingestion (humans)	Major route is inhalation, but speculation has been made that Goya licked his paint brushes, providing an additional exposure route
Lead shields for radiation barriers	N/A				
Lead undercoating	"Red lead" (Pb_3O_4)	Flakes and dust from weathering or mechanical cleaning	Soil and water	Inhalation (humans); inhalation and ingestion (other biota)	Large structures such as bridges coated with Pb compounds may contaminate large areas when coating is detached
Leaded paint, commercial and domestic use	Lead chromate ($PbCrO_4$), as well as white and red lead	Flakes and dust from weathering, sanding, and erosion	Soil and water	Inhalation (humans); inhalation and ingestion (other biota)	Major source for young children through mouthing toys and furniture, also pica; serious problem even for adults when sanding old

Source	Chemical species	Release medium	Environmental compartment	Exposure route	Comments
					paint; major source of soil contamination especially in urban areas and in older housing (Lead in paint for domestic use is now regulated.)
Roofing material	Oxidation products and soluble salts of lead	Water from runoff and from rooftop cistern collection systems for water		Possible ingestion (humans); aquatic ecosystems	Very limited concern, but with acidic precipitation may enhance the mobilisation of Pb
Primary smelting of lead and other metallic ores	Oxidation products of lead	Air: particulates, including dust Soil: ash and other solid waste products	Soil and water	Inhalation and possibly ingestion (humans); inhalation and ingestion (other biota)	Older Pb-Zn smelters emitted rather large amounts of Pb, affecting livestock and wildlife
Secondary lead smelting	Oxidation products of lead	Same as the primary smelter	Soil and water	Inhalation and ingestion (humans)	Modern operations tend to be much cleaner
Tetraethyl lead production[a]	Tetraalkyl lead	Water (local accidental release)		Ingestion and dermal exposure (aquatic biota)	Very local and unusual problem, but has been documented
Tetraethyl lead combustion in engines	Oxidation products	Air (fine particles)	Soil and water	Inhalation (humans); inhalation and ingestion (other biota)	A major source of environmental lead contamination on a widespread scale between the 1940s and the 1980s
Batteries	Metallic Pb if discarded; oxidation products if recycled in secondary smelters (see above)	Landfill	Water and soil	Ingestion by biota	
Gunshot and fish sinkers	Metallic Pb	Lake sediment		Ingestion by water birds	Attempts are being made to control source of exposure by substituting other metals

[a] See Section 6.5.1.

293

Patterson also contended that the high Pb levels in some ancient strata of ice cores do not represent Pb deposition at the time but are the result of contamination of samples during collection and analysis (Patterson and Settle, 1976). This information is included here not just to show the uncertainty involved in making assessments of historical contamination but also to illustrate some of the possible technical problems involved in trace metals analysis, as referred to in Section 6.3.7. In contrast, recent analyses of Pb in the bones of preindustrial and contemporary Alaskan sea otters showed that the respective concentrations of Pb were similar. However, their isotopic compositions were distinct. The older bones had ratios typical of the natural Pb of the region, whereas the modern bones had ratios characteristic of industrial Pb from elsewhere (Bunce, 1994). This comparison illustrates how the technique of determining isotopic ratios, as we saw also in Figure 6.10, can contribute to source identification for contaminants.

Regardless of the situation concerning natural versus anthropogenic sources, or historical versus contemporary exposure to Pb, there is no doubt that humans are exposed to Pb from their environment and that Pb is toxic. Humans are exposed to lead through the inhalation of dust or other particles and through ingestion of contaminated food and water. Direct occupational exposure is omitted from this account, although some of the dose-response relationships that are used to assess safe levels have been established from occupational models.

The atmosphere is the primary mode of transport, and urban air has higher concentrations of Pb than does the air in rural or remote sites. Thus, it is generally true to say that urban dwellers are exposed to more lead than are those living in rural settings. Nevertheless, in particular environmental situations, such as those related to local contamination from industry, old plumbing systems, or local anomalies, exceptions occur. Concerning dietary ingestion of Pb, oral intake via food is quantitatively more significant than ingestion via water, but, for environmental exposure, an important source for intake via ingestion is the mouthing of objects contaminated by Pb in dust or soil. These last two sources are of paramount concern for young children not only because of their habits and their proximity to surfaces such as the floor and the ground but also because of their susceptibility to and their high rate of absorption of ingested Pb. Dermal exposure is of much less significance than inhalation or ingestion because absorption is very limited from this route.

Human exposure to lead can be assessed over the long term, as described earlier, by its concentration in bone. It also accumulates in other calcium (Ca)-based tissues such as teeth and eggshells. There is a relationship between Pb and Ca in metabolism, and this is shown quite significantly in the enhancement of Pb toxicity when the diet is deficient in Ca. Bone, however, is not a target organ in that it shows no obvious symptoms. Recent exposure to Pb can be more reliably assessed by measuring the concentration of Pb in blood. The convention for this measure is expressed in micrograms per decilitre ($\mu g\,dL^{-1}$). The value of this parameter has become a standard for determining exposure. It integrates the absorbed lead dose from all sources and is used to determine recent exposure. Furthermore, it is quantitatively related to effects. Although subject to the usual errors of sampling error and contamination, it is a fast reliable method for screening populations and alerting physicians and toxicologists to potential problems if used properly. The most contentious component is, however, the actual value that represents a "safe" condition.

Lead has long been known to have toxic effects on humans. Frank lead poisoning (plumbism), which results from acute exposure to high levels of lead, is manifest by effects of the central nervous system, with stupor, coma, and convulsions. Palsy in the form of wristdrop (the inability to hold the hand in line with the arm) is one of the classical and easily recognised manifestations of lead poisoning. These types of symptoms, associated with blood Pb levels of $70–100\,\mu g\,dL^{-1}$, are extremely rare in modern medicine, especially from environmental exposure. The more subtle effects of lower exposures are now the subject of research and speculation, as well as being of concern for regulatory agencies.

Until the 1970s, a blood Pb of $25\,\mu g\,dL^{-1}$ was the benchmark for protection against the adverse effects of environmental exposure to Pb. More recently, with research on the so-called subclinical effects of Pb, medical specialists, especially paediatricians, as well as toxicologists, have questioned whether the value of $25\text{-}\mu g\,dL^{-1}$ Pb in blood is sufficiently protective. The subclinical effects, by definition, occur in the absence of any symptoms of overt Pb poisoning (Rodricks, 1993). These symptoms are, for the most part, behavioural effects in children, including hyperactivity, poor attention span, and IQ (intelligence quotient) deficits. Such effects are difficult to measure and are not specific to Pb. They may also be confounded by other environmental factors such as socioeconomic and nutritional status, which are themselves often statistically related to lead exposure. Furthermore, there is limited information on the long-term persistence of these subtle effects. Nevertheless, passionate expositions have been made concerning the subtle effects of exposure to levels of Pb previously considered to be safe. The fact that the foetus and children under 6 years of age are the prime targets for subclinical effects makes the issue all the more compelling.

The dilemma of determining a safe or acceptable concentration of Pb is paralleled for a number of other contaminants and has relevance to the entire concept of a threshold for effect. In general, the degree of proof that is generally required for a scientifically defensible decision is not going to be available prior to developing guidelines for exposure to toxic substances. The alternative is to apply a risk assessment approach, which is discussed in Chapter 10. For Pb, there have been recommendations to use a level of lead in blood between 5 and $10\,\mu g\,dL^{-1}$ as a guideline for potential adverse effects in children. Such low levels are fairly challenging to measure and may also be difficult to achieve in urban populations.

A number of physiological or biochemical processes are affected by lead. Lead interferes with the haem system in a number of ways. Levels of protoporphyrin in serum erythrocytes increase when Pb concentrations are elevated and the enzymes ferrochelatase and δ-aminolaevulinic acid dehydratase inhibited by lead. ALAD has proven to be a useful biochemical indicator (biomarker) for this metal because the enzyme activity can be readily determined. The effects of Pb on the human haemopoietic system result in decreased haemoglobin synthesis, leading to anaemia with pallor as a classic symptom of Pb poisoning. Anaemia has been observed in children with blood Pb above $40\,\mu g\,dL^{-1}$. However, the biochemical changes in the haem system begin at much lower lead exposures than this. There is still no consensus as to the functional/clinical significance of these biochemical changes that occur at very low lead levels; this exemplifies a problem common to a number of biochemical indicators, which has been examined in Section 4.4.1.

Some recent studies on adult humans have implicated Pb in the elevation of blood pressure, but a causal relationship has not been firmly established.

In summary, lead has a wide range of effects on humans, from subcellular through physiological to behavioural; the effects depend on the type of exposure, its duration, and its level. Challenges still remain in terms of assessing the risk of very low levels of exposure to Pb, particularly for young children. Closely connected to this is the question of whether or not there is a threshold for effects (World Health Organisation, 1995).

OTHER BIOTA

Even though the greatest emphasis in studies of Pb toxicity has been on human health, other biota, notably terrestrial plants, fish, benthic biota, and wildlife, especially waterfowl, are exposed to this metal in their environment. The extensive knowledge base of lead toxicology for humans has proven valuable for understanding its effects on other biota. For example, the inhibition of ALAD by Pb has been shown to occur in a wide range of vertebrates, and the classic "wrist drop" of severe plumbism has been tragically mimicked by the necks of swans exposed to Pb through ingestion of lead in shot or fishing weights. Indeed the highest exposure for nonhuman biota to Pb appears to have resulted from the direct ingestion of Pb in objects such as these.

Laboratory tests have been made on the effects of lead on a number of organisms; many studies have addressed birds and fish, and recently some attention has been paid to invertebrates, particularly benthos. Effects on terrestrial plants have also been examined in the context of direct atmospheric deposition of Pb as well as the contamination of soils via atmospheric Pb. Lead-tolerant strains of plants have been isolated from Pb-contaminated habitats, suggesting that at least some part of the Pb in the substrates is biologically available. From the point of view of the environmental toxicology of Pb, laboratory studies can provide important information in terms of mechanisms of effects of Pb as well as for risk assessment. But caution has to be used in that environmental exposures are often much lower, up to two orders of magnitude lower, than the doses used in laboratory studies; thus, the effects elicited in laboratory experiments may have little relevance in the real world. Furthermore, the biological availability of Pb is very strongly affected by environmental conditions, and laboratory experiments often use salts of Pb that are highly available. A third cautionary point has to do with the behaviour of Pb in the food chain. As already described, this behaviour leads to decreases in concentration of Pb with increasing trophic levels. Overall, evidence for the risks from Pb to aquatic biota and to organisms feeding in the aquatic system appear to be relatively limited.

Aquatic biota are potentially at risk from exposure to Pb from indirect and direct releases into their environment. Reference has already been made to the deposition of airborne Pb into water bodies that are soft and acidified or acid-sensitive. Comparisons have been made of identical fish species of identical age across a spectrum of pH, and consistently the organs of fish from softwater lakes have statistically higher Pb concentrations than those in harder water. In contrast to mercury, much of the accumulated Pb in fish is not in the muscle but in organs. This distinction would be important if the fish were to be consumed by humans; however, wildlife predators normally consume the whole fish. The implication of the accumulation of Pb in fish from acidic waters in terms of effects is not clear. Fish from higher trophic levels and fish-eating birds will be exposed to this Pb in their diet, but with the tendency of Pb to biominify (see Section 6.5.2), the dietary concentrations are unlikely to be high or to result in harmful effects.

Toxic effects of Pb to aquatic life via the aqueous phase are probably quite unusual in the real world and are likely to occur mainly when there is direct release into water, such as in industrial effluents or from accidental spills, in combination with water of low pH and low hardness. Acute toxicity of Pb to fish in laboratory experiments has been recorded at concentrations from 0.1 to $500\,mg\,L^{-1}$ (ppm), but the amount of dissolved Pb in hard waters is typically of the order of less than $1\,\mu g\,L^{-1}$ (see Section 6.5.1). However, in laboratory tests, death by asphyxia may result from Pb-induced production of mucilage, which has adverse effects on gill function. Even insoluble Pb in colloidal form can have harmful effects on fish gills. The foregoing refers to inorganic Pb. Harmful effects on fish have resulted from releases of tetraethyl lead at the site of its manufacture. Interestingly, even though fish show the same inhibition of the enzyme δ-aminolaevulinic acid dehydratase that mammals show as a result of exposure to inorganic Pb, tetraethyl lead does not elicit this response (Hodson, 1986). Because of the decrease in use and production of tetraethyl lead in the early to mid 1980s, this route of exposure for aquatic biota has become of much less concern.

Rooted aquatic plants and benthic organisms are exposed to Pb that has been deposited into sediments, and bioaccumulation of Pb by these organisms certainly occurs, indicating that at least part of the lead is biologically available. However, the relationship between Pb accumulation and harmful effects is not clear-cut, especially in field studies. Benthic organisms, both freshwater and marine, accumulate very high concentrations of Pb in various tissues with no obvious adverse effects. Harmful effects of Pb on sediment-dwelling biota are difficult to demonstrate in the field, and laboratory studies indicate that rather high concentrations of Pb are required to elicit a toxic response. Reviews of sequential extraction (Section 6.2.4) of Pb from sediments show that the dominant fraction with which Pb associates is usually the iron/manganese oxide one. Studies on marine bivalves have shown that extractable iron (Fe), along with organic carbon, tends to decrease Pb availability. Although more detailed studies are required, it is generally true to say that the availability of Pb from sediments appears to be quite limited, and that, in part because of this limited availability, toxic effects of Pb on benthic biota are usually not considered to be a major problem.

The literature contains many reports of Pb poisoning in waterfowl and more rarely in raptors. Most poisonings of wild birds from Pb have been attributed to the inadvertent ingestion of Pb shot that accumulates in sediments. Other sources of Pb poisoning for wild birds, such as mine waste, are more difficult to determine but have been discussed. The biochemical indicators of Pb exposure that have been described for humans are also seen in birds, thus protoporphyrin and ALAD are useful indicators of recent Pb exposure for these organisms, although not for total body burden. Dietary lead accumulates primarily in the bones of birds, as well as in soft tissues, notably the kidneys.

Laboratory studies of the chronic toxicity of dietary Pb to avian species show a wide interspecies range of susceptibility. Interestingly, in a comparison between Pb ingested as metal shot versus Pb as acetate, the effects on test mallards were similar, suggesting similar availability (Beyer et al., 1988). The harmful effects of chronic Pb exposure on birds includes growth impairment and subtle reproductive effects such as reduced clutch size and inadequate Ca deposition in eggshells. As with humans, dietary Ca has been shown to influence the toxic effects of dietary Pb.

6.6   Cadmium

6.6.1   The occurrence, sources, and properties of cadmium

Cadmium, a relatively rare trace metal, has assumed importance as an environmental contaminant only within the past 60 years or so. It is a common by-product of mining and smelting operations for zinc, lead, and other nonferrous metals but is also released during coal combustion, refuse incineration, and steel manufacture. It is used in the electroplating of steel, as a component of various alloys, and in plastics manufacture, as a pigment and in nickel-cadmium batteries. Worldwide production in 1935 totalled 1,000 metric tonnes per year and today is in the order of 21,000 metric tonnes per year. Approximately 7,000 metric tonnes of cadmium are released into the atmosphere annually as a result of anthropogenic activity (mainly nonferrous mining) compared with 840 metric tonnes from natural sources (mainly volcanic activity). Direct release of Cd into the aquatic environment totals about 7,000 metric tonnes annually, distributed approximately 3 : 1 between natural weathering and industrial discharge.

Atmospheric Cd may be a significant source of the metal for humans and other mammals. It is volatile and may be present as vapour, particularly in industrial settings. Otherwise, it may be inhaled in particulate form. Cadmium is present as a fine aerosol in cigarette smoke, and daily consumption of a package of cigarettes will approximately double Cd intake. In nonsmokers, respiratory absorption of Cd is generally between 15 and 30%.

Cadmium, atomic number 48, atomic weight 112, appears below zinc in the second row of transition elements in the periodic table (Figure 6.3) and shares some common properties with Zn. Like Zn, only the 2+ oxidation state is reflected in its compounds. Its substitution for Zn, an essential trace element, in certain enzymes probably accounts for some of its toxicity. Even though Cd is not generally regarded as having any physiological function, it has been shown to ameliorate Zn deficiency successfully in some diatom species. This appears to be the only demonstration of an apparent nutritive function for Cd.

Bioavailable Cd is best predicted by the free cadmium ion. The tendency for the metal to form chlorocomplexes in saline environments renders the metal less available from solution and may largely explain the inverse relationship between cadmium accumulation and salinity in the estuarine environment. Generally, the formation of soluble inorganic or organic complexes of Cd reduces cadmium uptake by aquatic organisms, although there have been some reports of increased Cd uptake in the presence of organic ligands. In such cases, it seems likely that the attachment of the organic complex to the membrane in some way facilitates the transfer of the Cd to the membrane receptor site. In most cases, it seems unlikely that Cd traverses biomembrane in complexed form, although the recent demonstration of enhanced Cd (and Zn) uptake in the presence of citrate (Errecalde et al., 1998) suggests that "accidental" uptake of the Cd-citrate complex is possible via a specific permease.

6.6.2 The physiological and ecological behaviour of cadmium

There have been several demonstrations of the ability of Cd to compete with calcium for a variety of receptor sites in organisms. Thus, it seems likely that transepithelial Cd transport may occur via calcium channels or through the mediation of specific calcium transport mechanisms such as Ca ATPases (Figure 3.1). Within cells, Cd competes with Ca for specific binding proteins such as calmodulins and may be accumulated in calcified tissue including skeletal elements.

Cadmium has a strong tendency to concentrate in aquatic animals, particularly bivalve molluscs where bioconcentration factors of $>10^6$ have been reported. Plants, too, concentrate Cd to a large degree, and this may be greatly enhanced by the agricultural application of commercial sludge, which may contain up to $1,500\,mg\,Cd\,kg^{-1}$. The U.S. Food and Drug Administration has recommended that sewage sludge applied to land should contain no more than $29\,mg\,Cd\,kg^{-1}$. In humans, dietary intake represents the largest source of Cd exposure, with average daily intakes in populations from Europe and North America being approximately $20–40\,\mu g$. In communities living on a predominantly seafood diet, the intake may be much higher. In open ocean water, the close correlation between the depth profile of Cd and a nutrient such as phosphate has been taken as evidence that soft tissues act as a major vector for Cd transport in the ocean environment.

6.6.3   The toxicity of cadmium

In workplace situations, inhalation of Cd fumes may result in pulmonary oedema and pneumonitis. Respiratory Cd exposure has also been linked to increased risks for both lung and prostate cancer, and the metal has now been classified by the International Agency for Research on Cancer as a Category 1 (human) Carcinogen.

The most serious cases of cadmium poisoning have been reported from Japan, and first came to light in the Jintsu River basin near the city of Toyama. Effects were traced to the consumption of rice grown in water that had been contaminated since the 1920s by Zn, Pb, and Cd from a mining and smelting operation. Symptoms were initially reported in 1955, although the condition probably predated this by at least 20 years. Patients were mainly postmenopausal women, long-term residents of the area, who presented multiple symptoms including severe renal dysfunction, osteomalacia, and osteoporosis. Skeletal deformities were accompanied by severe pain, which gave the condition its name, *Itai-Itai*, loosely translated as ouch-ouch disease. Pregnancy, lactation, vitamin D deficiency, and other dietary deficiencies were thought to be cofactors in the disease. Kidney malfunctions included decreased renal absorption of phosphate, glucose (glucosuria), and protein (proteinuria).

Cadmium has a long half-life in the human body (20–30 years), about 20% occurring in the liver and about 30% occurring in the kidneys. The kidneys appear to be the most important site of intoxication following chronic Cd exposure with damage resulting from the inhibition of enzymes responsible for resorption processes in the nephric tubules. Other enzymes inhibited by Cd include δ-aminolevulinic acid synthetase, arylsulfatase, and alcohol dehydrogenase, even though it enhances the activity of δ-aminolaevulinic acid dehydratase, pyruvate dehydrogenase, and pyruvate decarboxylase.

Cadmium accumulation in the liver is related to its association with low-molecular-weight metal-binding proteins and metallothioneins, which are synthesised in the liver and other tissues as a result of exposure to the metal. Other metals responsible for metallothionein induction include Hg, Cu, Zn, and possibly Ag. MTs probably perform a detoxifying function for all these metals but may have additional regulatory capability for essential trace metals such as Cu and Zn. Cd has also been shown to induce the synthesis of low-molecular-weight compounds called phytochelatins in plant tissue. Like MTs, these probably serve to protect enzymes and other receptors from the toxic effects of the metal.

In crustaceans, the main site of Cd storage is the hepatopancreas. This organ has a variety of digestive and metabolic functions, which may be damaged by prolonged Cd exposure. De Nicola et al. (1993) described a variety of pathological effects in the hepatopancreas of the marine isopod *Idotea baltica* exposed to Cd, including swollen mitochondria and picnotic nuclei. They also recorded a proliferation of granular inclusions, which they considered to be sites of Cd storage and, therefore, a detoxification mechanism. However, exposure concentrations employed by these workers considerably exceeded those that might be expected in the marine environment. Sublethal cadmium exposure during embryonic and larval development reduced long-term survival and growth, with males showing greater sensitivity than females. Differential survival of Cd-exposed genotypes of *I. baltica* indicated that chronic exposure to sublethal Cd levels may have the potential to induce long-term changes in population structure.

6.7   Copper

6.7.1   The occurrence, sources, and properties of copper

Copper is a rosy-pink-coloured transition metal, of atomic weight 63.54 and atomic number 29. In metallic form, it is very malleable and ductile and an excellent conductor of heat and electricity, melting at 1,083°C. It has two valance states: Cu^{+1}, cuprous, and Cu^{+2}, cupric. In the lithosphere, it occurs in trace quantities in metallic form and as Cu compounds in ores, including copper pyrites ($CuFeS_2$) and malachite [$CuCO_3 \cdot Cu(OH)_2$]. It has been known to humans for at least 6,000 years, being readily reduced to metallic form from malachite ore by heating with carbon. It has long been used in pure metallic form or alloyed with other metals, as bronze (copper-tin) and brass (copper-zinc). Copper is easily worked, but the alloys of Cu are not as soft as the pure metal. In metallic form, pure or alloyed, Cu has been used to make tools, ornaments, statuary, jewellery, coins, and vessels for food and beverages, as well as for electrical wiring, in plumbing, and for electroplating. Salts of Cu have been used as fungicides, molluscicides, and algicides.

The major processes that result in the mobilisation of Cu into the environment are extraction from its ore (mining, milling, and smelting), agriculture, and waste disposal. Soils have become contaminated with Cu by deposition of dust from local sources such as foundries and smelters, as well as from the application of fungicides and sewage sludge. Aquatic systems similarly receive Cu from the atmosphere, as well as from agricultural runoff, deliberate additions of copper sulphate to control algal blooms, and direct discharge from industrial processes.

Copper salts are moderately soluble in water: the pH-dependence of sorption reactions for copper compounds means that dissolved concentrations of Cu are typically higher at acidic to neutral pH than under alkaline conditions, all other things being equal. Copper ions tend to form strong complexes with organic ligands, displacing more weakly bound cations in mixtures. Complexation facilitates Cu remaining in solution, but usually decreases its biological availability. Copper also forms strong organometal complexes in soils and in sediments. Again, its availability is normally decreased by complexation.

6.7.2   The physiological and ecological behaviour of copper

Copper is an essential trace nutrient for all known living organisms. It is required for the functioning of more than 30 enzymes, all of which are either redox catalysts (e.g., cytochrome oxidase, nitrate reductase) or dioxygen carriers (e.g., haemocyanin). Prokaryotic and eukaryotic organisms have mechanisms for transporting Cu into cells. The process of cellular copper transport and its insertion into metalloenzymes results from an elaborate interplay between specific metal ion transport proteins and associated chaperone proteins. Such an orchestrated sequence of copper transfers between proteins functions to protect the organism from the toxic effects of inappropriate copper binding and to deliver the metal accurately to the correct enzyme (Pufahl et al., 1997).

Although required as a nutrient, Cu, as shown in Figure 6.1, can become toxic at concentrations exceeding that which is required. It has already been pointed out in the introduction that the concentration per se in the environment is not sufficient to determine the biological effect of Cu, whether stimulatory or harmful. In general, Cu conforms to the FIAM. The chemical form and, thus, the availability of Cu are influenced not only by organic matter but also by pH (whose influence interacts with that of organic matter) and redox potential. Water hardness, as calcium and magnesium carbonates, affects the toxicity of Cu, but this trait almost certainly acts through the protective effects of Ca ions on and in living cells rather than on the chemical speciation of Cu. The protective role of Ca is so consistent that water quality criteria for the protection of aquatic life are given different values for hard and soft water.

Copper accumulates in living organisms but does not biomagnify in food webs. In aquatic systems, copper tends to be in rather low concentrations in water, ranging from 0.001 to $0.1 \mu g\,L^{-1}$ in uncontaminated fresh waters and from 0.03 to $0.6 \mu g\,L^{-1}$ in ocean waters. Even in contaminated waters, Cu tends to leave the water column and, like most contaminants, accumulates in sediments. Copper is of particular concern in estuarine river sediments. Elevated Cu concentrations (e.g., $50–100 \mu g\,L^{-1}$ in water and more than $7,000 \mu g\,g^{-1}$ in sediments) can occur in aquatic ecosystems such as lakes that receive mine tailings or where copper sulphate has been added to control algal blooms.

In terrestrial ecosystems, Cu that is added from anthropogenic activities accumulates in soil and behaves in a manner similar to that in sediments (i.e., it may be bound by organic matter and its biological availability is strongly affected by pH, redox potential, and organic matter).

6.7.3   The toxicity of copper

The toxicity of Cu for mammals is rarely of concern; most mammals tested can tolerate high concentrations of Cu in their diets. Indeed, in some places, domestic animals are provided with Cu supplements in their feed. Similarly, Cu in water and beverages does not represent a problem, and standards for Cu in drinking water are based on taste, not on the risk of toxicity. The reason for the lack of Cu toxicity for mammals is twofold: primarily the explanation is that mammals have a biochemical system that can detoxify Cu in the liver and kidney. Exceptions arise only in the case of specific metabolic disorders, such as Wilson's disease where the patient's metabolic system cannot detoxify dietary Cu. The other reason concerns the low availability of Cu in many foods because of its tendency to bind with organic matter.

Terrestrial systems are somewhat susceptible to copper toxicity, although crop plants are more likely to be copper deficient than toxified. Soil microbial activity has been shown to be adversely affected by contamination by metals, particularly at pH values below neutrality, and copper is one of the metals that has been implicated. With its known antifungal and algicidal properties, it is not surprising that other microorganisms are affected by Cu. Functional changes such as decreased carbon mineralisation in soils, a rather general indicator of soil microbial activity, have been correlated with increased cupric ion activity. The nitrogen cycle has been investigated in terms of the effects of Cu. In pure culture in the laboratory, Cu has been shown to affect the microbial process of nitrification adversely, but this relation has not been reported consistently for nitrification or ammonification in the soil environment.

The major concern for environmental impacts of copper concerns the aquatic system. Fish and crustaceans are generally 10 to 100 times more sensitive to Cu than are mammals, and algae are up to 1,000 times more sensitive (Förstner and Witmann, 1979; Hodson et al., 1979). Copper was one of the metals for which some of the earliest studies on factors affecting availability and toxicity were carried out, and a good body of literature based on laboratory tests addresses these factors. In laboratory tests, planktonic invertebrates such as daphnids are even more sensitive to copper than are fish, but for both groups the range of toxic thresholds is quite wide. Sensitivity depends upon biological factors such as species, size, and life stage, as well as on chemical factors affecting availability. In the field, copper rarely exists as the sole metallic contaminant; furthermore, copper pollution is frequently accompanied by low pH, which may itself be toxic, in addition to exerting an effect on the form of copper. Some of the more easily interpreted field studies are those in which copper sulphate has been added over a number of years, for example to farm ponds.

Copper accumulates in sediments and in sediment-dwelling organisms. When the binding of Cu to various sediment fractions (see Section 6.2.4) has been investigated for "real" sediments, Cu is predominately in the Si-bound and organically bound fractions. This is consistent with the chemical properties and behaviour of Cu. The toxicological effects of Cu in sediments on sediment-dwelling organisms are more difficult to quantify than is bioaccumulation. Values for toxicity of Cu in sediments to benthic organisms vary widely, from the hundreds to the thousands of micrograms per gram of Cu in sediment on a dry weight basis. Laboratory tests tend to yield lower values for toxicity than do field tests, in part at least because

Cu availability is not known in the field. The effect of sulphide binding (see AVS in Section 5.4) of copper as well as other metals may explain the apparently low toxicity of copper in sediments. In fact, the availability of copper in sediments, and thus its effect, is influenced profoundly by the chemistry of the sediments. As seen in Case Study 4.1, there were benthic communities that did not differ significantly from the control sites, yet the total copper values exceeded the sediment-quality criteria.

The so-called spiked sediment approach is one of the approaches used to establish sediment-quality guidelines for metals; it involves adding known amounts of the contaminant to sediments and using test organisms to bioassay the toxicity. The value for acute toxicity is then modified by a factor of 10 to account for chronic toxicity and often by a further safety factor to account for uncertainty. Ontario's guideline for Cu is $16 \mu g \, g^{-1}$ for the lowest effect level and $110 \mu g \, g^{-1}$ for the severe effect level (SEL). Yet the literature records examples of sediments with $1,000 \mu g \, g^{-1}$ Cu that had no apparent adverse effects on benthic invertebrates. With the present state of understanding concerning the toxicity of sediment-bound Cu, it seems prudent to use a site-specific approach. The criteria refer to ideal situations, and the guidelines are, as the name suggests, for guidance.

Tolerance or resistance to copper has been reported quite frequently, for higher plants, algae, fungi, and bacteria. In most examples, it is assumed that tolerance has been selected for by very high concentrations of Cu in the environment, typically from mining activities or from repeated fungicide or algicide applications.

In summary, Cu certainly continues to be of concern for aquatic life, especially in acid-sensitive or soft waters. However, in both freshwater and saltwater, its bioavailability (and toxicity) is largely controlled by the DOC content of the water. Copper tends to form complexes with dissolved humic acid and other organic compounds.

6.8   Nickel

6.8.1   The occurrence, sources, and properties of nickel

Nickel (Ni) is a group II transition metal which is commonly used for electroplating, stainless steel manufacture, and in nickel-cadmium batteries. Like other divalent transition metals, it exists as the hexaquo ion $[Ni(H_2O)_6]^{+2}$ and dissolved salts in natural waters. Pristine streams, rivers, and lakes contain $0.2–10 \mu g \, L^{-1}$ total dissolved nickel, and surface water near nickel mines and smelters contain up to $6.4 \, mg \, L^{-1}$. Seawater contains approximately $1.5 \mu g \, L^{-1}$ of which approximately 50% is in free ionic form.

Ni enters the atmosphere from fossil fuel burning, smelting, and alloying processes, waste incineration, and tobacco smoke. Ni exists in the atmosphere primarily as water-soluble $NiSO_4$, NiO, and complex metal oxides containing nickel. The average ambient total concentration in air in the United States is $0.008 \, mg \, m^{-3}$. The current limit set by the U.S. Occupational Safety and Health

Administration (OSHA) for airborne Ni is $1 \, mg \, m^{-3}$. However, evidence of high Ni accumulation in lung tissue and recent findings linking exposure to small amounts of particulate Ni compounds with DNA damage have led to proposals to reduce this limit substantially.

6.8.2   The physiological and ecological behaviour of nickel

In mammals, including humans, the primary route for nickel uptake is by inhalation. A small amount is ingested with food and water, but most is eliminated in the faeces. In aquatic organisms, nickel uptake is influenced by water hardness such that nickel toxicity decreases with increasing water hardness. Water-soluble Ni salts can enter an organism by simple diffusion or through Ca^{2+} channels (Fletcher et al., 1994) and accumulate in the cytosol of cells until they reach equilibrium with the external Ni concentration or saturate cell membrane binding sites (Azeez and Banerjee, 1991; Fletcher et al., 1994). Within the cytosol, Ni^{2+} becomes bound to proteins and low-molecular-weight ligands including amino acids such as cysteine and histidine. Binding to albumin in the blood of vertebrates precedes elimination in the urine. The half-life of water-soluble nickel salts in humans is about 1–2 days. Because these compounds cycle through organisms quickly, they do not penetrate the nuclear envelope of cells and tend to have low toxicity. Some enzymes, such as ureases, contain Ni, and it has been shown to be an essential element in many vertebrates, invertebrates, and cyanobacteria.

6.8.3   The toxicity of nickel

Nickel has long been considered to be relatively nontoxic when compared to other heavy metals. However, several different pieces of evidence have caused this view to change somewhat in the last decade. A study by Kszos et al. (1992) indicated that nickel toxicity to freshwater species tends to be masked by the presence of other metals in water polluted by mining, electroplating, and stainless steel production. Other studies have shown that chronic effects of nickel exposure such as the inhibition of growth and reproduction in invertebrates occur at Ni concentrations that are up to two orders of magnitude lower than acute levels (e.g., Azeez and Banerjee, 1991; Kszos et al., 1992). Additionally, several studies linking exposure to small amounts of particulate Ni compounds with DNA damage have increased concern about the carcinogenicity of Ni in mammals. As a result of studies such as these, the U.S. Environmental Protection Agency (1994) placed nickel on its Categories of Concern list.

Nickel toxicity is highly dependent on the form in which it is introduced into cells. Nickel compounds can be divided into three categories of increasing acute toxicity:

1. Water soluble nickel salts [$NiCl_2$, $NiSO_4$, $Ni(NO_3)_2$, and $Ni(CH_3COO)_2$]
2. Particulate nickel [Ni_3S_2, NiS_2, Ni_7S_6, and $Ni(OH)_2$]
3. Lipid-soluble nickel carbonyl [$Ni(CO)_4$]

Nickel carbonate ($NiCO_3$) has attributes of groups I and II. It dissolves readily in cell culture and distributes within cells as a water-soluble salt, but its high uptake by cells and acute toxicity are more like those of particulate nickel. Acute nickel toxicity to humans is generally inversely proportional to Ni solubility in water. Nickel carbonyl is more toxic than particulate nickel because the kinetics of gas phase delivery of $Ni(CO)_4$ to lung tissue are much faster than incorporation of particulates into cells. In terms of long-term toxicity and carcinogenicity, however, the order is water-soluble salts < $Ni(CO)_4$ < particulates.

Although Ni toxicity increases with decreasing water solubility of the Ni compound, free Ni^{2+} is the ultimate toxic form inside the cell. This apparent contradiction is resolved if one considers routes of nickel uptake and elimination by cells. Soluble Ni compounds have a fast biological turnover and low toxicity, and particulate Ni compounds enter cells via phagocytosis. Lysosomes attach to the phagocytic vesicles and assist in Ni dissolution, although the mechanism of this is unknown. The vesicles then aggregate around the nuclear envelope, at which point Ni^{2+} enters and reacts directly with DNA resulting in fragmentation and cross-linking (Nieboer et al., 1988). In mammals, carcinogenicity is a greater concern because particulate Ni is a long-term source of Ni^{2+} that is delivered directly to DNA upon dissolution. The half-life of pools of particulate Ni in the human lung and nasal tissue is 3–4 years.

Nickel carbonyl, a gas, is an occupational hazard specific to nickel refining and the use of nickel catalysts. It is a highly lipid soluble and rapidly crosses the alveolar-blood barrier in both directions. A significant amount is exhaled as $Ni(CO)_4$, and the remainder is localised in tissues (primarily the lung) in free ionic form. Acute toxicity is due to the inhibition of enzyme activity in the lung by subsequent respiratory failure.

The U.S. Environmental Protection Agency's chronic water quality criterion (C) for Ni is calculated using the following equation:

$$C(\mu g\, L^{-1}) = e^{(0.76[L_n(\text{hardness})]+4.02)}$$

6.9　Selenium

6.9.1　The occurrence, sources, and properties of selenium

Selenium is a metalloid that exists in three oxidation states in surface waters: Se(−II), selenide; Se(IV), selenite; Se(VI) selenate. In addition it occurs in the elemental form Se(0), which is insoluble in water. It appears between sulphur and the metal tellurium in group VIA of the periodic table (Figure 6.3) and shares properties of both these elements.

Selenium enters the environment from many different industrial situations including coal-burning, fly-ash piles, and the manufacture of glass, paint, petroleum, textiles, and electrical components. Historically, it has been occasionally employed as a pesticide, and, although its use has been largely discontinued, it is still retained in some shampoos used to control human dandruff and canine mange. Elevated concentrations in aquatic systems may occur where the soil is naturally rich in selenium. Normally soils contain about $5\,mg\,kg^{-1}$ Se, although levels up to $80\,mg\,kg^{-1}$ may be found.

Selenate [Se(VI)] is stable in a well-oxidised environment and selenate salts are highly water soluble. As such, Se is freely available to aquatic animals and to plants. However, dietary Se is considered to be a more important source of the element than the aqueous form. Selenite [Se(IV)] is favoured under mildly oxidising conditions. It is less water soluble than selenate and is largely unavailable to plants in this form. But selenite is readily incorporated into organic compounds such as the amino acid derivatives Se-methylselenomethionine and Se-methylselenocysteine, which are taken up by plants. These compounds may substitute for amino acids in proteins causing conformational and functional changes resulting in toxicity. Similarly, protein degradation may result from bonding between inorganic Se accumulated in tissues and sulfhydryl groups on amino acid constituents of intracellular proteins. Selenomethionine is embryotoxic and teratogenic in developing birds when supplied in laboratory feed at levels exceeding $4\,mg\,kg^{-1}$, causing symptoms similar to those seen in field studies. Hydrogen selenide is a toxic gas existing in reducing environments.

6.9.2   The physiological and ecological behaviour of selenium

Vertebrates such as rats are able to methylate selenate and selenide to compounds such as dimethylselenide and dimethyldiselenide. Methylation of selenium is mediated by bacteria. Other bacteria are capable of oxidation of elemental Se to selenite (SeO^{2-}_3).

Selenium has been demonstrated to be an essential micronutrient for several organisms, although it appears that there is a very narrow window between beneficial and toxic selenium levels. Selenium is a component of the enzyme glutathione peroxidase and numerous proteins including haemoglobin, myosin, cytochrome c, and several ribonucleoproteins. Hodson and Hilton (1983) reported Se deficiency symptoms in fish at tissue levels $<0.1\,\mu g\,L^{-1}$ dry weight, but evidence of toxicity at $>10\,\mu g\,Se\,g^{-1}$. Selenium deficiency in mammals can result in disorders of the liver, pancreas, and other tissues. Cattle from the Florida Everglades region deficient in Se showed symptoms of anaemia and reduced growth and fertility.

6.9.3   The toxicity of selenium

There is epidemiological and experimental evidence that Se can act as an anticarcinogen and antimutagen at low levels but is capable of genotoxicity at higher concentrations (Bronzetti and della Croce, 1993). Several studies have also indicated an ameliorative effect of Se on Hg and cadmium toxicity in humans and aquatic animals. Additions of Se to Swedish lakes have been made in an attempt to accelerate the reduction of mercury concentrations in fish.

The current recommended water quality criterion for Se in the United States is $5\,\mu g\,L^{-1}$. This was primarily based on a study carried out in Belews Lake, North Carolina, where decreased fish survival and reproduction were linked to Se concentrations in the water column as high as $10\,\mu g\,L^{-1}$. Laboratory-based tests have tended to indicate somewhat higher levels for acute Se toxicity, often above $1\,mg\,L^{-1}$. However, the use of acute toxicity data to evaluate Se toxicity has been questioned in view of the importance of Se bioaccumulation in the response of organisms seen in the Belews Lake study (Lemley, 1985).

6.10　Phosphorus

6.10.1　The occurrence, sources, and behaviour of phosphorus

Phosphorus (P) is a luminous (phosphorescent) element, with an atomic weight of 30.98, an atomic number of 15, and valence states of (III) and (V). The element does not occur free in nature but compounds such as calcium phosphate $[Ca_3(PO_4)_3]$ and apatite $[CaF_2 \cdot 3Ca_3(PO_4)_2]$ occur naturally and abundantly. Phosphorus is an essential element; both inorganic and organic compounds of P are part of many biological systems. Phosphate is notably an integral part of the energy transfer system of all living organisms.

From the point of view of environmental toxicology, the major concerns for sources of P are its use in detergents and in agricultural fertilisers. Synthetic detergents until the last two decades were phosphate-based, and some continue to contain relatively large concentrations of P. Domestic sewage, therefore, contains P not only from human waste but also from laundry and other detergents that are put into drains. The P enters water bodies either directly or via sewage treatment plants (STPs). Agricultural crops and, to a lesser extent, forest systems require additions of phosphate as fertiliser, which is produced commercially. Excess of the fertiliser that is added to farm fields enters water bodies from runoff and drainage. Farm manure similarly provides a source of P to water bodies. Phosphates are also mobilised from weathering and dissolution of rocks and from soils. Softer rocks and the soils derived from them release more P to the environment than do harder rocks such as granites. Soils from which vegetation cover has been lost will tend to erode and P is released into water bodies as a consequence. There are also diffuse supplies of P for aquatic systems from atmospheric deposition. Some of these are of natural and some of anthropogenic origin.

6.10.2　The physiological and ecological behaviour of phosphorus

Living organisms and thus all ecosystems require a continuous supply of P, and the element is frequently the limiting nutrient for plants. The ecological significance of this can be quite profound. The law of limiting factors states that the rate of a particular process is controlled by the factor (nutrient) that is in the lowest supply relative to demand. The primary producers of aquatic systems require phosphate, and freshwater aquatic systems are most frequently P-limited. The various species of algae and blue-green bacteria that comprise the phytoplankton communities in different types of freshwater and saltwater are adapted to the phosphorus concentrations typical of their particular environment. Some planktonic algae have mechanisms to take up excess P when conditions are unusual and P is in excess supply (termed luxury uptake); this is likely to occur during turnover of a lake or when physical conditions are turbulent and P is released from sediments or from the watershed. Reserves are stored inside the cell as polyphosphate bodies, and P can be released enzymatically during periods of P starvation.

Surface waters have been generally categorised into low-nutrient, unproductive systems, described as oligotrophic systems, rich in nutrients and productive, described as eutrophic, and an intermediate condition known as mesotrophic. Like most classification systems, these are not in fact discrete categories, but rather a continuum exists. Any condition in terms of nutrient status can exist naturally and support a functioning ecosystem. Over time, it is probable that a slow progression from oligotrophic through mesotrophic to eutrophic may occur naturally, but this could take many thousands of years. Thus, a deep oligotrophic lake set in geological conditions of hard insoluble bedrock and nutrient-poor soils in a cool temperate region is likely to remain oligotrophic for a very long time. A short-term pulse of a nutrient over the course of a relatively steady set of conditions is unlikely to cause more then a brief increase in productivity. However, if there is a major shift in the nutrient status of a water body, not only will productivity increase, but biomass and species composition will also change. Initially, the effect of an increase in a limiting nutrient will be most obvious in the phytoplankton community in the form of an algal bloom, but if such an increase is sustained, overall ecosystem effects soon follow.

Eutrophic waters are not inherently less stable or viable than other types. Indeed, very productive waters have their own characteristic flora and fauna and provide excellent fisheries. The problem referred to as eutrophication, more properly cultural eutrophication, arises when excess nutrients result in overfertilisation. Oligotrophic and mesotrophic waters are more susceptible to this process than are waters that are already eutrophic because communities in the latter category are already adapted to high nutrient concentrations.

The nutrient that is most frequently, although not invariably, responsible for cultural eutrophication is P because it is most frequently the limiting nutrient and because human activities have enhanced the addition of P to surface waters. The series of interrelated processes that typify cultural eutrophication illustrate the interdependence of the various components of an ecosystem. Algal blooms are themselves natural phenomena in freshwater and marine systems, and the mechanisms are still poorly understood, but typically they involve excessive and unusual production in one or a few species of planktonic algae, whose populations crash after blooming. The bloom may itself be harmful because of natural toxins produced by the species that is blooming. When the trigger is overfertilisation by P, green or blue-green algae may grow into visible mats or sheets in the open water and along shorelines. As they die, they give large amounts of organic matter to the sediments as well as leave dead material along the shoreline; the latter may change the physical habitat for littoral biota; all types of dead material stimulate bacterial decomposition, which makes unusual oxygen (O) demands. The resulting depletion of O can be very detrimental to organisms in the deeper parts of the lake and in the sediments. These regions, particularly during periods of stratification, are not replenished with O and may become anoxic. In the water column, fish and planktonic invertebrates, particularly those that normally occupy oligotrophic systems and require high oxygen concentrations, often die. Death is the result of O depletion rather than the effect of any toxic substance.

In the sediments of overfertilised water bodies, anoxia often prevails, with changes in the species composition of invertebrates, leading to trophic effects through change in the food supply for the consumers of benthic invertebrates. Furthermore, aerobic bacteria, normally dominant in surficial sediments, are replaced by anaerobic forms, which produce hydrogen sulphide (H_2S), itself a toxic gas. The characteristic "rotten egg" odour of anaerobic decomposition is normal for certain systems at certain periods of the year, but for a previously O-dominated system, the production of (H_2S) is a symptom of overfertilisation.

Other ecosystem-level effects that can result from overfertilisation arise from changes in species composition in the phytoplankton. Species of diatoms, chrysophytes, and chlorophytes are the most important algae in oligotrophic systems: these are normally P-limited. When P becomes available in excess and is no longer limiting, according to the law of limiting factors, some other factor will become limiting. For the phytoplankton, this is usually nitrogen. In the absence of a ready supply of nitrate or other forms of N that the algae can use, these forms are at a competitive disadvantage. In contrast, cyanophytes (blue-green bacteria) can utilise gaseous nitrogen (N_2), which means that under conditions of N limitation, such nitrogen fixers will become dominant. Blue-green bacteria, although normal components of phytoplankton, are frequently not the preferred food for herbivores, being poorly grazed or not assimilated; indeed, they may even be toxic. Thus, herbivorous zooplankton and planktivorous fish may be adversely affected by limited or unsuitable food. Loss or changes in these communities may also affect the higher trophic levels. In the most extreme cases, some or all of the communities may become so altered that there is no longer a viable system. More frequently, however, there are profound changes from the original condition, but a viable, albeit less valuable or stable, system continues to function.

Chemical changes are part of cultural eutrophication. Anoxia and lowering of pH as decomposition of the large organic load occurs result from bacterial activity altering the chemistry of the sediments. Nutrients, including P itself, and the metals iron and manganese in the sediments, may undergo chemical changes that favour their release into the water column and alter their availability in this or some other manner. Certainly, there is a reservoir of P in the sediments of overfertilised water bodies, and, during periods of anoxia, this P can be released into the water column (Jacoby et al., 1982). This means that even if the external supply of excess P is identified and controlled, under some conditions the eutrophied water body may have an internal supply of P, and the process of eutrophication will continue even after abatement.

Although no absolute pattern can be said to typify cultural eutrophication, a picture emerges of a very complex series of primary and higher order changes, any or all of which may occur as a result of overfertilisation. The impacts go far beyond a scientific interest in an ecosystem: Drinking water, fisheries, and recreational use of the water as well as aesthetic considerations have made this a topic of major socioeconomic importance, and these concerns almost certainly stimulated research, as they have for other topics in environmental toxicology. Cultural eutrophication can now be said to exemplify a problem for which we have a rather complete understanding of causes and effects, and of mechanisms, from the organism to the ecosystem level. But this was not always so. The problem attracted considerable attention in the 1960s, when the condition of certain large water bodies, notably Lake Erie of the North American Great Lakes and some rivers and lakes in Europe, as well as shallow marine systems (e.g., Chesapeake Bay and Boston Harbor) were observed to have deteriorated in the manner described in Case Studies 4.5 and 9.1.

Identification of the causal agent did not follow immediately. These large water bodies were typically the subject of many perturbations, including overfishing, dredging, and major changes in land use in their watersheds. During the 1950s, a fivefold increase in oxygen demand was seen in the sediments of Lake Erie, and in the central basin increasingly anoxic conditions denied refuge to several cold water fish species such as lake trout, blue pike, and lake whitefish. Accelerating land clearance in the catchment area of this and other Great Lakes contributed to the deposition of clay, which compacted at the sediment surface to form a hard impervious layer having poor oxygen exchange and was unsuitable for fish egg survival.

In the course of investigations in the 1960s, because of the multiple stresses on many water bodies, excess fertilisation by P was not consistently identified as the obvious cause of, for example, fish kills. Furthermore, considerations other than scientific ones were involved. The implication of detergents in a phosphate-based model for eutrophication was not easily accepted by the manufacturers of detergents. And in common with almost any ecosystem-based problem, laboratory experiments alone, although often providing evidence of the pivotal role of P, were not definitive in terms of identifying the cause and the mechanisms of fish kills, algal blooms, and anoxia.

Whole system manipulations have been referred to (Section 4.7.2) in the context of approaches to environmental toxicology. A whole-lake manipulation experiment provided some of the most convincing proof, at least for the public, that P was the nutrient of concern for cultural eutrophication. Schindler (1990) and his group used the approach with considerable success. The Experimental Lakes Area of north-western Ontario has been referred to already in Section 4.7.2 and Case Study 4.7. Over a relatively long period of time beginning in 1969, several large-scale experiments on nutrients were carried out at the ELA. Lake 226, a small oligotrophic lake, was divided by a curtain into two parts, one of which received additions of P, N, and C while the other received only N and C. The result was an algal bloom in the part that had received P, N, and C but not in the part that received only N and C. A photograph of an aerial view of the lake (Plate 4.1), showing the two distinct sections with the conspicuous algal bloom only on the side that had received P, was published as a dramatic indication that P, and not N or C, was the limiting nutrient. Many more detailed experiments followed, in Lake 226 as well as several others in the ELA, addressing the mechanisms and the biological changes in the treated as compared with control lakes. More detail is provided in a series of publications, including Schindler and Fee (1974).

The quantitative relationship between P concentration and primary productivity in lakes has been modelled by Dillon and Rigler (1974) who showed a statistical relationship between chlorophyll-*a* and the concentration of total P. Chlorophyll-*a* is the pigment common to all groups of algae and, thus, provides an index of the standing crop of phytoplankton. Thus, as discussed in Section 4.2, chlorophyll-*a* can be used as an indicator of the nutrient status of a water body. Used with caution, chlorophyll-*a* may be used as a cheaper and more convenient parameter for assessing the nutrient status of a water body than other more complex biological variables such as community structure, especially for monitoring purposes.

Cultural eutrophication is still an important environmental issue, but, for the most part, gross overfertilisation is now much less common. Through regulation, detergents have lower phosphate content; furthermore, many STPs have tertiary treatment that removes most of the P from the finished water before it is released from the treatment plant. Agricultural runoff and atmospheric deposition from non-point sources remain as major sources of P to water bodies. Prevention is, of course, the ideal situation. The potential for nutrient loading from changes in land use and the installation of septic tanks for developments that are close to water bodies need to be controlled, and, for the most part, the planning process in developed countries attempts to take this into account.

Eutrophication is theoretically reversible; indeed, there are examples of aquatic systems that have recovered from overfertilisation. The simplest type of recovery is reversal of the effects by removal of the source of excess nutrients. Other cases may require more active intervention to attain an ecologically viable system. Recovery is revisited in Chapter 11.

In summary, the element phosphorus, although not a toxic substance in the conventional sense, is an inorganic contaminant that can be of great significance in aquatic ecosystems.

6.11 Fluorine

6.11.1 The occurrence, sources, and behaviour of fluorine

Fluorine is the thirteenth most common element on Earth and is the lightest and most reactive of the group VIII halogen elements (Figure 6.3). It exists principally as the monovalent fluoride ion, which forms a major component of minerals such as fluorspar (CaF_2), cryolite (Na_3ALF_6), and fluorapatite [$(Ca_{10}F_2)(PO_4)_6$]. It is a common component of soils, averaging approximately 200 mg kg^{-1} worldwide.

Fluoride has limited water solubility. In Europe and North America, natural waters contain 0–10 mg L^{-1} with an average of about 0.2 mg L^{-1}. Seawater has a remarkably constant fluoride concentration, approximately 1.2 mg L^{-1}.

Fluoride is discharged into the environment from a variety of sources. Gaseous fluoride (largely HF) is emitted through volcanic activity and by several different industries. Gaseous and particulate fluoride are major by-products of coal burning (coal contains 10–480 mg kg^{-1} F; average 80 mg kg^{-1}) and are released during steel manufacture and smelting of nonferrous metals. Aluminium manufacture involves the use of cryolite, fluorspar, and aluminium fluoride and is often a potent source of environmental fluoride. Fluoride-containing minerals are also raw material for glass, ceramics, cement, and phosphate fertilisers. The following simplified equation illustrates how hydrogen fluoride is released through the acidification of phosphate rock during the manufacture of phosphate fertiliser:

$$3[Ca_3(PO_4)_2]CaF_2 + 7H_2SO_4 \rightarrow 3[Ca(H_2PO_4)_2] + 7CaSO_4 + 2HF \quad (6.5)$$

6.11.2 The toxicity of fluoride

Fluoride has pathological effects on both plants and animals.

PLANTS

As a plant pathogen, fluoride has caused widespread damage to a variety of crops within the atmospheric fallout radius of many of the industries mentioned previously. The degree of plant damage may be tracked to specific point sources using climatological data such as prevailing wind direction and the amount of precipitation.

Fluoride is principally accumulated by plants in gaseous form (HF) through the stomata of the leaves. It dissolves in the aqueous phase of the substomatal cavity and is transported in ionic form in the transpirational stream to the leaf tips and margins. Some enters the leaf cells and accumulates within the subcellular organelles.

The effects of fluoride on plants are complex because they are involved with many biochemical reactions. The general injury symptoms are tip and marginal chlorosis (yellowing) and necrosis (burning) of the leaves. There also appears to be a general reduction in growth and reduced seed germination. One of the early symptoms of fluoride damage in plants is a loss of chlorophyll, which seems to be related to the destruction of chloroplasts. This inhibits photosynthesis. Fluoride has also been shown to have a direct effect on enzymes associated with glycolysis, respiration, lipid metabolism, and protein synthesis. Examples include phospho-glucomutase, pyruvate kinase, succinic dehydrogenase, pyrophosphatase, and mitochondrial ATPase.

ANIMALS

Although fluoride has only moderate acute toxicity to animals and is not regarded as a serious threat to wildlife, it can pose a significant threat to humans and domestic animals under certain conditions. Fluorides have been shown to cause chromosome damage and mutations in plant and animal cells, thereby implying a potential carcinogenic effect, but the most serious problems associated with fluoride exposure are generally skeletal disorders.

Airborne pollutants containing fluoride have probably caused more widespread damage to livestock in industrialised countries than any other pollutants. Regardless of the nature of the fluoride compound, the symptoms are the same:

- Abnormal calcification of bones and teeth;
- Stiff posture, lameness, rough hair coat;
- Reduction in milk production;
- Weight loss.

The latter symptoms are directly related to poor nutritional status caused by the skeletal defects, which inhibit the ability of animals to feed properly. Cattle and sheep are the most susceptible, while pigs and horses are much less so, and poultry are the most resistant.

Fluoride ingestion principally occurs through eating contaminated forage and feed (particularly of high phosphate content), and, over several decades, courts have made numerous monetary awards to livestock farmers for losses caused by ingestion of fodder containing excessive amounts of fluoride.

A typical case was that of Rocky Mountain Phosphates, Inc. (a superphosphate plant) versus Garrison, Montana, USA. where fluorosis in cattle, and an attendant fall in milk and calf productivity, appeared shortly after operations began in 1963. Measured fluoride concentrations in grass forage near the plant ranged as high as 10,000 mg kg^{-1}. Bone fluoride levels in livestock with serious fluorosis can reach >2,000 mg kg^{-1} on a dry fat-free basis. Instances such as this have led to the development of the following suggested standards for fluoride content of forage:

- Not over 40 mg kg^{-1} dry weight per year.
- Not over 60 mg kg^{-1} dry weight for more than two consecutive months.
- Not over 80 mg kg^{-1} dry weight for more than one month.

Occupational fluorosis has been diagnosed in factory workers, particularly in aluminium smelters and phosphate fertiliser plants, where symptoms take the form of clinical osteosclerosis. Fluoride exposure in affected workers is principally from airborne sources. Skeletal fluorosis in humans is often associated with a bone fluoride level of $2,000 \, mg \, kg^{-1}$ dry fat-free bone.

Evidence of cancer in communities exposed to high fluoride levels is conflicting. An excess of respiratory tract cancers has been reported in fluorspar miners in Newfoundland, Canada. Although these cancers have been attributed to radon and radon daughter products, the cancer incidence in these miners was five times greater (per unit of radon exposure) than in Colorado. Fluorspar is postulated to have a cocarcinogenic role.

6.12 Questions

1. Elements may be required as macronutrients or as micronutrients. Other elements may have no known biological role. Distinguish among typical organismal responses to these three types of effects, by diagram and text, providing at least two examples for each type.

2. All elements of the periodic table can be classified as metals or nonmetals, with the exception of a small number of so-called metalloids. List the main physical and chemical properties that chemically distinguish metals from nonmetals.

3. Define ligand and give examples of naturally occurring and synthetic ligands that are known to bind with metals. Explain what is meant by complexes, and provide an equation for a general type of metal-ligand interaction. Illustrate what is meant by multidentate and chelation.

4. Explain what is meant by bioavailability, and provide a list of the main factors that are believed to determine the bioavailability of a metal or metalloid. Provide a simple example of an experiment that would demonstrate the concept and determine the availability of a contaminant or a nutrient for a plant system.

5. List the recent technical advances (at least six) for determining metal species, and, for each approach, provide an example.

6. Prepare an account of the major properties of mercury in the context of its environmental toxicology.

7. Give an historical account of cadmium toxicity to humans including specific pathological effects. Describe factors affecting the bioavailability of cadmium in the aquatic environment and ways employed by organisms to detoxify this metal.

8. Compare and contrast the environmental behaviour, route(s) of exposure, mode of toxic action, and concern for the release into the environment of lead and copper.

6.13 References

Agranoff, D. A., and Krishna, S. (1998) Metal ion homeostasis and intracellular parasitism, *Molecular Microbiology*, **28**, 403–12.

Azeez, P. A., and Banerjee, D. K. (1991) Nickel uptake and toxicity in cyanobacteria, *Toxicological and Environmental Chemistry*, **30**, 43–50.

Beyer, W. N., Spann, J. W., Sileo, L., and Franson, J. C. (1988) Lead poisoning in six captive avian species, *Archives of Environmental Contamination and Toxicology*, **17**, 121–30.

Bronzetti, G., and della Croce, C. (1993) Selenium: Its important roles in life and contrasting aspects, *Journal of Environmental Pathology Toxicology and Oncology*, **12**, 59–71.

Bunce, N. (1994) *Environmental Chemistry*, 2nd ed., Wuertz Publishing, Winnipeg.

Campbell, P. G. C. (1995) Interactions Between Trace Metals and Aquatic Organisms: A Critique of the Free-Ion Activity Model. In *Metal Speciation and Bioavailability*, eds. Tessier, A., and Turner, D. R., pp. 45–102, John Wiley & Sons, New York.

Chiaudani, G., and Vighi, M. (1978) The use of *Selenastrum capricornutum* batch cultures in toxicity studies, *Mitt. Internat. Verein. Limnol.*, **21**, 316–29.

Clarkson, T. E. (1995) Mercury Toxicity: An Overview. In *Proceedings, National Forum on Mercury in Fish*, EPA 823-R-95-002, pp. 91–3, U.S. EPA Office of Water, Washington, DC.

Coale, K. H., and Flegal, R. H. (1989) Copper, zinc, cadmium and lead in surface waters of Lakes Erie and Ontario, *Science of the Total Environment*, **87/88**, 297–304.

Daly, H. R., Campbell, I. C., and Hart, B. T. (1990) Copper toxicity to *Paratya australiensis*: I. Influence of nitrilotriacetic acid and glycine, *Environmental Toxicology and Chemistry*, **9**, 997–1006.

De Nicola, M., Cardellicchio, N., Gambardella, C., Guarino, S. M., and Marra, C. (1993) Effects of cadmium on survival, bioaccumulation, histopathology, and PGM polymorphism in the marine isopod *Idotea baltica*. In *Ecotoxicology of Metals in Invertebrates*, Special SETAC Publication, eds. Dallinger, R., and Rainbow, P. S., Lewis, Boca Raton, FL.

Dillon, P. J., and Rigler, F. H. (1974) The phosphorus-chlorophyll relationship in lakes, *Limnology and Oceanography*, **19**, 763–73.

Di Toro, D. M., Mahoney, J. D., Hansen, D. J., Scott, K. J., Hicks, M. B., Mayr, S. M., and Redmond, M. S. (1990) Toxicity of cadmium in sediments: the role of acid-volatile sulfide, *Environmental Toxicology and Chemistry*, **9**, 1487–502.

Drever, J. I. (1997) *The Geochemistry of Natural Waters: Surface and Groundwater Environments*, Prentice Hall, Upper Saddle River, NJ.

Environment Canada. (1985) Status report on compliance with chlor-alkali mercury regulations, 1982–1983. Report EPS1/HA/1. Environment Canada, Environmental Protection Service, Ottawa.

Errecalde, O., Seidl, M., and Campbell, P. G. C. (1998) Influence of a low molecular weight metabolite (citrate) on the toxicity of cadmium and zinc to the unicellular green alga *Selenostrum capricornutum*: An exception to the free-ion model, *Water Research*, **32**, 419–29.

Fletcher, G. G., Rossetto, F. E., Turnball, J. D., and Nieboer, E. (1994) Toxicity, uptake and mutagenicity of particulate and soluble nickel compounds, *Environmental Health Perspectives*, **102**, Supp. 3, 69–79.

Forstner, U., and Witmann, G. T. W. (1979) *Metal Pollution in the Aquatic Environment*, Springer-Verlag, Berlin.

Harrison, R. M., and de Mora, S. J. (1996) *Introductory Chemistry for the Environmental Sciences*. Cambridge Environmental Chemistry Series 7, Cambridge University Press, Cambridge.

Hodson, P. V. (1986) The Effects on Aquatic Biota of Exposure to Lead. In *Pathways, Cycling, and Transformation of Lead in the Environment*, ed. Stokes, P. M., pp. 203–24, The Royal Society of Canada, Commission on Lead in the Environment, Toronto.

Hodson, P. V., and Hilton, J. W. (1983) The nutritional requirements and toxicity to fish of dietary and waterborne selenium, *Ecological Bulletins*, **35**, 335–40.

Hodson, P. V., Borgmann, U., and Shear, H. (1979) Toxicity of copper to aquatic biota. In *Biogeochemistry of Copper Part II. Health Effects*, ed. Nriagu, J. O., pp. 307–72, John Wiley and Sons, New York.

Horowitz, A. J., Elrick, K. A., and Colberg, M. R. (1992) The effect of membrane filtration artefacts on dissolved trave element concentrations, *Water Research*, **26**, 753–63.

Hutchinson, T. C., Fedorenko, A., Fitchko, J., Kuja, A., Van Loon, J. C., and Li, J. (1975) Movement and compartmentation of nickel and copper in an aquatic ecosystem. In *Environmental Biogeochemistry*, 2nd International Symposium, Environmental Biogeochemistry, March 1975, vol. 2, ed. Nriagu, J. O., pp. 565–85, Ann Arbor Science Publishers, Ann Arbor, MI.

Jacoby, J. M., Lynch, D. D., Welch, E. B., and Perkins, M. A. (1982) Internal phosphorus loading in a shallow eutrophic lake, *Water Research*, **16**, 911–19.

James Bay Mercury Committee. (1993) Report of the activities of the James Bay Mercury Committee 1992–1993. Montreal.

Jaworski, J. F. (1978) Effects of lead in the environment – 1978: Quantitative aspects. NRCC No. 16736, pp. 181–262, Associate Committee on Scientific Criteria for Environmental Quality. National Research Council of Canada, Ottawa.

Jensen, S., and Jervelov, A. (1969) Biological methylation of mercury in aquatic organisms, *Nature*, **223**, 753–4.

Kszos, L. A., Stawart, A. J., and Taylor, P. A. (1992) An evaluation of nickel toxicity to *Ceriodaphnia dubia* and *Daphnia magna* in a contaminated stream and in laboratory tests, *Environmental Toxicology and Chemistry*, **11**, 1001–12.

Lee, B.-G., Griscom, S., Lee, J.-S., Choi, H. J., Koh C.-H., Luoma, S. N., and Fisher, N. S. (2000) Influences of dietary uptake and reactive sulfides on metal bioavailability from aquatic sediments, *Science*, **287**, 282–4.

Lemley, A. D. (1985) Toxicology of selenium in a freshwater reservoir: Implications for environmental hazard evaluation and safety, *Ecotoxicology and Environmental Safety*, **10**, 314–38.

Mason, R. P., Fitzgerald, W. F., and Morel, F. F. M. (1994) Global mercury model. The biogeochemical cycling of elemental mercury: Anthropogenic influences. *Geochim. Cosmochim. Acta*, **58**, 3191.

Minnesota Pollution Control Agency. (1994) *Strategies for Reducing Mercury in Minnesota*, Minnesota Pollution Control Agency, Minneapolis.

Morel, F. M. M., and Hering, J. G. (1993) *Principals and Applications of Aquatic Chemistry*, Wiley and Sons, New York.

Myers, G. J., Davidson, P. W., Cox, C., Shamlaya, C., Cernichiari, E., and Clarkson, T. W. (2000) Twenty-seven years studying the human neurotoxicity of methylmercury exposure, *Environmental Research, Section A*, **83**, 275–85.

Nieboer, E., and Richardson, D. H. S. (1980) The replacement of the nondescript term "heavy metals" by a biologically and chemically significant classification of metal ions, *Environmental Pollution*, **1**, 3–26.

Nieboer, E., Rossetto, F. E., and Menon, R. (1988) Toxicology of nickel compounds. In *Metal Ions in Biological Systems: Nickel and Its Role in Biology*, vol. 23, eds. Sigel, H., and Sigel, A., Marcel Dekker, New York.

Nriagu, J. O. (1994). Origin, long-range transport, atmospheric deposition and associated effects of heavy metals in the Canadian Environment. A report prepared for Atmospheric Environment Service, Environment Canada. December 30, 1994.

Ontario Ministry of the Environment. (1992) Ontario Sports Fish Monitoring Program. 125 Resources Road, Etobicoke, Ontario.

Patterson, C. C. (1965) Contaminated and natural lead environments of man, *Archives of Environmental Health*, **11**, 344–60.

Patterson, C. C., and Settle, D. M. (1976) The Reduction of Orders of Magnitude Errors in Lead Analyses of Biological Materials and Natural Waters by Evaluation and Controlling the Extent and Sources of Industrial Lead Contamination Introduced During Sample Collection, Handling and Analysis. In *Accuracy in Trace Analysis: Sampling, Sample Handling, and Analysis*, 7th IMR Symposium, NBS Special Publication 422, ed. LaFleur, P. D., pp. 321–51, U.S. National Bureau of Standards, Washington DC.

Pufahl, R. A., Singer, C. P., Peariso, K. L., Lin, S.-J., Schmidt, P. J., Fahmi, C. J., Cizewski Cullota, V., Penner-Hahn, J. E., and O'Halloran, T. V. (1997) Metal ion chaperone function of the soluble Cu(1) receptor At_x1, *Science*, **278**, 853–6.

Rasmussen, P. (1996). Trace metals in the environment: A geochemical perspective. GSC Bulletin 429, Geological Survey of Canada, Ottawa.

Rodricks, J. R. (1993) *Calculated Risks. The Toxicity and Human Health Risks of Chemicals in Our Environment*, Cambridge University Press, Cambridge.

Schindler, D. W. (1990) Natural and Anthropogenically Imposed Limitations to Biotic Richness in Fresh Waters. In *The Earth in Transition. Patterns and Processes of Biotic Impoverishment*, ed. Woodwell, G. M., pp. 425–62, Cambridge University Press, Cambridge.

Schindler, D. W., and Fee, E. J. (1974) Experimental Lakes Area: Whole-lake experiments in eutrophication, *Journal of the Fisheries Research Board of Canada*, **31**, 937–53.

Schut, P. H., Evans, R. D., and Scheider, W. A. (1986) Variation in trace metal export from small Canadian watersheds, *Water, Air, and Soil Pollution*, **28**, 225–37.

Stokes, P. M., and Hutchinson, T. C. (1975) Copper toxicity to phytoplankton as affected by organic ligands, other cations and inherent tolerance of algae to copper. In *Workshop on the Toxicity to Biota of Metal Forms in Natural Water. Duluth, October 1975*, eds. Andrew, R. W., Hodson, P. V., and Konasewich, D. E., pp. 159–86, Standing Committee on the Scientific Basis for Water Quality Criteria of the International Joint Commission's Research Advisory Board, Duluth.

Sunda, W. G., and Gillespie, P. A. (1979) The response of a marine bacterium to cupric ion and its use to estimate cupric ion activity in seawater, *Journal of Marine Research*, **37**, 761–77.

Thornton, I. (1995) *Metals in the Global Environment: Fact and Misconceptions*, International Council on Metals in the Environment, Ottawa.

U.S. Environmental Protection Agency. (1993) Water Quality Guidance for Great Lakes System and Correction: Proposed rules. *Federal Register*, April 16, 1993.

U.S. Environmental Protection Agency. (1994) New Chemicals Program (NCP). Categories of Concern (revised August 1994). Washington, DC.

Voldner, E. C., and Smith, L. (1989) Production, usage and atmospheric emissions of 14 priority toxic chemicals. Proceedings of a workshop on Great Lakes Atmospheric Deposition, International Air Quality Advisory Board of the International Joint Commission for the Great Lakes, Scarborough, Ontario, October 29–31.

Watmough, S. A., Hughes, R. J., and Hutchinson, T. C. (1999) $^{206}Pb/^{207}Pb$ ratios in tree rings as monitors of environmental change, *Environmental Science and Technology*, **33**, 670–3.

Wesolowski, D. J., and Palmer, D. A. (1994) Aluminum speciation and equilibria in aqueous solution: V. Gibbsite solubility at 50°C and pH 3–9 in 0l1 molal NaCl solutions, *Geochimica et Cosmochimica Acta*, **58**, 2947–70.

Winfrey, M. R., and Rudd, J. W. M. (1990) Environmental factors affecting the formation of methylmercury in low pH lakes, *Environmental Toxicology and Chemistry*, **9**, 853–69.

Wixson, B. G., and Davies, B. E. (1993). Lead in soil: Recommended guidelines. Science Reviews, Society for Environmental Geochemistry and Health, Northwood, Middlesex, England.

World Health Organisation (WHO). (1995) Inorganic lead, *Environmental Health Criteria* 165. United Nations Environment Programme, International Labour Organisation and the World Health Organisation, Geneva.

Wren, C. D., McCrimmon, H. R., and Loescher, B. R. (1983) Examination of bioaccumulation and biomagnification of metals in a Precambrian Shield Lake, *Water, Air, and Soil Pollution*, **19**, 277–91.

6.14 Further reading

The biological availability of metals: National Research Council of Canada (NRCC). 1988. *Biologically available metals in sediments*. Campbell, P. G. C., Lewis, A. C., Chapman, P. M., Crowder, A. A., Fletcher, W. K., Imber, B., Luoma, S. N., Stokes, P. M., and Winfrey, M. Associate Committee on Scientific Criteria for Environmental Quality. NRCC No. 27694. National Research Council of Canada, Ottawa.

The biological behaviour of metals: Hare, L. 1992. Aquatic insects and trace metals: Bioavailability, bioaccumulation and toxicity, *Critical Reviews in Toxicology*, **22**, 372–89.

Mercury: Clarkson, T. W. 1997. The toxicology of mercury, *Critical Reviews in Clinical Laboratory Sciences*, **34**, 369–409.

Lead poisoning and decline of the Roman Empire: Gilfillan, S. C. 1990. *Rome's Ruin by Lead Poison*. Wenzel Press, Long Beach, CA.

Eutrophication (general): Simpson, R. D., and Christensen, J. L., Jr. 1997. *Ecosystem Function and Human Activities*, Chapman and Hall, New York.

Valiela, I. 1995. *Marine Ecological Processes*, 2nd ed., Springer-Verlag, New York.

Eutrophication (United States): Bricker, S. B., Clement, C. G., Pirhalla, D. E., Orlando, S. P., and Farrow, D. R. G. 1999. *National Estuarine Eutrophication Assessment. Effects of Nutrient Enrichment in the Nation's Estuaries*, National Ocean Service, National Oceanic and Atmospheric Administration, Silver Spring, MD.

Appendix: Properties of selected metals and metalloids

This appendix provides some basic facts about the uses, geochemistry, and biological effects of the metals and metalloids that are important in the context of environmental toxicology. Some emphasis has been placed on the uses of these elements because their mobilisation in the environment is frequently linked to their use by humans. A selection of information on regulatory aspects is included, and this will serve to show that sometimes there are large discrepancies among jurisdictions. For more comprehensive coverage of criteria, guidelines, standards, and the like, the reader is referred to the web sites for the various regulatory agencies.

The following 14 elements are listed: aluminium, arsenic, cadmium,* chromium, copper,* iron, lead,* manganese, mercury,* nickel,* selenium,* silver, vanadium, and zinc.

* These elements are dealt with in more detail in the text.

1.   Aluminium

Metal/metalloid. Aluminium, Al.

Atomic number. 13.

Atomic weight. 26.98.

Uses. Aluminium metal is used in cans and foils, for kitchen utensils, for outdoor building decoration, and for industrial applications where a strong, light, easily constructed material is needed. Its electrical conductivity is only about 60% that of copper; nevertheless, it is used in electrical transmission lines because of its lightness and its relatively low price. Aluminium is combined with other metals to make strong, lightweight alloys, useful for household and industrial purposes and of particular importance in the construction of modern aircraft and rockets. Aluminium compounds are also used as water softening agents, as flocculating agents in sewage and water treatment facilities, and in pharmaceuticals such as antacid preparations.

Forms in the natural environment. The third most abundant element in the Earth's crust and the most abundant metallic element, aluminium occurs in rocks, soils, clay, and other minerals, primarily in the form of aluminosilicates. It is mined in very large amounts as bauxite ($Al_2O_3 \cdot 2H_2O$). Aluminium normally occurs naturally in insoluble forms, including complexes with hydroxyl, fluoro-, organic, and phosphato-ligands. Naturally occurring aluminium compounds are insoluble at pH 5.5–6.0, but their solubility increases rapidly with decreasing pH. Organic ligands also affect the chemical speciation of the element in natural waters. In waters and soil solutions of pH < 5.0, the most prevalent form is "free" Al^{3+}. Naturally occurring aluminium is mobilised and solubilised by weathering and especially as a result of acidification.

Anthropogenic forms/conversions. Pure aluminium is obtained from bauxite through the Bayer process, which involves chemical conversion to Al_2O_3 followed by electrolysis to produce metallic aluminium. The major anthropogenic modification of the aluminium cycle, especially from a toxicological point of view, is the conversion of insoluble to soluble species of aluminium in water and in soils, typically as a result of acidification.

Sources for biota. Industrial processes may provide sources of aluminium for biota, with secondary smelters of nonferrous metals as a major point source. But more commonly the source for biota is a geological one that has been mobilised by human activity. Other direct sources for ingestion include medications and cooking utensils.

Exposure pathway(s) for humans. The primary routes of exposure for humans are through ingestion of drinking water, beverages, food, and medications. Aluminium compounds are often added deliberately to food; furthermore, acidic foods and beverages may be inadvertently contaminated from cooking utensils or cans. Occupational exposure to aluminium occurs in the primary refining industry and in manufacturing processes that use aluminium.

Effects on humans. Aluminium appears to accumulate primarily in the bone and brain tissue, but it may also accumulate in the liver and spleen. Several bone diseases have been associated with exposure to the element, notably osteomalacia. This condition often occurs in patients who are on long-term haemodialysis treatment for chronic kidney failure: The patients are given high oral doses of aluminium hydroxide to control excess phosphate. Aluminium exposure in such patients may also occur intravenously via the dialysis fluid. Neurological conditions, particularly patients suffering from Alzheimer's disease and dialysis dementia, have unusually high concentrations of aluminium in brain tissue. To date, the element has not been causally related to neurological disease in humans.

Pathways to ecosystems. The most well studied phenomenon in the context of aluminium and ecosystems is the release of soluble forms of the metal into aquatic ecosystems, from the watershed or from sedimented material. Terrestrial systems, notably those with poorly buffered forest soils, may also undergo release of soluble forms of aluminium. Managed soils and well-buffered aquatic systems rarely represent important pathways.

Effects on nonhuman targets. Aluminium is not considered to be an essential trace nutrient for plants or animals. Plants are, in general, much more sensitive to aluminium than are mammals and lower organisms. Aluminium in its ionic form is particularly phytotoxic at pH below about 4.5. It interferes with phosphorus metabolism within the plant so that aluminium toxicity is, symptomatically, like phosphorus deficiency; it also interferes with calcium uptake. In naturally aluminium-rich systems, local plants are tolerant to unusually high concentrations of the element, and some plants hyperaccumulate aluminium. Aluminium is of concern for fish and invertebrates in acidified systems. However, because the elevation is related to acidic pH, in the real world it is not possible to uncouple the effects of aluminium from those of acidification per se. Aluminium compounds may precipitate on gill membranes of fish and result in asphyxiation. Biota accumulate aluminium, and although there is no evidence for food chain biomagnification, there have been suggestions that high-level predators such as passerine birds may be ingesting aluminium from food sources such as aquatic invertebrates.

Regulatory aspects. In North America, the general guideline for aluminium in drinking water is $100 \mu g L^{-1}$. The water quality objectives range from 5 to $100 \mu g L^{-1}$. The Canadian interim water quality guidelines for aluminium are $5 \mu g L^{-1}$ in waters of pH greater than or less than 6.5, and $100 \mu g L^{-1}$ for waters of pH > 6.5. For Ontario, the water quality objective for waters with pH 4.5–5.5 the interim objective is $15 \mu g L^{-1}$ and for waters of pH 6.5 or greater is $75 \mu g L^{-1}$ in a clay-free sample. Ontario's interim Provincial Water Quality Objective (PWQO) also states that "at pH > 5.5 to 6.5, no condition should be permitted which would increase the acid soluble inorganic aluminium concentration in clay-free samples to more than 10% above natural background concentrations for waters representative of that geological area of the Province that are unaffected by man-made inputs".

Reference. Nieboer, E., B. L. Gibson, A. D. Oxmand, and J. R. Kramer 1995. Health effects of aluminum: A critical review with emphasis on drinking water, *Environmental Reviews*, **3**, 29–81.

2.   Arsenic

Metal/metalloid. Arsenic, As.

Atomic number. 33.

Atomic weight. 74.92.

Uses. Historically, arsenic has been used as a pesticide and rodent poison, in the production of pigments and chemical weapons (e.g., mustard gas), in semiconductors, in wood preservatives, and as a growth promoter for poultry and pigs. Because of widespread health concerns, however, the use of arsenic nowadays is primarily restricted to metallurgical applications and to the manufacture of wood preservatives.

Forms in the natural environment. Arsenic occurs in nature as both a free element and a part of different compounds. Over 200 minerals containing arsenic have been identified, the most common of which is arsenopyrite. Arsenic is typically found in minerals containing sulphide mineralizations, often associated with other metal ores. Arsenic concentrations in minerals may range from a few parts per million to percentage quantities. Arsenic can exist in four oxidation states (−III, 0, III, V) and forms a triprotic acid, arsenic acid (H_3AsO_4), in aqueous solution. Trivalent (As III) and pentavalent (As V) are the most common oxidation states. Of these, As V is the most prevalent and the least toxic. Arsenic may occur in either organic or inorganic forms, although the inorganic forms are more toxic. Release of arsenic from minerals is enhanced by microbial action, and volatile methylated forms of arsenic may be produced by bacteria and algae.

Anthropogenic forms/conversions. Even though many minerals such as arsenopyrite have commercially extractable levels of arsenic, today most arsenic is produced as a by-product of the extraction of copper, lead, gold, and silver from their ores. Significant amounts of arsenic are released in liquid effluent from gold-milling operations using cyanide and in stack gases from roasting of gold ores. Arsenic may be mobilised from soils that are flooded (e.g., for creation of reservoirs), and the application of phosphate fertilisers to agricultural lands can mobilise otherwise tightly bound arsenic in soil.

Sources for biota. Arsenic is present in the environment naturally as a result of weathering and erosion of rock and soil, forest fires, ocean gas releases (e.g., bubble bursting), microbial activity, and volcanism. Some groundwater has naturally high arsenic concentrations. The major anthropogenic sources of the element to air include emissions from pesticide manufacturing facilities and coal-fired power generators. Runoff and leachate from domestic and industrial waste sites, agricultural fields, and metal mines and smelters may be major sources of arsenic to groundwater and surface water bodies.

Exposure pathway(s) for humans. Humans can take up arsenic by ingestion, by inhalation, and through skin or mucous membranes. Well water in regions of the world that have naturally high levels of arsenic in the groundwater will represent a route of exposure through drinking water for humans as well as for livestock. Food and ambient air are other significant vectors of arsenic exposure to humans.

Effects on humans. Cases of chronic arsenic poisoning from drinking contaminated well water have been reported in local populations of areas such as the west coast of Taiwan, Bangladesh, and Region Lagunera of northern Mexico. Common symptoms of arsenic intoxication include skin lesions, hyperpigmentation, keratosis, and a peripheral vascular affliction known as blackfoot disease. Arsenic is a carcinogen in humans, and cancers may develop from exposure to arsenic at concentrations much lower than those associated with chronic As poisoning. Lung cancers in workers employed at smelters and arsenical pesticide manufacturing facilities have been related to airborne arsenic levels, and cancers of other sites including the stomach, colon, liver, and urinary system may be associated with occupational exposure to arsenic.

Pathways to ecosystems. Water bodies may become contaminated with arsenic from domestic and industrial waste waters, runoff or leachate from agricultural fields and metal mines and smelters, and atmospheric fallout. In aerobic soils and sediments, the element is strongly associated with clays and iron oxides, which may limit its mobility and bioavailability. Dissolution of oxides, when sediments become reducing (anaerobic), can cause a flux of arsenic back to the water column; similarly, poorly drained soils may release arsenic to porewater and groundwater following flooding. Plants can absorb arsenic through roots or foliage; the element is absorbed by fish through the gut or the gills.

Effects on nonhuman targets. The role of arsenic as a nutrient is still controversial, but it appears to be an essential ultratrace nutrient for red algae, chickens, rats, and pigs, where a deficiency results in inhibited growth. The element is best known, however, for its adverse effects. Arsenic has a high acute toxicity to aquatic life, birds, and terrestrial animals. Algae are some of the most sensitive groups of organisms to arsenic and show decreases in productivity and growth when exposed to arsenic at very low concentrations (e.g., $<5\,\mu g\,L^{-1}$). The chronic toxicity is high for aquatic life but more moderate for birds and land animals. Organic forms of arsenic are less toxic than inorganic forms. Arsenic is not biomagnified in aquatic or terrestrial food chains. Various aquatic organisms including algae, crustaceans, and fish can bioaccumulate arsenic; bioconcentration factors of up to several thousand have been reported. Marine crustaceans and molluscs accumulate more arsenic than do fin fish.

Regulatory aspects. In 1992, the U.S. EPA listed arsenic as a hazardous air pollutant and recommended a permissible exposure limit of $10 \mu g \, m^{-3}$. The U.S. EPA also lists inorganic arsenic in Group A, designated as having high carcinogenic hazard. The World Health Organisation (WHO) guideline for arsenic in drinking water is $50 \mu g \, L^{-1}$. The Canadian guideline for arsenic in water for the protection of freshwater aquatic life is $5 \mu g \, L^{-1}$ and the Ontario interim PWQO is also $5 \mu g \, L^{-1}$. The Ontario sediment quality guideline is $6 \mu m \, g^{-1}$ (dry weight) as the lowest effect level. The Canadian interim guideline for arsenic and marine life is $12.5 \mu g \, L^{-1}$.

Reference. Environment Canada. 1993. Priority Substances List Report: Arsenic and Its Compounds, EN40-215/14E. Environmental Canada, Ottawa.

3.   Cadmium

Metal/metalloid. Cadmium, Cd.

Atomic number. 48.

Atomic weight. 112.4.

Uses. Globally, cadmium is used in a number of major applications, including nickel-cadmium batteries, coatings (e.g., electroplating), pigments (cadmium sulphide is a yellow pigment), plastic stabilisers, and alloys. It is also used in some control rods and shields in nuclear reactors and in phosphors for both black-and-white and colour television tubes. In the past, cadmium was used as a fungicide in turf grass production, although its use in this application was banned in Canada in 1990.

Forms in the natural environment. The average concentration of cadmium in the Earth's crust is estimated to be about $0.1 \, mg \, kg^{-1}$, although anomalously high concentrations (i.e., $0.3–980 \, mg \, kg^{-1}$) can occur in certain sedimentary rocks, especially black shales and phosphorites. Soils derived from parent material rich in the element may also have very high concentrations of cadmium. It occurs in all zinc ores, principally in the mineral sphalerite, and cooccurs with zinc and lead in other sulphide minerals. In soil solution and in aquatic systems, cadmium is primarily found in the 2+ valence state, either as the free (hydrated) ion (Cd^{2+}), or as inorganic complexes [e.g., $CdOH^+$, $CdCl_2$, $Cd(SO_4)_2$]. Cadmium also reacts with organic ligands such as humic and fulvic substances to form organic complexes. Cadmium may precipitate in sediments and soils as sulphides under reducing conditions and commonly forms oxides and carbonate compounds (CdO, $CdCO_3$) in strongly toxic environments. In the atmosphere, inorganic Cd compounds (e.g., CdO) are commonly associated with particulate matter and have relatively short tropospheric residence times (<4 weeks) due to rapid removal from air through wet and dry deposition.

Anthropogenic forms/conversions. Fertilisers manufactured using Cd-rich phosphates are a ubiquitous source of cadmium in agricultural soils. Sewage contains cadmium from human excretion, industrial effluents, and disposal of domestic waste containing zinc. The application of sewage sludge as a fertiliser may also contribute to relatively high cadmium concentrations in arable soil and potentially increase cadmium uptake in crop plants. The mobility and bioavailability of cadmium may be affected by physical and chemical characteristics in different environmental media. Cadmium bioavailability in aquatic environments and in soil solution is enhanced under conditions of low pH. Anthropogenic stressors such as acid deposition may, thus, contribute to cadmium mobility and availability in both freshwater and terrestrial environments.

Sources for biota. Anthropogenic sources of cadmium include base metal smelting and refining (primarily lead-zinc), fossil fuel combustion, solid waste disposal, and application of sewage sludges to agricultural soils. The majority of anthropogenic releases are to the atmosphere. Cadmium may also enter the environment naturally, through weathering of rocks and soils, volcanic emissions, and forest fires.

Exposure pathway(s) for humans. Humans are potentially exposed to cadmium in air, water, soil, and food, although ingestion of cadmium-contaminated food is probably the most significant uptake route. Cigarette smokers have substantially higher exposure to cadmium than do nonsmokers because of the relatively high amount of the metal in cigarettes (approximately $0.19\,\mu g\,Cd$/cigarette) and the availability of inhaled Cd compounds. In Canada, it is estimated that people living in the vicinity of base metal smelters receive a Cd exposure that is one to two orders of magnitude higher than members of the general population receive.

Effects on humans. The most well-documented case of cadmium intoxication in humans occurred in Japan after the Second World War. Pollution from a zinc smelter resulted in high levels of cadmium in soil and rice, and many local residents (particularly postmenopausal women) developed a disease termed *Itai-Itai* (ouch-ouch), characterised by osteomalacia, osteoporosis, and kidney dysfunction. Inorganic cadmium compounds have been classified as "probably carcinogenic" to humans when taken up through inhalation.

Pathways to ecosystems. Cadmium may enter groundwater and surface water systems indirectly via run off from solid waste disposal sites and agricultural fields, or directly through effluent from sewage facilities, mines, and smelters. The majority of anthropogenic releases of Cd are to the atmosphere; emissions from base-metal smelters and refuse incinerators may cause elevated levels of Cd in ecosystems, even those relatively distant from the source.

Effects on nonhuman targets. Cadmium has no known biological function. Cadmium toxicity has been demonstrated in all orders of plants and animals, in both terrestrial and aquatic systems. Planktonic and benthic invertebrates appear to be the most sensitive groups in both marine and freshwater systems, with critical levels as low as $0.2 \, \mu g \, Cd \, L^{-1}$. Cadmium sensitivity in soil organisms has been well documented, with the scale of effects ranging from the level of individual (e.g., decrease in growth) to community (e.g., altered populations of soil microbes). Cadmium bioavailability, however, is modified by the physical and chemical characteristics of the environmental media in which the metal is present. In freshwater and soil solutions, the mobility and bioavailability of cadmium are enhanced under conditions of low pH, altered redox, low hardness, and low organic matter content. In marine and estuarine systems, cadmium speciation is dominated by chloride complexation: The toxicity and accumulation of the element are inversely related to salinity. Although the element tends to bioaccumulate in organisms, there is no evidence of biomagnification of Cd in food chains.

Regulatory aspects. The Canadian interim guideline for the protection of aquatic life from Cd is $0.017 \, \mu g \, L^{-1}$. The fact that water hardness affects the toxicity of cadmium in freshwater systems is incorporated into the Ontario water quality objectives: For soft water, the interim PWQO is $0.1 \, \mu g \, L^{-1}$, and in hard water systems it is $0.5 \, \mu g \, L^{-1}$. In the U.S. EPA's Ambient Water Quality Criteria, the maximum acceptable average concentration of cadmium is $0.66 \, \mu g \, L^{-1}$ in soft water ($50 \, mg \, L^{-1}$ as $CaCO_3$) and $2 \, \mu g \, L^{-1}$ in hard water ($200 \, mg \, L^{-1}$ as $CaCO_3$). The Canadian guideline for Cd in sediment is $0.6 \, \mu g \, g^{-1}$ (dry weight), which coincides with the lowest effect level for Ontario's sediment quality guidelines. The Canadian water quality guideline for marine life is $0.12 \, \mu g \, L^{-1}$.

Reference. Environment Canada. 1994. Priority substances list assessment report: Cadmium and its compounds, EN40-215/40E. Environmental Canada, Ottawa.

4.   Chromium

Metal/metalloid. Chromium, Cr.

Atomic number. 24.

Atomic weight. 51.99.

Uses. Chromium is used to harden steel, to manufacture stainless steel, and to make various alloys, as well as to anodise aluminium, notably in the aircraft industry. It is used in plating to provide an attractive lustrous surface and to prevent corrosion. Dichromates such as $K_2Cr_2O_7$ are of major importance in tanning leather.

Forms in the natural environment. The seventh most abundant element on Earth, chromium occurs naturally primarily in ultrabasic and basic rocks. Although the metal can occur in nine different oxidation states, the two valence states – trivalent Cr(III), the most stable, and hexavalent Cr(VI) – are the most important. Trivalent chromium is a hard acid that forms strong kinetically inert complexes with a variety of ligands. Dissolved Cr(III) has a tendency to adsorb to surfaces. The chemistry of pentavalent chromium is quite different from the trivalent. It forms a number of stable oxyacids and anions, of which chromate (CrO_4^{2-}) is the most significant for the environment. The interconversion of Cr(II) and Cr(VI) occurs in such a way that the reduction of Cr(VI) to Cr(III) occurs readily by a variety of reducing agents in water and soil, but few oxidants in natural waters can oxidise Cr(III) to Cr(VI).

Anthropogenic forms/conversions. The principle sources of chromium to the atmosphere are cooling towers, releases from chromium plating, and the incineration of fossil fuel, municipal waste, and sewage sludge, as well as other metallurgical and chemical industries. Nonferrous metal smelters and refineries and pulp and paper plants contribute chromium to the aquatic environment through liquid effluents. The discharge of chromium in waste, including landfills, has resulted in groundwater contamination.

Sources for biota. Natural mobilisation of chromium involves windblown dusts, volcanic emissions, and sea-salt aerosols. Anthropogenic activities already described result in local or more widespread contamination of air, water, soils, and sediments by chromium. Because the pentavalent form of the element is far more toxic than the trivalent form, the chemical distinction between the two has profound biological significance. Because of the chemistry referred to earlier, especially with respect to interconversion, even when released as Cr(VI), much of the chromium burden in the environment is in the Cr(III) form.

Exposure pathway(s) for humans. Occupational exposure typically involves inhalation, and this would also apply to atmospheric chromium as nonoccupational exposure. Dermal and digestive tract pathways occur in occupational exposure.

Effects on humans. Chromium is an essential trace element for humans, having a role in insulin and glucose metabolism. In common with other essential trace nutrients, toxicity also occurs at higher concentrations of the element. Inhalation of Cr(VI), as chromate dust or as chromic acid mist, can lead to ulceration of the nasal mucosa and perforation of the nasal septum. There is an occupational risk for pulmonary cancer through inhalation of dust contaminated with chromium for workers in industries such as chrome plating, chromate production, and chrome pigment production. Other harmful effects, also occupational, occur if large amounts of Cr(VI) are absorbed through the skin or the digestive tract, which can be lethal. Information on effects of the element on humans at more typical (i.e., much lower) environmental concentrations is lacking.

Pathways to ecosystems. Air, water, soil, and sediments all potentially contribute chromium to ecosystems. Although biota can accumulate the element (e.g., benthic biota can concentrate chromium from sediments), tissue concentrations are often low: Fish can depurate the element, and aquatic macrophytes and algae generally show low concentrations in tissue. There is no evidence of biomagnification in food chains.

Effects on nonhuman targets. Chromium has been shown to be essential for a number of species of laboratory vertebrates, but the database for other organisms is very limited. It has been shown to be essential for mammals in its role in the regulation of carbohydrate metabolism. It also has been shown to counteract the harmful effects of cadmium in rats and vanadium in chickens. It does not appear to be an essential element for plants, although it may have beneficial effects, referred to as stimulation. Nor has it been proven as essential for aquatic organisms. The main concern related to exposure to chromium is for its adverse effects. Aquatic organisms are the most well studied and show acute and chronic toxic effects from chromium exposure, with Cr(VI) being far more toxic than Cr(III). Mammals and birds show altered reproductive rates as a result of high dietary or injected chromium, but most of the data are from laboratory experiments; the extent of chromium effects in the field is less well understood. Excess chromium causes abnormal morphology and coloration in citrus plants when exposed to high concentrations of chromium in soil.

Regulatory aspects. The U.S. EPA Ambient Water Quality Criteria for chromium are as follows: For protection of freshwater aquatic life, the criterion for total recoverable hexavalent chromium is $0.29 \mu g L^{-1}$, while for salt water the criterion is $0.18 \mu g L^{-1}$, both as 24-hr averages. The U.S. EPA criterion for trivalent chromium is hardness-related: total recoverable trivalent chromium should not exceed $2,200 \mu g L^{-1}$ in soft water ($50 mg L^{-1}$ as $CaCO_3$), and for hard water ($200 mg L^{-1}$ as $CaCO_3$) total recoverable trivalent chromium should not exceed $9,900 \mu g L^{-1}$. The Canadian water quality guidelines are $8.9 \mu g L^{-1}$ for trivalent chromium (interim guideline) and $1.0 \mu g L^{-1}$ for hexavalent chromium. The Ontario PWQO is $100 \mu g L^{-1}$ and the Ontario sediment quality guideline is $16 \mu g g^{-1}$ (dry weight) for the lowest effect level. The Canadian marine water quality guidelines are $56 \mu L^{-1}$ for trivalent chromium (interim guideline) and $1.5 \mu L^{-1}$ for hexavalent chromium.

Reference. Nriagu, J. O., and E. Nieboer (eds.). 1988. Chromium in the Natural and Human Environments. In *Advances in Environmental Science and Technology*, vol. 20, John Wiley and Sons, New York.

5.   Copper

Metal/metalloid. Copper, Cu.

Atomic number. 29.

Atomic weight. 63.54.

Uses. Metallic copper has long been used in pure metallic form or alloyed with other metals, as bronze (copper-tin) and brass (copper-zinc). The alloys of copper are less soft than the pure metal. In metallic form, pure or alloyed, copper has been used to make tools, ornaments, statuary, jewellery, coins, and vessels for food and beverages, as well as for electrical wiring, in plumbing, and for electroplating. Salts of copper have been used as fungicides, molluscicides, and algicides.

Forms in the natural environment. The element has two valance states: I, cuprous, and II, cupric. It occurs in trace quantities in metallic form and as copper compounds in ores, including copper pyrites ($CuFeS_2$) and malachite ($CuCO_3 \cdot Cu(OH)_2$). Copper salts are moderately soluble in water: The solubility of most copper compounds is greater at acidic to neutral pH than under alkaline conditions. Copper ions tend to form strong complexes with organic ligands, displacing more weakly bound cations in mixtures. Complexation facilitates copper remaining in solution but usually decreases its biological availability. Copper also forms strong organometal complexes in soils and in sediments.

Anthropogenic forms/conversions. The major processes that result in the mobilisation of the element into the environment are extraction from its ore (mining, milling, and smelting), agriculture, and waste disposal.

Sources for biota. Water and food for humans and livestock may contain copper from distribution systems, containers, and pesticides. Soils have become contaminated with copper by deposition of dust from local sources such as foundries and smelters, as well as from direct application of fungicides and sewage sludge, but uptake into plants is limited. Food chain contamination from soil is rarely a problem, and copper does not biomagnify in terrestrial or aquatic food chains. Aquatic systems receive copper from the atmosphere, as well as from agricultural runoff, deliberate additions of copper sulphate to control algal blooms, and direct discharge from industrial processes, all of which represent potential sources for aquatic biota.

Exposure pathway(s) for humans. Occupational exposure to the metal can occur through inhalation of contaminated dusts. For the general population, ingestion of food and drink represent the major routes of exposure.

Effects on humans. Copper is an essential trace element for humans, but deficiency is uncommon. The toxicity of copper for humans is rarely of concern; humans can tolerate high concentrations of the element in their food and in drinking water. Reasons for the lack of copper toxicity for humans as for other mammals is twofold: Because of its high affinity for organic matter, dietary copper is rarely bioavailable, and, most importantly, mammals have a biochemical system that can detoxify copper in liver and kidney. Exceptions arise only in the case of specific metabolic disorders, such as Wilson's disease where the patient's metabolic system cannot detoxify dietary Cu.

Pathways to ecosystems. Atmospheric deposition to water and to soil, and runoff and leaching from soils into water bodies, represent major pathways for terrestrial and aquatic biota. Water is the most important vector for copper to biota. Copper accumulates over time and persists in undisturbed soils and particularly in sediments, providing potential exposure routes for plant roots, microflora, microfauna, and benthic organisms. The affinity of copper for organic matter limits its availability through these routes.

Effects on nonhuman targets. Deficiency of the nutrient element can occur in livestock and can be induced in laboratory animals. Indeed, domestic animals are sometimes provided with copper supplements in their feed. Crop plants are more likely to be copper deficient than toxified. Copper is one of several metals that has been implicated as adversely affecting soil microbial activity. Even though terrestrial organisms are somewhat susceptible to copper toxicity, the major harmful effects of copper are for aquatic organisms. Fish and crustaceans are very susceptible to copper intoxification, generally 10 to 100 times more sensitive than are mammals and algae. Copper is less toxic in hard than in soft water, a fact that is recognised by regulators and that is often built into standards and objectives.

Regulatory aspects. The U.S. EPA 24-hour average criterion for protection of freshwater aquatic life is $5.6\,\mu g\,L^{-1}$ total recoverable copper, with never-to-be-exceeded levels on a hardness-related scale of $12\,\mu g\,L^{-1}$ for soft water ($50\,mg\,L^{-1}$ as $CaCO_3$) and $43\,\mu g\,L^{-1}$ for hard water ($200\,mg\,L^{-1}$ as $CaCO_3$), as total recoverable copper. The Canadian guidelines for copper in fresh water are $2\,\mu g\,L^{-1}$ for soft water (0–$120\,mg\,L^{-1}$ $CaCO_3$), $3\,\mu g\,L^{-1}$ for water of hardness 120–$180\,mg\,L^{-1}$ $CaCO_3$, and $4\,\mu g\,L^{-1}$ for hard water ($>180\,mg\,L^{-1}$ $CaCO_3$). The Ontario interim PWQOs for copper are $1\,\mu g\,L^{-1}$ for soft water and $5\,\mu g\,L^{-1}$ for water with hardness greater than $10\,mg\,L^{-1}$ $CaCO_3$. Standards for Cu in drinking water are based on taste, not on the risk of toxicity, and for Ontario Canada the drinking water objective (stated as "Not Health Related") is $1\,mg\,L^{-1}$. The Ontario sediment quality guideline is $16\,\mu g\,g^{-1}$ (dry weight) for the lowest effect level.

Reference. Nriagu, J. O. (ed.). 1980. *Copper in the Environment*, pp. 357–81, John Wiley, New York.

6.   Iron

Metal/metalloid. Iron, Fe.

Atomic number. 26.

Atomic weight. 55.84.

Uses. The use of iron has a very long history, being the final of the Stone-Bronze-Iron age sequence. It is a major component of carbon steels, alloy steels, and wrought iron and is incorporated into various other alloys. Products in which these iron-containing substances are used are extremely numerous and include automobiles, ships, land-based construction, machinery, and containers.

Forms in the natural environment. The fourth most abundant element in the Earth's crust, iron occurs in most rocks and soils. Important ores are oxides and carbonates. Iron can exist in divalent [Fe(II) ferrous] and trivalent [Fe(III) ferric] forms. From the point of view of environmental chemistry, the single most important aspect of the chemical forms of iron is the respective properties of ferrous and ferric iron and their interconversions. Ferrous iron is the more soluble and more toxic form, but, under aerobic conditions, it is readily converted to ferric iron, which is much less soluble and much less toxic.

Anthropogenic forms/conversions. Human activities that mobilise iron include smelting and refining of metals, steel manufacturing, and metal plating. Waste disposal, particularly landfills, may result in the mobilisation of ferrous iron when anaerobic and/or acidic conditions prevail. Acid mine drainage typically contains high concentrations of dissolved iron.

Sources for biota. Airborne particulates containing iron are deposited onto soils and into water bodies. Geological sources also supply iron to soils and water. But although present in soils and water, iron is often not biologically available. Food items represent the main sources for biota.

Exposure pathway(s) for humans. Oral intake, both natural and supplemental, represent the most common routes of exposure for the general population. Blood transfusions also supply large amounts of iron to the involved patients.

Effects on humans. Iron is a required nutrient for humans, having a role in many redox reactions in metabolic functions and being a component of haemoglobin. For human health, the literature reveals that deficiency of iron is a major concern, and that toxicity is rare except in unusual and very well-defined situations. Acute toxicity of iron is known only from exposure to therapeutic iron preparations, whereas chronic iron toxicity ("iron overload") is invariably related to excessive dietary iron intake, typically from consuming large quantities of red meats or dietary supplements.

Pathways to ecosystems. Most of the iron in soils is insoluble. Plants can only take up this element in dissolved form. Sediments of aquatic systems contain high (percentage) concentrations of iron. Weathering and erosion mobilise particulate forms of iron as well as release soluble iron, all of which can enter the surface water directly or indirectly. Much of the iron in water bodies is in the form of suspended particulates, and this is the form in which iron is typically transported through water bodies. Leaching of iron from soils can result in its transport in soluble form to groundwater or to surface water.

Effects on nonhuman targets. Iron is an essential element, required for many metabolic functions, especially in redox reactions. It is a core component of chlorophyll and an integral component of haemoglobin and other oxygen-carrying pigments. Concern for iron and aquatic plants is more likely to be for deficiency of the element than for its toxicity, related to the low solubility of inorganic compounds of ferrous iron. Toxicity has been demonstrated in the laboratory for aquatic organisms, but experimental design is typically confounded because of the conditions required to dose the biota with dissolved iron. Evidence for iron toxicity to aquatic organisms in the field is still controversial. Sediment-dwelling biota may be adversely affected by high iron concentrations, but, because of confounding factors, field data are difficult to interpret. The literature indicates that iron toxicity for terrestrial plants is usually encountered only in waterlogged (and oxygen limited) or extremely acidic soils.

Regulatory aspects. The Canadian water quality guideline for iron is $0.3\,\mathrm{mg\,L^{-1}}$. Ontario's drinking water objective for iron is an aesthetic objective, at $0.3\,\mathrm{mg\,L^{-1}}$. The Ontario PWQO is also $0.3\,\mathrm{mg\,L^{-1}}$. There is no Canadian sediment quality guideline; the Ontario sediment quality guideline is 2% for the lowest effect level.

References. Lauffer, R. B. 1992. *Iron and Human Disease*. CRC Press, Boca Raton, FL. Subcommittee on Iron. 1979. NRC Committee on Medical and Biologic Effects of Environmental Pollutants: Iron. Division of Medical Sciences Assembly of Like Sciences National Research Council, University Park Press, Baltimore.

7.   Lead

Metal/metalloid. Lead, Pb.

Atomic number. 82.

Atomic weight. 207.

Uses. Lead is now the fifth most commonly used metal in the world. The use of the metal in plumbing has continued from ancient times to the present day. Although lead for plumbing and for construction has now been largely discontinued, older lead-based systems still remain in use. The largest recent use of lead, and probably the most significant from the point of view of environmental contamination, is in tetraethyl lead, which has been, and in some jurisdictions continues to be, used as an additive to gasoline.

Forms in the natural environment. Lead occurs naturally in trace quantities. Weathering and volcanic emissions account for most of the natural processes that mobilise the element. The usual valence state of lead is II, in which state it forms inorganic compounds. Lead can also have a valence state of IV, forming covalent compounds.

Anthropogenic forms/conversions. Older uses of lead, in glazes, in paints, and in crystal tableware have now been partially or completely discontinued, but the legacy of lead contamination, particularly in soil, remains as a potential human health or environmental problem. Lead is released into air during primary and secondary smelting of lead ores and wastes, and smelting of ores in which lead is a contaminant. These practices continue, although there have been improvements in terms of decreasing emissions in recent decades. The practices release lead into the atmosphere, where it falls at relatively short distances from the respective sources into soil and water.

Sources for biota. Lead in air, in food, in water, and in soils and sediments are all potential sources for biota. Lead in air may be a direct source for biota, but the air is also a vector by which lead is transported to other media. Terrestrial plants incorporate very little lead from the soil via root uptake, but surface contamination of root and leafy vegetables can result in ingestion of lead. Dissolved lead and lead in sediment represent sources for aquatic biota. There may be some limited food chain transfer of the element, but lead is not biomagnified in food chains.

Exposure pathway(s) for humans. Humans are exposed to lead through the inhalation of dust or other particles. For environmental exposure, an important source for intake via ingestion is the mouthing of objects contaminated by lead in dust or soil, the latter being particularly important for younger children. Humans are also exposed to lead through ingestion of contaminated food and water. Oral intake of lead is more significant via food than via water.

Effects on humans. Plumbism results from acute exposure to high levels of lead and is manifest by effects of the central nervous system, with stupor, coma, and convulsions. Palsy in the form of wristdrop is one of the classical manifestations of lead poisoning. These types of symptoms, associated with very high blood lead levels, are extremely rare in modern medicine, especially from environmental exposure. A number of physiological or biochemical processes are affected by lead. Lead interferes with the haem system. Levels of protoporphyrin in serum erythrocytes increase when lead concentrations are elevated and the enzyme δ-aminolevulinic acid dehydratase is inhibited by lead. The effects of lead on the human haemopoietic system result in decreased haemoglobin synthesis, leading to anaemia with pallor as a classic symptom of lead poisoning. The more subtle effects of lower exposures include behavioural effects in children, including hyperactivity, poor attention span, and IQ deficits. These effects are not always easy to relate to lead exposure causally because of difficulties in measurement and the existence of confounding factors. Low-level effects of lead are now the subject of a great deal of research, as well as a concern for regulatory agencies.

Pathways to ecosystems. Nonhuman biota, notably terrestrial plants, fish, benthic biota, and wildlife, especially waterfowl, are exposed to this metal in their environment. Deposition of airborne lead and the resulting soil contamination represent potential exposure routes for terrestrial organisms. Ingestion of lead fishing weights and lead shot by waterfowl represents some of the highest exposures to lead for nonhuman biota. Aquatic biota are potentially at risk from exposure to lead from indirect and direct releases into their environment, particularly in acidified systems.

Effects on nonhuman targets. Laboratory tests have been made on the effects of lead on a number of organisms; many studies have addressed birds and fish, and recently some attention has been paid to invertebrates, particularly benthos. Aquatic biota are potentially at risk from exposure to lead from indirect and direct releases into their environment; however, toxic effects of lead to aquatic life via the aqueous phase are probably quite unusual in the real world and are likely to occur mainly when there is direct release of lead into water, as for example in industrial effluent or from accidental spills, in combination with water of low pH and low hardness. Lead in water bodies that are soft and acidified or acid-sensitive appears to be more available to fish than that in harder waters, as judged by the relative accumulation of lead in tissues. However, this accumulation is seen as an indicator of exposure rather than representation of any toxic effect. Effects on terrestrial plants have also been examined, in the context of direct atmospheric deposition of lead as well as the contamination of soils via atmospheric lead. Lead-tolerant strains of plants have been isolated from lead-contaminated habitats, suggesting that at least some part of the lead in the substrates is biologically available.

Regulatory aspects. The Canadian water quality guidelines for lead range from 1 to $7 \mu g L^{-1}$, depending on water hardness, with the lowest, $1 \mu g L^{-1}$, for very soft water ($0–60 mg L^{-1}$ as $CaCO_3$) and the highest, $7 \mu g L^{-1}$, for very hard water ($>180 mg L^{-1}$ as $CaCO_3$). The USA criteria are very similar. Ontario's revised interim PWQOs for lead are $1 \mu g L^{-1}$ for very soft water ($<30 mg L^{-1}$ as $CaCO_3$), 3 $\mu g L^{-1}$ for water of hardness $30–80 mg L^{-1}$ as $CaCO_3$, and $5 \mu g L^{-1}$ for water of hardness $>80 mg L^{-1}$ as $CaCO_3$ $mg L^{-1}$ as $CaCO_3$. Ontario's drinking water objective for lead is $10 \mu g L^{-1}$ and the U.S. drinking water criterion for lead is $50 \mu g L^{-1}$. The Canadian interim sediment quality guideline for lead is $35 \mu g g^{-1}$ (dry weight), which is almost the same as the Ontario lowest effect level of $31 \mu g g^{-1}$ (dry weight).

Reference. Bunce, N. 1994. *Environmental Chemistry*, 2nd ed., Wuertz Publishing, Winnipeg.

8.   Manganese

Metal/metalloid. Manganese, Mn.

Atomic number. 25.

Atomic weight. 54.94.

Uses. The element is used in the manufacture of ferroalloys, batteries, glass, inks, ceramics, rubber, and wood preservatives. A very recent use of manganese is as an anti-knock additive to gasoline in the form of methylcyclopentadienyl manganese tricarbonyl (MMT).

Forms in the natural environment. Manganese is an abundant crustal element, often present in igneous rocks. Weathering of crustal rocks, ocean spray, and volcanic activity are the main natural processes that mobilise manganese. The common ores are pyrolusite (MnO_2), rhodochrisite ($MnCO_3$), and psilomelane ($BaMn^{2+}Mn_g^{4+}O_{16}$). Oxidation states include Mn(II), Mn(III), and Mn(IV). Changes among these oxidation states have profound effects on the environmental chemistry of the element. The chemical speciation of manganese is altered by a number of factors including pH and redox potential. Acidification of surface waters has increased the concentration of dissolved manganese.

Anthropogenic forms/conversions. Mining, metallurgical processing, and steel manufacturing mobilise manganese and release it, mainly as airborne particles, but also, from tailings ponds, water can become contaminated. Fuel combustion and waste incineration also release manganese, although on regional or even global scales, the quantities from these are far less significant than those from industrial sources. Most airborne manganese falls close to the source. MMT is a particularly toxic manganese product; its manufacture and use require stringent precautions.

Sources for biota. Under normal conditions, the main source of manganese as a nutrient is dietary for all animals. Airborne manganese represents a major source for workers in occupational settings. Less is known about environmental sources to humans, but urban air may represent a significant source to humans, particularly for young children. Sources for other biota, apart from diet, include soil or surface waters, with increased availability in acidified systems.

Exposure pathway(s) for humans. Inhalation is the major exposure route for persons working in mines and other industries, and this route would also apply to any environmental exposure. The exception is MMT, which is absorbed through the skin, and via this route accesses the central nervous system.

Effects on humans. Manganese is essential for all types of organisms, with a role in nucleic acid synthesis and the metabolism of carbohydrates, as well as involvement with the maintenance of the nervous system, the formation of blood clotting factors, and the function of female sex hormones and thyroid hormones. Deficiencies for humans occur rarely when diets are inadequate but more frequently from metabolic disorders. In higher doses, typically from occupational exposure, it is toxic. Acute manganism results in neurological dysfunction, with disorientation, memory impairment, and physical disabilities. Patients with Parkinson's and other neurological diseases have elevated manganese, but to date no causal connection between brain lesions associated with these nonoccupational neurological diseases, and manganese exposure, has been established.

Pathways to ecosystems. Deposition of manganese from the atmosphere into water and soil as well as deliberate addition of fertilisers represent potential pathways for direct uptake into plants, animals, and microorganisms. Food chain transfer apparently does not occur in terrestrial systems, but results from experimental approaches with aquatic systems suggest that weak biomagnification may occur in aquatic food chains.

Effects on nonhuman targets. Deficiency of the element can be a problem, particularly for agricultural plants. The relationship between manganese and iron uptake is probably related to competition between the two cations. Deficiency of manganese is generally not related to the absolute concentration of the element in soils but rather to availability: The divalent cation Mn^{2+} is the most important form for plant uptake. Deficiency symptoms are characteristic and can be relieved by foliar spraying. Manganese toxicity to terrestrial plants is poorly understood. It is rare in natural field conditions but occurs in cultivated plants. Toxicity is also related to the chemical species rather than to the total concentration of manganese. Manganese toxicity to aquatic organisms has been shown under experimental conditions, but its significance as a toxic substance to aquatic biota in the field remains poorly understood.

Regulatory aspects. There appear to be no water quality criteria or guidelines in Canada or the United States for manganese. The Ontario lowest effect level for sediments is $460 \mu g\, g^{-1}$ (dry weight).

Reference. Stokes, P. M., P. G. C. Campbell, W. H. Schroeder, C. Trick, R. L. France, K. J. Puckett, B. Lazerte, M. Speyer, and J. Donaldson. 1988. *Manganese in the Canadian Environment*, NRCC No. 26193, Associate Committee for Scientific Criteria for Environmental Quality, National Research Council of Canada, Ottawa.

9.   Mercury

Metal/metalloid. Mercury, Hg.

Atomic number. 80.

Atomic weight. 200.59.

Uses. Use of cinnabar as a pigment for red ink in China goes back 3,000 years and is still in current use. Currently, mercury is used in instruments to measure temperature and pressure and is still widely used in certain types of batteries and in fluorescent tubes and other electrical appliances. Biocidal uses such as slimicides, fungicides, antiseptics, and internal medications are now almost entirely absent from developed countries. The use of mercury in dental amalgam is beginning to be supplanted in parts of North America and Europe, following the demonstration that increased tissue burdens of the metal are related to the number of faces of fillings in the mouth. Whether or not this type of burden represents a health risk is still controversial.

Forms in the natural environment. The three most important forms of mercury can be distinguished as: elemental, metallic mercury (Hg^0); divalent inorganic mercury (Hg^{2+}); and methylmercury [CH_3Hg^+ and $(CH_3)_2$]. Geological sources of mercury include elemental and combined forms in soils, rocks, and water. Notably, mercury is relatively concentrated in fossil fuels such as oil and coal, with implications for atmospheric emissions when the fuel is burned. In aquatic systems, mercury has a high affinity for particulate matter and, thus, is found predominately in suspended or bed sediment rather than in the dissolved phase. In the natural environment, mercury undergoes chemical and biochemical transformations, the most significant of which is microbial methylation, which occurs primarily in freshwater and saltwater ecosystems. Monomethylmercury is the form that has the most harmful effects on living organisms and the form that accumulates in tissues, notably in the muscle of fish.

Anthropogenic forms/conversions. A number of human activities have directly and indirectly altered the cycle of mercury, resulting in exposure of humans and ecosystems to various forms of the element. These forms include the combustion of coal and oil and the incineration of municipal and medical waste and sewage sludge. The mining and smelting of mercury, as well as copper and lead smelting, release the element into the atmosphere. Until recently, chlor-alkali plants released mercury into water bodies. Other human activities that result in the mobilisation of mercury are the manufacture of batteries and the preparation of dental amalgam, paint, industrial instruments, and electrical appliances, especially fluorescent lamps.

Sources for biota. Mercury enters the atmosphere through direct and indirect pathways, from soils, rocks, water, and sediment, as well as from living systems. Aside from occupational situations, the main sources for humans and other biota is ingestion of mercury-contaminated food, the most notable of which is fish.

Exposure pathway(s) for humans. Occupational exposure to mercury is mainly through inhalation of elemental mercury vapours or contaminated dusts. Misuse of mercurial fungicides can lead to contamination of foodstuffs and subsequent ingestion of mercury compounds, but the main risk from mercury for the general population is from consumption of contaminated fish. Biomagnification results in concentrations of methylmercury in fish muscle that render the fish unfit for human consumption even though the fish show no adverse effects. Recent studies have indicated that mercury released from dental amalgam can increase the body burden of the element.

Effects on humans. The syndrome in humans resulting from exposure to methylmercury has been called Minamata disease, named for the first recorded widespread epidemic of environmental methylmercury poisoning, in the Japanese fishing village of Minamata, beginning in 1953. Methylmercury is a neurotoxin, and, through its action on the brain, symptoms include aphasia, ataxia, convulsions, and death.

Pathways to ecosystems. Mercury is deposited into terrestrial and aquatic systems from the atmosphere, as wet and dry deposition, and may travel long distances from its original source as a consequence of the volatility of the element and some of its compounds. Once in terrestrial or aquatic systems, mercury undergoes a series of chemical and biological transformations, the most significant of which, as noted, lead to the formation of methylmercury. In the aquatic system, mercury alone of the metals is biomagnified through the food web.

Effects on nonhuman targets. Mercury has no known essential biological function. Methylmercury is the form of major concern for adverse effects. Organisms at the greatest risk are high-level consumers, such as domestic animals and wildlife. For mammals, methylmercury is a neurotoxin, and, through its action on the brain, symptoms include aphasia, ataxia, convulsions, and death. At doses higher than those eliciting neurological symptoms, kidney, cardiovascular, and digestive systems among others can be affected. Lower forms of life are less susceptible to mercury poisoning than are higher vertebrates, but the "processing" of mercury through the ecosystem means that the higher trophic levels such as predatory wildlife populations are particularly at risk.

Regulatory aspects. The current U.S. EPA criterion for mercury in freshwater is $0.012 \mu g \, L^{-1}$, and the U.S. potable water criterion is $<2.0 \mu g \, g^{-1}$. The Canadian water quality guideline for mercury is $0.1 \mu g \, L^{-1}$, whereas Ontario's PWQO is $0.2 \mu g \, L^{-1}$ in a filtered water sample. For Ontario, Canada, the maximum acceptable concentration of mercury is $1 \mu g \, L^{-1}$. Canada's interim sediment quality guideline is $0.13 \mu g \, g^{-1}$ (dry weight), and Ontario's lowest effect level is $0.2 \mu g \, g^{-1}$ (dry weight). The food intake criteria (all foods, in $\mu g \, kg^{-1}$ fresh weight) are <500 for Canada, $<1,000$ for the United States, and <30 for Benelux countries.

Reference. Watras, C. J., and J. W. Huckabee (eds.). 1994. *Mercury Pollution Integration and Synthesis*, Lewis, Boca Raton, FL.

10.   Nickel

Metal/metalloid. Nickel, Ni.

Atomic number. 28.

Atomic weight. 58.69.

Uses. Nickel is commonly alloyed with other metals for use in various applications, such as the manufacture of stainless steel and electronic equipment, and in the shipbuilding, automotive, and aeronautical industries. Nickel compounds are also used in electroplating, in pigments and glazes for ceramics, in Ni/Cd batteries, and in jewellery.

Forms in natural environment. Nickel occurs naturally in the Earth's crust at an average concentration of approximately $75 \mu g g^{-1}$. It is the 24th most abundant element in the Earth's crust. Nickel occurs as a trace constituent in a number of minerals, particularly those with high iron and magnesium contents, including pyroxenes and olivine. The mineral pentlandite [$(Ni, Fe)_9S_8$] is an important commercial source of the element. Nickel is most commonly found in the environment in the +II oxidation state. In natural waters with high sulphate concentrations (e.g., those which have received historically high levels of acid deposition), nickel sulphate may be the predominant soluble species. In aquatic systems, nickel is distributed between dissolved and particulate forms, depending on the pH, redox potential, suspended sediment load, and the like of the water body. Nickel in surface water and in soil solution is generally present in dissolved form at pH < 6.5. Under reducing conditions in sediment, insoluble nickel sulphide, which is less bioavailable than soluble nickel species, may be formed. In soils, nickel may be adsorbed by clays and oxides of iron and manganese but is mobilised in soil solution at lower pH.

Anthropogenic forms/conversion. Nickel released to the atmosphere as a result of smelting processes is likely to be in the form of nickel sulphate, nickel subsulphide, and/or nickel oxide, the majority of which is water-soluble and may be removed from the atmosphere via wet and dry deposition. Acid rain may mobilise nickel from soils and rocks in watersheds, increasing its availability to both terrestrial and aquatic biota.

Sources for biota. Nickel enters the environment through natural processes such as weathering and erosion of rocks and soils, forest fires, volcanic emissions, sea spray, and particle exudates from vegetation. Human activities such as base metal mining, smelting, refining, fossil fuel combustion, and solid waste disposal/incineration are also significant sources of the metal to the environment.

Exposure pathway(s) for humans. Humans may be exposed to nickel through inhalation of both gaseous nickel compounds (e.g., nickel carbonyl) and particulate nickel (in dust and tobacco smoke) and through ingestion of food and water. Human studies have shown that approximately 35% of inhaled nickel is absorbed, compared with less than 10% absorption of ingested nickel. Studies have suggested that nickel might be an essential element for humans; however, there have been no reports of nickel deficiencies in humans, and its status as a nutrient for humans remains equivocal.

Effects on humans. Occupational studies of smelter employees have demonstrated a relationship between airborne nickel concentrations and the incidence of lung and nasal passage cancer. Other nonmalignant disorders of the respiratory tract such as asthma and tissue damage have been associated with inhalation of nickel carbonyl. Cigarette smoking can increase inhaled nickel exposure by as much as 4 µg nickel per pack of cigarettes.

Pathways to ecosystems. Nickel compounds are relatively soluble; thus, the element is rather mobile and widely distributed in the environment. Nickel may be deposited in ecosystems remote from point sources via airborne transport of particulates. Nickel-laden leachate from waste rock and tailings generated through mining operations may contribute to groundwater and surface water contamination, and large amounts of the element may be transferred to aquatic systems through municipal sewage effluent containing industrial wastes (e.g., from electroplating or steel mills).

Effects on nonhuman targets. Nickel can be bioconcentrated in aquatic organisms, and bioconcentration factors from approximately 100 to 5,000 on a dry weight of tissue to water basis. There is no evidence to suggest that the metal is biomagnified in either terrestrial or aquatic food chains. Nickel is an essential element for plant growth but can cause deleterious effects at high concentrations. General signs of nickel toxicity in plants include stunted growth, deformation of plant parts, and chlorosis. Animals vary in their sensitivity to nickel. Certain taxa (e.g., some fungi, microorganisms, invertebrates) have been observed to be relatively tolerant to nickel and may control their body burdens either by limiting uptake or through efficient excretion. Other taxa such as certain molluscs, crustaceans (e.g., *Daphnia*) are relatively sensitive to nickel and may be unable to survive; therefore, they are absent from nickel-contaminated sites. Observed nonlethal responses of fish and invertebrates to high nickel exposure (chronic and acute) include decreased growth, genotoxicity, and tissue damage.

Regulatory aspects. The Canadian water quality objective for nickel in fresh waters ranges from 25 to 150 µg L^{-1} depending on water hardness, with the higher values for harder waters. The U.S. criteria are similar and are also hardness-related. The Ontario PWQO is 25 µg L^{-1}. There are no Canadian sediment quality guidelines for nickel.

References. Chau, Y. K., and O. T. R. Kulikovsky-Codeiro. 1995. Occurrence of nickel in the Canadian environment. *Environmental Reviews*, **3** (1), 95–120.

Environment Canada. 1994. Priority Substances List Report: Nickel and its compounds. Environment Canada, Ottawa.

11.   Selenium

Metal/metalloid. Selenium, Se.

Atomic number. 34.

Atomic weight. 78.96.

Uses. Selenium is used in electrical rectifiers, in photocopying, as a photographic toner, and as an additive for stainless steel. Its photovoltaic action makes it useful in the production of photocells, exposure meters, and solar ells. Selenium is used in the manufacture of glass to decolourise it as well as in making ruby-coloured glass and enamels.

Forms in the natural environment. Selenium commonly occurs in sedimentary rocks, frequently combined with sulphide minerals or with silver, copper, lead, and nickel minerals. The element exists in three oxidation states in surface waters: −II, selenide; IV, selenite; VI, selenate. In addition, it occurs in the elemental form Se^0, which is insoluble in water. Selenate (VI) is stable in a well-oxidised environment, and selenate salts are highly water-soluble. As such, Se is freely available to aquatic animals and to plants. Selenite (IV) is favoured under mildly oxidising conditions. It is less water-soluble than selenate and is largely unavailable to plants in this form. Hydrogen selenide is a toxic gas existing in reducing environments.

Anthropogenic forms/conversions. Selenium enters the environment from many different industrial situations including coal-burning, fly-ash piles, and the manufacture of glass, paint, petroleum, textiles, and electrical components. Historically, it has been employed occasionally as a pesticide and, although its use has been largely discontinued, it is still retained in some shampoos used to control human dandruff and canine mange.

Sources for biota. Seleniferous soils (soils that are naturally enriched with the element) are a source for the accumulation of the element by locoweeds (*Astragalus* spp.), without any toxic effects to the plants. Livestock feeding on these weeds receive potentially toxic doses of selenium.

Exposure pathway(s) for humans. Absorption through the skin is poor. Humans can be exposed to selenium in drinking water, typically in inorganic form and occasionally the ground water can become contaminated with selenium from abandoned mining areas or old waste disposal sites. Meat from animals fed on selenium rich grains can contain organic selenium, which is absorbed through the human digestive system.

Effects on humans. Based on epidemiological and experimental evidence, it appears that selenium can act as an anticarcinogen and antimutagen at low levels but is capable of genotoxicity at higher concentrations. Several studies have also indicated an ameliorative effect of selenium on mercury and cadmium toxicity in humans. Chronic exposure to selenium in some village populations in the People's Republic of China caused effects such as deformed nails, brittle hair, and loss of feeling in arms and legs.

Pathways to ecosystems. Selenium contamination of water has been shown to lead to accumulations of selenium in fish tissue, which is then passed on to piscivorous wildlife.

Effects on nonhuman targets. Selenium has been demonstrated to be an essential micronutrient for several organisms. The element is a component of the enzyme glutathione peroxidase, as well as numerous proteins including haemoglobin, myosin, cytochrome c and several ribonucleoproteins. Deficiency of selenium has been shown for some fish. Selenium deficiency in mammals can result in disorders of the liver, pancreas, and other tissues. Cattle from the Florida Everglades region deficient in Se showed symptoms of anaemia and reduced growth and fertility. The doses of selenium that elicit toxic responses are not very much greater than those considered as nutrient concentrations. Selenite is readily incorporated into organic compounds such as the amino acid derivatives Se-methylselenomethionine and Se-methylselenocysteine, which are taken up by plants, substituting for amino acids in proteins causing configurational and functional changes resulting in toxicity. Similarly, protein degradation may result from bonding between inorganic Se accumulated in tissues and sulfhydryl groups on amino acid constituents of intracellular proteins. Seleneomethionine is embryotoxic and teratogenic in developing birds when supplied in laboratory feed at levels exceeding $4\,mg\,kg^{-1}$, causing symptoms similar to those seen in field studies. Selenium appears to have an ameliorative effect on mercury and cadmium toxicity.

Regulatory aspects. The current recommended water quality criterion for selenium in the United States is $5\,\mu g\,g^{-1}$. The Canadian water quality guideline is $1\,\mu g\,L^{-1}$. The Ontario PWQO is $100\,\mu g\,L^{-1}$, and Ontario's maximum acceptable concentration in drinking water is $10\,\mu g\,L^{-1}$. The U.S. drinking water criterion for selenium is $<500\,\mu g\,L^{-1}$.

Reference. American Society of Agronomy and Soil Science Society of America. 1989. *Selenium in Agriculture and the Environment.* SSSA Special Publication No. 23. Madison, WI.

12.   Silver

Metal/metalloid. Silver, Ag.

Atomic number. 47.

Atomic weight. 107.9.

Uses. Silver is used to manufacture many products, including coins, jewellery, ornamental containers, sterlingware, batteries, and mirrors. Silver is also present in photographic film and paper, drinking water, disinfectants, prophylactic agents (silver nitrate), cloud seeding preparations (silver iodide), solder, alloys, and electrical contacts. Silver is used in medicine as a germicide, in dental amalgams, and for the treatment of burns.

Forms in natural environment. Silver has four oxidation states: 0, I, II, and III; the first two are most common in the environment. Silver is normally present at trace concentrations ($<1\,ng\,L^{-1}$) in aquatic systems because of its tendency to form insoluble complexes (e.g., sulphides), and so it tends to accumulate in sediment. Silver is naturally occurring in the Earth's crust at trace concentrations ($0.1\,mg\,kg^{-1}$). In rock and soil, silver is mainly found in sulphide compounds, in association with lead, iron, tellurides, or gold; it is rarely found in its metallic state.

Anthropogenic forms/conversions. Coal burning releases silver in particulate form to the air and may result in silver accumulation at points distant from the source. Mining and smelting of silver produces silver-enriched waste rock and slag. Silver, which is released to the aquatic environment from the photographic industry, is primarily in sulphide and chloride forms, which are generally unavailable to biota.

Sources. Major anthropogenic sources of this element to the environment include mining and smelting (of silver and/or associated metals) and solid and liquid wastes from the photographic industry and other silver-related industries (e.g., silver plating).

Exposure pathway(s) for humans. Humans may receive occupational exposure to silver in the air and may ingest the element, which is present in food and drink, from eroded tableware or cutlery.

Effects on humans. Silver is not an essential element; is excreted in faeces, urine and perspiration; and may accumulate in the hair and bones. The most commonly reported effect of silver in humans is argyria (skin discoloration), which is cosmetic rather than toxic.

Pathways to ecosystems. Silver is released to the air in particulate form, from ore processing and the burning of fossil fuels, especially coal. The main input of this element to the environment is through disposal of silver-containing solid waste, including domestic waste, sewage sludge, and industrial wastes. Silver enters the aquatic environment primarily in the form of sulphides (particulate or colloidal form), chlorides, and soluble organic complexes.

Effects on nonhuman targets. Silver toxicity in aquatic systems is related to the free ion (Ag^+) activity. The Ag^+ ion is surface active and does damage to aquatic organisms by binding to specific sites on or in the gills. Ligands such as humic substances, chloride, and sulphide ions bind silver and decrease the activity of the free ion and, therefore, the toxicity of the metal. Many silver complexes are insoluble (e.g., Ag_2S), and most silver that enters aquatic systems accumulates in the sediment. Silver toxicity (death, low productivity, altered reproduction) has been demonstrated in laboratory experiments with both plants and animals.

Regulatory aspects. The Environment Canada guideline for silver for the protection of aquatic life is $0.1 \mu g L^{-1}$. The Ontario PWQO is the same. The U.S. criteria for silver for the protection of aquatic life are related to hardness. At hardness $50 mg L^{-1}$ and $500 mg L^{-1}$ as $CaCO_3$, the criteria for silver are 1.2 and $13.0 \mu g L^{-1}$, respectively.

Reference. Purcell, T. W. and J. J. Peters. 1998. Sources of silver in the environment, *Environmental Toxicology and Chemistry* **17** (4), 539–46.
Beads: Environmental Toxicology and Chemistry.

13.   Vanadium

Metal/metalloid. Vanadium, V.

Atomic number. 23.

Atomic weight. 50.94.

Uses. The structural strength of vanadium makes it useful in nuclear applications. About 80% of the vanadium now produced is used as ferrovanadium or as a steel additive. It is used in rust-resistant springs and steel tools. Vanadium foil is used as a bonding agent for binding titanium to steel. Vanadium is also used in dyes, inks, and paints, and the pentoxide V_2O_5 is used in ceramics and as a catalyst in the production of plastics and sulphuric and nitric acids.

Forms in natural environment. Vanadium is present naturally in the Earth's crust at an average concentration of $150 mg kg^{-1}$, which gives it the same abundance as nickel, copper, zinc, and lead, although the element has a far more even distribution than these. Vanadium in rock is commonly associated with titanium and uranium ores. Crude oil and coal contain high levels of vanadium (average 50 and 25 mg kg^{-1}, respectively, with a range of 1–1,400 mg kg^{-1}) which can be released to the atmosphere during fuel combustion.

Anthropogenic forms/conversions. During refining and distillation of crude oil, much of the vanadium remains in the residual oils because of its low volatility and, as a result, becomes more concentrated than the original crude oil. Residual oils are mainly used for heating and electric power generation, and during fuel combustion, much of the contained vanadium is released as part of the fly-ash particulates.

Sources for biota. The majority of vanadium in the atmosphere results from human activity, particularly in the combustion of oil and coal and as a by-product of petroleum refining. Vanadium enters water bodies from deposition of airborne particulates, weathering of vanadium-rich soils, and direct dumping. Most of the vanadium in aquatic systems, however, is natural and is a result of soil runoff.

Exposure pathway(s) for humans. The major route of human exposure to vanadium is through inhalation. The majority of fly-ash particles released from oil combustion are of respirable size (i.e., diameter less than 10 μm), and particulates may be deposited and accumulate in the lungs.

Effects on humans. Vanadium that is present in food is poorly absorbed in the gastrointestinal tract (<10%) although gastrointestinal distress had been linked with industrial exposure to vanadium. Vanadium is excreted by the kidneys. Respiratory irritation may arise from exposure to high vanadium concentrations (0.05 mg m^{-3}) in fumes, and the element may accumulate in the lungs with age but does not appear to cause toxic effects.

Pathways to ecosystems. In comparison with many other elements, vanadium in soil is highly mobile under neutral or alkaline conditions. Its mobility is decreased under acidic and oxidising conditions. In reducing environments, the element is nearly immobile in soil. Estuarine muds may have particularly high concentrations of vanadium as a result of precipitation of soluble vanadium when freshwater mixes with seawater. Vanadium is present in the atmosphere in particulate form, primarily as a result of anthropogenic activities. Natural sources may also contribute to the concentrations in the atmosphere, resulting from the weathering of soils, marine aerosols and volcanic emissions. Anthropogenic sources of this element to the air, however, are far more important in urban areas, where vanadium concentrations can be very high (>11 ng m^{-3}), resulting from the combustion of vanadium-rich fossil fuels. In nonpolluted areas, concentrations in air average 0.02 μg m^{-3}.

Effects on nonhuman targets. Vanadium appears to increase the growth rate of certain algae when present at trace ($1–10\,\mu g\,L^{-1}$) concentrations in the growth media. Higher concentrations ($20–40\,\mu g\,L^{-1}$) have been shown to decrease species diversity of phytoplankton. Vanadium is essential in rats and chickens and is required by some bacteria. Vanadium toxicity has been demonstrated in laboratory experiments using both aquatic and terrestrial plants and animals, although concentrations used in these experiments are generally higher than those found in the environment.

Regulatory aspects. Due to its apparent low toxicity and poor absorption in humans, vanadium concentrations in the environment are not regulated by government agencies.

Reference. Van Zinderen Bakker, E. M., and J. F. Jaworski. 1980. Effects of Vanadium in the Canadian Environment. National Research Council Pub. No. 18132, Ottawa.

14.   Zinc

Metal/metalloid. Zinc, Zn.

Atomic number. 30.

Atomic weight. 65.29.

Uses. Zinc metal is used in batteries, for roof covering, and to protect iron from corrosion (galvanising). Zinc is also used in coins. Alloys that incorporate zinc include brass, nickel silver, commercial bronze, spring brass, soft solder, and aluminium solder. Oxides of zinc are used in the manufacture of paints, rubber products, cosmetics, pharmaceuticals, floor coverings, plastics, printing ink, soap, textiles, electrical equipment, and ointments.

Forms in the natural environment. Zinc is the 23rd most abundant element in the Earth's crust. Some soils have naturally high zinc concentrations. The principal zinc minerals are sulphides such as sphalerite and wutzite (ZnS), which usually occur in association with other ores including copper, gold, lead, and silver. In natural waters, zinc is in the form of the divalent cation Zn^{2+} (hydrated Zn^{2+} at pH between 4 and 7) as well as in the form of fairly weak complexes.

Anthropogenic forms/conversions. Zinc is mobilised into air and water in the mining and primary smelting of zinc and of other metals, in electroplating, in fossil fuel combustion, and in solid waste incineration. From the atmosphere, deposition onto soil and into water occurs, with increased dissolution in these media if acidification occurs.

Sources for biota. Sources of zinc for humans in occupational situations are mainly in the atmosphere in the form of dust; however, for the general population, zinc in food, originating from natural uptake and, to a lesser extent, from natural or anthropogenic origins, are the typical sources. Similarly for other animals, food and water are the main sources. Oral supplements constitute a source for selected members of human populations. Natural or anthropogenic zinc in soil is the source for terrestrial plants. Zinc in soil is an indirect source, via runoff, for aquatic organisms. Deposition of zinc from air into water, especially if the water is acidified, is a major source for aquatic biota, as is naturally occurring and anthropogenic zinc in sediments, which is a primary source for benthos and potentially a secondary source for organisms in the water column.

Exposure pathway(s) for humans. Workers in certain industries are exposed through inhalation. For the more general population, zinc is typically ingested in food, especially organ meats, seafood, and finfish, all of which accumulate zinc naturally during life.

Effects on humans. Zinc is an essential trace nutrient, a constituent of many enzymes, and a coenzyme for a number of systems. It is involved in the synthesis of RNA and proteins. Zinc deficiency occurs in humans, and oral supplements are available. There is some evidence that specific conditions, including deficient immune systems, prostate problems, diabetes, and macular degeneration, will respond to zinc supplements. Zinc toxicity is manifest in the occupational disorder called metal fume fever, related to the inhalation of zinc dust, but the long-term health effects of inhaling zinc dust are not well understood. Toxicity of the metal is not generally of concern for humans in the general population: Homeostasis is achieved through a system of binding zinc to small proteins (metallothioneins) in the kidney.

Pathways to ecosystems. Zinc from the atmosphere enters ecosystems via wet and dry deposition, whence it becomes incorporated into soils, surface waters, and sediments. Zinc is transferred in the aquatic food chain, although not biomagnified, so for this pathway can be significant heterotrophs.

Effects on nonhuman targets. Zinc is a required trace element, and terrestrial plants are known to show zinc deficiency. Like other trace nutrients, zinc can be toxic at concentrations higher than those needed for nutrition. Zinc in soil can be toxic to terrestrial plants, for example on metal contaminated sites, and such sites evoke the selection of zinc-tolerant plant types. Zinc toxicity to terrestrial animals has been demonstrated in laboratory studies, but these studies often require doses much higher than environmental concentrations. Aquatic biota are probably of most major concern related to zinc toxicity, with fish and invertebrates at risk from environmentally realistic concentrations of the metal. Most organisms that have been examined have the metal-binding protein that appears to control zinc concentrations in the over-all organism. Zinc accumulates in various tissues but is not biomagnified.

Regulatory aspects. The U.S. EPA, the Canadian Department of the Environment, and Ontario all have the drinking water guideline for zinc as $5\,mg\,L^{-1}$, and in all cases this is based on taste (i.e., is an aesthetic, not health protection, objective). The Canadian guideline for the protection of freshwater life is $30\,\mu g\,L^{-1}$, while Ontario's revised PWQO is $20\,\mu g\,L^{-1}$. The U.S. criteria for zinc, to protect aquatic life, are based on total recoverable zinc and are related to hardness. In soft water, $50\,mg\,L^{-1}$ as $CaCO_3$, the criterion is $180\,\mu g\,L^{-1}$, and in hard water, $500\,mg\,L^{-1}$ as $CaCO_3$, the criterion for zinc is $570\,\mu g\,L^{-1}$. The Canadian interim sediment quality guideline is $123\,\mu g\,g^{-1}$ (dry weight), and Ontario's lowest effect level is $120\,\mu g\,g^{-1}$ (dry weight).

Reference. National Academy of Sciences. 1979. *Zinc*, University Park Press, Baltimore.

7

○ ○

Organic compounds

7.1   The nature of organic compounds

All organic compounds contain carbon, and nearly all contain hydrogen and have
a least one carbon-hydrogen bond. The simplest organic compounds contain only
carbon and hydrogen and are called hydrocarbons. Alkanes are hydrocarbons in
which C and H are linked by single covalent bonds consisting of two shared electrons. They may exist as straight chains, branched chains, or cycloalkanes, ring-
like structures. The general formula for straight-chain or branched-chain alkanes
is C_nH_{2n+2}, and it is C_nH_{2n} for cycloalkanes. Some alkanes and alkane derivatives
are shown in Figure 7.1. Alkenes, also called olefins, contain double bonds consisting of four shared electrons (e.g., ethylene, $H_2C=CH_2$, Figure 7.2). Alkynes
have triple C bonds consisting of six shared electrons (e.g., acetylene, $HC\equiv CH$).

It is possible for two or more compounds to have the same chemical formula
yet have the constituent atoms arranged in different ways. Different configurations
of similarly formulated compounds are called isomers. For example, structural
isomers result from straight-chain or branched alkanes containing the same number
of carbon atoms (e.g., pentane, 2-methyl butane, Figure 7.1). In cis-trans isomers,
the order of the constituent atoms doesn't change although the spatial relationship
between them is altered (e.g., dichloroethylene, Figure 7.2).

Aromatic or aryl compounds are based on a carbon ring structure, usually the
6C benzene ring, which unusually has strong bonding between adjacent carbon
atoms and a low (1:1) hydrogen to carbon ratio. The bonding between carbon and
adjacent hydrogen atoms may be pictured as an equilibrium state between two
resonant forms shown in Figure 7.3 and conventionally portrayed as a circle within
a hexagon. Although the benzene ring is highly stable, the component hydrogen
atoms can be replaced by individual atoms (e.g., chlorine) or by multiatomic sub-
stituents (e.g., hydroxyl or alkyl groups) (Figure 7.4). Both alkanes and aromatics
often contain other elements such as oxygen (O), nitrogen (N), chlorine (Cl), and
sulphur (S), which are incorporated into specific atomic configurations or functional groups (e.g., hydroxyl, carboxyl, aldehyde, and amine).

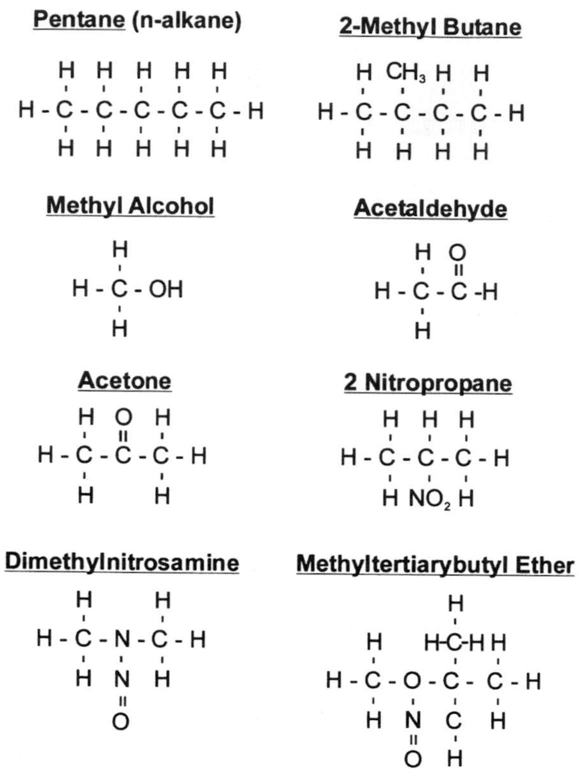

Pentane (n-alkane)

H H H H H
| | | | |
H - C - C - C - C - C - H
| | | | |
H H H H H

2-Methyl Butane

H CH₃ H   H
| | | |
H - C - C - C - C - H
| | | |
H H H H

Methyl Alcohol

H
|
H - C - OH
|
H

Acetaldehyde

H O
| ||
H - C - C -H
|
H

Acetone

H O H
| || |
H - C - C - C - H
| |
H   H

2 Nitropropane

H H H
| | |
H - C - C - C - H
| | |
H NO₂ H

Dimethylnitrosamine

H   H
| |
H - C - N - C -H
| | |
H   N   H
     ||
     O

Methyltertiarybutyl Ether

H
|
H   H-C-H H
| | |
H - C - O - C - C - H
| | |
H   N   C   H
           || |
           O H

Figure 7.1 Structural formulas for alkanes and derivatives.

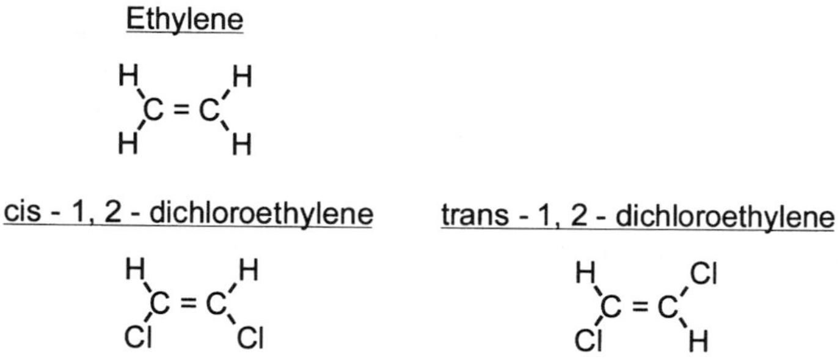

Ethylene

H   H
 \   /
  C = C
 /   \
H   H

cis - 1, 2 - dichloroethylene

H   H
 \   /
  C = C
 /   \
Cl   Cl

trans - 1, 2 - dichloroethylene

H   Cl
 \   /
  C = C
 /   \
Cl   H

Figure 7.2 Examples of alkenes showing cis-trans isomers.

Benzene itself is a colourless highly flammable liquid with carcinogenic properties (see Section 2.3.1) and is used in the manufacture of a wide variety of other compounds including polystyrene plastics, alkyl-benzene surfactants, and pesticides. Substitution of a hydrogen atom by hydroxide forms the aromatic alcohol phenol, which, despite its early use as an antiseptic, is highly toxic in a number of ways. Phenol causes necrosis of skin tissue and is rapidly absorbed through the skin causing renal malfunction, cardiac arrhythmias, and convulsions. Phenol

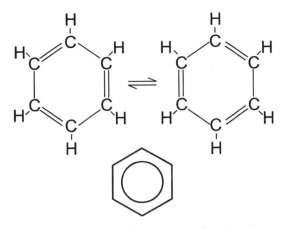

Figure 7.3 Schematic for benzene ring structure.

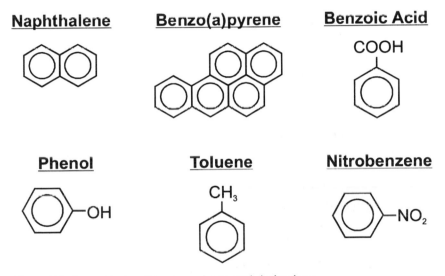

Figure 7.4 Aromatic (aryl) hydrocarbons and derivatives.

continues to be an important industrial product used widely in a variety of chemical syntheses and polymer manufacture. Nitro (NO_2) and chloro (Cl) substitutions on the aromatic ring greatly influence the chemical and toxicological properties of phenolic compounds.

Single-ring aromatic compounds such as benzene, toluene (methylbenzene), ethyl benzene, and phenol are all common environmental contaminants whose lipophilicity and small molecular size facilitate their rapid absorption through biological membranes. Multiring structures, called polynuclear aromatic hydrocarbons, tend to have lower volatility and lower acute toxicity than single-ring aromatics but may exhibit a high degree of chronic toxicity (see Section 7.6).

Halide-substituted hydrocarbons have a wide range of industrial uses. Many lower molecular-weight chlorine-substituted compounds are solvents [e.g., dichloromethane (alkyl), trichloroethylene (alkenyl), and monochlorobenzene (aryl)]. Monochloroethylene (vinyl chloride), a gaseous carcinogen, is the basic raw material for polyvinylchloride plastic. Low-molecular-weight compounds saturated with combinations of Cl and fluorine (F) (e.g., CCl_3F, C_6ClF_5), known as chlorofluorocarbons, have been widely used to manufacture of lightweight foam products and as refrigerants for air conditioning units.

Organooxygen compounds, those with oxygen-containing functional groups, including alcohols, aldehydes, carboxylic acids, epoxides, ethers, and ketones, have an enormous range of properties. Many are readily metabolised by biota, although in certain molecular configurations functional groups such as epoxides may be carcinogenic.

Aldehydes can be formed by the photooxidation of hydrocarbons in polluted air, although some are commonly used in industry. Formaldehyde is the most notorious environmental aldehyde and by far the most common. It accounts for about half the total aldehydes in polluted air and is used to manufacture phenolic resins, as a tissue preservative, and in the processing hides and textiles. Like most aldehydes, it is extremely irritating to mucous membranes, particularly the nose and eyes. Studies with rats have implicated formaldehyde as a nasal carcinogen, although extrapolation to other species seems uncertain, and no direct evidence is available for humans. Nevertheless, formaldehyde has been described as a probable human carcinogen.

Esters of the carboxylic acid, phthalic acid (Figure 7.5), are truly ubiquitous environmental contaminants. Widely used as plasticisers in the manufacture of polyvinyl chloride, they may constitute up to 50% by weight of some products such as PVC tubing and are very common contaminants of any substance found in plastic containers, tubing, and the like. As such, they appear in many chromatograph traces, sometimes in surprisingly high concentrations. In 1985, 0.8 million tonnes of phthalate esters were produced in the United States, approximately one quarter of the world's production. About 40% of this total ends up in refuse dumps each year, by far the largest reservoir of these compounds. Approximately 1% of this production is vaporised. For many years, phthalate esters were considered relatively harmless as environmental contaminants, although their recent association with the peroxisome proliferase receptor and their implication as possible hormone disrupters have renewed interest in this common class of compounds (see Section 7.5).

Most ethers have minor importance as environmental contaminants, although some concern has been expressed over chlorinated derivatives of diphenyl ether, which have been used as heat exchangers and are by-products of chlorophenol production. Polybrominated diphenyl ethers are used as fire retardants. Methyltertiarybutyl ether (Figure 7.1) is the principal substitute for tetraethyl lead as an anti-knock additive in gasoline.

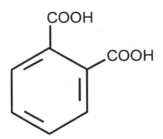

Figure 7.5 Phthalic acid.

Most nitro (NO_2-substituted) compounds of environmental concern are aromatic derivatives (e.g., nitrobenzene trinitrotoluene, TNT). Dinoseb (4,6-dinitro-2-*sec*-butylphenol) is a substituted phenolic compound used as a herbicide and has been implicated in causing birth defects through maternal exposure early in pregnancy and sterility in exposed men.

Nitrosamines are a group of *N*-nitroso compounds (NOCs) of particular environmental significance in that most have been found to be carcinogenic. They are formed by the reaction of nitrites with secondary and some tertiary amines. Nitrites originate as food preservatives or via nitrate from drinking water and some foods. The best studied nitrosamine, dimethylnitrosamine (Figure 7.1), is formed by the reaction of nitrite with demethylamine, an important compound used in soap and rubber manufacture.

7.1.1 Behaviour and transport

Organic chemicals enter the environment in a variety of ways and in a variety of forms. In environmental toxicology, we are particularly concerned with their bioavailability and their toxicity, although having information on their sources, mode of release, and means by which they are transported and partitioned among different compartments of the ecosystem is important to understand the epidemiology associated with toxic effects. Organic compounds may be released into the environment as solvents and by-products of a variety of manufacturing processes including fuel production. Agrochemicals and other pest control products are deliberately released into the environment either to promote production or to selectively destroy competitors and consumers of favoured species.

Size and shape of organic molecules and features such as the degree of chlorination will affect fate, transport, bioaccumulation, and toxicity in complex, interactive ways. Particularly over the last 10 years, it has been determined that relatively small organic molecules with low water solubility and high vapour pressure may be transported in the atmosphere. Several studies have documented the presence of organochlorines in snow, ice, and fauna of the Canadian Arctic, far removed from potential sources (e.g., Bidleman et al., 1989; Gregor and Gunner, 1989) and in moss and lichens from the Antarctic (Focardi et al., 1991). Exchange

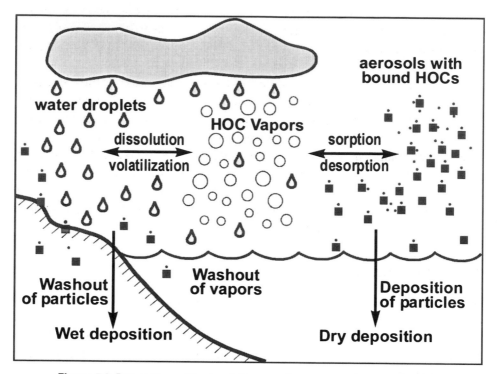

Figure 7.6 Deposition of hydrophilic organic compounds onto the Earth's surface. After Poster and Baker (1997).

of polynuclear aromatic hydrocarbons and polychlorinated biphenyls between water and atmosphere has been demonstrated to play a major part in determining the input of these chemicals to water bodies such as the Great Lakes and Chesapeake Bay (Baker et al., 1997). Figure 7.6 illustrates atmospheric depositional processes that influence the deposition of natural and anthropogenic hydrophilic organic compounds onto the Earth's surface. The degree of chlorination affects the persistence of organic compounds in the environment since more highly chlorinated compounds tend to be more refractory. The pattern of chlorination in PCBs may also be critical in determining their differential partitioning between water, sediment, and biota as well as their toxicity (see Section 7.3). Generally speaking, small organic molecules with three or fewer benzene rings (e.g., dichlorodiphenyltrichlorethane, DDT; PCBs) exhibit acute toxicity associated with their ability to disrupt membranes as a result of their planar shape. Larger molecules such as PAHs have low acute toxicity but may share more chronically toxic, sometimes carcinogenic, properties with smaller planar molecules such as dibenzodioxins and dibenzofurans.

The nonpolar nature of many organic compounds confers on them hydrophobic properties that cause them to become associated with particulate matter, although the strength of this association may, again, be dependent on the degree of chlorination. In the aquatic environment, particulate matter is operationally defined as that which is trapped by a 0.45-μm filter, although, in fact, it is composed of an enormous size range of both organic and inorganic material (e.g., clays, oxides, decomposing plant and animal debris, and live algae) having a wide range of sorptive sites or phases. It may take the form of suspended solids or bulk sediment. Important processes influencing the distribution and transport of organic contaminants in the aquatic environment are the exchanges occurring between particulates and interstitial water, exchanges occurring between suspended particles and open water, and exchanges at the sediment-water interface (see Section 5.3.6). The association of many organics with nonaqueous phases means that food chain transfer plays a major role in their bioaccumulation, and some compounds have been found in increasing concentrations at higher trophic levels. This phenomenon is called biomagnification (see Section 6.2.6).

7.2   Pesticides

Worldwide, an estimated 5 billion tonnes of pesticides are applied annually. Although the majority are used in agriculture, bactericidal and fungicidal agents are also employed in heavy industry where they are added to cooling fluids used with metal machining operations. Agricultural pesticides may be broadly categorised as herbicides (70% of total by weight), insecticides (20%), and fungicides (10%). Since 1945, synthetic pesticide use in the United States has increased 30-fold, and the toxicity and effectiveness of pesticides has increased by as much as 20-fold during the same period. The annual cost of pesticide application in the United States in 1994 was approximately $4 billion, resulting in an estimated saving of $12 billion in crops that would otherwise have been destroyed.

Between 1964 and 1998, wheat production in India increased approximately fivefold in the face of a doubling of arable land. Much of this increased efficiency is attributed to the use of chemical pesticides, and similarly impressive figures are available from many other parts of the world.

The domestic market for pesticides is also very large, particularly in developed countries. Very little is known about the contribution of nonagricultural uses of pesticides and fertilisers to surface water runoff, although as much as one third of all pesticide use may be for nonagricultural purposes.

Most insect and related anthropod species are beneficial and play essential roles in natural ecosystems. Only 1% (about 9,000 species) are pests of humans and agriculture and are responsible for the worldwide destruction of about 13% of food and fibre crops. Negative aspects of pesticide use have focused on their persistence in the biosphere and their chronic toxicity to nontarget species such as birds. Subtle, but highly important effects from the agricultural standpoint concern the destruction of species would otherwise exercise biological control over pests and of insect pollination vectors. In the United States, approximately $1 billion in crops are ruined each year through the overapplication or inappropriate use of pesticides, and many countries are currently adopting aggressive initiatives designed to reduce pesticide usage while maintaining crop yields. For example, between 1986 and 1998, Sweden achieved a 75% reduction in annual average agricultural pesticide use compared with the period from 1981 to 1985.

Most synthetic agricultural insecticides used since the Second World War may be categorised as chlorinated organics, organophosphates, carbamates, and pyrethroids. Several well-known compounds that have been determined to be hazardous to health have been banned from domestic use in recent decades. However, many are still exported. For example, under current U.S. law, there are no restrictions to the foreign sale of toxic chemicals. Legal action may still arise, however, if the hazard is insufficiently communicated. Sterility occurred in banana plantation workers in Costa Rica resulting from handling the nematocide dibromochloropropane (DBCP) without suitable protective clothing. Lawsuits charged the American manufacturer with improper labelling including labelling only in English.

7.2.1   Chlorinated organics

DDT

Of all the enormous variety of chlorinated organic pesticides produced in the last 60 years, none has received as much notoriety as DDT. Dichlorodiphenyltrichloroethane or 2,2-bis-(p-chlorophenyl)-1,1,1-trichloroethane was first synthesised in 1874 but was not marketed as an insecticide until 1942 when it was widely used by the U.S. Army as a fumigant for typhus and malaria prevention. In later years, its use was expanded by the World Health Organisation to combat a variety of insect-borne diseases.

DDT is a small planar molecule, which may be metabolised to DDD [1,1-dichloro-2,2-bis(p-chlorophenyl)ethane] or DDE [1,1-dichloro-2,2-bis(p-chlorophenyl)ethene] through mixed function oxidase activity (Figure 7.7). DDD itself has been marketed as a pesticide. All these compounds act as sodium channel blockers, although they produce a variety of physiological effects. For example, DDT was the first pesticide discovered to have estrogenic properties, and, like many similar compounds, it can initiate a variety of reproductive effects. Early reports of songbird mortality following the use of DDT to control Dutch elm disease (Hickey and Hunt, 1960) were followed by widespread records of egg-shell thinning and adverse effects on reproductive performance, particularly in raptors (Hickey and Anderson, 1968). Many of these effects were associated with DDE, which is a relatively stable DDT metabolite and has itself been used as a pesticide. Reproductive failure in the brown pelican (*Pelicanus occidentalis*) has been attributed to DDE. Such symptoms, together with the persistence of DDT metabolites, led to their ban in the United States, Canada, and many other countries in the early 1970s, although, even before its ban, they were being slowly rendered redundant because of the development of genetic resistance in several target organisms (Dover and Croft, 1986). Methoxychlor was developed as a substitute for DDT and was

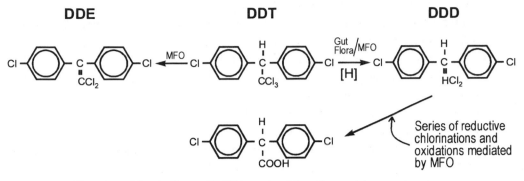

Figure 7.7 Metabolism of DDT by mixed function oxidase system.

designed to be rapidly metabolised to harmless by-products but was subject to the same genetic resistance as developed toward DDT.

Despite its widespread ban in several countries, DDT remains the insecticide of choice for malaria control in many tropical regions from Mexico to Vietnam in spite of the fact that its use is illegal in some of these areas. As recently as 1993, DDT residues in excess of $100\,mg\,kg^{-1}$ were reported in dried fish taken from rice fields in Bangladesh. In most areas, DDT residues in fish, birds, and mammals have shown a decline since the 1970s. One issue of concern in bioaccumulation studies has been the identification of possible sources of the parent compound. Because of the rapid breakdown of DDT relative to DDE, the ratio of these two compounds is often used as an indicator of recent exposure. For example, a DDE:DDT ratio of <50 is thought to indicate significant concentrations of DDT relative to DDE and, therefore, recent exposure to the parent compound.

Even though DDT and its derivatives are increasingly regarded as being of historical interest only, their persistence in the environment may still cause problems, often in conjunction with other refractory chlorinated pesticides and PCBs (Section 7.3). For example, Case Study 4.2 dealt with a variety of pathological conditions seen in benthic fish exposed to DDT- and PCB-contaminated sediments. Prominent among concerns are reproductive failure and carcinogenesis, and such fears are multiplied when mammalian populations are affected. One such population that serves as an interesting case study is the St. Lawrence estuary Beluga whale population, which shows signs of multiple symptoms perhaps resulting in part from exposure to several different chlorinated compounds concentrated in the sediments of their native habitat (Case Study 7.1).

CYCLODIENES

The cyclodienes aldrin, dieldrin, and heptachlor (Figure 7.8) as well as chlordane and endrin were developed largely for control of the Japanese beetle and fire ants in the late 1950s and early 1960s, although endrin was also used as a rodenticide. Like DDT, they resulted in adult bird mortalities, and, to a lesser degree, they were

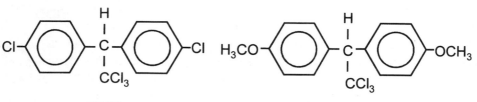

DDT   **Methoxychlor**

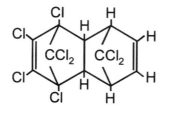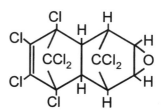

Aldrin   **Dieldrin**

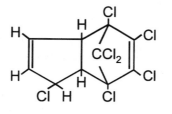

Heptachlor

Figure 7.8 Some chlorinated pesticides.

implicated in egg-shell thinning. Heptachlor, which was widely used as a seed treatment, was responsible for mortalities among nontarget mammalian species such as foxes through food chain accumulation involving prey species such as pigeons. Additionally, cyclodienes have been implicated as carcinogens, and restrictions in their manufacture and use during the 1970s and early 1980s have largely removed them from the food chain. However chlordane, represented by several isomers and metabolites, was still used for subterranean termite control until its deregistration in 1987. It remains in common use in Latin America and Africa for termite and ant control and may be used in Ontario, Canada, by special permit. Technical chlordane is a mixture of chlordane isomers and nonachlor isomers that form oxychlordane as a metabolite in biota. Heptachlor is still used in several South American countries, notably Argentina and Brazil.

HEXACHLOROCYCLOHEXANE

Hexachlorocyclohexane HCH (Figure 7.9) was developed as a replacement for heptachlor as a seed treatment. Technical HCH consists of several stereoisomers

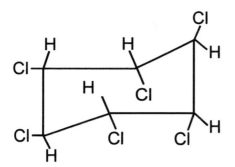

Figure 7.9 Hexachlorocyclohexane (Lindane).

of which the dominant insecticide is the gamma form, lindane, still widely used in North America. Its classification as a carcinogen has restricted its use around poultry and human habitation, although it is rapidly metabolised to harmless water-soluble products and remains an effective insecticide with toxic properties similar to those of DDT.

Technical HCH and lindane are still heavily used in China, India, and elsewhere in the developing world. They are a major contaminant in cold Arctic regions, which seem to act as condensation traps for compounds having intermediate volatility.

MIREX AND CHLORDECONE (KEPONE)
These two heavily chlorinated compounds with a boxlike structure (Figure 7.10) were marketed for the eradication of fire ants in the southern United States. Discharge of 53,000 kg of Kepone into the James River, Virginia, during the 1970s and the development of neurological symptoms among workers at the Hopewell Kepone manufacturing plant led to extensive toxicological studies and concern over the persistence of the compound. The plant was abruptly closed in 1975, and the incident proved to be a test case for the newly enacted (1972) amendments to FIFRA and highlighted flaws in occupational health regulation. Allied Chemical Company, which was held liable in the case, cancelled Kepone production in 1977.

Widespread Mirex and photomirex contamination of Lake Ontario probably resulted from production at the Hooker Chemical Plant on the Niagara River until the early 1970s (see Case Study 7.1).

PYRETHROIDS
Extracts of dried chrysanthemum or pyrethrum flowers containing the insecticidal esters pyrethrins, jasmolin, and cinerin have been used for centuries as insecticides, although the insecticidal components pyrethrin I and pyrethrin II were not isolated and characterised until 1924 (Figure 7.11). The most important source of natural

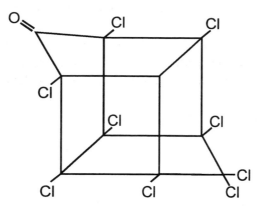

Kepone (Chlordecone)
(Mirex has 2 Cls in place of =O)

Figure 7.10 Chlordecone and mirex.

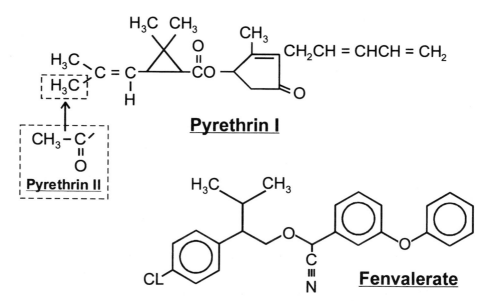

Pyrethrin I

Pyrethrin II

Fenvalerate

Figure 7.11 Natural pyrethrins and fenvalerate (an artificial pyrethroid).

pyrethrins is Kenya, which produced half of the world's supply in 1965. In recent years demand far outstripped supply and led to the development of a family of synthetic pyrethroids (e.g., fenvalerate; Figure 7.11). Pyrethroids are neurotoxins, causing a variety of neurophysiological symptoms including repetitive neuronal discharges, similar to the action of DDT, and disturbances in synaptic transmission. Pyrethroid esters such as permethrin have also been shown to inhibit Ca-Mg

ATPase, resulting in an increase in intracellular calcium in neurones and enhanced neurotransmitter release.

These properties combine to elicit rapid knock-down and mortality in insects, with the result that these compounds are in common use as agricultural insecticides on crops such as cotton and are extremely popular as domestic and horticultural insecticides. They currently account for nearly half of worldwide insecticide usage. Their popularity as household insecticides is enhanced by their low acute and chronic toxicity to mammals. Pyrethroids are rapidly metabolised by the mixed function oxidase system, and certain formulations contain MFO inhibitors such as piperonyl butoxide to potentiate the action of the insecticide. Although there have been isolated reports of neurological symptoms developed by workers through occupational exposure to pyrethroids, these have resulted from inappropriate handling and application.

7.2.2   Organophosphate pesticides

The first organophosphate pesticide, tetraethylpyrophosphate (TEPP) was developed in Germany during the early 1940s and was soon followed by parathion in 1944. Parathion (Figure 7.12) is a phosphorothionate ester, which, after ingestion, is metabolically converted (activated) to paraoxon by the substitution of oxygen for sulphur. Paraoxon, a structural analogue of acetylcholine, attaches to the active site of the enzyme acetylcholinesterase. Unlike acetylcholine, which binds to the enzyme in a reversible manner, paraoxon forms a stable covalent complex from which it is difficult to regenerate the original enzyme (Figure 7.13). Irreversible binding to acetylcholine makes compounds such as paraoxon extremely toxic to nontarget organisms such as birds and mammals. Their toxicity is exacerbated by their ability to be absorbed through the skin. These properties have been utilised in the development of nerve gases such as sarin, which is lethal at doses as low as $0.01\,mg\,kg^{-1}$ and was used to deadly effect in the Japanese subway bombing incident in 1994.

It is, therefore, important to develop organophosphorus compounds as pesticides that can be metabolised by mammals to less toxic forms. The best known example is the phosphodithioate insecticide malathion, which is hydrolysed to relatively harmless derivatives by a carboxylation reaction. Another example of an organophosphate capable of being inactivated by mammalian metabolism is dichlorvos. This pesticide has a structure similar to paraoxon and is used in pesticide strips because of its relatively high volatility.

It is generally assumed that organophosphate pesticides degrade sufficiently quickly in the environment to pose little threat to ecosystems, although several cases of unintentional poisoning have been reported. In Essex, England, a spill of the organophosphate chlorpyrifos killed all the arthropods and 90% of the fish over a 37 km stretch of the Roading River (Raven and George, 1989). Determination of brain acetyl cholinesterase (AChE) activity has been widely used to diagnose and evaluate the effects of organophosphate in birds, and to a lesser extent in

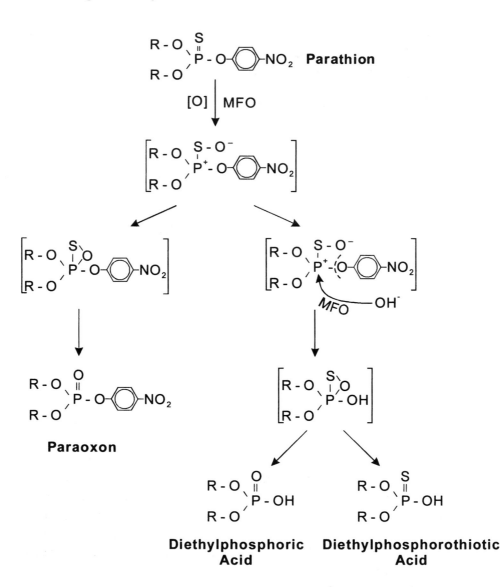

Figure 7.12 Intoxication (activation) and detoxification (deactivation) of parathion via mixed function oxidase system.

amphibians, reptiles, small mammals, and fish. In birds, a reduction of AChE of 50% is generally regarded as life-threatening. Reproductive effects in birds have been traced to a decrease in luteinising hormone and steroid hormones.

7.2.3   Carbamate pesticides

The carbamate pesticides are derivatives of carbamic acid and, like organophosphate pesticides, they act as acetylcholinesterase inhibitors. However, they have a

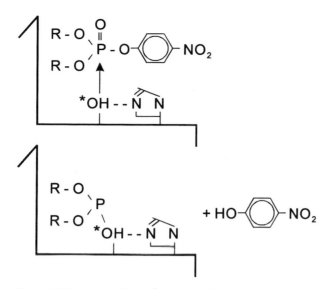

Figure 7.13 Inactivation of acetylcholinesterase by paraoxon.

much lower dermal toxicity than most organophosphates and are relatively easily degradable. Two carbamate pesticides in common usage on grasses and other plants are carbaryl and carbofuran (Figure 7.14). In Canada and the United States, both are commonly employed to control grasshopper populations in the western plains. As with organophosphates, instances of accidental bird poisonings, resulting in convulsion and reproductive failure, have been reported.

In many passerine species, carbofuran has an LC_{50} close to $1 \, mg \, kg^{-1}$ body mass. Preening, ingestion of poisoned insects, and inhalation are all significant sources. Young birds may be up to $100\times$ more sensitive than adults. Although some field deaths have been reported for gulls, owls, and pigeons, the lethal dose for most species generally exceeds that which might be expected to occur through the ingestion of contaminated food.

Organophosphate and carbamate pesticides can also affect immune system competence. Malathion, methyl malathion, parathion, and dichlorvos have all been reported to suppress humoral immune response in laboratory mice, and other compounds produce changes in autoimmune function such as hypersensitivity and sensitisation.

7.2.4 Phenoxyacid herbicides

2,4-Dichlorophenoxyacetic acid (2,4-D) and 2,4,5-trichlorophenoxyacetic acid (2,4,5-T) (Figure 7.15), along with their esters and salts, have been used in several formulations of herbicide as plant growth regulators. These compounds are moderately water-soluble, are readily degraded to tricarboxylic acids in the environment, and pose relatively few environmental problems in normal application.

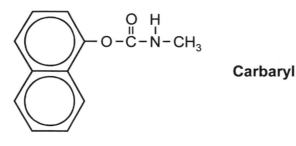

Carbaryl

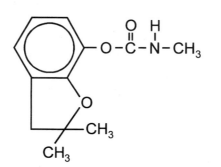

Carbofuran

Figure 7.14 Carbamate pesticides.

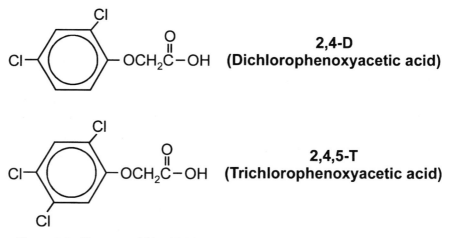

2,4-D
(Dichlorophenoxyacetic acid)

2,4,5-T
(Trichlorophenoxyacetic acid)

Figure 7.15 Phenoxyacid herbicides.

Their most notorious usage was as Agent Orange, an approximately 1:1 mixture of 2,4-D and 2,4,5-T, which was employed in large quantities as a defoliant by the American military in Vietnam during the late 1960s and early 1970s.

Silberner (1986) reported a link between the handling of 2,4-D by farmers and the incidence of non-Hodgkin's lymphoma, a finding that appears to be supported by more recent evidence (Hardell et al., 1994). Occupational exposure to

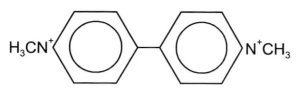

Figure 7.16 Paraquat, a bipyridilium herbicide.

2,4-D was also implicated as a cause of necrosis and abnormal development of sperm in farm workers (Lerda and Rizzi, 1991). Carcinogenicity and mutagenicity associated with exposure to phenoxyacid herbicides is thought to be associated with the low concentration of tetrachlorodibenzodioxin (TCDD), which is a by-product of the 2,4,5-T manufacture, rather than the phenoxyacids themselves. The evidence remains controversial, however, and this is reflected in the litigation surrounding awards made to Vietnam veterans who were exposed to Agent Orange. The implication of 2,4,5-T in TCDD production has led to the banning or strict control of this compound in most developed countries, whereas 2,4-D continues in widespread use as a domestic and agricultural pesticide. The main epidemiological evidence concerning 2,4,5-T and the high prevalence of cancer (e.g., soft tissue sarcomas) comes from studies of forestry workers in Scandinavian countries.

7.2.5 Bipyridilium herbicides

The two most important members of this group of compounds are diquat and paraquat (Figure 7.16). Paraquat, which was registered for use in 1965, is still the most widely used postemergent herbicide in Latin America, the Caribbean, and many other tropical areas. Bipyridilium compounds quickly destroy plant cells and have a high binding affinity for inorganic minerals such as clay, which rapidly reduces their bioavailability. This property enables rapid replanting following their application.

Accidental exposure to paraquat has resulted in hundreds of human deaths. Ingestion, inhalation, or dermal exposure may result in severe toxicity and death within 24 hr. Symptoms include necrosis of the lung, liver, kidneys, adrenals, and heart. Lungs are particularly susceptible even with nonpulmonary exposure. Survivors of paraquat poisoning often develop progressive pulmonary fibrosis within 5 to 10 days of exposure.

7.2.6 Triazine herbicides

The herbicidal properties of the *s*-triazines were discovered during research begun in 1952 by the J. R. Geigy Chemical Company of Basel, Switzerland. Since then, several major products including atrazine, simazine, cyanazine, and prometryn have been developed for the control of broad-leaved annual and perennial weeds

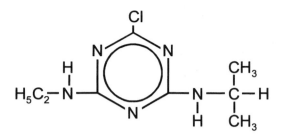

Figure 7.17 Atrazine, a triazine herbicide.

in corn (>80% of use), sorghum, and sugar cane. Atrazine (Figure 7.17) accounts for more than 75% of triazine usage, and in 1994 approximately 30×10^6 kg of atrazine were applied in North America, principally in the midwestern United States.

Atrazine enters plants through both roots and leaves and acts by inhibiting photosynthesis via the blockage of electron transport in photosystem II. This leads to the destruction of chlorophyll and an inhibition of carbohydrate synthesis. Because of its highly specific mode of action, atrazine and other triazines are regarded as toxic threats to nontarget plants rather than animals, and this is borne out by the relatively high acute LC_{50}s seen in most animal species tested.

7.3   Polychlorinated biphenyls

7.3.1   Chemistry and effects

Polychlorinated biphenyls are small planar molecules consisting of two benzene (phenyl) rings linked by a carbon-carbon bond and having chlorine atoms substituted at one or more of the ten available carbons in the molecule (Figure 7.18). They theoretically comprise 209 isomers and congeners. They have relatively low water solubility and high lipid solubility. These properties increase with the degree of chlorination. Octanol-to-water coefficients increase from a $\log K_{OW}$ of 4.6 for monochloro forms to a $\log K_{OW}$ of 8.4 for the decachloro form. This hydrophobicity also means that they rapidly and strongly adsorb to sediments. These characteristics dictate that particulate ingestion and food transfer play major roles in their bioaccumulation, and that they have a high capacity for biomagnification as they are passed to higher trophic levels.

By 1976, worldwide production of PCBs totalled approximately 6×10^8 kg. Over 90% was manufactured in the United States by the Monsanto Corporation under the trade name Aroclor. In Japan, they were marketed as Kanechlor, and in Germany, as Clophen. Aroclor mixtures were given numerical designations which indicated the percent chlorine in the formulation. For example, Aroclor 1232 had 32% chlorine. Initially employed largely as plasticisers, their use increased dramatically in the 1960s, particularly in the electrical industry. Because of their high

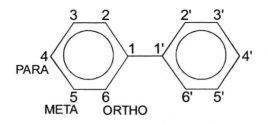

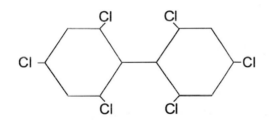

3, 3', 4, 4', 5 -PENTACHLOROBIPHENYL

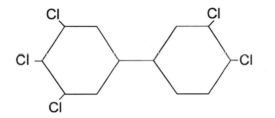

2, 2', 4, 4', 6, 6'-HEXACHLOROBIPHENYL

Figure 7.18 Structure and nomenclature of PCBs.

boiling point, they were in great demand as cooling fluids for high-temperature processes. Their high dielectric constant and electrical resistivity made them ideal for heat transfer in circuitry and electronics, and their chemical stability contributed substantially to the low maintenance of capacitors and transformers.

During the 1960s evidence that directly linked PCBs to a number of human and environmental health problems began to accumulate. In 1966, the Swedish chemist Jensen discovered PCBs as trace contaminants in chromatograms of environmental samples being screened for DDT. Since then, they have been detected in almost every kind of living organism, even in abyssal fish collected from the Arctic Ocean. In 1968, attention was focussed on an outbreak of PCB-related poisoning, which became known as Yusho (rice oil disease). A leak in a factory heat exchanger resulted in rice oil contamination with Kanechlor 400, a 48% chlorinated biphenyl. The contaminated oil was used for food preparation by more than 1,200 victims. The chief symptom was chloracne, skin eruptions usually found on the face and back. Adipose tissue from patients contained up to $75\,mg\,kg^{-1}$ of PCBs. Other clinical signs included respiratory and chronic bronchitis-like symptoms, decrease in conduction velocity in peripheral sensory nerves, diminished growth in young males and pigmented skin of newborn babies. During the same year, chicken feed prepared with contaminated rice oil killed over 400,000 chickens. The chickens experienced laboured breathing, droopiness, and decreased egg production.

In the early 1960s, mink ranchers in the north central United States and Canada began to notice reproductive complications and excessive newborn mortality among their stock. By 1967, an 80% increase in newborn mortality was found; this increase showed a significant correlation with the percentage of coho salmon in the mother's diet. PCBs were implicated as the cause of the excessive newborn mortality for several reasons: (1) high PCB residues in Lake Michigan coho salmon, (2) the sensitivity of mink to PCBs determined from controlled experiments, and (3) bioaccumulation of PCB residues in mink tissue. Subsequent investigations demonstrated substantial contamination of food chains associated with the Great Lakes. A characteristic of many of these studies was the very high degree of biomagnification of PCB achieved, particularly at the lower end of the food chain. Relative to ambient water Safe (1980) recorded bioconcentration factors of 2×10^4 for plankton, 2×10^5 for catfish, and 7×10^7 for herring gulls from the Lake Ontario ecosystem.

In addition to the Laurentian Great Lakes ecosystem, there are several, well-documented instances of PCB pollution at specific sites. Notable among these is the case of PCB contamination of the upper Hudson River resulting from the manufacture of electrical equipment; 0.8 million kg of PCBs were released into the river over two decades through discharges regarded as acceptable industrial practice at the time. Most of this material was trapped locally in sediments until the removal of a dam at Fort Edward, New York, in 1973, which resulted in the contamination of much of the river. Discovery of PCB levels in some fish in excess of $300\,\mu g\,kg^{-1}$ led to the closure of the striped bass (*Morone saxatilis*) fishery in the Hudson River. The incident also prompted the passage of the U.S. Toxic Substances Control Act of 1976.

Although the uses of PCBs were restricted in the United States in 1974, and their production was halted altogether in 1977, large quantities remain in industrial use. Recent declines in PCB levels have been reported in fish from the United States and Europe, although their persistence in the environment still poses problems in several areas. Despite extensive information on health problems associated with PCB ingestion and PCB's implication as both a carcinogen and a promoter, toxicological evidence has always been complicated by the fact that investigators are always dealing with a variable mixture of compounds having different properties. It is now widely assumed that some of the toxicity historically attributed to PCBs was due to trace amounts of contaminants such as polychlorinated dibenzo-*p*-dioxins (PCDDs) and dibenzofurans (PCDFs). These contaminants result from partial oxidation of PCBs during their manufacture and use. For example, Yusho rice oil was found to have a relatively high degree of contamination with PCDFs. Recent improvements in analytical methods have also permitted detailed study of relatively few PCB congeners, specifically coplanar molecules, which, it is now assumed, may be responsible for most PCB toxicity. It has been shown that PCBs having four or more chlorine atoms at both para and meta positions in the biphenyl rings, but no chlorine atoms in the ortho positions, attain isostereomerism with PCDDs and PCDFs. Hence, some coplanar PCBs (e.g., 3,3′,4,4′,5-pentachlorobiphenyl and 3,3′,4,4′,5,5′-hexachlorobiphenyl share similar properties with PCDDs and PCDFs, both in terms of the symptoms elicited and their ability to induce microsomal cytochrome P450 enzymes of the 1A family (see Section 7.9).

7.3.2   Evidence of decline in environmental PCBs

Aerial transport of PCBs has always been an important vector for the dispersal of these and other hydrophobic organic contaminants. In 1983, Thomann and DiToro reported atmospheric PCB input to Lake Superior to be as high as 7,550 kg yr^{-1}. Historical measurements of PCB concentrations in the atmosphere over the Great Lakes region and the Chesapeake were in the 1–3 ng m^{-3} range, although recent atmospheric measurements in both these regions indicate a tenfold decrease in atmospheric PCBs over the last 15 years (Leister and Baker, 1994). Between the late 1960s and late 1980s, atmospheric PCBs in rural England declined by a factor of 50 (Jones et al., 1992).

Since 1975, in Canada, the Ontario Ministry of the Environment has used young-of-the-year spot tail shiners (*Notropis hudsonus*) as biomonitors for near-shore waters of the Great Lakes. The near-shore environment is regarded as best reflecting anthropogenic chemical input, and the spot tail shiner is an important forage species in these areas. A 1990 survey showed that, of 36 sites that had been sampled since the mid 1970s, 16 showed a significant decline in PCB tissue residues with the most pronounced drop occurring during the late 1970s. Nevertheless, 15 of the 37 sites samples in 1990 had PCB concentrations in whole fish exceeding the International Joint Commission (IJC) aquatic life guideline (100 ng g^{-1}). The highest concentrations (>1,500 ng g^{-1}) were found in fish downstream from aluminium and auto manufacturing plants on the St. Lawrence River, although several sites in Lake Ontario and the Niagara River also had elevated levels. Among the latter was Cayuga Creek, site of the abandoned Love Canal hazardous waste site.

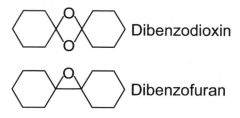

Figure 7.19 Dibenzodioxin and dibenzofuran.

7.4 Dibenzodioxins and dibenzofurans

Polychlorinated dibenzo-*p*-dioxins and dibenzofurans are a group of toxic halogenated aryl hydrocarbons (HAH) formed as by-products in a variety of manufacturing and incineration processes. They have the general formulae shown in Figure 7.19. The number of possible chlorine substitutions on the dioxin molecule may vary from one to eight, giving a possible 72 positional isomers. Substitution at the lateral (2, 3, 7, and 8) positions is associated with a particularly high degree of toxicity, and much of the available toxicological information derives from studies of the 2,3,7,8 tetrachloro-*p*-dioxin (TCDD) isomer.

Historically PCDDs and PCDFs were notable as by-products in the manufacture of PCBs and chlorophenol herbicides such as 2,4,5-T. In 1970, as much as one half of the environmental release of these compounds came from the use of chlorophenol pesticides and herbicides, and their massive application as Agent Orange in the Vietnam conflict in the late 1960s and early 1970s formed the basis of ongoing controversy over the effects on exposed Vietnam veterans and civilians. Currently, the principal sources of PCDD and PCDF emissions are from medical and municipal waste incineration. Other important sources include steel mills, metal processing plants, and the industrial burning of polychlorinated phenol-treated wood chips. PCDDs and PCDFs are also formed in the pulp and paper industry through the bleaching of wood pulp with chlorine (Section 9.3.3). The release of pulp mill effluent contributes approximately 5% of the total PCDD and PCDF emissions in the United States and has prompted concern over the bioaccumulation of these compounds in fish inhabiting receiving waters. Ingestion of contaminated fish and shellfish is regarded as a major source of PCDDs and PCDFs for consumers such as birds and mammals.

In vertebrates, PCDDs, PCDFs, and related HAHs cause a broad spectrum of symptoms including weight loss; reproductive failure; thymic atrophy; hepatotoxicity; porphyria, chloracne, and other types of dermal lesion; immunotoxicity; and induction of phase I and II xenobiotic-metabolising enzymes. Through the mediation of the Ah (aromatic hydrocarbon) receptor, TCDD influences DNA transcription and alters the function of both regulatory and structural genes. Regulation of MFO (P450) activity via the Ah receptor is shown in Figure 7.20 and described in greater detail in Section 7.8.2. The overall results include (a) increased

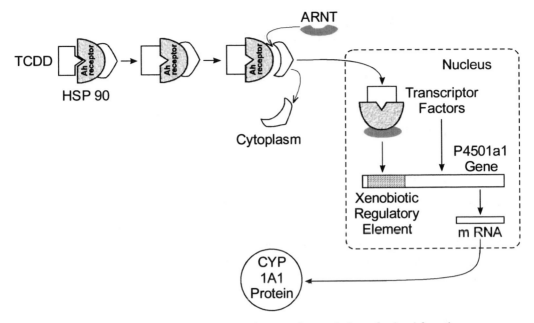

Figure 7.20 Role of Ah receptor in genetic regulation of mixed function oxidase enzyme CYP1A1. ARNT, Ah receptor nuclear translator; HSP90, heat shock protein 90 (MW).

cytochrome P450 and cellular metabolism, (b) alteration of membrane growth factors and cell differentiation, and (c) increased incidence of mutagenesis. Although it is suspected that many of the aforementioned symptoms result in some way from the action of TCDD in tandem with the Ah receptor, the exact mechanism is poorly understood. For example, the powerful antioestrogen activity of the TCDD-Ah complex is thought to result from interference with estrogen production via the estrogen receptor (ER), which is analogous to the Ah (see Section 7.5.2). Safe et al. (1991) suggested at least five alternative pathways for this inhibition to occur.

It is now understood that a broad range of polychlorinated compounds having more or less similar structure to TCDD may bind with the Ah receptor (Figure 7.20) and cause a variety of adverse effects. These compounds include polynuclear aromatic hydrocarbons (PAHs), PCBs, polychlorinated naphthalenes, polyhalogenated diphenyl ethers, polychlorinated anisole, polychlorinated fluorenes, and polychlorinated dibenzothiophenes. This has led to a structure activity approach to assigning toxicity of complex mixtures of such compounds. Toxicity is expressed as TCDD equivalents (TCDD-EQ), also referred to as toxic equivalency factors (TEFs), which are used to calculate the Ah receptor activity contributed by individual compounds in a mixture. This model has been demonstrated to predict the degree of enzyme induction, embryo mortality, or teratogenicity accurately (Safe, 1990).

7.5   Organic chemicals as environmental estrogens (endocrine disrupters)

7.5.1   Rationale

A sustained increase in breast cancer over the last several decades has prompted recent concern over the ability of low-molecular-weight (usually chlorinated) organic compounds to mimic the effects of estrogens in mammals and other vertebrates. About 40% of all cancers in women are mediated by hormones, and a large body of experimental and epidemiological evidence indicates a pivotal role for estrogen in breast cell proliferation and the hypertrophy of other secondary sex organs. The incidence of abnormal sexual development in other vertebrate groups including birds (Fry, 1995), reptiles (Guillette et al., 1994), and fish (Jobling and Sumpter, 1993) has also been interpreted as evidence for the presence of so-called environmental estrogens or xenoestrogens. More recently, the more conservative term *endocrine disrupter* has been employed and seems a more appropriate descriptor.

It has long been known that certain natural compounds have hormone-like properties (Colborn et al., 1993), and the presence of weakly estrogenic compounds in plant foodstuffs and in essential oils has also been demonstrated. In synthetic organic chemicals, too, hormonal activity has been recognised for some time. As early as 1949, it was known that crop dusters handling DDT had reduced sperm counts (Patlak, 1996). Bitman et al. (1968) demonstrated that DDT produced characteristic estrogen responses in the reproductive tracts of rats and birds, and in a more recent study, a decrease in the alligator population in Lake Apopka, Florida, was linked to a previous spill of DDT and dicofol (Guillette et al., 1994).

There is now a long list of chemicals implicated as potential endocrine disrupters including triazine herbicides, several PAHs, cyclodienes, phenoxyacids, hydroxy-PCBs, and dioxins. More recent additions to this inventory are plasticisers (e.g., bisphenol A, phthalate esters), surfactants (e.g., alkyl phenol polyethoxylates), and solvents, indicating a large structural diversity in such chemicals. This diversity, in turn, would indicate that endocrine disruption may operate through diverse mechanisms. However, many of these compounds or their primary oxidative metabolites possess a common structural relationship to the phenolic A ring in estradiol (Figure 7.21).

Endocrine-disrupting compounds are believed to exert their influence by (1) mimicking the effects of endogenous hormones such as estrogens and androgens, (2) antagonising the effects of endogenous hormones, (3) altering the pattern of synthesis and metabolism of normal hormones, and (4) modifying hormone receptor levels (Soto et al., 1995).

7.5.2   Proposed mechanism for the action of estrogenic compounds

Gonadotropin-releasing hormone released from the hypothalamus stimulates release of luteinising hormone (LH) and follicle-stimulating hormone (FSH) from

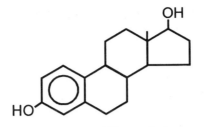

17β-Estradiol

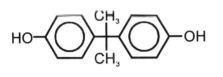

Bisphenol A

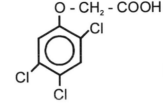

2,4,5-T

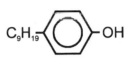

p-nonylphenol

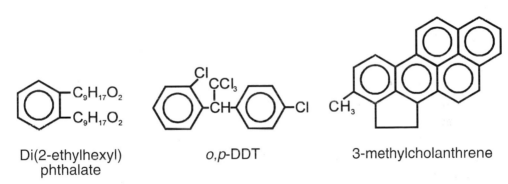

Di(2-ethylhexyl) phthalate

o,p-DDT

3-methylcholanthrene

Figure 7.21 Examples of environmental estrogens (xenoestrogens).

the pituitary of mammals and birds and gonadotropin I and II from the pituitary of lower vertebrates. Either FSH or gonadotropin II stimulates follicle growth and estradiol synthesis in the ovary, but the other gonadotropin is also important in stimulating ovarian development. In oviparous vertebrates, the release of gonadotropin-releasing hormone and follitropin, from the hypothalamus and pituitary glands, respectively, stimulates ovarian follicle growth in the female along with estradiol synthesis. In oviparous vertebrates, the release of estradiol from the ovary causes the liver to produce large amounts of vitellogenin, a high-density lipoprotein, which is a precursor for egg yolk. This is lipoprotein transported from the liver in the bloodstream and incorporated into developing oocytes. Any perturbation of this sequence through hormonal disruption could have detrimental effects not only on oogenesis, but also fecundity, embryonic development, egg

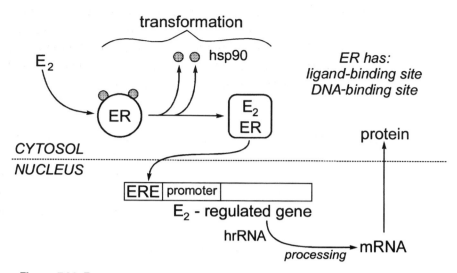

Figure 7.22 Estrogen receptor binding sequence. E_2, 17β-estradiol; ER, estrogen receptor; ERE, estrogen response element of regulated gene; hsp90, heat shock protein.

hatching and larval survival (Anderson et al., 1996). Compounds that mimic the effect of estrogen have been shown to stimulate vitellogenin production artificially. Circulating levels of vitellogenin have, therefore, proven to be a useful means of screening for exposure to estrogen-like chemicals in the environment.

Endocrine disrupters are usually small molecules that regulate the activity of estrogen-responsive genes in a manner similar to that postulated for TCDD binding to the Ah receptor in cytochrome P450 induction. The estrogen receptor, like the Ah receptor, is normally complexed with two molecules of the heat shock protein hsp90, which dissociate upon binding of the estrogen to the receptor. Similar to the Ah receptor, the estrogen receptor (ER) has a ligand binding site and a site that binds to the DNA molecule. It is impossible to overlook the similarity between the overall sequence triggered by Ah binding (Figure 7.18) and ER binding (Figure 7.22). As we learn more about these processes, it is likely that they will converge into a more unified scheme for ligand-activated gene transcription. However, as it now stands, the binding configuration of the ER seems quite different from the Ah binding and has no equivalent of the Ah/ARNT heterodimer that is a feature of Ah-DNA binding. Most xenoestrogens have potencies between 5×10^{-2} and 5×10^{-4} that of the natural estrogen 17β-estradiol or the synthetic estrogen diethyl-stilbestrol (DES), which are used as standards.

In view of the large range of man-made chemicals that are capable of endocrine disruption, this topic has become one of the most controversial environmental issues in recent years. A large research effort has focussed on the development of effective screening techniques for quantifying estrogenic activity. Earlier assays concentrated on the proliferation of the female genital tract as determined by an increase in the weight of the uterus in laboratory rats. More recent, in vitro assays include measurement of the proliferative effect of estrogens on human breast cancer (MCF7) cell cultures and the production of the phospholipoprotein vitellogenin (VTG) in rainbow trout (*Oncorhynchus mykiss*) hepatocytes. VTG production has also been employed as a biomarker of estrogenic exposure in salmonids such as rainbow trout collected from the field (see Case Study 7.3).

7.5.3 Effect of organic chemicals on male reproductive health

Male reproductive health is considered by some to be decreasing worldwide. Several studies have indicated that, in general, sperm concentration and semen quality have dropped significantly, and the incidence of testicular cancer, cryptorchidism (undescended testicles), and various deformities of the reproductive tract has increased. Some of these symptoms are correlated. For example, it has been shown that boys who are born with cryptorchid testes exhibit a tenfold increase in the incidence of testicular cancer.

Testicular cancer is the most common form of cancer in young men, and its occurrence has approximately doubled in the past 50 years. Like cryptorchidism, testicular cancer affects different nationalities and ethnic groups at different rates. For example, in Denmark, the rate of testicular cancer is approaching 1% of the male population, about four times that observed in Finland (Toppari et al., 1996), and Caucasian males are approximately three times more likely to be affected than are black males in the United States.

One of the best known and most often cited reports on semen quality is that of Carlsen et al. (1992), which includes data from approximately 15,000 men and over 60 studies. The authors report a steady decline in semen quality (42% decrease in mean sperm count; 18% decrease in semen volume) between 1940 and 1990. Several more recent studies have given inconsistent results, e.g., some show significant declines in sperm count; others indicate no such decline. However, in general, it appears that urban areas have a declining trend in sperm concentrations, whereas rural areas show consistent sperm counts. These results support the hypothesis that semen quality reflects environmental factors rather than genetic alterations in the general population.

It has been postulated that changes in the male reproductive system may be mediated by xenoestrogens acting on the male foetus in utero. Sexual differentiation occurs between weeks six and nine of foetal development when the SRY gene on the short arm of the Y-chromosome is activated and induces the expression of Müllerian Inhibiting Substance (MIS), which is released by the Sertoli cells. MIS inhibits development of the Müllerian ducts, which would otherwise form the uterus and the upper portion of the vagina. MIS thereby facilitates the formation of the Wolffian duct, which develops into the vas deferens and epididymus. The Sertoli cells are primarily responsible for spermatogenesis. Their proliferation and function is under the control of follicle-stimulating hormone, which is in turn regulated by negative feedback by estrogens. This means that FSH functions at the normal level when the correct concentrations of estrogens are available, but when estrogens are present in abundance, the function of FSH is impaired. Inhibition of FSH lowers the number of Sertoli cells, thereby limiting MIS secretion and the quantity of sperm produced by the testes. Failure of MIS to arrest the formation of the Müllerian ducts completely has been associated with cryptorchidism. These events provide a logical scenario for impairment of the male reproductive system and feminisation through endocrine disruption.

Evidence for this sequence has been sought through studies of chemicals designed specifically as estrogens and those whose estrogenic properties are incidental. Diethylstilbestrol is a xenoestrogen prescribed to more than 5 million pregnant women worldwide from the late 1940s to the late 1960s. Its purpose was to prevent abortions and complications associated with pregnancy. However, its use was discontinued after many girls who had been exposed to DES in utero developed a rare cancer called clear cell carcinoma. DES was also used in livestock farming and so an unknown number of people worldwide may have been exposed to the drug. Studies of males exposed to DES in utero have shown a significantly increased incidence of abnormalities of the reproductive system and reduced sperm concentration compared with unexposed males. The potential for sustaining these abnormalities through several generations is a particularly troubling aspect of these findings.

7.5.4   Environmental influences on breast cancer

Overall incidence of breast cancer rose 45% between the years 1960 and 1985. Dozens of hypotheses have emerged in an effort to explain this trend, and women's reproductive habits, diet, medication, and their environment have all been examined as possible culprits. Scientists have investigated links between cancer and various aspects of reproductive history including age at childbirth, oral contraception, and induced or spontaneous abortion. However, a lack of consensus in these studies and the fact that only 5% of breast cancer cases can be attributed to the recently discovered BRCA1 gene have led to investigation of chemical influences such as medications and environmental estrogens.

Some studies have implicated the drug diethylstilbestrol as a causative agent (see foregoing discussion). There are several conflicting reports, and Giusti et al. (1995), in a review of the literature, agreed with the hypothesis that a link existed between diethylstilbestrol and breast cancer. Although DES may explain only a very small proportion of cases, it demonstrates that a foreign estrogen can result in increased risk of breast cancer and raises the question whether environmental estrogens could possibly explain the increasing numbers of breast cancer victims.

Studies such as that of Wolff et al. (1993) have shown a significant relationship between DDE exposure and the incidence of breast cancer in women, although other organochlorines such as PCBs have yielded inconsistent results and the role of environmental estrogens in causing breast cancer remains speculative. Laboratory experiments on rats using DDT have suggested a cancer-promoting role for

Table 7.1. Examples of peroxisome proliferases

Plasticising agents	Phthalate esters
Solvents	Trichloroacetic acid
	Perchloroethylene
Herbicides	Lactofen
	Fomasafen
	2,4-D
	2,4,5-T
Hypolipidemic drugs	Chlofibrate
	Finofibrate
Antidiabetic drugs	Thiazolidinediones
Human adrenal steroid and nutritional supplement	Dehydroepiandrosterone

this compound. Rats exposed to the cancer initiator 2-acetamidophenanthrene (AAP) and fed a diet contaminated with DDT developed (primarily mammary) tumours significantly faster than rats not fed DDT. However, rats fed only DDT did not develop tumours.

7.5.5 Peroxisome proliferases

Peroxisomes are subcellular organelles found in most plant and animal cells that perform a variety of metabolic functions including cholesterol and steroid metabolism, β-oxidation of fatty acids, and hydrogen peroxide-derived respiration. Peroxisome proliferators are a structurally diverse group of environmental contaminants including herbicides, solvents, plasticisers, and several drugs and dietary supplements (Table 7.1). As their name suggests, they cause a large increase in both size and number of peroxisomes in exposed organisms. At the tissue level, symptoms include enlargement of the liver resulting from hypertrophy and hyperplasia of hepatocytes. Prolonged exposure to peroxisome-proliferating agents results in the increased incidence of liver tumours in laboratory animals, although they are not considered to be genotoxic agents. Such carcinogenic activity is, therefore, thought to derive from indirect DNA damage. A possible causative agent is hydrogen peroxide produced as a by-product of the cyclic oxidation of fatty acids. A further contributory factor is the increase in cell proliferation that results from exposure to peroxisome-proliferating agents. Peroxisome proliferators are, therefore, considered to function as cancer promoters (see Section 2.3.1).

There is also evidence to suggest that peroxisome proliferators may act as endocrine disruptors, and several metabolic and nuclear hormone receptor pathways have been implicated in this regard. Examples include aromatase, which converts testosterone to estradiol and androstenedione to the less active estrone and 17β-hydroxysteroid dehydrogenase. A potential mechanistic basis for peroxisome

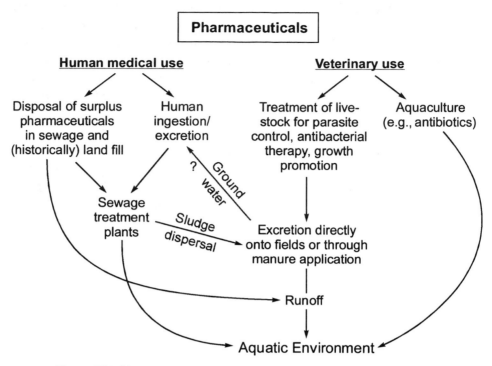

Figure 7.23 Pharmaceuticals in the environment.

proliferators acting as endocrine disrupters may be the cross-recognition of the estrogen receptor by the estrogen-responsive elements (ERE) of estradiol-responsive genes.

7.5.6   Pharmaceuticals in the environment

Until recently, little attention had been paid to the release of medical substances or pharmaceuticals into the environment. However, the recognition of xenobiotics as potential endocrine disruptors at very low concentrations has refocussed attention on prescribed or over-the-counter medicines or dietary supplements deliberately designed to elicit a biological effect. These include drugs and metabolites belonging to a wide range of groups such as beta blockers, antiphlogistics, lipid-regulating agents, antibiotics, and analgesics that find their way into the environment via manufacturing waste, the disposal of unused product, or the excretion of unchanged or metabolised compound. Pharmacokinetic data indicate that the human excretory rate of unchanged drugs exceeds 50%, and detectable levels of several such chemicals have been found in sewage and river water. A scheme describing pathways for environmental fate and effects of pharmaceuticals is shown in Figure 7.23.

Some, such as metabolites of the lipid-regulating agent clofibrate, are very persistent and have the potential to be retained in a biologically active form in the environment. Clofibric acid was detected in Berlin tapwater at concentrations between 10 and $165 \, ngL^{-1}$ and from several groundwater samples close to the city (Stan et al., 1994). A study conducted by Richardson and Bowdon (1985) on the River Lee in England indicated that approximately $170 \, ngL^{-1}$ were released into the river annually in quantities of >1 tonne (1,000 kg) giving predicted concentrations in excess of $0.1 \, \mu gL^{-1}$. Several studies have reported high levels (up to $5 \, mg kg^{-1}$ dry weight) of antibiotics from sediments associated with fish farms.

In evaluating the possible effects of pharmaceuticals on humans and other species, particular attention is being paid to the potential for endocrine disruption, as described in the foregoing sections. However, another line of investigation concerns the possibility that antibiotics could quantitatively and qualitatively alter natural bacterial assemblages with ramifications for a variety of bacterially mediated chemical transformations. The impact of such pharmaceuticals is still uncertain, although their environmental release is substantial. For example, in a small country such as Denmark (population 5.2 million), with a relatively large agricultural economy, over 100 tonnes of antibiotics are used annually as agricultural growth promoters. It is estimated that 30–90% is excreted in the urine as biologically active substances.

7.6 Polynuclear aromatic hydrocarbons

The carcinogenic potential of several aromatic hydrocarbons derived from coal tar has been known since the 1930s, and the etiology of diseases such as lung cancer has long been principally concerned with occupational factors such as exposure to volatile components of petroleum compounds and particulate and gaseous products in tobacco smoke. Rodents are particularly sensitive to PAHs, and subcutaneous injection of a wide variety of anthracene derivatives results in the development of sarcomas in mice and rats. Carcinogenicity usually requires activation, mediated by the mixed function oxidase system, culminating in the formation of dihydrodiol epoxide derivatives. Such metabolites bind covalently to DNA to form adducts (see Section 2.3.1).

PAHs are released into the environment principally through the burning of fossil fuels and as a result of the manufacture and use of petroleum and wood products. Of the $1,000 \, tonne \, yr^{-1}$ PAHs emitted into the atmosphere from UK sources, over 95% originates from unregulated fires, vehicle emissions, and domestic coal consumption. Over 75% of this is exported to continental Europe by westerly winds. Natural sources of PAHs include forest fires, oil seepage, and volcanoes, although it is clear that anthropogenic sources dominate PAH input to the environment. Total global PAH inputs to water from all sources, both natural and anthropogenic, have been estimated at $>80,000 \, tonne \, yr^{-1}$, and studies of dated sediment cores have shown that fluxes of PAHs to sediments have increased five- to tenfold in the last hundred years but are currently in decline.

In the atmosphere, partitioning of PAHs between the particulate and gaseous phase depends on the number, type, and size of particles present; the ambient air

Table 7.2. Relative importance of sources of organic contaminants to Chesapeake Bay

Organics	Susquehanna River load (kg/yr)[a]		Atmospheric deposition load (kg/yr)[b]
	Dissolved	Particulate	
ΣPCBs	97	68	37
Fluorene	37	85	27
Phenanthrene	63	388	155
Fluoranthene	108	1,020	189
Pyrene	104	925	184
Benz(a)anthracene	12	364	44
Chrysene	15	316	114
Benzo(a)pyrene	5	436	53

[a] Annual loads entering Chesapeake Bay via the Susquehanna River, measured at Conowingo, MD, between February 1994 and January 1995.
[b] Total atmospheric deposition loads directly to the surface of the Chesapeake Bay below the fall lines ($1.15 \times 10^{10} \, m^2$), as measured by the Chesapeake Bay Atmospheric Deposition Study, 1990–1992.

temperature; and the vapour pressure of the individual compounds. Removal from the atmosphere may result from dry particle deposition, vapour exchange between the atmosphere and terrestrial surfaces and water bodies, and scavenging through precipitation. Nitro-PAHs may be formed by the photoactivated reaction of oxides of nitrogen (NO_x). Nitropyrene, for example, is a significant carcinogen in urban air and remains volatile compared with higher molecular weight PAHs that tend to bind to particulates.

Table 7.2 compares annual atmospheric deposition of PAHs on the Chesapeake Bay with input from the Susquehanna River, which supplies about 60% of the water to the bay. Atmospheric input is approximately equal to the dissolved riverine fraction entering Chesapeake Bay, although overall input is dominated by the suspended particulate fraction, much of which is in the form of suspended sediments. Because of their high octanol-water (K_{OW}) and organic carbon (K_{OC}) adsorption partition coefficients, most PAHs tend to adsorb to aquatic sediments and bioaccumulate in organisms such as crustaceans and bivalve molluscs, which have poorly developed mixed function oxidase systems for their metabolism.

Although concern over PAHs in the environment is often articulated in terms of their potential carcinogenicity to man and other mammals, most studies of PAHs in the aquatic environment have concentrated on natural resources. Both laboratory and field studies in the aquatic environment have indicated that PAH exposure affects steroid levels, gonadal development, fecundity, fertilisation and hatching success, although in many of these studies the toxicology is not differentiated from accompanying organic contaminants such as polychlorinated organic compounds. Similarly, genotoxic effects have been demonstrated in environments heavily contaminated with PAHs, although in view of the coincidence of compounds such as PCBs, these studies tend to adopt a weight-of-evidence, correlative approach. Typical examples include the high incidence of neoplasms in bottom-dwelling fish from coastal areas where sediments have high levels of PAHs and other organic compounds: hepatocellular and cholangiocellular carcinomas in livers of winter flounder (*Pseudopleuronectes americanus*) from Boston Harbor, English sole (*Pleuronectus vetulus*) from Puget Sound, and mummichogs (*Fundulus heteroclitus*) from the Elizabeth River, Virginia. Histogenesis of these neoplasms in English sole is considered to be similar to tumour development in rodents, although a clear linkage between carcinogenicity in aquatic animals and mammals has yet to be made. However, a national survey of cancer incidence in the United States indicated that mortality in human females resulting from ovarian and other cancers of the reproductive system at sites near the Maine and Indian Rivers, Florida, were significantly higher than the natural average and coincided with a high incidence of tumours in local catfish.

Table 7.3 (Law et al., 1997) is a compilation of estimated safe levels for individual PAHs obtained through a variety of extrapolations. When the European Union Maximum Admissible Concentration of $200\,\mathrm{ng\,L^{-1}}$ total PAH (ΣPAH) was applied to surface water samples collected from around the United Kingdom, 56/177 (32%) were in excess of this value, which was established for the protection of human consumers of seafood. An ecotoxicological assessment of these samples suggested that only 15% posed any toxic threat to aquatic organisms.

7.7 Petroleum hydrocarbons

Crude oil is a complex mixture of thousands of chemical compounds formed from the burial and transformation of marine organisms over hundreds of millions of years. The transformation process involves heat and pressure and may be catalysed by mineral constituents of the Earth's crust such as aluminosilicates. As it is formed, oil migrates to the surface of the Earth's crust, where it seeps out naturally or becomes trapped in sedimentary rock formations beneath an impervious substrate. These underground reservoirs have been tapped for energy since the first commercial well was drilled in Pennsylvania in 1859. Today, petroleum accounts for approximately 40% of the world energy needs, about half of this being required for transportation.

Even today, estimates of reservoir gas and oil reserves are remarkably inexact and are probably only accurate to within approximately one or two orders of magnitude. Hunt (1995) has estimated that only about 1% of oil and 0–5% of gas ever make their way into producible reservoirs. Of the remainder, about 50% is lost at the surface and the rest is trapped in rocks out of reach of current extraction technology. Recent evidence suggests that oil and gas fields may be continually

Table 7.3. Extrapolated "safe" levels (mg^{-1}) for aquatic organisms exposed to PAHs in water, developed by various countries and international organisations (Based on Law et al. 1997)

Organics	Country/organisation					
	Netherlands (maximum permissible concentration)	Canada (draft interim water quality criteria)	United States (environmental quality criteria)	Denmark (water quality criteria)	European Union (maximum admissible concentration)	Oslo and Paris Commission (provisional ecotoxicological assessment criteria)
Anthracene	0.2	0.12*		0.01		0.005–0.05
Benzo(a)anthracene	0.1					
Benzo(a)pyrene	0.1	0.008				0.01–0.1
Benzo(k)fluoranthene	0.02					
Benzo(ghi)perylene						
Fluoranthene	0.5					0.05–0.5
Naphthalene	2.0	11.0	4.6	1.0		1.0–10.0
Phenanthrene		0.8				
ΣPAH			0.03*		0.2*	

* To protect human consumers of aquatic life collected from PAH-contaminated waters.

recharging themselves from deeper in the Earth's crust (Whelan et al., 1993), and that hydrocarbon seepage from the ocean floor may be far larger than had previously been imagined.

Crude oil itself varies widely in chemical composition and properties according to its source. For example, a typical North Sea crude oil is much lighter than a Venezuelan crude oil. Chemical characteristics of crude oils, such as density and sulphur content, dictate their marketability and, like many other minerals, this affects the cost benefits of their extraction from the ground.

Of the crude oil constituents, 98% are organic. The three main hydrocarbon groups are aliphatics, cycloalkanes, and aromatics.

> *Aliphatics (paraffins).* These include *n*-alkanes (straight chain) and *iso* alkanes (branched chain) in which carbon atoms are attached to hydrogen or other carbon atoms (Figure 7.1). They comprise 60 to >90% of the hydrocarbon content of crude oils. Saturated hydrocarbons with fewer than five carbons are gases at room temperature. Those with 5 to 17 or 18 carbons are liquids. Paraffins with 20 to 35 carbon atoms per molecule are solids and are referred to as waxes. They are present in solution at elevated temperatures but may solidify at lower temperatures.
>
> *Cycloalkanes (naphthenes).* These saturated compounds are similar to alkanes but are formed into rings (Figure 7.24). Each ring may have one or more paraffinic or alkyl side chains.
>
> *Aromatics.* These compounds contain one or more aromatic rings, which have a stable alternate double-bond configuration. Single-ring aromatics in crude oil include benzene, toluene, ethylbenzene, and xylenes, the so-called BTEX compounds. Naphthalene, anthracene, pyrene, and coronene have two, three, five, and six rings, respectively, and belong to the polynuclear aromatic group of compounds (PAH) some of which are carcinogenic (see Chapter 2). In addition to pure hydrocarbons, some organic compounds in crude oil contain small amounts of oxygen, nitrogen, and sulphur. Examples of such heteroatomic organics include phenols and thiophenes. Two important heteroatomic groups are resins (MW, <700–1,000) and asphaltenes (MW, 1,000–10,000).

Crude oil is of little practical use in its unrefined state and is subjected to a process of fractional distillation to separate the lighter fractions, which are used as fuels. Gasoline, for example, comprises only C-4–C-10 compounds and may contain up to 30% olefins which are double-bond, straight-chain, or cyclic hydrocarbons produced by a catalytic cracking process. The presence of olefins and branched chain hydrocarbons much improves the performance of gasoline fuels and minimises knocking. It is important to remember that the dispersive properties of oil and its toxicity are greatly affected by the relative proportion of refined products. Lighter oils are more easily dispersed but are more toxic than heavier

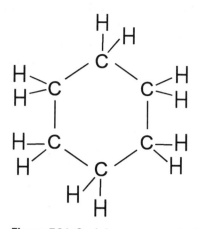

Figure 7.24 Cyclohexane, a cycloalkane.

oils. This has an important bearing on the methods used to treat an oil spill (see Chapter 9).

7.8   Organotins

While inorganic tin compounds pose little toxicological risk due to their low solubility and poor absorption, some organotin compounds, notably the alkyl tins, have demonstrated a variety of toxic effects to aquatic organisms at very low concentrations. Biocidal applications of these compounds include their use as selective molluscicides, agricultural pesticides, and antifouling agents on ships. Tributyltin proved to be particularly effective as an antifouling agent and, through the 1960s, became increasingly popular as the active ingredient in highly effective antifouling paints. These paints demonstrated superior effectiveness and longevity compared with copper-based paints, with significant savings in vessel downtime and fuel consumption (related to drag). Between 1965 and 1980, worldwide organotin consumption increased from 5,000 to 35,000 metric tonnes.

By the late 1970s and early 1980s, several reports implicating high TBT concentrations in the vicinity of boat moorings as the cause of unnatural shell thickening and poor reproductive performance in oysters began to appear (Alzieu and Portman, 1984). Laboratory experiments demonstrated that TBT levels as low as 2–$8\,ngL^{-1}$ had detrimental effects on oysters, and it was concluded that coastal TBT contamination contributed to a decline in the native oyster (*Ostrea edulis*) fishery on the east coast of England during the 1970s. Reports of problems with bivalve molluscs were followed by demonstrations that TBT caused a condition known as imposex in the dog whelk (*Nucella lapillus*). Imposex is characterised by the development of a vas deferens and a penis in females leading to the blockage of the female genital duct (Gibbs and Bryan, 1986). In this case, TBT seems to operate as an endocrine disrupter leading to increased testosterone levels. Although the exact mechanism of endocrine disruption remains unclear, there is some evidence to suggest competitive inhibition of cytochrome P450, which is involved in the conversion of testosterone into 17β-estradiol (Beltin et al., 1996). Other evidence suggested that TBT can also inhibit the formation of sulphur conjugates of testosterone and its metabolites thereby inhibiting their excretion (Ronis and Mason, 1996). Although the genus *Nucella* is particularly sensitive to TBT, other neogastopods are affected to varying degrees. In more sensitive species, imposex is initiated at TBT levels as low as $2\,ng\,L^{-1}$. At the peak of TBT use, the highest seawater concentrations found in harbours and marinas approached $1,000\,ng\,L^{-1}$ ($1\,\mu g\,L^{-1}$), although levels between 10 and $100\,ng\,L^{-1}$ occurred in the open waters of estuaries and bays. TBT had significant effects on some marine ecosystems, and in the late 1980s restrictions on its use began, first in France and the United Kingdom and later in other countries including the United States and Canada. TBT regulation affects the TBT content of antifouling paints, their leaching rate, and the size of the vessel to which TBT may be applied. Larger ships (>25 m) are exempt from mandatory TBT and current TBT restrictions are by no means worldwide. For example, in southeast Asia, TBT is restricted in Japan but few other countries in the region. Where restrictions have been enforced, subsequent recovery of sensitive species have been reported. For example, following TBT regulation in France, dramatic increases in oyster settlement (*Crassostrea virginica*) have been recorded (Alzieu, 1986). Until quite recently, it was assumed that TBT was the only organotin compound able to induce imposex. However, Horiguchi et al. (1997) demonstrated experimentally that triphenyltin was also able to promote the development of imposex in the whelk *Thais clavigera*. Even though TPT has also been used in antifouling paints, it is more commonly used as an agricultural fungicide.

7.9 Metabolism of organics

7.9.1 Introduction

Throughout their evolution, both bacteria (prokaryotes) and complex organisms (eukaryotes) have had to deal with the problem of detoxifying and eliminating a variety of organic chemicals accumulated from the environment (exogenous) or produced by the organism itself (endogenous). Foreign, exogenous compounds, termed xenobiotics, have multiplied many thousandfold by man's manufacturing activities: Probably over 200,000 man-made organic compounds are in existence. It is, therefore, a tribute to the enzyme systems involved that the great majority of these new compounds are successfully metabolised and detoxified. Transformation of nonpolar organic chemicals occurs through two different kinds of reactions. Phase I reactions are catabolic in nature and involve the addition of a functional group to the parent compound, thereby creating a more reactive intermediate. Phase II reactions are catabolic or biosynthetic in nature and involve the conjugation of the intermediate with water-soluble moieties that are more mobile and that may be excreted in the bile or urine.

7.9.2   Phase I reactions

The addition of a polar functional group (Phase I reaction) is principally achieved by controlled enzymatic oxidation following the generalised reaction:

$$RH + NADPH + O_2 + H^+ = ROH + NADP + H_2O$$

where RH is the substrate, and ROH is its hydroxylated form. The enzymes responsible for this reaction are called monooxygenases. Two major monooxygenase groups have been identified: the flavoprotein monooxygenases (FMO), also known as the amine oxygenase system, and the cytochrome P450 family of enzymes, also called the mixed function oxidase system. Both groups are found in soluble form in prokaryotes and membrane-bound in eukaryotes. FMOs, which have FAD as a prosthetic group, have been identified from mammals, fish, and molluscs. Substrates include phosphines, thiols, and a wide range of aliphatic amine and secondary and tertiary aromatic amines.

The MFO system, discovered in the 1940s, is probably the most important enzyme system for xenobiotic metabolism. It is found in all phylogenetic levels of organisms, including plants, and is probably at least 450 million years old. In mammals, about 70% of MFO activity is in the liver, although significant levels are found in most tissues. MFO activity is very low in foetal tissue – about 1% of the adult level. After birth, MFO activity rises to near adult levels within a few months. There are now well over 400 different P450 proteins classified into at least 36 families according to their amino acid sequences; hence, the classification is still evolving. At present, it includes the prefix CYP followed by a numerical term for family (40% identical amino acid sequence) and an Arabic letter for subfamily (60% identical amino acid sequence). Many P450 enzymes are given a final identifying number (e.g., CYP1A1, CYP1A2), although there is some ambiguity associated with this. In some cases, the last number is specific to a particular species; in others, the nomenclature applies to many species. Particular attention has been given to the enzyme CYP1A1, which is induced through a sequence of events beginning with the binding of a xenobiotic chemical to a cytoplasmic Ah receptor (Figure 7.20).

REGULATION OF CYTOCHROME P450 INDUCTION

Poland and co-workers (e.g., Knutson and Poland 1982), working with 2,3,7,8-tetrachlorodibenzo-*p*-dioxin, identified a high affinity receptor molecule in the cytosol of liver and other tissues. This Ah receptor, which is normally associated with the heat shock protein hsp90, recognises the inducing compound, and the resulting TCDD-Ah receptor complex passes to the nucleus where it releases the hsp90 and crosses the nuclear membrane with the aid of a specific protein, the Ah receptor nuclear transferase (ARNT). Now in its transcriptionally active form, the TCDD-Ah receptor complex interacts with a specific region of the CYP1A1 gene known as the xenobiotic regulatory element. This step allows access by other proteins to the promoter region of the gene which, in turn, initiates synthesis of the

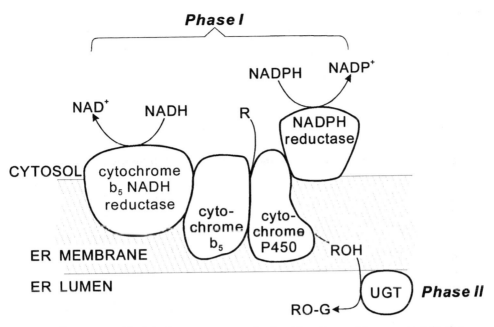

Figure 7.25 Model of components of mixed function oxidase system in the membrane of the endoplasmic reticulum. ER = endoplasmic reticulum; NAD = nicotinamide adenine dinucleotide; NADP = nicotinamide adenine dinucleotide phosphate; UGT = UDP glucuronosyltransferase; R = nonpolar substrate; ROH = hydroxylated substrate; RO-G = glucuronide.

messenger RNA (mRNA) responsible for transcribing the appropriate sequence for CYP1A1 synthesis. The cytochrome P450 molecule is then incorporated into the membrane of the endoplasmic reticulum.

STRUCTURE AND FUNCTION

The cytochrome P450 system consists of a closely aligned complex of molecules embedded in the lipid membrane of the endoplasmic reticulum. Some accounts differ with respect to the exact configuration of this unit, but most recognise the cytochrome P450 enzyme which can accept electrons from either one of two molecules. One is cytochrome P450 NADPH reductase, which contains two subcomponents, flavin adenine dinucleotide (FAD) and flavin mononucleotide (FMN). The other is cytochrome b_5, which may, in turn, receive electrons from cytochrome b_5-NADH reductase. A model of these components is shown in Figure 7.25, which also indicates the relative position of the Phase II (conjugative) enzyme UGT. This utilises products from the P450 system as substrates. Although each component is shown singly, in fact, the cytochrome P450 NADPH reductase molecule is associated with 10 to 20 cytochrome P450 enzyme molecules and 5 to 10 molecules of cytochrome b_5.

The nonpolarity of the substrate causes it to associate with the lipid bilayer of the membrane and gain access to the active site of the cytochrome P450 enzyme.

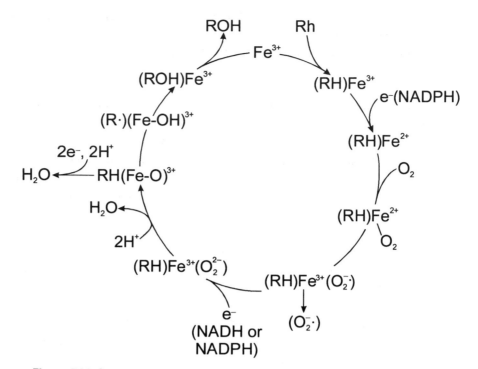

Figure 7.26 Sequence of events in mixed function oxidase activity on nonpolar organic compounds.

This event causes a change in the absorption spectrum of the haeme moiety of the enzyme and triggers a flow of electrons from NADPH via NADPH-cytochrome-P450 reductase to the haeme ion, reducing it from Fe(III) to Fe(II). The strong affinity of oxygen for Fe(II) results in the binding of molecular oxygen to cytochrome P450. A second electron is then transferred from NADH via NADH reductase. The reduced electrons are added to molecular oxygen to produce the superoxide anion, which is an extremely powerful oxidising agent. The substrate is then spontaneously oxidised. One oxygen atom is inserted into the substrate, probably with the abstraction of hydrogen and recombination of the transient hydroxyl and carbon radicals, to give the product. The other oxygen combines with hydrogen ion to produce water. This sequence of events is shown in Figure 7.26.

The MFO system catalyses a wide variety of reactions. Although all these reactions involve, at some point, the fixation of oxygen, this may appear either in the product or the by-product (e.g., aldehyde by-product of dealkylation). Based on structure activity studies, some general rules may be applied that predict the site and degree of oxidation of substrates by the MFO:

- The compound must be sufficiently lipid-soluble to gain access to the active site.
- The compound must have an oxidisable atom (e.g., C, N or S).

- Oxidisable atoms with a high density of reduced electrons favoured for MFO oxidation.
- Oxidation is blocked by steric interference.

STERIC INTERFERENCE

Carbon atoms are more difficult to oxidise when bonded to chlorine rather than hydrogen atoms due to electronegative repulsion between chlorine and oxygen atoms. Chlorine atoms also withdraw electrons from carbon atoms in the same ring, reducing the ease with which these carbon atoms can be oxidised. Hence, the proximity of a chlorine atom may inhibit the oxidation of an adjacent carbon atom. This explains the refractory nature of several highly chlorinated hydrocarbons [e.g., chlordecone (kepone)] (Figure 7.9).

7.9.3   Important mixed function oxidase reactions

Many important environmental transformations are carried out by the reactions summarised in Figure 7.27. There a several variants of these reactions. In hydrolytic dehalogenation, for example, the halogen atom is replaced by —OH, not H. In alkyl halides, this involves the addition of water; in aryl halides, O_2 is added. Dechlorination reactions are involved in the metabolism of many chlorinated pesticides, of which DDT is, perhaps, the best-known example (Figure 7.7). The more rapid metabolism of methoxychlor (Figure 7.8) involves both dechlorination and demethylation reactions.

MFO-mediated metabolism may result in products that are more or less toxic than the parent compound. Oxidation of the thion moiety of the phosphorothionate pesticide parathion is required for activation to its toxic oxygen analog, paraoxon. However, this process shares a common intermediate with the deactivation of parathion through cleavage of the phosphorus-aryl bond (Figure 7.11).

Aromatic hydroxylation is thought to have an intermediate epoxidation step (Figure 7.28). Most epoxides are extremely reactive and toxic, yet they are usually only transient intermediates in such reactions. Their hydrolysis to diols, therefore, represents a detoxification process. Enzymes catalysing this reaction are termed epoxide hydrolases. The toxicity of epoxides is enhanced where the molecular configuration retards their metabolism. A typical case is illustrated by the carcinogenic activation of benzo(a)pyrene (Figure 7.29), where the formation of a diol is an intermediate step in the production of the compound 7,8-dihydrodiol-9,10-*trans*-epoxide. The hydrolysis of the 9,10 epoxide is retarded by the proximity of the adjacent benzene ring in the so-called bay area of the molecule. Another group of compounds requiring MFO activation to carcinogenic forms are the nitrosamines, formed by the reaction of nitrite ions with secondary and tertiary amines.

Stimulation of genetic activity associated with the mixed function oxidase system or with changes in endocrine activity has proven to be a powerful indicator of possibly detrimental physiological change relating to potential carcinogenicity. Furthermore, MFO enzymes have been increasingly implicated in

S - dealkylation : $R - S - CH_3 \longrightarrow \left[R - S - CH_3OH \right] \longrightarrow R - SH + CH_2O$

N - dealkylation : $R - NH - CH_3 \longrightarrow \left[R - NH - CH_3OH \right] \longrightarrow R - NH2 + CH_2O$

O - dealkylation : $R - O - CH_3 \longrightarrow \left[R - O - CH_3OH \right] \longrightarrow R - OH + CH_2O$

deamination : $R - \underset{NH_2}{CH} - CH_3 \longrightarrow \left[R - \underset{NH_2}{\overset{OH}{C}} - CH_3 \right] \longrightarrow R - \overset{O}{C} - CH_3 + NH_3$

aliphatic hydroxylation : $R - CH_2 - CH_2 - CH_3 \longrightarrow R - CH_2 - CHOH - CH_3$

aromatic hydroxylation :

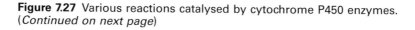

reductive dehalogenation : $R - CCl_3 \longrightarrow R - CHCl_2 + HCl$

Figure 7.27 Various reactions catalysed by cytochrome P450 enzymes. (*Continued on next page*)

changes in reproductive potential. Altered genetic activity and changes in associated enzyme levels have proven to be useful biological indicators of exposure to specific groups of chemicals, and this so-called biomarker approach to determining the degree of risk from such chemicals is discussed in Sections 4.4.1 and 4.4.2. In addition to its induction (de novo synthesis) by dioxins, PCBs, and other chlorinated compounds, there is a growing acceptance of an association between CYP1A1 and endogenous steroid metabolism, with implications for hormonal activity and reproductive function (Section 7.5). CYP1A1 has been shown to affect fertilisation, reproductive success, and gonadal development.

The induction of CYP2-like isozymes has been associated with the presence of peroxisome-proliferating agents (Section 7.5.6) and has the potential for significant adverse organism health effects including developmental abnormalities, reproductive changes, and changes in fatty acid metabolism (Haasch et al., 1998).

Oxidations

N - Oxidation (e.g. aniline → N - hydroxyaniline)

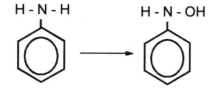

S - Oxidation (e.g. aldicarb)

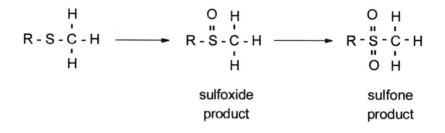

| sulfoxide | sulfone |
| product | product |

P - Oxidation (e.g. paraoxon)

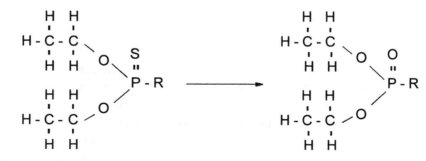

Figure 7.27 *(continued)*

Dehalogenations

Reductive dehalogenation

<pre>
   Cl Cl   Cl Cl
   | | | |
R- C - C - Cl   ------>   R- C - C - H
   | | | |
   Cl H   Cl H
</pre>

Oxidative dehalogenation

<pre>
   Cl Cl   Cl Cl
   | | | |
R- C - C - Cl   ------>   R- C - C = O
   | | |
   Cl H   Cl
</pre>

Dehydrodehalogenation

<pre>
    Cl Cl   Cl   Cl
    | | \   /
Cl- C - C - Cl   ------>   C = C
    | | /   \
    Cl H   Cl   H
</pre>

Dealkylations

O - Dealkylation (e.g. methoxychlor)

<pre>
      H   [H   ]   H
      | | | | |
R - O - C - H   ------> | R-O-C-H | ------> R - OH + C = O
      | | | | |
      H   [OH   ]   H
</pre>

N - Dealkylation (e.g. carbaryl)

<pre>
      H   [H   ]   H
      | | | | |
R - N - C - H   ------> | R-N-C-H | ------> R - NH2 + C = O
      | | | | | |
      H   [H OH]   H
</pre>

S - Dealkylation (e.g. methylmercaptan)

<pre>
      H   [H   ]   H
      | | | | |
R - S - C - H   ------> | R-S-C-H | ------> R - SH + C = O
      | | | | |
      H   [OH   ]   H
</pre>

Figure 7.27 *(continued)*

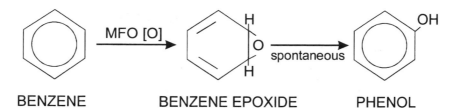

Figure 7.28 Hydroxylation of benzene by mixed function oxidase system.

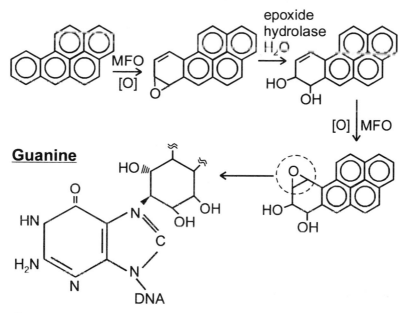

Figure 7.29 Carcinogenic activation of benzo(a)pyrene.

Alternatively, decreases in CYP2-like proteins have been associated with the presence of estrogenic compounds (Arukwe et al., 1997). Therefore, the regulation of CYP2-family isozymes plays a pivotal role in the well-being of the organism.

MFO INHIBITION

The mixed function oxidase system may be inhibited by carbon monoxide and cyanide, which bind to the Fe(II) of reduced cytochrome P450, blocking access to oxygen. Other substances act as inhibitors by occupying the active site of cytochrome P450 yet undergoing only very slow oxidation (e.g., SKF 525-A, piperonyl butoxide). Chlorinated alkanes (e.g., CCl$_4$) and alkenes (e.g., vinyl chloride) may be transformed by the MFO system into reactive metabolites, which in turn bind covalently to the haeme moiety of cytochrome P450, thereby inactivating the enzyme.

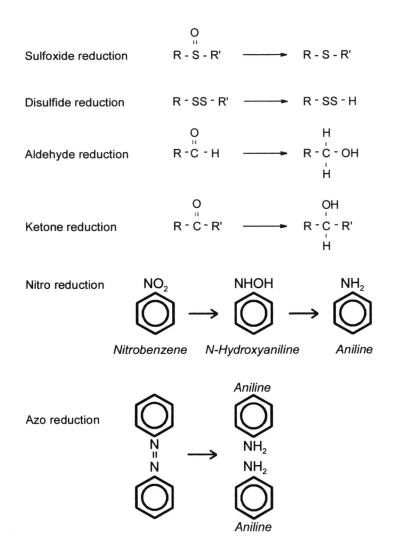

Figure 7.30 Reductions mediated by mixed function oxidase system.

7.9.4 Reductions

Under low-oxygen conditions, electrons produced by the MFO system are also capable of carrying out reductions. Many reductions of xenobiotic compounds are mediated by gut microflora, and some examples are shown in Figure 7.30. These include azo- and nitro-reductions. In a broader environmental context, the carcinogenicity of many aromatic amines formed in part by nitrate contamination of groundwater has prompted a good deal of interest in the ability of free-living bacteria to metabolize such compounds under field conditions (Gorontzy et al., 1993). Both reductive and oxidative dehalogenations are catalysed by cytochrome P450. An example of reductive dechlorination is the transformation of DDT to DDD shown in Figure 7.7 (further dechlorinations may subsequently occur), although the

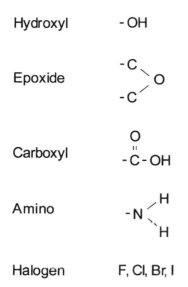

Hydroxyl	- OH
Epoxide	$-C \overset{\diagdown}{\underset{\diagup}{O}}$
Carboxyl	$\overset{O}{\overset{\|}{-C-OH}}$
Amino	$-N \overset{H}{\underset{H}{\diagup}}$
Halogen	F, Cl, Br, I

Figure 7.31 Functional groups for phase II (conjugation) reactions.

more common pathway for DDT degradation is a dehydrochlorination catalysed by cytochrome P450 and glutathione and resulting in the ethylene form, DDE.

7.9.5 Phase II reactions

Phase II reactions are biosynthetic in nature, leading to the formation of a conjugate. Conjugation is the joining together of two compounds and is an energy-consuming process. In such reactions, the conjugating agent is an endogenous compound, and the substrate is a xenobiotic chemical having the requisite functional groups for metabolism by conjugating (Phase II) enzymes. These functional groups are shown in Figure 7.31. Some xenobiotics may already possess such functional groups; others require prior action by Phase I enzymes to add or unmask reactive functional groups. The major conjugating agents for nucleophilic (electron-rich) compounds are glucuronides, whereas the principal conjugating agent for electrophilic (electron-deficient) compounds is glutathione.

Glucuronides are the most common endogenous conjugating agents. The sugar acid, glucuronic acid, is conjugated to O, N, and S heteroatoms after initial activation to uridine diphosphate glucuronic acid (UDPGA) (Figure 7.32). An example of an *o*-glucuronide is illustrated by the conjugation of pentachlorophenol with UDPGA, which is catalysed by UDP glucuronyl transferase found in the microsomal membrane (Figure 7.33). The resulting conjugate is much more polar (and therefore more water-soluble) than pentachlorophenol. Examples of S- and N-glucuronides are 2-mercaptothiazole and aniline glucuronide, respectively. Carboxylic acids and alcoholic oxygen may also be glucuronidated. Specific transport proteins facilitate the rapid excretion of glucuronides into urine and bile.

$$D\text{-glucose-1-PO}_4 \ + \ UTP \xrightarrow[\text{pyrophosphorylase}]{\text{UDPG}} UDP\text{-}a\text{-}D\text{-glucose} \ + \ P_2O_7^{4-}$$

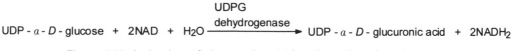

$$UDP\text{-}a\text{-}D\text{-glucose} \ + \ 2NAD \ + \ H_2O \xrightarrow[\text{dehydrogenase}]{\text{UDPG}} UDP\text{-}a\text{-}D\text{-glucuronic acid} \ + \ 2NADH_2$$

Figure 7.32 Activation of glucuronic acid for phase II conjugation.

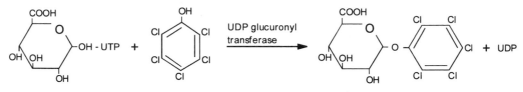

Figure 7.33 Conjugation of pentachlorophenol with UDPGA.

Glutathione (GSH) is a tripeptide that is a critically important conjugating agent for reactive electrophiles, which would otherwise attack vital macromolecules. It forms conjugates with a wide variety of xenobiotic compounds including aromatic hydrocarbons, aryl halides, alkyl halides, aromatic nitro-compounds, alkenes, alkyl epoxides, and aryl epoxides. Glutathione transferases are a family of enzymes that catalyse these conjugations. The glutathione conjugates are transported to the bile by specific transport proteins. However, they are rarely excreted directly and usually undergo further enzymatic conversion to mercapturic acid (Figure 7.34), a readily excreted polar compound.

Sulphate conjugation involves substantial energy input from ATP but provides an effective means of eliminating a variety of compounds in the urine since the sulphur conjugates are highly water-soluble. The types of compounds conjugated by sulphur are alcohols, phenols, PAHs, and aromatic amines. Enzymes that catalyse these reactions are called sulphotransferases. They exist in a variety of forms and specificities. Conjugation of phenol with sulphur is shown in Figure 7.35. Sulphation is a competing pathway with glucuronidation which is much more common due to the much higher concentrations of glucuronide, available as a conjugating agent.

Other pathways play minor roles in conjugation reactions and involve only a few compounds. For example, amino acid conjugation is largely confined to carboxylic acids such as the phenoxyacids 2,4-D and 2,4,5-T.

7.10 Environmental mobility of organic compounds

We have attempted in the chapter to provide students with at least background knowledge of several organic environmental contaminants and some aspects of

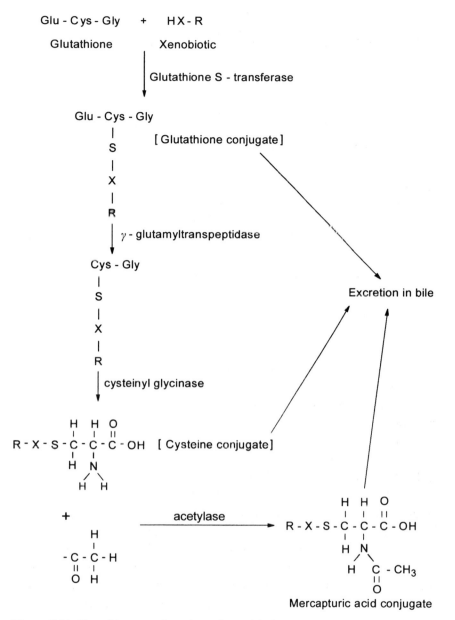

Figure 7.34 Glutathione conjugation of xenobiotic.

their metabolism. It is useful at this point to summarise the most important physiocochemical properties affecting the fate and transport of these compounds although some mention of these has been made in Chapter 5.

Compound mobility is affected by the following elements:

- Molecular weight
- Melting point

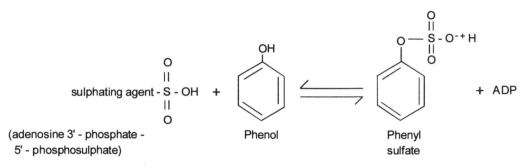

Figure 7.35 Conjugation of phenol with sulphate.

- Boiling point
- Water solubility – Mobile chemicals tend to have high water solubilities ($>30\,\mathrm{mg\,L^{-1}}$).
- Distribution coefficient between solid and liquid phases – Mobile chemicals tend to have low (<2) soil/water or sediment/water distribution coefficients, K_d, where

$$K_d = \frac{[\text{Chemical in aqueous phase}]}{[\text{Total chemical}] - [\text{Chemical in aqueous phase}]}$$

- Organic binding constant (K_{oc}) – Mobile chemicals tend to have K_{oc} values <500 where

$$K_{oc} = \frac{K_d}{\%\ \text{Organic carbon in (soil/sediment) sample}}$$

- Volatility; determined as Henry's law constant (K_H) – Mobile (volatile) compounds have K_H values of $<10^{-2}\,\mathrm{atm\,m^{-3}\,mol^{-1}}$.
- Polarity – Mobile compounds in soil and sediment tend to be negatively charged and, therefore, repelled by negatively charged clay particles. Positively charged compounds tend to be adsorbed to clay, although this may be blocked by H^+. Therefore, low pHs (high $[H^+]$) may increase chemical mobility in soils and sediments.

RATE OF DEGRADATION

The rate of degradation of a compound is dependent not only on the properties of the compound itself but also on the ambient conditions (i.e., whether the compound is in solution or adsorbed to particulates, the organic content of sediment or soil, the redox potential, and the pH, among others). Chemical persistence in sediment or soil is often recorded empirically as a range of chemical half-lives ($t_{1/2}$) under different physicochemical conditions. Compounds having a half-life of <5 days in sediment or soil are regarded as nonpersistent, whereas those with a $t_{1/2} > 60$ days are designated as persistent. In the United Kingdom, environmental regulators use the groundwater ubiquity score (GUS) to assess the potential for a particular compound to contaminate groundwater:

$$GUS = (1g \text{ solilt } t_{1/2}) \cdot [4 - (1 g K_{oc})]$$

Bacterial degradation of xenobiotic chemicals in ecosystems is of critical importance in determining their fate and often their degree of effect on other organisms. Relative to gut microflora, comparatively little is known about bacterially mediated chemical transformation in the natural environment, although the situation is rapidly changing. Although such a subject is beyond the scope of the text, fortunately there are some excellent reviews of the subject, notably that of Neilson (1994). Neilson stresses the large number of alternative pathways used by prokaryotes in biotransformation, and although his primary focus is on bacterially mediated metabolism of organic compounds, his extensive coverage of fungal metabolism provides examples of models that have proven to be good models of metabolic pathways utilised by mammals.

7.11 Case studies

Case Study 7.1. Pathology of beluga whales in the St. Lawrence estuary, Quebec, Canada

An isolated population of beluga whales (*Delphinapterus leucas*) has existed for probably several thousand years in the St. Lawrence estuary and in the Gulf of St. Lawrence. Increased hunting pressure reduced numbers from as many as 10,000 individuals at the beginning of the twentieth century to no more than 500 by 1979, when the Canadian government intervened and completely banned their hunting.

Since then, the population has not expanded, and reasons have been sought for the unexpected lack of increase in the population. Low reproduction rates and habitat degradation by hydroelectric power projects have been cited as contributory factors, but there is also evidence that toxic chemicals may have had a detrimental effect on the health of the beluga whale population.

The summer distribution of the animals is close to the confluence of the St. Lawrence River, draining the Great Lakes and the Saguenay River, which drains the Lake St. Jean area on the Canadian preCambrian Shield.

Between 1983 and 1984, more than 175 belugas were found dead in the St. Lawrence estuary, and these have formed the basis of an epidemiological study to determine the relationship between chemical contamination and pathological condition. Over that period, 73 of the carcasses were subjected to full autopsy. Specimens examined covered the full age range from 0 to 30 years. Adult belugas range from 3.6 m (females) to 4.5 m (males). Comparisons were made with a sample of 171 belugas, killed in Alaska and examined in the early 1980s, and with other species of whales and seals from the St. Lawrence.

St. Lawrence belugas had higher or much higher concentrations of mercury, lead, Mirex, DDT, and PCBs than Arctic belugas and lower levels of cadmium. No PCDDs and only low levels of a few PCDFs were found in St. Lawrence belugas, similar to their Arctic counterparts. PAHs and associated metabolites were comparable in both beluga populations with the exception of benzo(a)pyrene, which was much higher in the St. Lawrence animals.

The very high levels of total DDT and PCBs in St. Lawrence beluga whales (>10 × Arctic belugas) were consistent with the chronic contamination of Great Lakes fauna by these compounds during the 1970s. However, there was no indication of a decline in DDT and PCB levels, as seen in many fish species since the 1980s. Unlike other St. Lawrence cetaceans, DDT and PCB concentrations in the belugas were not inversely related to size and continued to increase with age, although substantial loads were passed on from females to young via the placenta or through lactation. Particularly high levels of Mirex and other organochlorines could be traced principally to Lake Ontario, from where migrating fish species (e.g., eels) and suspended particulates have been suggested as major chemical vectors to the whale population (Béland et al., 1993). Elevated benzo(a)pyrene levels in St. Lawrence belugas may be due to several different sources: The St. Lawrence and Saguenay Rivers support a variety of chemical manufacturing industries, many of which burn fossil fuels, and Saguenay River sediments are locally highly contaminated with benzo(a)pyrene from aluminium smelters.

Pathological examination of 73 St. Lawrence belugas indicated a variety of symptoms including an unusually high incidence of tumours (40%), 14 of which were cancerous. Tumours are extremely rare in free-ranging cetaceans and were not found in the Arctic belugas and other whales and seals from the St. Lawrence. More than half of the neoplasms in St. Lawrence belugas were found in tissues of the digestive system. Of the adult females, 45% possessed lesions to the mammary glands, which correlated with compromised milk production.

Reproductive dysfunction in St. Lawrence belugas was suggested by the fact that, unlike Alaskan belugas, no females above 21 years old were discovered to be pregnant or had recently given birth. This and other measures of fecundity indicated that a significant portion of the St. Lawrence population was not as productive as Arctic belugas.

The isolated St. Lawrence beluga whale population has provided a unique opportunity to study the epidemiology of chemical-related pathology in a mammalian species occupying a position in the food chain similar to humans. Even though the cause-and-effect relationship is equivocal in some respects, the study supports a causal relationship between several toxic compounds and the health and fecundity of the population.

Case Study 7.2. Recovery of double-crested cormorants (*Phalacrocorax auritus*) in the Great Lakes

Great Lakes populations of this large fish-eating bird fell dramatically throughout the 1960s, and, by 1970, the species had declined to only about 10% of its population level in the 1950s. There were several reasons for this. Double-crested cormorants were seen by many people as a nuisance species, competing with humans for valuable fish resources, and were killed by shooting and by destruction of eggs and nesting sites. Stocks of commercial fish declined rapidly during the 1960s and, no doubt, contributed to declines in piscivorous bird species. High levels of both DDE and PCBs in eggs and developing chicks were also implicated in this regard, and egg-shell thickness declined by approximately 30% during the 1960s, leading to a critically high egg destruction rate during incubation. Fecundity declined from a "normal" level of approximately 2 chicks per pair to an average of less than 0.2 chicks per pair. Poor reproductive performance was accompanied by a variety of deformities in developing young, including club feet, extra digits, crossed bills, and eye and skeletal deformities. By some accounts, no more than 100 breeding pairs existed in the Great Lakes region in 1970, and during the first part of that decade, several states placed the species on the endangered species list.

In the mid 1970s, cormorant numbers began an even more dramatic recovery, and, from 1978 to 1991, the average annual rate of population increase was 35%. Over 38,000 breeding pairs were recorded in the Great Lakes region in 1991. By that time, egg-shell thickness had recovered, and, in Lake Ontario, the hatching rate was up to 1.9 per breeding pair.

The recovery of the species has been attributed to a reversal in all factors that had contributed to the species' demise: Human persecution of cormorants has fallen to a low level as the perception of these birds as nuisances has declined. The emergence of the forage fish (*Alosa pseudoharengus*), also known as alewives, as the preferred prey of cormorants, particularly during the breeding season, clearly improved reproductive condition. Decline in PCB, and particularly DDF, concentration in adult birds and eggs was associated with an increase in egg-shell thickness to precontamination levels.

It is clear that chlorinated compounds were just one of a cumulative set of factors that contributed to cormorant population fluctuations; this case study might equally belong in Chapter 9. Nevertheless, there is a general consensus that their influence was probably critical, and that the extent of the observed recovery would not have been sustained without regulation of these chemicals.

Case Study 7.3. Feminisation of fish in English rivers

Vitellogenin is a precursor of egg yolk that is synthesised in the liver of developing female salmonids under the control of endogenous estradiol. It circulates in high concentration in the blood of maturing female fish and is incorporated into growing oocytes where it stimulates their rapid development prior to ovulation. Vitellogenin may be determined in the blood of fish using a specific radioimmunoassay.

In male rainbow trout (*Oncorhynchus mykiss*), circulating levels of endogenous estradiol are below detectable limits, and there is normally no expression of the gene responsible for VTG production. If exposed to exogenous sources of estrogens, however, male fish are able to synthesize VTG in quantities approaching those of a female fish.

Using this assay, it has been shown that the majority of effluents from sewage treatment plants in the United Kingdom are estrogenic to fish. Figure 7.36 shows mean

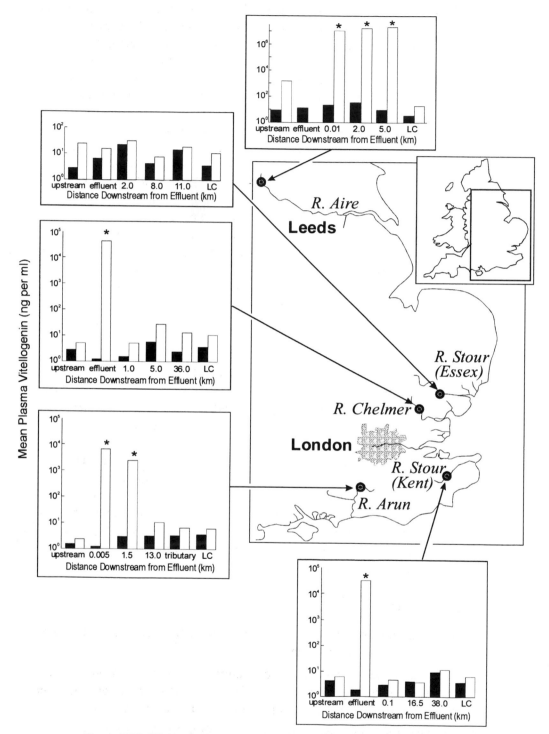

Figure 7.36 Effect of sewage treatment plant effluents on plasma vitellogenin levels in caged rainbow trout (*Oncorhynchus mykiss*) held in British rivers. The *x*-axis shows distances (km) downstream from effluent outfall. Results from fish held upstream from the effluent and from laboratory controls (LC) are also shown.

plasma VTG concentrations in caged male rainbow trout before and after being held in five English rivers for 3 weeks during the summer of 1994 (Harries et al., 1997). Fish were entrained in the effluent from selected sewage treatment plants on these rivers and at three downstream stations and one station upstream from each STP. Four of these rivers showed highly significant ($p < .001$) elevations in plasma VTG relative to control fish held in the laboratory and to fish held upstream from the STP.

The discharge from the Bures STP on the River Stour was the only effluent that failed to stimulate an increase in VTG and was the only negative result from this and an earlier effluent survey (Harries et al., 1996). The largest response in terms of both VTG levels and geographical extent was obtained from the River Aire in northern England where highly significant effects were recorded 6 km downstream from the Marley STP (Figure 7.36, top graph). Plasma VTG concentrations at the 5-km station were no different from fish held only 10 m from the Marley outfall and represented a millionfold increase over controls. High VTG levels were associated with highly significant inhibition of testicular growth in these fish (recorded as a low gonadosomatic index) coupled with an increase in liver weight relative to total body weight, the hepatosomatic index (Figure 7.37). The absence of data from fish exposed to undiluted effluent from the Marley site (Figure 7.37) was due to the death of those fish.

Implicated as hormone disrupters in this study were the alkylphenol polyethoxylates (APEOs), industrial surfactants having a worldwide production of 300,000 tonnes during the 1980s. However, natural estradiol and estrone and synthetic ethnylestradiol were also found in sewage effluents, and the relative importance of natural and artificial hormones and hormone disrupters of industrial origin is still being debated.

In extending their study to the roach (*Rutilus rutilus*), a cyprinid, Jobling et al. (1998) were able to demonstrate that hormonal changes were associated with intersexuality, the development of both male and female gonadal material in the same fish. Using a seven-point scale in describing the relative degree of disappearance of the sperm duct and the appearance of primary and secondary oocytes in genetically male fish, they were able to correlate the intersex condition with the fishes' proximity to the sewage outfall as summarised in Figure 7.38.

7.12 Questions

1. Give an account of the principal factors influencing the mobility of organic chemicals in the environment.

2. Define bioconcentration, bioaccumulation, and biomagnification in the context of inorganic substances. Explain the difference in application of these concepts for inorganic and organic substances respectively.

3. Describe the most important factors affecting the bioaccumulation and toxicity of chlorinated organic compounds, giving specific examples.

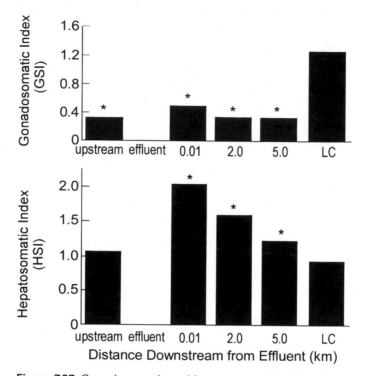

Figure 7.37 Gonadosomatic and hepatosomatic indices in rainbow trout (*Oncorhynchus mykiss*) upstream and downstream from the Marley sewage treatment plant on the River Aire, England (see Figure 7.36).

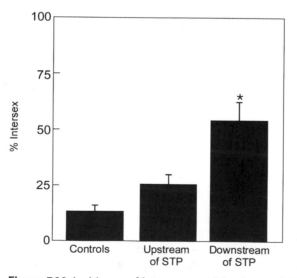

Figure 7.38 Incidence of intersex condition in roach (*Rutilus rutilus*) relative to proximity to sewage outfalls in English rivers.

4. Write an essay on the structure, mode of action, and environmental toxicology of the following classes of pesticide: (a) organophosphates, (b) phenoxyacid herbicides.

5. Give an account of the toxicity, history of use, regulation, and environmental consequences of environmental legislation on (a) polychlorinated biphenyls and (b) tri(n)butyltin.

6. Describe the toxicity and metabolism of tetrachlordibenzodioxin.

7. Give a critical account of the epidemiological and environmental evidence that organic compounds may disrupt endocrine function in humans and other animals. Include specific putative endocrine disrupters.

8. Describe the principal components and catabolic activity of the mixed function oxidase (monooxygenase) system. Give examples of some common metabolic reactions carried out by this enzyme system.

7.13 References

Alzieu, C. (1986) TBT Detrimental Effects on Oyster Culture in France – Evolution Since Antifouling Paint Regulation. In *Oceans '86 Proceedings International Organotin Symposium*, pp. 1130–4, Institute of Electrical and Electronics Engineers, New York.

Alzieu, C., and Portman, J. E. (1984) The effect of tributyltin on culture of *C. gigas* and other species, *Proceedings of Annual Shellfish Conference*, **15**, 87–101.

Anderson, M. J., Olsen, H., and Matsumura, F. (1996) *In vivo* modulation of 17β-estradiol-induced vitellogenin synthesis and estrogen receptor in rainbow trout (*Oncorhynchus mykiss*) liver cells by β-naphthoflavone, *Toxicology and Applied Pharmacology*, **137**, 210–18.

Arukwe, A., Förlin, L., and Goksøyr, A. (1997) Xenobiotic and steroid biotransformation enzymes in Atlantic salmon (*Salmo salar*) liver treated with an estrogenic compound, 4-nonylphenol, *Environmental Toxicology and Chemistry*, **16**, 2576–83.

Baker, J. E., Poster, D. L., Clark, C. A., Church, T. M., Scudlark, J. R., Ondor, J. M., Dickhut, R. M., and Cutter, G. (1997) Loadings of atmospheric trace elements and organic contaminants to the Chesapeake Bay. In *Atmospheric Deposition of Contaminants to the Great Lakes and Coastal Waters*, ed. Baker, J. E., pp. 171–94, SETAC Press, Pensacola, FL.

Béland, P., DeGuise, S., Girand, C., Lagacé, A., Martineau, D., Michaud, R., Muir, D. C. G., Norstrom, R. J., Pelletier, E., Ray, S., and Shugart, L. R. (1993) Toxic compounds and health and reproductive effects in St. Lawrence beluga whales, *Journal of Great Lakes Research*, **19**, 766–75.

Beltin, C., Oehlmann, J., and Stroben, E. (1996) TBT-induced imposex in marine neogastropods is mediated by an increasing androgen level, *Helgoländer Meeresuntersuchungen*, **50**, 299–317.

Bidleman, T. F., Patton, G. W., Walla, M. D., Hargrave, B. T., Vaso, W. P., Erickson, P., Fowler, B., Scott, V., and Gregor, D. J. (1989) Toxaphene and other organochlorines in Arctic Ocean fauna: evidence for atmospheric delivery, *Arctic*, **42**, 307–13.

Bitman, J., Cecil, H. C., Harris, S. J., and Fries, G. F. (1968) Estrogenic activity of *o,p'*-DDT in the mammalian uterus and avian oviduct, *Science*, **162**, 371–2.

Carlsen, E., Giwereman, A., Keiding, N., and Skakkebaek, N. E. (1992) Evidence for decreasing quality of semen during last 50 years, *British Medical Journal*, **305**, 609–13.

Colborn, T., von Saal, F. S., and Soto, A. M. (1993) Developmental effects of endocrine-disrupting chemicals in wildlife and humans, *Environmental Health Perspectives*, **101**, 378–84.

Dover, M. J., and Croft, B. A. (1986) Pesticide resistance and public policy, *Bioscience*, **36**, 78–85.

Focardi, S., Gaggi, C., Chemello, G., and Bacci, E. (1991) Organochlorine residues in moss and lichen samples from two Antarctic areas, *Polar Record*, **162**, 241–4.

Fry, D. M. (1995) Reproductive effects in birds exposed to pesticides and industrial chemicals, *Environmental Health Perspectives*, **103**, 165–71.

Gibbs, P. E., and Bryan, G. W. (1986) Reproductive failure in populations of the dog-whelk, *Nucella lapillus*, caused by imposex induced by tributyltin from antifouling paints, *Journal of Marine Biological Association of the United Kingdom*, **66**, 767–77.

Giusti, R. M., Iwamoto, K., and Itatch, E. E. (1995) Diethylstilbestrol revisited: A review of long-term health effects, *Annals of Internal Medicine*, **122**, 778–88.

Gorontzy, T., Küver, J., and Blotevogel, K.-H. (1993) Microbial reduction of nitroaromatic compounds under anaerobic conditions, *Journal of General Microbiology*, **139**, 1331–6.

Gregor, D. J., and Gunner, W. D. (1989) Evidence for atmospheric transport and deposition of organochlorine pesticides and polychlorinated biphenyls in Canadian Arctic snow, *Environmental Science and Technology*, **23**, 561–6.

Guillette, J. L. J., Gross, T. S., Mason, G. R., Matter, J. M., Percival, H. F., and Woodward, A. R. (1994) Developmental abnormalities of the gonad and abnormal sex-hormone concentrations in juvenile alligators from contaminated and control lakes in Florida, *Environmental Health Perspectives*, **102**, 680–8.

Haasch, M. L., Henderson, M. C., and Buhler, D. R. (1998) Induction of CYP2M1 and CYP2K1 lauric acid hydroxylase activities by peroxisome proliferating agents in certain fish species: Possible implications, *Marine Environmental Research*, **46**, 37–40.

Hardell, L., Eriksson, M., and Degerman, A. (1994) Exposure to phenoxyacetic acids, chlorophenots or organic solvents in relation to histopathology, stage and anatomical localization of non-Hodgkin's lymphoma, *Cancer Research*, **54**, 2386–9.

Harries, J. E., Sheahan, D. A., Jobling, S., Mattiessen, P., Neall, P., Routledge, E. J., Rycroft, R., Sumpter, J. P., and Taylor, T. (1996) A survey of estrogenic activity in United Kingdom inland waters, *Environmental Toxicology and Chemistry*, **15**, 1993–2002.

Harries, J. E., Sheahan, D. A., Jobling, S., Mattiessen, P., Neall, P., Sumpter, J. P., Tylor, T., and Zaman, N. (1997) Estrogenic activity in five United Kingdom rivers detected by measurement of vitallogenesis in caged male trout, *Environmental Toxicology and Chemistry*, **16**, 534–42.

Hickey, J. J., and Anderson, D. W. (1968) Chlorinated hydrocarbons and eggshell changes in raptorial and fish-eating birds, *Science*, **162**, 271–3.

Hickey, J. J., and Hunt, I. B. (1960) Initial songbird mortality following Dutch elm disease control program, *Journal of Wildlife Management*, **24**, 259.

Horiguchi, T., Shiraishi, H., Shimizu, M., and Morita, M. (1997) Effects of triphenyltin chloride and five other organotin compounds on the development of imposex in the rock shell, *Thais clavigera. Environmental Pollution*, **95**, 85–91.

Hunt, J. M. (1995) *Petroleum Geochemistry and Geology*, W. H. Freeman, New York.

Jobling, S., Nolan, M., Tyler, C. R., Brighty, G., and Sumpter, J. P. (1998) Widespread sexual disruption in wild fish, *Environmental Science and Technology*, **32**, 2498–506.

Jones, K. C., Sanders, G., and Wild, S. R. (1992) Evidence for the decline of PCBs and PAHs in rural vegetation and air in the United Kingdom, *Nature*, **356**, 137–41.

Jobling, S., and Sumpter, J. P. (1993) Detergent components in sewage effluent are weakly oestrogenic in fish. An *in vitro* study using rainbow trout (*Oncorhyncus mykiss*) hepatocytes, *Aquatic Toxicology*, **27**, 361–72.

Knutson, J. C., and Poland, A. (1982) Response of murine epidermis to 2,3,7,8-tetrachlorodibenzon-p-dioxin: Interaction of the Ah and Nr loci, *Cell*, **30**, 225–34.

Law, R. J., Dawes, V. J., Woodhead, R. J., and Matthiessen, P. (1997) Polycyclic aromatic hydrocarbons (PAH) in seawater around England and Wales, *Marine Pollution Bulletin*, **34**, 306–22.

Leister, D. L., and Baker, J. E. (1994) Atmospheric deposition of organic contaminants to the Chesapeake Bay, *Atmospheric Environment*, **28**, 1499–520.

Lerda, D., and Rizzi, R. (1991) Study of reproductive function in persons occupationally exposed to 2,4-dichlorophenoxyacetic acid (2,4-D), *Mutation Research*, **262**, 47–50.

Neilson, A. H. (1994) *Organic Chemicals in the Aquatic Environment: Distribution, Persistence and Toxicity*, Lewis, Boca Raton.

Patlak, M. (1996) A testing deadline for endocrine disrupters, *Environmental Science and Technology*, **30**, 540a–4a.

Poster, D. L., and Baker, J. E. (1997) Mechanisms of atmospheric wet deposition of chemical contaminants. In *Atmospheric Deposition of Contaminants to the Great Lakes and Coastal Waters*, ed. Baker, J. E., pp. 51–72. SETAC Press, Pensacola, FL.

Raven, P. J., and George, J. J. (1989) Dursban spill on the River Roading and its impact on arthropods, *Environmental Pollution*, **59**, 55–70.

Richardson, M. L., and Bowdon, J. M. (1985) The fate of pharmaceutical chemicals in the aquatic environment – A review, *Journal of Pharmacy and Pharmacology*, **37**, 1–12.

Ronis, M. J. J., and Mason, A. Z. (1996) The metabolism of testosterone by periwinkle (*Littorina littorea*) in vitro and in vivo: Effects of tributyltin, *Marine Environmental Research*, **42**, 161–6.

Safe, S (1980) Metabolism, Uptake, Storage and Bioaccumulation. In *Halogenated Biphenyls, Terphenyls, Napthalenes, Dibenzodioxins and Related Products*, ed. Kimbrough, R. D., Elsevier, North-Holland, New York.

  (1990) Polychlorinated biphenyls (PCBs), dibenzo-*p*-dioxins (PCDDs), dibenzofurans (PCDFs) and related compounds: environmental and mechanistic considerations which support development of toxic equivalency factors (TEFs), *Critical Reviews in Toxicology*, **21**, 51–88.

Safe, S., Astroff, B., Harris, M., Zacharewski, T., Dickerson, R., Romkes, M., and Biegel, L. (1991) 2,3,7,8-Tetrachlorodibenzo-*p*-dioxin (TCDD) and related compounds as antioestrogens: Characterization and mechanism of action, *Pharmacology & Toxicology*, **69**, 400–9.

Silberner, J. (1986) Common herbicide linked to cancer, *Science News*, **130**, 167.

Soto, A. M., Sonnenschein, C., Chung, K. L., Fernandez, M. F., Olea, N., and Serrano, F. T. (1995) The e-screen assay as a tool to identify estrogens: An update on estrogenic environmental pollutants, *Environmental Health Perspectives*, **103**, 113–22.

Stan, H.-J., Herberer, T., and Linkerhagner, M. (1994) Occurrence of clofibric acid in the aquatic system – Is the use in human medical care the source of the contamination of surface, ground and drinking water?, *Vom Wasser*, **83**, 57–68.

Thomann, R. V., and DiToro, D. M. (1983) Physico-chemical model of toxic substances in the Great Lakes, *Journal of Great Lakes Research*, **9**, 474–96.

Toppari, J., Larsen, J. C., Christiansen, P., Giwereman, A., Grandjean, P., Guillette, L. J., Jegou, B., Jensen, T. K., Jouannet, P., Keiding, N., Leffers, H., McLachlan, J. A., Meyer, O., Muller, J., Rajpert-Demeyts, E., Sheike, T., Sharp, R., Sumpter, J., and Skakkebaek, N. E. (1996) Male reproductive health and environmental xenoestrogens, *Environmental Health Perspectives*, **104**, 741–803.

Whelan, J. K., Kennicult, M. C., II, Brooks, J. M., Schumacher, D., and Eglinton, L. B. (1993) Organic geochemical indicators of dynamic fluid flow processes in petroleum basins, *Advances in Organic Geochemistry*, **22**, 587–615.

Wolff, M. S., Toniolo, P. G., Lee, E. W., Rivera, M., and Dublin, N. (1993) Blood levels of organochlorine residues and risk of breast cancer, *Journal of the National Cancer Institute*, **85**, 648–52.

8

○ ○

Ionising radiation

8.1   Introduction

This chapter introduces students to the basic toxicology associated with ionising radiation, beginning with some definitions of underlying processes and the terms that we use to describe them. The unique nature of radionuclides requires quite a different approach to their toxicology from that applied to stable elements, although certain shared concepts, such as threshold toxicity, are noted and cross-referenced with other chapters. In this chapter, we are concerned with the potentially toxic nature of the radionuclides themselves, while recognising that they share the same chemical toxicity as the respective stable element. For example, at the molecular level, radioactive ^{210}Pb has the same toxicity as its stable counterpart ^{207}Pb. This fact, of course, forms the rationale for the use of radioisotopes as tracers in biochemistry, physiology, and ecology. With the assumption that the radioactive element behaves in the same way as its stable isotopes, the passage and the dynamics of numerous substances through various components of the ecosystem can be traced. This aspect of radionuclides is not dealt with in this chapter.

Before the adverse health effects of radiation exposure were fully recognised, deliberate exposure to radioisotopes often formed part of therapies for a variety of ailments. It was not until after intense nuclear weapons testing programmes in the 1950s and 1960s together with accumulated information from victims of the two wartime nuclear detonations that the full extent of harmful effects began to be fully understood. Even then, this realisation took many years to develop. Thus, we note that, in the 1970s and 1980s, some aspects of what might be termed the nuclear industry, notably disposal, were poorly regulated. This oversight was particularly apparent in some military establishments where the deployment of weapons was given a higher priority than certain regulatory details. For example, in the former Soviet Union, low- and medium-level radioactive wastes were routinely dumped on the ground with few adequate safeguards. In countries that paid some attention to containment of waste radiation materials, it is clear that earlier methods were often critically inadequate and leaks common. As a society, we are still recipients of a legacy of past inadequacies of containment and disposal. Several well-publicised, major accidents at nuclear facilities, beginning as far back as the late 1950s, have served as sharp reminders to be extremely vigilant at all stages of what has come to be known as the nuclear cycle. Intensive studies of miners and mill workers have indicated that this vigilance should be extended to the ore extraction and refinement stages. Almost all the available information concerning the toxicology of radionuclides is, therefore, derived from a number of inadvertent "experiments" on living systems stemming from accidental or even deliberate radiation exposure. As such, the subject is dominated by the consequences of human/mammalian radiation, and much less is known about the effects of radiation on other organisms and ecosystems as a whole.

The roles of nuclear weapons programmes and electricity generation as potential radiation hazards have strong political implications. This affects choices for methods of power generation as a whole, which is discussed in Sections 9.4.2 and 9.4.4.

8.2   Definitions

8.2.1   What is ionising radiation?

When matter is subdivided, the molecule is the smallest amount that has the same properties as the original material. Molecules can be further subdivided into atoms, each type of atom being characteristic of an element. When atoms are divided, the constituents found are electrons, protons, and neutrons. These can be further broken down to yet smaller components, the nature of which is highly theoretical.

The nucleus of an atom consists of positively charged protons and electrically neutral neutrons. Together they form the nucleon, which comprises 99% of the atomic mass of the atom. The nucleus is surrounded by a cloud of negatively charged electrons. Normally, the negative electrical charge associated with the electrons balances the positive charge on the protons. Atoms of the same element all have the same number of protons in the nucleus, but, within an element, the total number of neutrons may vary. Elements may, therefore, exist in two or more different forms having slight differences in mass. These different forms are called isotopes. For example, 99% of naturally occurring carbon in the environment (^{12}C) contains six neutrons and six protons in the nucleus. However, 1% of naturally occurring carbon (^{13}C) has an extra neutron in the nucleus. Both of these isotopes are stable, and, although they have virtually identical chemical properties (as determined by their electron configuration), their slightly differing mass causes tiny but discernible differences in how they are partitioned in various components of the ecosystem. Isotopes of the same element may be differentiated analytically using a mass spectrometer (see Section 6.2.4).

Many naturally occurring isotopes are unstable and may emit radiation as a result of their nuclear instability. These are known as radioisotopes or radionuclides. All elements above bismuth in the periodic table are naturally radioactive, and several of the lighter elements have one or more radioactive forms. Radioisotopes are also formed as a result of human activity.

Energy is released as a radioisotope tends to revert to a less energetic state. This process is called radioactive decay. Each radioisotope has highly specific decay characteristics in terms of both speed of decay and the type of energy released, and these factors are utilised in detecting specific radioisotopes in minute quantities. Natural ionising radiation takes three principal forms, alpha (α) particles, beta (β) particles, and electromagnetic photons consisting of gamma (γ) rays and X-rays. Neutron radiation is only produced from synthetic radionuclides.

Alpha particles. Alpha decay produces alpha particles, which consist of two neutrons and two protons and carry a +2 charge, the equivalent of a helium nucleus. As a result of alpha decay, the resulting atom or daughter element has an atomic number two units lower and an atomic weight four units lower than that of the parent radionuclide. In other words, it drops two places in the periodic table. Although alpha particles have very high ionising capability, their energy is quickly dissipated in their passage through matter. They will only travel through a few centimetres of air and will be stopped by a piece of paper or the outer epithelium of an organism. Alpha emitters are of concern only if taken into the body by inhalation or ingestion.

Beta particles. Beta decay produces beta particles, which are negatively charged electrons formed when an excess of neutrons in the nucleon causes a neutron to be changed into a proton. If nuclear imbalance results in an excess of protons in the nucleon, the emitted particle is the positively charged equivalent of an electron known as a positron. This causes a proton to be changed to a neutron. Alternatively, the proton may be converted to a neutron with the capture of a satellite electron, a process known as electron capture. β emitters vary widely in their energy output, although β particles are stopped by a few millimetres of plastic or aluminium, or about a centimetre of tissue.

Electromagnetic photons. Electromagnetic photons released by radioisotopes consist of gamma rays or X-rays, which differ in their source. The γ rays originate from the atomic nucleus, whereas X-rays are emitted from the orbital electrons. Additionally, γ rays are emitted at energy levels that are highly specific to a particular radionuclide, a further diagnostic feature when identifying unknown γ emitters. They are essentially similar to photons of visible light and travel at the same speed as light. Their energy is inversely proportional to their wavelength, and highly energetic photons with very short wavelengths are capable of passing completely through a human body. Some properties of different types of photomagnetic radiation are shown in Figure 8.1. Both γ and X-rays are strongly ionising, and a thick protective layer of dense material such as lead or concrete is required to absorb all their energy. Absorption characteristics have been derived for a variety

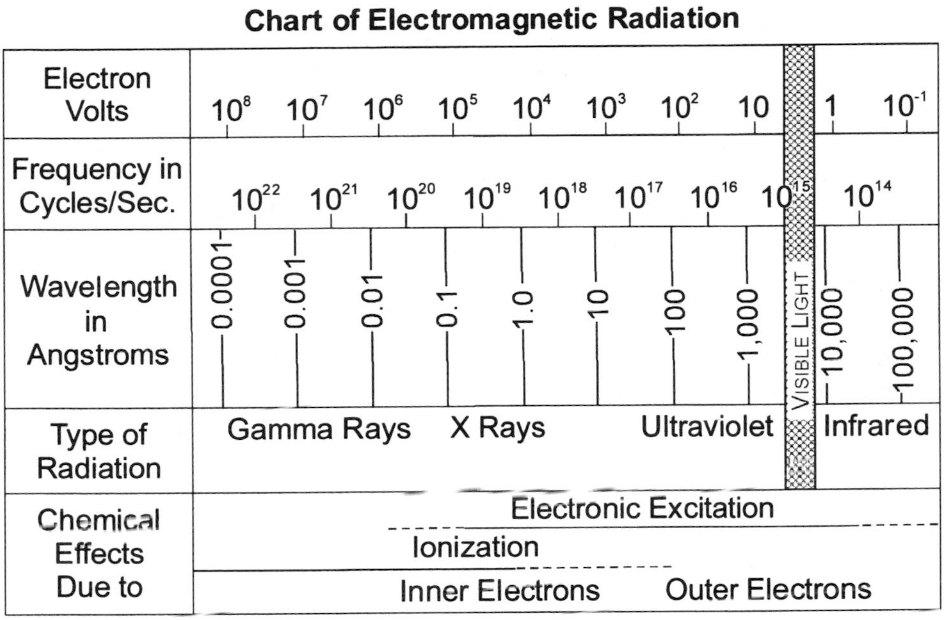

Chart of Electromagnetic Radiation

Electron Volts	10^8	10^7	10^6	10^5	10^4	10^3	10^2	10		1	10^{-1}
Frequency in Cycles/Sec.	10^{22}	10^{21}	10^{20}	10^{19}	10^{18}	10^{17}	10^{16}	10^{15}		10^{14}	
Wavelength in Angstroms	0.0001	0.001	0.01	0.1	1.0	10	100	1,000	VISIBLE LIGHT	10,000	100,000
Type of Radiation	Gamma Rays		X Rays				Ultraviolet			Infrared	
Chemical Effects Due to	Electronic Excitation / Ionization / Inner Electrons / Outer Electrons										

Figure 8.1 Properties of different types of electromagnetic radiation. After Grosch (1963).

of materials and have been summarised by Johns and Cunningham (1980). The attenuation of radiation is an exponential process, which is dependent on the type and energy of the emitter and the density of the absorbing material. For γ rays, attenuation can be expressed as

$$I = I_0 e^{-\mu d} \tag{8.1}$$

where I_0 is the initial measure of radiation (photons $cm^{-2} s^{-1}$), μ is the attenuation coefficient (cm^{-1}), and d is the thickness of attenuating material.

Each radionuclide has a characteristic decay rate, which is independent of temperature, physical form, or chemical combination. The rate of decay follows first-order kinetics such that

$$N = N_0 e^{-\lambda t} \tag{8.2}$$

where N is the number of atoms at time t, N_0 is the initial number of atoms at $t = 0$, and λ is the decay constant. The decay rate of any radionuclide is usually expressed in terms of its half-life ($t_{1/2}$), which is determined as

$$t_{1/2} = \frac{0.693}{\lambda} \tag{8.3}$$

The radioactive half-life is, therefore, the time taken for 50% of a particular radioisotope to decay to its daughter. Highly radioactive substances have short half-lives, whereas those with long half-lives have low radioactivity. Energy released

Table 8.1. Uranium radioactive decay series

Isotope	Half-life	Radiation
^{238}U	4.5×10^9 years	α, β
^{234}Th	24 days	β
^{234}Pa	1.17 min	β, γ
^{234}U	2.45×10^5 years	α, γ
^{230}Th	7.54×10^4 years	α, γ
^{226}Ra	1.6×10^3 years	α, γ
^{222}Rn	3.8 days	α, γ
^{218}Po	3.1 min	α
^{214}Pb	26.8 min	β, γ
^{214}Bi	19.9 min	β, γ
^{214}Po	164 μs	α, γ
^{210}Pb	22.3 years	β, γ
^{210}Bi	5 days	β
^{210}Po	138.4 days	α, γ
^{206}Pb (stable)		

as a result of radioactive decay is measured in millions of electron volts (MeV). α particles have energies between 3 and 6 MeV, β particles have energies between 0.005 and 3 MeV, and γ rays, 0.1 and 2 MeV.

The decay of one radioisotope may result in the formation of a daughter element, which may itself be unstable. This, in turn, will decay to a further element and so on through a decay chain until a stable element is formed. Probably the most important and best known decay chain is that of uranium-238, shown in Table 8.1.

Radiation from radioisotopes dissipates as it passes through matter (Equation 8.1). The absorption of this energy produces excited atoms by displacing electrons from one orbital to another or by the ejection of electrons with the subsequent formation of a charged ion, hence the term ionising radiation. Different forms of radiation produce ionisation by different mechanisms. Charged particles (α and β) act directly by transferring their kinetic energy to the neutral atom and dislodging an outer orbital electron. Electromagnetic radiation (γ and X-rays) acts indirectly by accelerating a secondary charged particle such as a free or loosely held electron, which produces the ionisation effect. Neutrons ionise by colliding with nitrogen and releasing a proton or by directly colliding with hydrogen. All processes have in common the production of an ion pair.

8.2.2   Units of measurement

When considering measurements of radioactivity, the situation is complicated by the fact that, while the traditional units of measurement are being phased out in favour of standard international (SI) units, the older units remain in active use in the United States. Thus, the old and new units continue to exist side by side (Table

Table 8.2. Units of measurement of radioactivity

Old unit	International (SI) unit
Measure of radioactive emission – Curie (Ci). Originally defined as the amount of radioactivity emitted by 1 g radium. Later standardised to 2.2×10^{12} nuclear disintegrations per minute (dpm).	Bequerel (Bq); one nuclear disintegration per second; $1\,Bq = 27.03\,pCi$
Measure of absorbed radioactivity – rad. Amount of radioactivity causing 1 kg tissue to absorb 0.01 J energy $(= 2.38 \times 10^{-6}\,cal\,g^{-1})$.	Gray (Gy); an energy absorption of $1\,J\,kg^{-1}$; $1\,Gy = 100\,rad$
Measure of destructive dose – rem. Equivalent of 1 rad of hard X-rays or 0.05 rad of α particles.	Sievert (Sv); replacement for rem as destructive dose equivalent; $1\,Sv = 100\,rem$

8.2). There are two types of radioactive measurement. One measures the amount of radioactivity emitted from a substance. The other is a measure of absorbed radioactivity, which, in turn, has been modified to take into account the destructive potential of different kinds of radiation.

Unlike nonradioactive chemicals, the damage from radioisotopes is usually related to their ionisation potential and the amount of radiation actually reaching the tissues. Therefore, for radioactive substances, it is common to use a unit of dose that represents the amount of energy deposited in tissues following exposure. The currently used SI unit for this is the Gray (Gy) (Table 8.2), which is equivalent to 1 Joule of energy per kilogram of tissue. Enough is known of the ionisation potential of different types of radiation to make further refinements to this concept of dose. When assessing radiation risk, it is important to consider that an α particle is capable of transferring to tissue 20 times more of its energy per unit distance travelled than a γ ray of similar energy. In other words, it has 20 times more linear transfer energy (LTE). Absorbed radiation doses are, therefore, normalised to dose equivalents by arbitrary factors known as quality factors or weighting factors (W_t), which take into account the LTE. The international measure of dose equivalent is the Sievert (Sv) (= 100 rem). Thus,

$$H(\text{equivalent dose in Sieverts}) = \text{Dose(Gy)} \times W_t$$

8.3 Effects of radiation at the molecular and cellular level

8.3.1 Molecular interactions

Energy absorbed by tissues exposed to radiation causes a variety of chemical changes in cells, resulting in a range of damaging effects. Some changes may be sufficiently severe that they cause tissue necrosis, which will lead to the death of the organism in extreme cases. Alternatively, exposure can lead to permanent changes in cellular metabolism, resulting in carcinogenesis or genetic disorders. These damaging effects originate at the molecular level and result from the ionisation of water molecules and the formation of the high energy radicals ·OH and ·H. The sequence or their formation follows:

Primary ionisation

$$H_2O \rightarrow H_2O^+ + e^-$$

Electron capture by neutral water molecule

$$H_2O + e^- \rightarrow H_2O^-$$

Breakdown of ionised water molecules

$$H_2O^+ \rightarrow H^+ + \cdot OH$$

$$H_2O^- \rightarrow \cdot H + OH^-$$

H^+ and OH^- combine to produce neutral water, resulting in the net reaction

$$H_2O \rightarrow \cdot H + \cdot OH$$

Although these free radicals have a very short lifetime in water, they are extremely reactive. They may react with each other to form water or two ·OH radicals may combine to form hydrogen peroxide:

$$\cdot OH + \cdot OH = H_2O_2$$

Hydrogen peroxide is a stable molecule but is a powerful oxidising agent. These radicals, together with H_2O_2, may interact with macromolecules such as nucleic acids, proteins, lipids, and large polymers. Large molecules may be broken up forming additional, organic radicals, which may initiate the formation of cross-linkages within the molecules or between molecules. In the latter case, molecular pairs, or dimers, may be formed, a process known as dimerisation. The proximity of oxygen to the initial site of chemical change increases the probability of damage. This may be due to increased H_2O_2 production, or through direct interaction with the organic radical. Interaction with proteins and lipids will cause a variety of adverse biochemical effects through the disruption of enzymes and membranes. It has been determined that 3,000–5,000 rad of absorbed dose is needed to rupture the plasma membrane of human cells. Disruption of lysosome membranes (in humans at doses between 500 and 1,000 rad) causes the release of damaging enzymes into the cytoplasm. Only a few thousand rad are required to disrupt mitochondrion function. At the microscopic level, they would appear swollen. Further examination would reveal that the internal structure has degraded, resulting in the disruption of ATP production for the cell. If the affected cell has insufficient ATP reserves, it cannot readily repair the damage, and death will ultimately result.

Radiation-induced DNA damage (e.g., point mutations, strand breakage, and chromosome damage) in somatic tissue will increase the incidence of carcinogenesis. In germinal tissue, such damage can lead to genetic disorders, which may be passed on to offspring or cause sterility.

Radiation may cause cell death in two different ways – mitotic death and interphase death. As its name suggests, interphase death occurs when the cell is not dividing. Not to be confused with necrosis, it is similar and perhaps identical to apoptosis, or programmed cell death (described in Section 2.3.1). Cells in the process of dividing (mitosis) are particularly sensitive to radiation. Such cells include bone marrow cells and stem cells of the gastrointestinal tract and skin, which are continually in a high state of replication in order to repair tissue damage. Young children and foetuses are particularly sensitive to radiation due to the high proportion of actively growing tissue. Mature, fully differentiated cells such as nerve and muscle cells are less sensitive to radiation. Even though blood cells such as erythrocytes and granulocytes are themselves resistant to radiation, their stem cells are not. This resistance eventually leads to a reduction in both red and white cell counts with attendant anaemia, reduction in oxygen-carrying capacity, and loss of immune function.

The lethal effect of radiation on replicating cells has been utilised in the treatment of malignant cancer cells, which have a high proliferation rate. Precisely directed doses of high-energy γ rays have been shown to destroy malignant tumours selectively. Damage to surrounding tissue is minimised by varying the angle of irradiation in controlled serial doses.

8.3.2 Effect of radiation on the immune system

The principal components of the mammalian immune system are lymphocytes, macrophages, and a series of immunological mediators, formed from macrophages, which regulate the activity of the lymphocytes. Macrophages are responsible for nonspecific immune reactions such as phagocytosis. Lymphocytes are highly specific in their immune response and are responsible for the recognition of foreign (nonself) antigens and the initiation and degree of the response. Small lymphocytes are functionally divided into two groups: B-cells produced by the bone marrow and T-cells produced by the thymus gland. B-cells are primarily responsible for humoral immunity and the production of circulating antibody. B-cells can differentiate into plasma cells. T-cells are mainly responsible for cellular immunity and hypersensitivity responses such as transplantation immunity and immunity to agents such as bacterial and viral antigens. Some subsets of T-cells known as helper or suppressor cells can modulate the effect of both T- and B-cells. A characteristic feature of T-cells is their high incidence of apoptosis prior to full differentiation.

For reasons that are not fully understood, some subpopulations of lymphocytes are extremely sensitive to radiation, but macrophages and plasma cells are very resistant. Generally speaking, radiation has a dose-dependent immunosuppressive

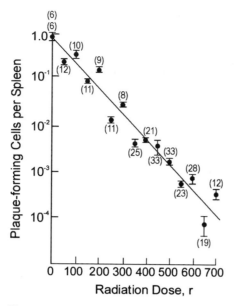

Figure 8.2 Effect of radiation dose on antibody production in mice. Modified from Kennedy et al. (1964).

effect. Figure 8.2 (from Kennedy et al., 1964) shows the effect of increasing doses of radiation on the ability of mice to produce an antibody to sheep red blood cells injected following irradiation. However, under certain experimental conditions, exposure to radiation may be associated with an increased immune response. Such a phenomenon appears to be related to the loss of a particular type of T-cell, which has a suppressive influence on the immune response.

8.4   Assessment of risk from radiation

In 1977, the International Commision on Radiological Protection (ICRP) introduced the term *effective dose equivalent* (H_t) to compare human carcinogenic and genetic risks from different tissue or whole-body radiation doses. The measure incorporates tissue weighting factors (w_t) in recognition of the fact that certain tissues are more susceptible than others to the effects of radiation. Individual tissue weighting factors are listed in Table 8.3. Table 8.4 shows estimated effects of one unit of low-dose, low-dose-rate irradiation on a population of one million people. Table 8.4 illustrates that, for an individual, the risk of harm depends not only on the dose but on the tissue exposed. The effective dose equivalent allows computation of risk from a mixture of whole-body and specific tissue exposures and is calculated as

$$H_t = w_t \times \text{Dose} \times W_t$$

Effective dose equivalents from different sources are assumed to be additive to calculate overall carcinogenic and genetic risk.

Table 8.3. Values for tissue-weighting factors (W_t) recommended by International Commission on Radiological Protection

Tissue/organ	Tissue weighting factor, W_t
Gonads	0.2
Colon	0.12
Lung	0.12
Red bone marrow	0.12
Stomach	0.12
Breast	0.05
Bladder	0.05
Liver	0.05
Oesophagus	0.05
Thyroid	0.05
Bone surface	0.01
Skin	0.01
Other tissues	0.05[a]

[a] Subject to modification if evidence of specific tissue irradiation exists.

Table 8.4. Estimated effects of low-level irradiation on different tissues[a]

Tissue	Effect	UNSCEAR[b] estimate (per 10^{-2} Gy)[d]	ICRP[c] estimate (per Sievert)[d]
Gonads	Mutations	63 (1st generation offspring of exposed parents)	10,000 (1st & 2nd generation offspring of exposed parents)
Body	All cancers	200	12,500
Breast	Fatal cancer	50 (population)	2,500 (workers)
Red bone marrow	Leukemia	20–50	2,000
Lung	Fatal cancer	25–50	2,000
Thyroid	Fatal cancer	10	500
Bone surfaces	Fatal cancer	2–5	500
Other tissues	Fatal cancer	35–93	5,000

[a] Risk is estimated as incidence of adverse effects per million persons.
[b] United Nations Scientific Committee on the Effects of Atomic Radiation.
[c] International Commission on Radiological Protection (1990).
[d] Units are as in original publications.

Estimates by the ICRP (1977) of human risk from radiation were based on data from Japanese atomic bomb survivors who had received doses between 0 and 4 Sv. In the earlier part of the study, observed effects were dominated by subjects who had received particularly high doses of radiation (>1 Sv). However, by 1990 attempts were being made to use the low-dose portion of the data to assess risks for a variety of different cancers (NAS, 1990). In this regard, toxicologists are faced with problems similar to those encountered with chemical toxicology at very low doses. Is there a toxic threshold, or is the dose-response linear? (See Section 2.3.4.) In the case of low-dose radiation, the possibility of protective or hormetic effects also had to be considered (see Section 2.3.5). For all cancers apart from leukemia, the data fitted a linear, nonthreshold dose response with no evidence to suggest that the slope in the lower dose portion of the curve was different from that based on the entire dose range. For leukemia there was some indication that the risk-versus-dose relationship might be better described by a linear quadratic curve where the slope in the low-dose range is shallower than in the high-dose range. However, the large confidence limits associated with these data mean that there is a high degree of uncertainty associated with low-dose risk.

To strengthen the assessment of risk from low-dose radiation, information has been gathered and analysed from a variety of industrial operations where workers face chronic exposure from external radiation. Probably the largest of these studies is that conducted at Hanford in southeast Washington State in the United States, which was established in the 1940s for plutonium production but which has since been expanded as a power generation and research facility. At the Hanford facility, over 32,000 workers with more than 6 months' work experience at the site have been monitored. Most of these workers received only very small occupational doses (<10 mSv), and only 10% received more than 50 mSv. This large data set gives no clear indication of cancer or genetic risks at these low exposures.

For the present, the relationship between radiation dose and harmful effects in humans is assumed by the ICRP to be linear with no threshold. This situation is true despite the fact that, in addition to epidemiological evidence that a threshold exists, several experimental studies with animals point to a threshold. Indeed, some studies suggest that organisms are capable of adaptive responses to very low doses of radiation, which may afford them some degree of protection against higher doses. In simple terms, the mechanism(s) by which such adaptation might be effected would be the induction of proteins involved in the protection from and/or repair of DNA damage caused by radiation. Hill and Godfrey (1994) provide a review of some of this evidence including data that suggest that the adaptive process might be triggered by stresses other than radiation (e.g., heat-induced radiation resistance). In light of such experimental evidence, the current ICRP assumption of a linear relationship between low radiation dose and deleterious effect remains controversial. Advocates of a linear dose response maintain that it errs on the conservative side (i.e., it overestimates risk). It has for its mechanistic basis the

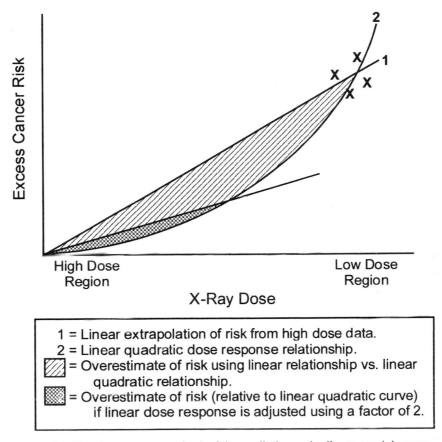

Figure 8.3 The dose-response for ionising radiation – the linear model versus the quadratic model.

one-hit model for carcinogenesis described in Section 2.3.1 and embodies the assumption that a molecular "hit" by a unit of radiation will result in a lesion. As contradictory evidence mounts, and more is learned about repair processes related to radiation, it may be necessary to modify this approach. The problem is illustrated in Figure 8.3, where we might make the assumption that the "true" dose-response is a linear quadratic represented by the curved line. The "currently accepted" linear relationship is based on noncontroversial high-dose data. Overestimated risk is, therefore, represented by the shaded area. Given all the uncertainties, future remedies may include the application of across-the-board correction factor. For example, it has been suggested by the ICRP that risk estimates might be halved when based on information from high doses at high dose rates. As more information is gathered on persons exposed to low-radiation doses, it is likely that risk estimates will be modified and may become more specific for different types of cancer and other radiation-induced disorders.

A special case arises where radionuclides are ingested and remain in the body as an irreversible radiation source subject to the decay half-life and the kinetics of elimination. The latter will depend on the particular tissues that accumulate the radionuclides. For example, those deposited in the bone will have a much longer residence time than those entering soft tissues. For this reason, the commited equivalent dose is used as the fundamental unit of risk for humans and takes into account the potential for a radiation dose to be delivered over a long period of time following ingestion. It is calculated on the basis of a "lifetime" of 50 years (70 years for children) and is equal to

$$H_{T50} = \int_{t0+50}^{t0} H_T dt$$

where H_{T50} = 50-year dose administered to tissue T as single intake at time T_0 and H_T = equivalent dose rate in tissue T at time t.

Radioisotopes of particular concern in this regard often include those produced by nuclear fission. For example, radium-226, a member of the uranium-238 decay series (see Table 8.1) with a half-life of 1,620 years, is an α and γ emitter having similar characteristics to calcium. As such, it is readily incorporated into bone. Strontium-90, a β emitter with a half-life of 29 days, is also largely deposited in bone. Caesium-137, a β emitter (half-life 30.2 years), replaces potassium in soft tissues such as muscle. Iodine-131 (β and γ emitter, half-life 8.1 days) is accumulated in the thyroid gland.

The ICRP has recommended an occupational exposure limit of 50 mSv (= 5 rem) yr^{-1} for workers in industries involving radionuclides but with stipulations relating to specific tissue exposures. For example, the ICRP's recommended annual dose limit for the large intestine is 15 mSv yr^{-1}. Detailed recommendations have been published on 240 different radionuclides that could be ingested or inhaled. For persons not exposed to radioactivity as a result of their profession, the recommended limit for annual whole-body dose is 5 mSv (= 0.5 rem) yr^{-1}. Risk factors published by the ICRP estimate the probability of adverse effect per mSv of dose received. Four categories are recognised.

Effect	Risk factor, mSv
Fatal cancer	5.0×10^{-5}
Nonfatal cancer	1.0×10^{-5}
Severe genetic effects	1.3×10^{-5}
Total deleterious effect	7.3×10^{-5}

Sensitivity to radiation varies widely among species. Figure 8.4, which compares acute lethal dose ranges for most taxa, indicates that mammals are the most radiosensitive group of animals yet studied. Of mammals studied so far, humans appear to be the most sensitive. Within taxonomic groups, it is generally found that younger developing organisms are more sensitive than older organisms.

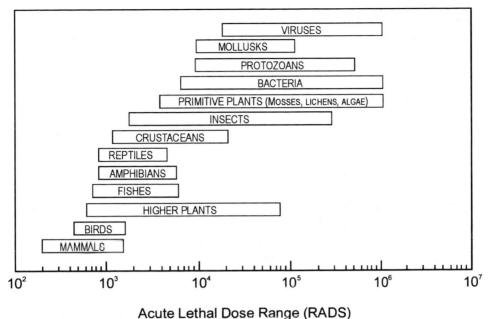

Figure 8.4 Effect of ionising radiation on different taxonomic groups.

8.5   Sources of radiation

8.5.1   Background radiation

Radioisotopes are present in all segments of the ecosystem, and radioactivity is distributed fairly homogeneously throughout the environment. Chemically, radioisotopes behave in the same way as their stable counterparts and are subject to the same uptake and physiological pathways. ^{40}K, for example, is concentrated in certain plants such as the coffee plant in the same way as the normal, stable isotope ^{39}K. High coffee consumption will, therefore, increase potassium intake and, therefore, exposure to ^{39}K. Uranium (principally occurring as ^{238}U) is a natural and ubiquitous component of rocks, particularly granite, and the degree of human exposure will be heavily influenced by relative proximity to different types of rock. Tritium (^{3}H) occurs principally in the atmosphere where it is produced by the reaction of cosmic ray protons and neutrons with atmospheric gases such as oxygen (O), nitrogen (N), and argon (Ar). Most tritium is produced by the reaction

$$^{14}N + n \rightarrow {}^{12}C + {}^{3}H$$

where n represents a $>4.4\,MeV$ neutron.

In terms of background radiation, cosmic radiation is approximately equal to natural terrestrial sources, although the most important source of natural radiation is through the inhalation of gaseous radioisotopes. The largest dose comes from short-lived daughters of radon (^{222}Rn) gas produced as part of the ^{238}U decay series.

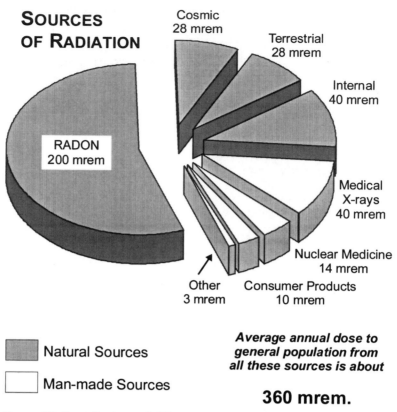

Figure 8.5 Contributions of different sources to background radiation.

Figure 8.5 shows the relative contribution of different sources to background radiation.

Nuclear energy is utilised principally for weapons production and the generation of electricity. Historically, release of radionuclides from nuclear weapons production and use (including testing) has substantially outstripped that from electricity generation. The advent of nuclear power has significantly altered the environmental input of radioisotopes, both quantitatively and qualitatively. For several elements, production related to anthropogenic activity greatly exceeds natural sources. For tritium, for example, production related to either electricity generation or the manufacture of weapons exceeds the global inventory from natural sources. While Plutonium-239 is a naturally occurring element, the largest global source is through nuclear weapons production.

8.5.2   Electricity production from nuclear power

Electricity is produced from nuclear energy principally through a chain reaction originated by the bombardment of the isotope ^{235}U with slow neutrons. Fission of the uranium atom results in a number of products of smaller molecular weight and the release of α,β particles and γ radiation together with high-energy neutrons. These collide with other ^{235}U nuclei to produce a chain reaction. The amount of energy produced per collision is 200 MeV (3.2×10^{-2} erg).

The nuclear fuel cycle is comprised of the following basic steps:

- Ore mining and milling
- Enrichment, conversion, and fabrication
- In-core fuel management
- Fuel reprocessing
- Waste disposal and storage

MINING AND EXTRACTION

The mode of extraction of ore-bearing rocks is determined by their depth and distribution. Underground mines typically harvest only about 50% of the total ore present, whereas surface mining can remove as much as 90–95% of the ore present in the deposit. Because the mean ^{235}U content of ore is 0.7% and the final product for commercial power generation must be 3–5% ^{235}U (much higher for military propulsion), considerable enrichment is required. This requires massive quantities of ore and the production of large amounts of mining overburden and mill tailings from the extraction process.

Mine tailings remain a source of uranium isotopes and decay products. Although legislation governing uranium mine and mill reclamation exists in the United States in the form of the Uranium Mill Tailings Act of 1978, contamination from tailings runoff and radioactive dust remain significant environmental issues, particularly where statutes are not rigorously enforced or in countries where no such laws exist. Even where masks and high-efficiency particle-arresting (HEPA) filters are used to trap fine particulates, miners and mill workers are not protected from gaseous radioisotopes such as radon. Several case studies have correlated the incidence of lung cancer with the inhalation of radon and daughter isotopes by uranium miners and mill workers. The actual carcinogen is not the inert radon gas itself but several of its particulate and short-lived strong α-emitting daughter decay products such as polonium-218. When such isotopes are inhaled and deposited in the lungs, they can transfer a significant amount of their energy through the thin bronchial epithelium to the underlying target cells.

ENRICHMENT, CONVERSION, AND FUEL FABRICATION

Uranium has only two significantly abundant natural isotopes, ^{235}U and ^{238}U, with natural abundances of 0.71 and 99.28%, respectively. The CANDU reactor and reactors using Magnox fuel use ^{238}U, whereas other commercial reactors must depend on fuel enriched 3–5% in ^{235}U. Currently, there are three known commercial isotopic enrichment processes: gas centrifugation, gaseous diffusion, and atomic vapour laser separation (AVLS). Enrichment by diffusion is based on diffusion kinetics at very high temperatures and uses gaseous UF_6 under pressure diffusing across a membrane. AVLS employs wavelength-tunable lasers to separate isotopes by differential photo-ionisation.

Wilson (1996) reported the existence of only 17 enrichment plants in 11 different countries, although roughly half of these have been shut down or operate at much reduced capacity.

IN-CORE FUEL MANAGEMENT

The two basic components of a nuclear fuel assembly are the nuclear fuel itself (either uranium, plutonium, or a mixture) and a protective (stainless steel) pressure vessel to contain the waste fissile material that is produced. The heat produced in the fission process is exchanged with water under high pressure, which is circulated at high velocity through the reactor. The rate of the chain reaction can be controlled by inserting boron control rods between the fuel rods, all of which are immersed in water. The control rods, or moderators, can be raised or lowered to increase or decrease the reaction rate. When 90% or more of the rods need to be removed to maintain the criticality, the nuclear fuel is considered to be "spent".

An operating reactor will produce about 80 radionuclides, with even more produced by decay chains yielding half-lives from a few milliseconds to billions of years. Of these, a subset of synthetic radionuclides has been the focus of particular concern due to their incorporation into biological systems and their significant longevity with respect to the total radiation emitted. These radionuclides follow:

$$\alpha \quad - \ ^{238}\text{U}, \ ^{232}\text{Th}$$

$$\alpha/\beta - \ ^{226}\text{Ra}$$

$$\beta \quad - \ ^{3}\text{H}, \ ^{14}\text{C}, \ ^{32}\text{P}, \ ^{90}\text{Sr}$$

$$\beta/\gamma - \ ^{40}\text{K}, \ ^{65}\text{Zn}, \ ^{131}\text{I}, \ ^{137}\text{Cs}$$

A brief account of some of these radionuclides is given in Section 8.5.3.

The release of some radioactivity during normal reactor operation is unavoidable. Some radioactive fission products leak through pinholes in the steel cladding surrounding the core into the cooling water. In addition to contamination by radioactive materials from the nuclear fuel, other radionuclides are formed by neutron activation of gases in the atmosphere, graphite (used as a moderator in some plants), and impurities in water in the vicinity of the reactor. Such neutron activation products include ^{76}As, ^{134}Cs, ^{137}Cs, ^{58}Co, ^{60}Co, ^{51}Cr, ^{64}Cu, ^{3}H, ^{85}Kr, ^{56}Mn, ^{32}P, ^{31}Si, and ^{65}Zn, which are often found in measurable quantities in liquid effluents from nuclear power plants. A nuclear reactor releases approximately 2×10^{11} Bq yr^{-1} of mixed fission and neutron activation products in liquid effluents. Larger amounts of ^{3}H, ^{85}Kr, ^{131}Xe, and ^{133}Xe are released as gaseous effluents. Routine radionuclide emissions from nuclear power plants are very small and are generally regarded as negligible in terms of their environmental effect.

REACTOR ACCIDENTS

Bodansky (1996) lists 11 electric nuclear reactor accidents involving containment breaches, although by far the most serious was the world's first complete reactor core meltdown at Chernobyl, Ukraine, in April 1986 (Case Study 8.1). This was estimated to have released up to 100 MCi (c. 370,000 TBq) of radioactivity over a large area of the northern hemisphere. Two other near-catastrophic accidents were the October 1957 fire at the plutonium-production reactor at Windscale (now Sellafield) on the northwest coast of England and the partial core meltdown at the Three Mile Island nuclear plant in Pennsylvania in the eastern United States in March 1979. The Windscale accident released approximately 7×10^{14} Bq (700 TBq) activity to the surrounding countryside and the Irish Sea. The partial meltdown at Three Mile Island, unlike Chernobyl, did not involve an explosion and catastrophic breach of the outer containment structure. Escape of radioisotopes was largely confined to the noble gases and of the biologically significant nuclides, only 15 Ci (c. 5.6×10^{11} Bq) of ^{131}I were released. Small increases in cancer rates (not fatalities) and thyroid tumour incidence were found nearer the plant than farther away.

FUEL REPROCESSING

Nuclear fuel is considered to be spent when c. 90% of the moderator/control rods must be removed to achieve a critical state. Spent fuel is then transferred from the reactor core to a holding tank where it is allowed to dissipate in terms of both heat and radioactivity before being shipped to a reprocessing facility where it is either disassembled and discarded or reprocessed. Although reprocessing in connection with plutonium production for nuclear weapons has taken place at U.S. Department of Energy weapons production sites, only three reprocessing plants exist worldwide for the regeneration of fuel for electric power production. Two are in France, and one (Sellafield) is in the United Kingdom.

In the repurification process, the fuel cladding is removed either by dissolution or stripping, followed by fuel dissolution in acid. Large amounts of NO_X, tritium, and noble gases are given off during fuel dissolution. The Purex process is almost universally used in the removal of unusable fission products from solution and solvent extraction of nitrate salts of the desirable ^{239}Pu and ^{235}U. Very large amounts of acid are used in the dissolution of cladding and fuel pellets. Equally large amounts of solvents such as diethyl ether, isobutyl ketone, dibutyl carbitol, and tributyl phosphate are used in extraction and extraction and repurification operations.

Reprocessing of uranium from irradiated fuel is different from processing natural uranium in that the additional isotopes ^{232}U and ^{236}U are present. Both must be removed from spent fuel. ^{232}U produces a strong γ emitter, and ^{236}U is a neutron flux poison. The latter creates an economic penalty for reenrichment of spent fuel. When compared with direct fuel disposal, recycling, if performed correctly, has many obvious environmental benefits including reduced waste volumes, reduced radioactive content of waste, and reduced toxicity of wastes. According to data from Fiori (1997) nuclear fuel reprocessing and recycling efforts have led to reductions in mining wastes (12.5% through reduced uranium demand) and in Pu toxicity of intermediate wastes (30% over a period of 10,000 years). The stream of long-lived thermally active high-level wastes has been reduced ninefold.

NUCLEAR WASTE AND ITS MANAGEMENT

There are two types of nuclear waste, high level and low level. High-level wastes encompass spent fuel and transuranic wastes and weapons processing effluents. Low-level wastes comprise the high-volume/low-radioactivity waste rock and milling from the uranium extraction process but also include protective clothing, tools laboratory research compounds, and other equipment associated with the operation and maintenance of nuclear facilities.

Plutonium is a major factor forcing the disposal rather than recycling of spent nuclear fuel. Although the Pu contained in spent fuel is of significant political, economic, and strategic military value (primarily as the critical trigger in constructing nuclear bombs that are easily human-portable), the spent fuel itself is, in effect, "self-protecting" with respect to the high degree of radiation and shielding necessary for its transport. There has been a certain amount of debate as to whether Pu metal itself is inherently toxic, or whether its toxicity derives from its strong α emission. Of course, we will never be certain because it is impossible to separate the element from its radioactivity.

LOW-LEVEL WASTE

Low-level wastes comprise 90% by volume of all radioactive wastes, yet they contain less than 1% of the total radioactivity disposed of. It has been estimated that a single, large, water-cooled nuclear reactor generates between 600 and 1,400 m^3 of low-level waste per year. The United States produces about 28,000 $m^3 yr^{-1}$ of commercial low-level wastes but four times this amount from its defence and research facilities. The latter are stored at the facility where they are generated. Different countries have adopted greatly diverse standards for the disposal of low-level waste. Until 1982, most countries dumped intermediate and low-level radioactive waste at deep-water oceanic sites beyond the continental shelf. Since then, land burial has been favored. Usually, the waste is packed in drums that are placed in shallow trenches covered with earth. However, some countries (e.g., Germany and Sweden) prefer deep geologic disposal similar to how they handle high-level waste. The United States relies on a system of landfills of varying degrees of packing and security depending on the type of waste. With groundwater and soil contamination the most important concerns, the most sophisticated of these landfills are thick concrete bunkers with remote monitoring devices. The Waste Isolation Pilot Project (WIPP) designed to permanently store low-level and transuranic military waste at an underground site in the salt flats of the Chihuahuan Desert is still being tested and has not been cleared for use.

Table 8.5. Radiological characteristics of some biologically important
radioactive isotopes

Isotope	Radiation	Specific activity, $Ci\,g^{-1}$	Physical half-life	Biological half-life
^{131}I	α, β, γ	1.23×10^5	8.05 days	138 days
^{90}Sr	α, β	40	28.1 yr	1.3×10^4 days
^{3}H	α, β	9,700	12.4 yr	5 days
^{226}Ra	α, γ	1.0	1,620 yr	900 days (bone, 12 yr)
^{137}Cs	α, β, γ	87.0	30 y	120 days

HIGH-LEVEL WASTE

Irradiated fuel rods and assemblies are even more radioactive than the new fuel
due to the accumulation of fission products, hence the term *high-level waste*. Fuel
rods remain dangerous for many thousands of years due to emissions from
radioisotopes with very long half-lives. The U.S. Environmental Protection Agency
standards for disposal of such wastes stipulate 10,000 years of isolation from the
biosphere as a requisite for storage. The current worldwide inventory of high-level
radioactive waste from nuclear power plant operation is about 40,000 metric tonnes
with approximately 8,000 tonnes being added annually. It is, therefore, estimated
that this total will reach 80,000 tonnes by the year 2025 if the current rate of nuclear
plant building and operation continues. Fuel rods account for about 95% of the
radioactivity from all sources, both civilian and military. The 1987 amendments to
the U.S. Nuclear Waste Policy Act led to the construction of a single high-level
waste site at Yucca Mountain in Nevada. Excavation of approximately 100 miles
of tunnels in volcanic rock 300 m below the surface of the mountain yet well above
the water table is still in progress. After completion, it is anticipated that high-level
waste will be placed in the repository over a 50-yr period after which time it will
be sealed and marked.

8.5.3   Radioisotopes of biological importance

Apart from the central importance of uranium and plutonium in electricity generation and weapons production, several other isotopes produced as by-products of
these processes have assumed practical importance in what might be called the
nuclear age, by virtue of their rapid incorporation into biological systems. These
were listed in Section 8.5.2, "In-core fuel management", and some of these are discussed in more detail here. A summary of radiological properties of biologically
important elements is provided in Table 8.5.

TRITIUM (^{3}H)

Tritium occurs naturally in the atmosphere, where it is produced by cosmic ray
protons and neutrons reacting with oxygen, nitrogen, and argon. Whenever water

is irradiated with a high flux of neutrons, tritium is produced according to the following equation:

$$2H + n \rightarrow 3H + \gamma$$

When released into the environment either by natural processes or through human activity, more than 99% appears as tritiated water, ^3H_2O. Tritium has a half-life of 12 yr. Most of the tritiated water released into the environment is found in surface waters. Surface water concentrations exceeded $200\,BqL^{-1}$ in surface water stations in the northern hemisphere following aboveground nuclear bomb tests in the 1960s relative to a background level of $0.6\,BqL^{-1}$. Total ^3H releases from weapons testing until 1973 were estimated as c. $3 \times 10^9\,Ci$ ($1.2 \times 10^{19}\,Bq$), a figure similar to the amount of ^3H released since that time as a result of nuclear power plant operation. As a result of human activity, ^3H dose estimates for people in the northern hemisphere are between 7 and 15× higher than those for people in the southern hemisphere.

RADIUM (^{224}Ra, ^{226}Ra, ^{228}Ra)

Radium is a luminescent metallic element found in minute quantities in uranium ores. The most common isotope, ^{226}Ra, has a half-life of 1,622 years, whereas ^{224}Ra has a half-life of only 3.62 days. Discovered by Marie Curie in the early years of the twentieth century, radium was rapidly incorporated into a number of medical treatments for a range of ailments from rheumatism to mental disorders. In Europe, ^{224}Ra therapy was successfully used for over 40 years to treat the pain associated with ankylosing spondylitis. For many years, Ra was used to paint luminescent watch faces, a practice now abandoned in favour of tritium-based organic paints, but Ra exposure has been linked to the development of osteogenic sarcoma, no such statistical relationship has been found with leukemia.

RADON (^{222}Rn)

Radon is a largely inert gas formed by the radioactive decay of radium (see uranium decay series, Table 8.1). The most stable isotope, ^{222}Rn, has a half-life of 3.82 days. There are now 11 large-scale studies that document the increased incidence of lung cancer in uranium miners. In view of this convincing evidence, the measurement of significantly increased levels of the gas in some modern homes has been a particular source of concern particularly within the last decade. While it has long been known that the gas is responsible for most of the background radiation received by humans, it is believed that this particular type of buildup is a consequence of better insulated homes, particularly in cooler climates such as Scandinavia and Canada where houses are often built on granite bedrock or constructed of granite-based materials. In assessing the possibility of cancer risk from domestic radon levels, toxicologists are currently evaluating the utility of data from the uranium industry to see whether high-dose models have any application in risk assessment at lower doses (see Section 8.3).

CAESIUM (^{134}Cs, ^{137}Cs)

Caesium (Cs) is an alkali metal like sodium (Na) and potassium (K) with soluble salts and a rapid turnover within an organism. Both isotopes have similar energies, although the half-life of ^{137}Cs is 30 years, whereas that of ^{134}Cs is 2.3 years. Although Cs has a low residence time within the body, it can be rapidly incorporated into some food chains where heavy contamination occurs. For example, ^{137}Cs was released in relatively large amounts as a result of the Chernobyl accident and was concentrated by fungi and lichens in Scandinavian countries. Contaminated lichens were consumed by reindeer to the extent that some animals acheived concentrations as high as 20,000 Bq kg f.w.$^{-1}$ (Forberg et al., 1992), which presented a potential hazard to humans consuming reindeer meat. Normal levels in reindeer meat are <200 Bq kg f.w^{-1}. Similar considerations caused the U.K. government to impose a strict monitoring program of sheep grazing on ^{137}Cs-contaminated pasture following the Chernobyl fallout.

STRONTIUM (^{90}Sr)

Strontium is an alkaline earth metal similar to calcium and magnesium and, as such, it is stored in the bone matrix. ^{90}Sr has a half-life of 28 years. Atmospheric concentrations of this fission product reached a peak during the period of above-ground weapons testing in the late 1960s and early 1970s but rapidly declined soon afterwards. Like other fission products, it became particularly concentrated in the northern hemisphere. In addition to its deposition in bone, Sr becomes concentrated in milk and, as such, may enter the human food chain as well as pass to suckling young.

IODINE (^{131}I)

Iodine-131 is one of the more hazardous fission products released as a result of weapons testing and nuclear accidents. Like strontium, it enters the pasture-livestock-milk food chain and becomes highly concentrated in the thyroid gland where it is a component of the thyroid hormone thyroxine. Although it has a relatively short biological half-life (c. 24 hr in sheep and cows) and a short radioactive half-life (8.05 days), it is a strong β/γ emitter with high specific activity, which rapidly transfers a large radiation dose to the thyroid. Doses of >500,000 Bq kg f.w.$^{-1}$ were found in the thyroid glands of sheep within a several hundred mile radius of the U.S. nuclear weapons testing site and the Windscale, U.K., reactor site in 1957 following the accident there, and in sheep from southern Europe more than 2 months after the Chernobyl accident. While livestock and other animals at long distances from the Chernobyl reactor were saved from serious health problems by the short radioactive half-life of ^{131}I (i.e., substantial decay had occurred before deposition and ingestion), much higher doses were received close to the plant and several incidences of thyroid cancer and birth defects have been reported from the area surrounding the power station.

8.6   Ecological effects of radiation

Ecological effects of radiation are difficult to identify, and there is no evidence of adverse effects resulting from exposure at or near background levels. More than 30 years of studies at the Windscale site after the 1957 accident failed to reveal any significant ecological effects from the fallout. Likewise, studies of invertebrates, macrophytes, and algae in ponds and streams within and beyond the perimeter of the Hanford site in Washington State in the northwest United States reveal some intersite population differences, but none that can be directly attributed to radiation exposure (Emery and McShane, 1980).

Although it has been possible to detect small localised changes in biota very close to some sites of heavy radioisotope usage and disposal where leakages have been reported, the overwhelming volume of information relating to ecosystem effects of radiation exposure comes from very large releases such as the Chernobyl accident and aboveground detonation of nuclear weapons. Where such investigations extend into risk assessment, the data base is expanded to include occupational radiation exposure. Such an approach is not unlike that used to characterize the environmental toxicology of nonradioactive substances where unintentional releases form the bases of field "experiments". In view of the risks involved, field experiments using radioisotopes are very rare. An interesting exception was the long-term experiment conducted at the Brookhaven National Laboratory in New York, USA, where a natural oak-pine forest was irradiated for 15 years (1961–76) by a 9,500 Ci ^{137}Cs source mounted on top of a 5-m tower. The principal aim of the investigators was to obtain chronic estimates of ecosystem change under controlled conditions. In general terms, the authors of the study described a two-phase effect: initially a thinning out of sensitive, larger bodied species such as trees in favor of smaller, more resistant ground-hugging species. This was followed by a "second sorting" of species wherein more uncommon or exotic species create new communities in conjunction with the survivors of the first sorting (Woodwell and Houghton, 1990). Investigators used a number of different parameters to compare the exposed site with reference sites. Depending on the end-point used, they were able to discern detectable differences due to radiation up to 50–100 m from the radiation source. Interestingly, they found simple diversity indices (irrespective of actual species found) to be less useful than parameters that included a quantitative comparative component such as the coefficient of community. The coefficient of community is the number of species shared by two communities expressed as a percentage of the total number of species in both. The coefficient I is expressed as

$$I = \frac{C}{A + B - C}$$

where C equals the number of species shared by the two communities, one of which has A species and the other B. This coefficient is essentially the same as that used

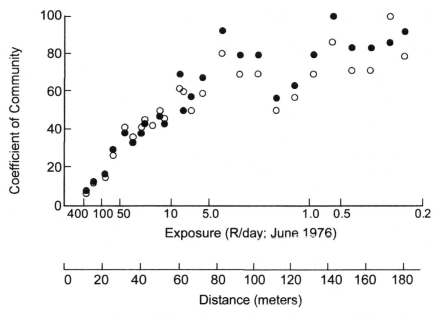

Figure 8.6 Changes in coefficient of community in plant communities exposed to a radiation gradient at Brookhaven National Laboratory, 1961–76. Measurements were made in 1976. Modified from Woodwell and Houghton (1990).

to compare animal communities upstream and downstream from a point source of pollution as described in Section 4.5.4 and Figure 4.12). Figure 8.6 describes changes in the coefficient of community for plant communities along the radiation gradient at the end of the experiment in 1976.

8.7 Case study

Case Study 8.1. The Chernobyl accident

On April 26, 1986, the world's largest reactor explosion occurred at unit four of the graphite moderated, 1,000 MW, boiling water pressure tube reactor at Chernobyl, 80 km north of Kiev in Ukraine, which was then part of the USSR. At the time of the accident, this type of reactor was one of 14 such reactors in the USSR, comprising more than half of the nuclear-powered electricity generation capacity of the country. The accident followed a dramatic and cumulative series of errors incurred while conducting a series of tests and included a blatant disregard of safety procedures, which were exacerbated by design faults of the RMBK reactor itself. Several safety systems were deliberately disabled. The single most important design problem is that steam generation in the fuel channels causes an increased number of neutrons to collide with the graphite leading to an increased rate of fission. Under such circumstances, a last desperate attempt to stop the reaction by inserting the control rods had the opposite effect, and within 4 s the reactor power reached more than 100 times its capacity. The resultant explosion sheared all 1,661 water pipes, blew the 1,000-tonne cap off the top of the core and ruptured the concrete walls of the reactor hall. Radioactive dust, including pulverised fuel, was thrown at least 7.5 km into the air and the intense heat started at least 30 fires. Radioactive release lasted about 10 days, and despite attempts to quench the reaction through air drops of boron carbide, sand, clay, and lead, there was a complete meltdown of the reactor core. Nine days following the initial explosion, the daily release rate of radioactive material was nearly as high as it was at the time of the initial release. As an immediate result of the accident, 31 people were killed, and another 65 of the original staff of the reactor had died by 1991. As a group, they experienced a death rate more than 100 times that of a comparable, unexposed population. In the months following the explosion, more than half a million people were involved with the construction of a sarcophagus around the melted core. In some areas, the radiation fields were as high as 66 rad min^{-1}. Despite the stated intent to limit workers to maximum doses of 20 rad (500 rad is regarded as a lethal dose), it can be appreciated that, in such an environment, many were probably exposed to considerably higher levels of radiation, and many probably received doses comparable to survivors of the Hiroshima and Nagasaki explosions. Including the 31 deaths, the USSR had a total of 237 cases of acute radiation sickness as a result of the accident.

Released radionuclides included essentially all the noble gases, volatile elements (i.e., ^{131}I and ^{134,137}Cs) and some refractory materials (i.e., ^{89,90}Sr, ^{141,144}Ce, ^{238,239,240}Pu) (Anspaugh et al., 1988). In Sweden, the fallout was identified as containing radioactive krypton, xenon, iodine, caesium, and cobalt (Megaw, 1987). Through integration of environmental data, it is estimated that some 100 petabequerel of ^{137}Cs (1 Pbq = 10^{15} Bq) were released during and subsequent to the accident. Total release of radioactivity exceeded 3×10^{18} Bq. Due to the prevailing weather patterns at the time of the accident, the spread of the fission product plume from Chernobyl was sustained for 11 days and affected many countries including Finland, Sweden, Denmark, Germany, France, Italy, Austria, Poland, the United Kingdom, and Czechoslovakia (Megaw, 1987). Japan, the United States, and Canada were slightly affected. In Poland, the closest country to Ukraine, the government banned the sale of milk from cows on pasture, and children were treated with potassium iodide to reduce their uptake of ^{131}I. In parts of Sweden where rain or snow had fallen since the accident, people were advised not to drink water, which might be 100 times more radioactive than normal. Although no acute effects occurred outside the USSR, the risk for lifetime expectation of fatal radiogenic cancer increased from 0 to 0.02% in Europe and 0 to 0.003% in the northern hemisphere (Anspaugh et al., 1988).

During the years following the Chernobyl accident, there have been numerous clinical studies of human populations exposed to radiation, some of which continue today. Several reveal symptoms both directly and indirectly related to the radiation. Distinct changes in the clinical picture of acute pneumonia were noted in patients subjected to constant, prolonged (1986–1990) effect of small doses of ionising radiation as a result of residing in the contaminated territory after the accident (Kolpakov et al., 1992). These changes included increased duration of the disease and the frequency of protracted forms as well as suppression of the immune system. The small radiation doses in Kiev, 80 km away from the accident, had a significant impact on the humoral immunity of the population (Bidnenko et al., 1992). Likhatarev et al. (1993) predicted a 1.4-fold increase in thyroid cancer morbidity (relative to spontaneous incidence) in children who lived in the heavily contaminated region of Ukraine in 1986.

Ecological effects

Many European species of mycorrhizal fungi, including several edible ones, were found to contain unacceptably high levels of ^{137}Cs (>1,000 Bq kg^{-1} dry wt) following the accident. ^{137}Cs concentrations in lichens and mosses in some areas were significantly elevated 5 years after the accident. The implications for food-chain transfer were discussed in Section 8.5.3.

A study of earthworm populations in a 30 km zone around the Chernobyl nuclear power plant following the accident indicated a significant, but temporary, depression in recruitment relative to control plots. Populations had recovered by the summer of 1988 (Krivolutzkii et al., 1992).

8.8   Questions

1. What are the principal forms of ionising radiation? Give specific examples of radionuclides responsible for each of these different types of emission.

2. Describe the effects of ionising radiation at the molecular level. Give two ways in which radiation may cause cell death. In what ways does radiation affect the immune system?

3. Give an account of the chronic effects of radiation exposure. What are the difficulties in assessing the risk from low levels of ionising radiation?

4. Write a short essay on how electricity is produced from nuclear power, including brief descriptions of the major steps in the nuclear fuel cycle.

5. Give an account of the treatment of waste products from nuclear power generation, including both reprocessing and waste management.

6. Give short accounts of the environmental significance of radon (^{222}Rn), caesium (^{137}Cs), strontium (^{90}Sr), and iodine (^{131}I).

7. Describe the immediate and long-term effects of the Chernobyl reactor accident. What were the lessons from this incident for the nuclear power industry?

8.9   References

Anspaugh, L. R., Catlin, R. J., and Goldman, M. (1988) The global impact of the Chernobyl reactor accident, *Science*, **242**, 1513–19.

Bidnenko, S. I., Nasarchuk, L. V., Fedorovskaya, E. A., Lyutko, O. B., and Openko, L. V. (1992) Follow-up antibacterial immunity in humans under altered radiation conditions, *Zhurnal Mikrobiologii Epidemiologii i Immunobiologii*, **1**, 33–6.

Bodansky, D. (1996) *Nuclear Energy: Principles, Practices and Prospects*, American Institute of Physics Press, Woodbury, NY.

Emery, R. M., and McShane, M. C. (1980) Nuclear waste ponds and streams on the Hanford site: An ecological search for radiation effects, *Health Physics*, **38**, 787–809.

Fiori, M. P. (1997). Accelerating cleanup: Focus on 2006. U.S. Department of Energy Report, Savannah River Operations Office, Aiken, SC.

Forberg, S., Odsjo, T., and Olsson, M. (1992) Radiocesium in muscle-tissue or reindeer and pike from northern Sweden before and after the Chernobyl accident – A retrospective study on tissue samples from the Swedish-Environmental-Specimen-Bank, *Science of the Total Environment*, **115** (3), 179–89.

Grosch, D. S. (1963) *Biological Effects of Radiation*, Blaisdel, Waltham, MA.

Hill, C. K., and Godfrey, T. (1994) Adaptive responses after exposure to low dose radiation. In *Biological Effects of Low Level Exposures: Dose-Response Relationships*, ed. Calabrese, E. J., pp. 255–68, Lewis, Boca Raton, FL.

International Commission on Radiological Protection (ICRP). (1977) *Recommendations of the International Commission on Radiological Protection*, report no. 26. Pergamon, New York.

International Commission on Radiological Protection (ICRP). (1990) *Recommendations of the International Commission on Radiological Protection*, report no. 60. Pergamon, New York.

Johns, J. E., and Cunningham, J. R. (1980) *The Physics of Radiology*, C. C. Thomas, Springfield, IL.

Kennedy, J. C., Till, J. E., Simonovich, L., and McCullough, E. A. (1964) Radiosensitivity of the immune response to sheep red cells in the mouse, as measured by the hemolytic plaque method, *Journal of Immunology*, **94**, 715–22.

Kolpakov, M. Y., Maltsev, V. I., Yakobchuk, V. A., Shatilo, V. I., and Kolpakova, N. N. (1992) Course of acute pneumonia in patients subjected to prolonged effect of small doses of ionizing radiation as a result of the Chernobyl atomic station accident, *Vrachebno Delo*, **7**, 11–15.

Krivolutzkii, D. A., Pogarzhevskii, A. D., and Viktorov, A. G. (1992) Earthworm populations in soils contaminated by the Chernobyl atomic power station accident, 1986–1988, *Soil Biology and Biochemistry*, **24**, 1729–31.

Likhatarev, I. A., Shandala, N. K., Gulko, G. M., Kairo, I. A., and Chepurny, N. I. (1993) Ukrainian thyroid doses after the Chernobyl accident, *Health Physics*, **64**, 594–9.

Megaw, W. J. (1987) *How Safe: Three Mile Island, Chernobyl and Beyond*, Stoddart, Ontario.

National Academy of Sciences (NAS). (1990) *Health Effects of Exposure to Low Levels of Ionizing Radiation*, National Academy of Sciences Report BEIR V, National Academy Press, Washington, DC.

Wilson, P. D. (1996) *The Nuclear Fuel Cycle: From Ore to Wastes*, Oxford University Press, Oxford.

Woodwell, G. M., and Houghton, R. A. (1990) The experimental impoverishment of natural communities: Effects of ionizing radiation on plant communities, 1961–1976. In *The Earth in Transition: Patterns and Processes of Biotic Impoverishment*, ed. Woodwell, G. M., pp. 3–24, Cambridge University Press, Cambridge.

9

○ ○

Complex issues

9.1   Introduction and rationale

When toxic substances cause or appear to cause problems in the environment or for human health, the issues are rarely simple, and the chemical and other potential causes of harm rarely occur in isolation. Typically, there is a great deal of complexity in the system of concern and uncertainty concerning the relationship between cause and effect. The reasons for this include the facts that (a) one typically encounters combinations of toxic substances, rather than single pollutants; (b) the system of concern may not in itself be clearly understood, even in the absence of chemical perturbation; and (c) data, particularly from field studies, are often incomplete. Yet, the environmental toxicologist is frequently challenged to address the effects of natural or anthropogenic perturbations, either after the fact or, ideally, in the planning stage of any human endeavour that might have an environmental impact.

Up to this stage in the text, although examples and case studies have been presented, we have dealt mainly with definitions, methods, and approaches and the environmental toxicology of individual substances or classes of substance. In the present chapter, we address a number of complex issues, exemplifying the impact of certain processes and also providing vehicles for integrating some of the basic concepts and information that have been covered so far.

Subsequent chapters on risk assessment, rehabilitation, and regulatory toxicology will show how the integration of the same kind of information can be applied to (a) assessment and planning in the face of uncertainty and in advance of any damage occurring (risk assessment, risk management), (b) rehabilitation or recovery of systems that have already sustained damage, and (c) responsible control and regulation of potentially toxic substances, normally prior to their release into or use in the environment, but sometimes after errors have been made.

9.2   The mining and smelting of metals

9.2.1   The issue

Metals and metallic products can be seen as fundamental to our contemporary way of life; indeed, progress in earlier and more primitive societies than ours was driven to a considerable degree by their discoveries and usage of metals such as iron, copper, and bronze. Precious metals have historically and in contemporary life provided aesthetic as well as practical and commercial enrichment to humans. Like other resource industries, the metals industries have shaped and continue to shape the social fabric of human societies. The industries influence the development of entire towns, with direct employment, secondary industries, and capital growth. These aspects are, for the most part, positive attributes of the industry. The negative aspects of metal mining and smelting include their potential to have adverse effects on the health of humans and of ecosystems. But these negative impacts are not inevitable. The more thorough the understanding of the mechanisms of these potential impacts, the better technology can be applied to minimise them.

In the past, the negative impacts of mining and the metallurgical industries appear to have been viewed as the necessary consequences of human development. Local damage, normally related to mining, was familiar to and, for the most part, accepted as normal by those who lived locally and usually depended upon the industry, whereas for the average member of society living elsewhere, the damage was scarcely visible. Metal processing, especially smelting, had more widespread effects, but even these were seen as relatively local and acceptable.

As with many environmental issues, public awareness of the environmental impacts of metal extraction and purification began to rise in the late 1960s. Attitudes changed quite rapidly, especially concerning smelting. One single factor appears to have been the most influential in bringing about these changes. In the 1970s, the issue of acid rain came to the fore, first in Europe, especially in the Nordic countries, and soon thereafter in North America. Tall stacks, whether related to the burning of fossil fuel or the high-temperature processing of sulphur-containing metal ores, release the precursors of sulphuric and/or nitric acids into the atmosphere. The long-range atmospheric transport of these gases is aided by the tall stacks from which the gases are initially discharged. In the 1960s and early 1970s, these same tall stacks had been advocated as a reasonable technology to improve local air quality by dispersing the gases to a greater distance and, in so doing, diluting the concentration in the air. This so-called pollution solution by dilution did not take into account the transport and transformation of the acidic gases, leading to acidic deposition many hundreds or thousands of kilometres from their respective sources.

Sensitive systems (i.e., those overlying granitic bedrock that provided little buffering capacity) were identified and monitored. Loss of buffering capacity can be dramatic, leading to lowering of pH in runoff and in surface water and to a chain of chemical and biological changes that represent ecosystem degradation. Research from private and public institutions came together, there was contention, the issue was highly publicised, pressures were brought to bear, and some regulatory changes led gradually to abatement of the emissions of acid-generating gases. There are many published studies of acidic deposition and its effects, some of which have already been referred to in other contexts. The topic is dealt with briefly in Section 6.6.2, and further reading is provided in Section 9.11. However, the purpose of raising it at this point is to demonstrate the pivotal significance of the issue in bringing about cleaner emissions from stacks, including those from smelters.

In the present climate of environmental protection and management, in most parts of the developed world when a new endeavour is proposed, extensive environmental impact assessments are required even before exploration and site development begins. In the past, this was not so, and there is a legacy of the environmental effects of metal mining and smelting from which a great deal can be learned.

9.2.2 Processes involved in the extraction and purification of metals

A life-cycle, or cradle-to-grave, review of metal purification properly begins at the exploration stage.

EXPLORATION

Mineral exploration initially employs noninvasive studies, including the use of existing geological maps, remote sensing from satellite or aircraft followed by greater refinement from low-flying aircraft and air photography, and eventually ground surveys. Geophysical exploration follows, involving seismic, gravity, resistivity, magnetic, and electromagnetic tools; radar; and induced polarisation (Ripley et al., 1996). Further exploration normally involves physical changes including digging trenches and drilling, which may be accomplished by hand or by hydraulic and blasting methods, all of which are to a greater or lesser extent invasive and likely to have some environmental impact.

Less invasive are biological methods. Geobotanical prospecting applies knowledge of the specific flora that grow on different types of mineralised soil (see also Section 4.4.3), so prospectors gain clues about mineral deposits from the native plants. Biogeochemical prospecting utilises plants in another manner, as "samplers", which is also a noninvasive approach. The roots of plants can be excavated, and, with the capacity for plants to bioaccumulate metals (see Sections 4.4.3 and 6.2.6) anomalously high concentrations of metals in the substrates can be detected by chemical analysis of the roots. Geozoology uses animals to locate areas of mineralisation.

Examples of development and application of modern techniques for exploration are provided in Ripley et al. (1996). Advances in engineering and other technology are providing more efficient as well as more environmentally benign methods for mineral exploration.

MINING

After mineral deposits have been identified and judged as economically viable, mines are constructed, either underground or on the surface. Following mining, concentration involving grinding and crushing is often required. Then the smelting and refining of metals includes pyrometallurgical processes, in which high temperatures are used; hydrometallugical processes, in which a solvent is used to leach metal from its ore; and electrometallurgical processes, when electrical energy is applied for the dissolution and collection of metal from aqueous solution, known as electrowinning. Every metallurgical operation has its own characteristics. Many combine the smelting or refining of a particular metal with other processes, and many operations are still undergoing modifications towards ends that include increasing efficiency, decreasing emissions and effluents, recycling and treating wastes, and linking and cross-linking products and waste materials. Waste management is properly incorporated into each stage of the processes listed here. The final activity is decommissioning, which is likely to involve reclamation or rehabilitation of the land and water that have been affected.

Metals are combined in various chemical forms in ores. Removal of the ore from the ground (i.e., mining) involves either underground mining, with drilling and blasting followed by removal of ore, or surface operations such as strip or opencast mining. The latter is more easily adapted to mechanical removal of ore, but often the depth and location of the ore body dictate a combination of underground and surface mining. Underground mining, while resulting in much more occupational risk than surface mining, provides the option of backfilling with the various waste products of mining and refining (see the following discussion of spoils and tailings). However, since metals may be mobilised in the waste products, the potential for groundwater contamination has to be addressed. Surface mining disturbs more surface than does underground mining, but it is often the more economically attractive procedure.

Nonentry mining is carried out *in situ*, typically by leaching the desired metal from its parent material by means of microbial or chemical treatment. Liquid is dispersed through the heap or other deposit of ore and the metal is recovered from solution. Nonentry mining is used mainly above ground, and typically for low-grade ores containing uranium and copper. The process obviates all the subsequent extraction and purification that is required for metals that are removed by conventional mining. In situ leaching avoids many of the environmental problems associated with extracting an ore and processing it by other methods, but it still runs the risk of contaminating groundwater if the leaching solution is not properly contained.

In many ores, the metal is not sufficiently concentrated to be smelted directly and may require beneficiation, which results in higher concentration of metal. Beneficiation involves grinding, concentration, and finally dewatering of the concentrate.

Whether underground or surface, a mine's operation produces two categories of waste: mine spoil (soil and rock removed when the mine is excavated) and mill tailings (the finely ground rock resulting from the benefication process). These materials vary greatly in chemical composition, but most have in common a paucity of plant nutrients and concentrations of metals that may range from nontoxic to highly toxic. The waste material may also be acid-generating (see following discussion). For any or all of these reasons, areas where mine and mill wastes are deposited and left untreated are usually barren and inhospitable to plant colonisation. The physical instability that results from lack of plant cover renders potentially toxic metal more mobile, and the ecological effect of lack of plant cover is clearly inhibitory for any succession to a functioning ecosystem.

SMELTING

The oldest form of extractive metallurgy is pyrometallurgy, of which the most widely used is some combination of roasting and smelting. Heat converts the metal sulphide or oxide into pure metal, and in the case of sulphide ores, sulphur dioxide is released.

For centuries in the United Kingdom and Europe and early in the twentieth century in the New World, roasting and smelting was carried out in open pits resulting in ground-level fumigation by acids and metal particulates. Workers, as well as adjacent ecosystems, were exposed to very high concentrations of contaminants, often with dramatic results. The practice of erecting tall stacks to disperse the volatile and other airborne waste products has improved local air quality, with beneficial effects for human health and for ecosystems. The challenge now is to limit the emissions, either by pretreatment of sulphur-containing ores or by scrubbing the waste stream of gas.

9.2.3 Substances of concern that are mobilised or formed and released during mining, smelting, and other purification processes

Many metals that are used in large quantities, such as iron (Fe), copper (Cu), nickel (Ni), lead (Pb), and zinc (Zn), occur in sulphide ore bodies, such as pyrite (FeS_2) and pyrrhotite ($Fe_{n-1}S_n$). The reduced sulphides, on exposure to air and water as they are brought to the surface, form sulphuric acid through a series of microbial and chemical oxidations. During mining, the resulting sulphuric acid dissolves iron and other metals from the ore and other geological material. Then the highly acidic metal-bearing water, often with pH as low as 2.0, drains into surface water bodies. This is known as acid mine drainage.

Sulphur dioxide from smelting is the single most problematic toxic substance from an environmental standpoint. Sulphur dioxide is extremely phytotoxic. And as described in the Section 9.2.1, acidic deposition results from chemical conversions and long-range atmospheric transport of the sulphur dioxide.

Particulate emissions from roasting and smelting of ore contain metals, usually in the form of oxides, which fall relatively close to the source, contaminating soils and surface water. If the soils or waters are naturally acidic, or if they are acidified by the sulphur emissions, then these metals become soluble, increasing their mobility and biological availability.

9.2.4   The environmental toxicology of metal mining and smelting

The exploration, site development, and mining of metals may be geographically separated from the refining of metals, and the respective impacts are distinct.

Table 9.1 summarises the toxic or potentially toxic effects on the environment and specifically on biota of the series of processes described earlier. Physical disruption that does not directly involve toxic substances and water consumption, other than contamination of water by toxic substances, clearly represents major environmental problems related to the metals industries, but these are outside the scope of this text.

MINING

From a toxicological point of view, acid mine drainage is the major problem in mining where sulphide ores are involved: research in engineering and applied biology continues to seek new solutions to this problem. The highly acidic water that drains from mines and mine-related activities is not only toxic because of its acidity per se but also dissolves and mobilises metals from ore and other geological material.

In addition to the acid mine drainage issue, the need to dispose of large volumes of waste material containing toxic substances but in relatively low concentrations is also a common problem. These so-called low-level types of waste, including mill tailings, can be of additional concern when the elements involved are radioactive, as in uranium mining and milling. Mill tailings, whether legacies from the past or in current production, are generally not amenable to destruction or removal, even though some fraction of them can be back-filled into worked-out mines. They represent challenges for ecosystem rehabilitation, a subject that is revisited in Chapter 11.

SMELTING

The atmospheric emissions from smelters represent the other major concern for the environmental impact of metal processing. Particulates in emissions generally contain oxides of metals, not only those metals that are to be won in the processing but also additional elements that co-occur in the ore but that, for various reasons, are not economically feasible to recover. These elements are all potentially toxic. They tend to fall relatively close to the source, and thus their effect is concentrated rather locally, typically over a radius of up to 100 km, depending on the height of the source and the topography. The environmental toxicology of a number of individual metals has been covered in Chapter 6, but this does not amply explain all the effects observed on ecosystems in the vicinity of smelters. When the smelted ores are predominately sulphide, as is frequently the case for copper, lead, nickel, uranium (U), and zinc, the effects of the metals are typically added to and exacerbated by acidification.

Even though surface waters undergo exchange or renewal through flow-through or flushing and, thus, do not normally accumulate metals with time unless input continues, soils and sediments do accumulate and may retain deposited materials. This type of contamination is normally concentrated at the surface of the soil or sediment, in contrast to metals of geological origin, which show enrichment near the bedrock or in any case lower in the soil profile. It should be apparent then that deposited metals will tend to affect the "active" zone in sediments, where most of the biological activity occurs, and similarly for soils, the rooting zone and the habitat of soil microflora and fauna will be the most severely impacted. Further complications can be anticipated since residues of metal in soil or sediment can be mobilised by physical, chemical, or biological processes, thus providing a secondary source of contaminants to water and air.

The aquatic and terrestrial flora and fauna in the vicinity of smelters have been well studied, and dramatic effects have been demonstrated, such as impoverishment of flora and fauna (e.g., Yan and Welbourn, 1990). However, the direct cause of observed effects on native biota or on managed systems such as forests and agricultural systems in the vicinity of smelters is often difficult to determine. The insults rarely occur singly; thus, several different toxic elements may co-occur. Generally, one may hypothesise that the effect is (a) due solely or primarily to a single toxic substance, (b) due to the additive effects of several toxic substances, or (c) due to synergistic effects of a combination of toxic substances. Occasionally, the symptoms of toxicity are sufficiently specific as to make identification of the cause rather simple. For example, there are some fairly specific symptoms of metal toxicity in plants (Antonovics et al., 1971), but the availability of such indicators in the field is rare. Add to the mixture of metals and other toxic substances the fact that sulphur dioxide fumigation and acidification frequently combine with metal contamination, and the puzzle is understandably complex. One must also consider that ecosystem level effects can result in part from indirect or secondary effects rather than from the initial response to a toxic substance.

The biota in surface waters close to sources tend to be quite impoverished, with shortened food chains, and frequently they show the complete absence of fish, the highest trophic level. Although such ecosystems may appear to be functional, it is apparent that they are less resilient in the face of additional stress than are the more diverse original systems (Yan and Welbourn, 1990). Comparable ecosystem level effects, mainly resulting from simplification of the structure of the biological system, have also been observed for terrestrial systems in the vicinity of smelters (Buchauer, 1971; Freedman and Hutchinson, 1980; Hutchinson and Whitby, 1974).

To provide some illustration of the combinations of metals that are mobilised in the extraction and processing of metals, several major and well-studied examples are shown in Table 9.2.

Table 9.1. Toxic substances released or mobilised in metal mining and smelting[a]

Activity	Process[a]	Substance(s)	Potential environmental effect[a]	Potential biological response[a]
Exploration	Overburden removed.	Various metals in dust and other particulates; exposure of metal sulphides	Soils and surface water contaminated with increased turbidity, acidification, and metals. May be more extensive than subsequent (mining) effects but for shorter time and less severe.	There are mainly short-lived effects: some occupational human health risks, local loss of sensitive species, possible selection of tolerant species, local change in community structure.
Mine site	Structures and waste disposal systems designed.	Various metals in waste materials, dust, and processing water	Ideally, release of metals and acids from wastes is very limited; site rehabilitation should be part of the design.	Ideally, adverse effects are minimised, and there are no irreversible effects.
Underground mining	Mine spoil produced in large volumes. Some may be backfilled into mine, but there is always excess requiring disposal. Ores are exposed to air and waster. Waste water often contains metals.	Metals, generally in low concentrations in spoil material, may leach or erode; oxidation of sulphide ore	Mine spoil heaps are infertile and may have toxic metals; soils and surface water are contaminated by metals; acid drainage flows into surface waters; groundwater may become contaminated.	Spoil heaps are infertile and often toxic, have poor plant cover and poor ecosystem recovery; long-term local impoverishment of flora and fauna; aquatic fauna and flora depleted or extinguished by acid mine drainage; possible toxicity in drinking or irrigation water if aquifer contaminated.
Surface mining	Mine spoil produced in large volumes. Ores exposed to air and water. There is high potential (greater than underground mining) for particulate emissions to atmosphere.	Metals, generally in low concentrations in spoil material, which may leach or erode; oxidation of sulphide ore to sulphuric acid; metals in airborne particulates	Overburden when replaced and mine spoil heaps are infertile and may have toxic metals; soils and surface water contaminated by metals from deposited particulates; acid drainage flows into surface waters.	Spoil heaps and overburden are infertile and often toxic, poor plant cover and poor ecosystem recovery; long-term local impoverishment of flora and fauna; biota in soils and surface water may be affected by toxic concentrations of metals from particulates; aquatic fauna and flora depleted or extinguished by acid mine drainage.
In situ leaching	Bacterial or chemical dissolution of metal from ore.	Dissolved metal, usually in acid solution	There is potential to contaminate groundwater if leachate not contained.	There is very little impact, with possible toxicity in drinking or irrigation water from aquifer if groundwater is contaminated.

Process	Description	Wastes and releases	Releases to environment	Environmental impacts
Beneficiation	Ore crushed or ground, followed by concentrating and dewatering. Large volumes of wet slurry deposited into open depressions know as tailing ponds.	Local deposition of metals in dust, in which mill tailings contain metals, often as sulphides	Local soils and water contaminated by metals in dust; metals and acid-generating material deposited into tailings ponds are very infertile and may be dispersed by wind and water.	Main problem is lack of vegetation for tailings ponds. Treatment to add nutrients and deacidity may be a continuing need after tailings become inactive.
Pyrometallurgic purification	Roasting and smelting	Slag (solidified waste rock) containing metals, notably iron; sulphur dioxide gas; metals in particulates; potential for wider dispersion or emissions from source than occurs for mine wastes, especially in the context of SO_2	Sulphur dioxide makes local fumigation; contributes to acidic deposition in rain-snow and fog, which can acidify sensitive surface water and soils; soils and surface water contaminated by metals in particulates.	Sulphur dioxide damage most severe to plant communities (plants more sensitive to SO_2 than animals), but there are major changes in species composition for all communities, usually impoverishment; damage to aquatic and terrestrial ecosystems distant from the source from acidic deposition; toxicity to local flora and fauna from metals in particulates, often combined with acidification.
Hydrometallurgic purification	Extraction of metals using water (sometimes in large volumes) and aqueous solutions of salts or microorganisms	Metals, acids, and process chemicals in process water	Minimal releases to the environment; process water is usually recycled. There is possible aquifer contamination.	There is very little impact, with possible toxicity in drinking or irrigation water from aquifer if groundwater is contaminated.
Electrometallurgic purification	Use of electrolytic methods to win metal from impurities – may be used alone or as a further refinement following smelting	Waste solutions containing metals and process chemicals	Waste disposal has potential to contaminate, but generally very little waste is released.	Indirect response only, if the process has high-energy requirements.
Decommissioning	Disturbed area is returned to original state or to some acceptable alternative state. Engineering solutions prearranged at planning stage.	Ideally, no toxic substances remain in available form	If off-site disposal is required, metallic, acid-generating and process chemicals may impact on other environments.	Ideally, there are no adverse effects; ecosystem may change by prior arrangement (e.g., from original natural woodland to recreational or ornamental park).

---

[a] Processes and environmental impacts that are *not* related to toxic substances have not been included.

Table 9.2. Selected examples of metal smelters

Name, location	Metal(s) recovered	Topic(s) of study	Effect(s)	Reference
Prescot, Merseyside, U.K. Complex of copper refineries and a brass foundry, operating for 60 years.	Copper, brass	Contamination of soils and vegetation, with distance from source	Widespread copper, cadmium, and zinc contamination of soils and vegetation with concentrations exponentially related to distance from sources. Metal in soils (Cu up to 2,000 ppm, Cd up to 15.4 ppm) mainly confined to surface.	Hunter and Johnson (1982)
Arnoldsein, Austria smelter	Lead and zinc	Metals in invertebrates	Thirteen species of ants all had accumulations of Pb, Cd, and Zn, related to distance from the smelter. In contrast, Cu, although emitted from the smelter and elevated in soils, appeared to be regulated by the ants.	Rabitsch (1995a, 1995b, 1995c)
INCO, Sudbury, Ontario, Canada – Other smelters, current and now closed, as well as older methods of metal extraction in the the area have had a complex series of effects	Nickel, copper, lesser amounts of other metals	Metal in deposition, soils, water, sediments, and biota; effects of smelter emissions on vegetation and aquatic life	Major contamination of all media by metals (Cu, Ni, Cd, Zn, Fe, etc.) and extensive acidification of soils and water, all related to distance from the smelter but not in a simple manner. The complex metallurgical and extraction history of the area makes precise identification of present-day effects difficult to determine. Vegetation has undergone major changes, with the original white pine forest largely eliminated, organic matter depleted, and in some cases soils completely lost. Lakes affected by metals and acidification; many lack fish.	Freedman and Hutchinson (1980); Hutchinson and Whitby (1974); Jeffries (1984); Matuszek and Wales (1992); Nriagu et al. (1982); Rose and Parker (1983); Taylor and Crowder (1983); Whitby and Hutchinson (1974)

The New Jersey Zinc Company (smelter), Palmerton, NJ, USA; closed in 1980	Zinc	Contamination of soils by metals; effects on arthropods and selected vertebrates	Litter layer highly contaminated with Cd (710 ppm), Pb (2,700 ppm), Zn (24,000 ppm), and Cu (<40 ppm). Soil invertebrates adversely affected, loss of amphibians related to the smelter; accumulations of metals in mammals including ungulates, but no clear adverse effects.	Beyer et al. (1985); Sileo and Beyer (1985)
The Hudson Bay Mining and Smelting Company, Flin Flon, Manitoba, Canada – The zinc operation in 1993 completed conversion to a hydrometallurgical operation, eliminating SO_2 emissions and most of the Hg and Cd emissions. The Cu smelter is being converted to a continuous reactor process, which will result in some improvements but not SO_2 containment.	Zinc and copper smelters since the 1930s	Metals in lake ecosystems, soils; effects on biota of metals and SO_2	Lead, arsenic, cadmium, zinc, and, to a lesser extent, copper, showed deposition patterns related to distance from the smelter; water, sediment, soils, and aquatic plants contaminated; fish affected by metals but not in a simple manner related to distance from the smelter (other factors also important); some evidence that smelter releases adversely affected voles but specific causal factors not clear.	Franzin et al. (1979); Franzin (1984); Franzin and McFarlane (1981); Harrison and Klaverkamp (1990)

9.3   Environmental impacts of pulp and paper mills

9.3.1   The issue

The paper industry is extremely valuable to many regional economies within the North American and Scandinavian economies. For example, it is the largest industrial employer in Canada, and the single most valuable export industry in that country. Annual paper and pulp production in the United States and Canada total about 80 million and 20 million metric tonnes, respectively.

Paper manufacture involves several different steps, many of which require large quantities of water. On average, the water requirement for paper products is about $100 \, m^3$ metric tonne^{-1} of paper, although different grades of paper require widely differing water volumes. Very briefly, the principal manufacturing processes are as follows:

- Bark separation – Bark comprises about 15% of the raw weight of wood. It is removed mechanically, releasing tannins, which are potentially toxic and contribute to biological oxygen demand (BOD).
- Chipping – Reduces wood to small pellets in preparation for pulping.
- Pulping – May be done chemically, mechanically (including use of heat), or by a combination of both approaches. Chemical pulping involves cooking the wood in either sulphurous acid (H_2SO_3) or a basic medium containing sodium hydroxide (NaOH) and sodium sulphide (Na_2S), the so-called Kraft process. Both remove over 90% of the lignin in the wood and result in the emission of noxious sulphur compounds (e.g., methyl mercaptan, dimethyl sulphide), which are responsible for the pungent odour associated with the operation.

The development in 1946 of chlorine dioxide bleaching in combination with the Kraft process led to a product with superior whiteness combined with strength and has resulted in this combination being the most popular pulping method. However, chlorine remains in widespread, though diminishing, use in bleaching operations. The Kraft process enables the efficient recycling of most of the cooking chemicals and results in the burning of most residual organics. The resulting effluent has a biological oxygen demand, which is only about 30% that of sulphite mill effluent. BOD from the paper industry remains a major concern and in North America has been estimated as equivalent to 75% of the BOD associated with untreated sewage.

Earlier studies of pulp and paper mill effluents were mainly concerned with chemical oxygen demand (COD), BOD, and the acute toxicity of resins, fatty acids, and chlorinated phenols. Depending on the pulping procedure, untreated mill effluents could contain levels of resin acids (principally dehydroabietic acid) as high as $80 \, mg \, L^{-1}$ and fatty acid concentrations as high as $22 \, mg \, L^{-1}$. Other concerns included the suffocating effect of deposited sludge on benthic fauna and physical clogging of fish gills resulting from high levels of suspended particulates.

More recently, problems associated with eutrophication and organochlorines, particularly dioxins and furans, have dominated the environmental agenda of the pulp and paper industry. Concerns also include the possible endocrine-disrupting properties of products such as phenolic compounds and phytestrogens such as β-sitosterol.

9.3.2   Substances of concern I: Nutrient enrichment from pulp mills

Growth of phytoplankton is limited by concentrations of phosphorus (P), nitrogen, or other required nutrients. Untreated mill effluents are enriched with these plant nutrients. In biologically treated mill effluents, although most of the nutrient content is removed, eutrophication of water bodies that receive these effluents is still observed. Priha (1994) studied the bioavailability of phosphorus in two types of mill effluent, activated-sludge-treated bleached Kraft mill effluent (BKME) and activated-sludge-treated paper mill effluent (PME), by measuring algal growth potential. Approximately 90% of the dissolved phosphorus and 45% of the particulate phosphorus in BKME were biologically available. On the other hand, less than 20% of the dissolved phosphorus was biologically available in PME.

In 1972, the start-up of a bleached Kraft pulp mill in Kamloops, British Columbia, Canada, resulted in a significant increase in algal standing crop in the Thompson River below Kamloops Lake. Since then, Thompson River has become the site for studies of the impacts of nutrient enrichment caused by BKME. Dubé and Culp (1996) have conducted experiments to determine the effects of increasing concentrations of BKME (0.25, 0.5, 1, 5, and 10%) on periphyton and chironomid growth in the Thompson River. Periphyton growth was significantly stimulated at all testing effluent concentrations as determined by increases in chlorophyll-a. Chironomid growth was also stimulated at lower effluent concentrations (less than or equal to 1%). These increases were attributed to the effects of nutrient and organic enrichment from BKME. This effluent served as a phosphorus source for periphyton and a carbon source for benthic insects grazing on the biofilm.

Algal growth in the Thompson River is thought to be phosphorus-limited and BKME serves as a phosphorus source, but mill effluents can also stimulate algal production in nitrogen-limited rivers (Bothwell, 1992). In the McKenzie River in Oregon, the atomic ratio of available N and P was 2:1 at the time of the study, indicating a nitrogen-limited system. The discharge of secondarily treated mill effluent to this river increased algal production. In summer, the concentration of dissolved inorganic nitrogen available for algal uptake was high enough to almost double algal production.

Mill effluents not only stimulate algal growth but also affect the structure and metabolism of algal community in receiving waters. Bourdier et al. (1990) found decreases in species richness and diversity and changes in taxonomic structure of the algal community in experimental streams receiving mill effluents. Decreases in diatom density were compensated by Chlorophyceae, which increased with increasing effluent concentration. Presence of effluent also led to a rise in heterotrophic biomass and activity.

9.3.3   Substances of concern II: Chlorinated products of paper pulp

ORGANOCHLORINES

The use of chlorine to bleach paper pulp results in the formation of organochlorines, which have the potential to cause harmful environmental effects. Chlorinated dioxins and furans are of particular concern due to their high toxicity and tendency to accumulate in biota. A study of 104 pulp and paper mills in the United States found mean 2,3,7,8-tetrachlorodibenzofuran levels in Kraft mill effluent and sludge of 0.5 and $806 \, \text{ng} \, \text{L}^{-1}$, respectively. Corresponding concentrations of 2,3,7,8-tetrachlorodibenzodioxin were 0.06 and $95 \, \text{ng} \, \text{L}^{-1}$. Bioconcentration factors of $>10^6$ for TCDF and TCDD have been reported from fish under field conditions. To minimise organochlorine formation, an initiative was begun in 1988 to substitute chlorine dioxide for chlorine during the pulp bleaching process.

Chlorophenols tend to accumulate in biota. Oikari et al. (1988) exposed juvenile lake trout to simulated BKME with added chlorophenols for 7 weeks. For pentachlorophenol, the concentration gradient between fish bile and water was 5.2×10^4. Lander et al. (1990) found that bioaccumulation factors of 4,5,6-trichloroguaiacol and metabolites from water ranged from 50 times for algae up to 700 times for invertebrates and fish.

Environmental monitoring of BMKE chlorophenol concentrations in a northern Canadian river system showed variation with seasonal river flows and mill process changes such as the substitution of chlorine dioxide for chlorine (Owens et al., 1994). At 100% chlorine dioxide substitution, concentrations of chlorophenols in effluents and receiving waters approached the analytical detection limit of 0.1–1 ppb. Chlorophenols were detected in the bile of mountain whitefish (*Prosopium williamsoni*) and longnose sucker (*Calostomus catostomus*) but were rarely detected in fillets. In the same study, contaminated fishes were held in uncontaminated water for 8 days, and a rapid decrease in chlorophenol levels was observed. This observation showed that chlorophenols were rapidly excreted by fish. Metabolites of chlorophenols were also detected in the bile of sand flathead (*Platycephalus bassensis*), an Australian marine fish, that was exposed to diluted chlorine dioxide bleached eucalypt pulp effluent for 4 days (Brumley et al., 1996).

CHLORATE

Substitution of Cl_2 by ClO_2 during the pulp-bleaching process reduces the concentrations of organochlorines; however, it generates another environmental concern. The use of ClO_2 as a bleaching agent results in the formation of the plant toxin chlorate (ClO_3^-). Chlorate is a close molecular analogue of nitrate (NO_3^-) and is taken up by the same transport mechanism. In higher plants, nitrate reductase can reduce chlorate to chlorite, which is highly toxic and causes general oxidative destruction of proteins. Nevertheless, because NO_3^- out-competes ClO_3^- for transport sites, additions of NO_3^- reduce the toxic effect of ClO_3^-.

Although ClO_3^- toxicity to some marine algae is known and has been implicated in environmental problems in the Baltic Sea, the addition of $500 \mu g L^{-1}$ of ClO_3^- to a low ambient NO_3^- ($10 \mu g\ NO_3^- L^{-1}$) river did not reduce the specific growth rates or change the taxonomic composition of the riverine diatom community (Perrin and Bothwell, 1992). This result indicated that ClO_3^- discharged from pulp mills into freshwater did not cause damage to the dominant algal producers such as diatoms.

9.3.4 The environmental toxicology of mill effluent

A variety of biochemical, histological, physiological, and ecological changes have been observed in aquatic organisms exposed to different mill effluents. Addison and Fraser (1996) studied the induction of hepatic mixed function oxidase in a benthic flatfish, English sole (*Parophrys vetulus*), sampled near coastal pulp mills in British Columbia, Canada. Their result showed significantly increased activities of MFO enzymes measured in fish collected from mill-effluent-contaminated sites compared with those from sites expected to be free of contamination. Indices of MFO activity in fish were well correlated with concentrations of chlorinated dioxins and related compounds in sediments at the sites and in marine invertebrates from these sites. Munkittrick et al. (1992) observed reproductive dysfunction such as reduced steroid levels, delayed maturity, and reduced gonad size; secondary sex characteristics; and fecundity in white sucker, longnose sucker, and lake whitefish exposed to BKME. Induction of hepatic MFO activity was also observed in the same study (see Section 7.8).

Numerous recent studies have demonstrated a link between reproductive dysfunction and exposure to BKME. Many have shown reduced hormone levels in exposed individuals, although the mechanisms controlling steroid levels, through synthesis or clearance, are complex. Interference may occur at several different levels within the steroid biosynthetic pathway. It has now been established that 17β-estradiol directly or indirectly influences cytochrome P450s associated with environmental contaminants by reducing enzyme activity. The phytoestrogen β-sitosterol, found in high concentrations in BKME, has been shown to lower plasma levels of reproductive steroids by reducing gonadal steroid biosynthesis in fish (MacLatchy and Van Der Kroak, 1994). Larsson et al. (2000) found a preponderance of male eelpout embryos (*Zoarces viviparus*) close to a pulp mill, which they ascribed to the presence endocrine disruptors in BKME.

A survival study was conducted with fingerlings of tilapia (*Oreochromis mossambicus*) to evaluate toxic effects of mill effluents on different bioenergetic parameters (Varadaraj and Subramanian, 1991). The 96-h LC_{50} was 6% of the effluent. Rates of feeding, food absorption, growth, metabolism, absorption efficiency, and conversion efficiency were all decreased due to the toxicity of mill effluents. The same fish was exposed to various sublethal mill effluent concentrations for 15 days and decreases in red and white blood cell counts, haemoglobin content, packed cell volume, and mean cellular haemoglobin concentration were observed (Varadaraj et al., 1993).

Abnormalities of the operculum were found in perch (*Perca fluviatilis*) from an area affected by mill effluent in the Gulf of Bothnia, Baltic Sea (Lindesjoeoe et al., 1994). Shortening of the distal part of the operculum and craterous formation on the operculum were observed. Maximum prevalences of abnormalities were found in 1983 and 1984. Decreases in abnormalities after 1984 coincided with a reduction in effluent concentrations.

A study was conducted in the Baltic Sea to verify general responses of adult and fry abundances and fish community structure, mainly *Perca fluviatilis* and *Rutilus rutilus*, to mill effluent (Sandstrom et al., 1991). Fish abundance was low close to the mill. However, at intermediary exposure, eutrophication caused very high fish biomass and a community dominated by cyprinids. Fry abundance was low, although studied areas provided potentially good recruitment habitats. These patterns were related to both organic enrichment and toxic pollution.

9.3.5   Mitigation: Means for minimising the impacts of pulp mills

The pulp and paper industry has made a great deal of progress in mitigating the adverse effects of discharges over the last 30 years. The initial impetus for these improvements was provided by legislation under the 1972 U.S. Clean Water Act and specific legislation passed by the Canadian Environmental Protection Service in 1972 dealing with liquid effluent discharges from pulp and paper mills.

Much like the treatment of sewage effluent, primary treatment of pulp and paper waste consists of initial filtration and pH adjustment, usually to within the 6.5–8.0 range, followed by the removal of suspended solids in a settling basin. In secondary treatment, up to 95% reduction in biological oxygen demand can be obtained by using activated sludge filters, although where space allows, better results are achieved with the use of biooxidation ponds having residence times of up to 10 days. In these ponds, aeration is employed to stimulate bacterial and protozoan catabolism of potentially toxic organic compounds. Following such treatment, effluents display dramatically reduced toxicity to aquatic organisms. The use of such lagoons also generates much less sludge than activated sludge treatment. Such treatments have resulted in at least a threefold reduction in BOD and suspended solids discharged per tonne of effluent since legislation was first enacted in the early 1970s.

More experimental approaches include the process employed at the Trenton, Ontario, Norampac plant, and some Quebec mills wherein the black liquor from the pulping process is not discharged to water at all. It still presents a disposal problem, however, and currently some of it is applied to gravel roads as a dust suppressant. This use of the liquor remains a controversial issue, because its application represents an aesthetic problem and its ecological effects are still of concern, particularly if the liquor is allowed to migrate from the road surface.

Recycling has become a major initiative in the paper industry and may be responsible for saving 4,000 km^2 of trees annually in addition to reducing pollution and water use associated with the industry. The use of unbleached products in the packaging industry points to a future commitment to a more conservative use of chlorine products and a cleaner image for the pulp and paper industry as a whole.

9.4 Electrical power generation

Electrical power can be generated in a number of ways. Steam to run generators is produced by burning fossil fuel such as coal or oil in steam-turbine plants or by harnessing nuclear energy in a nuclear reactor. Hydroelectric power plants harness the power of falling water. These three are quantitatively the most significant processes by which electricity is produced. Each may have environmental impacts in terms of mobilising or releasing potentially toxic substances to the environment, and these impacts are the subjects of this section. Impacts other than those related to toxic substances are not considered in detail here. Other methods, sometimes referred to as renewable energy sources, include the harnessing of wind power, solar power, and other nonconventional sources for the generation of electricity. These energy sources are not dealt with in detail here because they have minimal toxicological significance, but this is not to imply that in other ways they are entirely benign.

The subsequent transmission of electricity and its distribution is the same whether the ultimate source of energy is the combustion of fossil fuel, nuclear fission, water power, or unconventional sources.

The environmental toxicology of the three major types of electrical power generation are each treated briefly in a cradle-to-grave manner and then the more complex socioeconomic issues are considered.

9.4.1 The issue of producing electricity from fossil fuel

The extraction of fossil fuel from its geological sources involves the production of acid mine drainage from coal mines (see Section 9.2) and for oil and gas, there is a risk of contamination of terrestrial and aquatic systems during extraction from accidental leaks and spills, which can be major. The transportation of oil and gas, whether by pipeline or by tanker, raises similar concern because of the risk of accidental spills. Section 9.7 deals in more detail with the occurrence and impacts of oil spills. Oil and gas are, of course, used directly as fuels and not exclusively in the generation of electricity. The combustion of fossil fuel is the main cause of many atmospheric pollutants: CO_2, which is one of the greenhouse gases (see Section 9.5.2), and SO_2 and NO_x, which contribute to acidic deposition (Section 9.6.2). NO_x also contributes to changes in ozone levels (see Section 9.6.2).

Metallic contaminants in fossil fuels result from the tendency of living organisms to bioaccumulate contaminants natural or anthropogenic. When coal, oil, or gases are formed, the biological component of the fuel has already accumulated inorganic substances such as arsenic (As), mercury (Hg), and other metallic elements. These are locked into the fossil fuel deposits, only to be released by combustion or other types of processing. The more volatile elements, such as mercury and to a lesser extent cadmium (Cd), are released in gaseous form, whereas most other elements are released in particulate form. The latter elements tend to move less distance from the source than mercury. Fluoride is also released during the combustion of coal, once again because it has accumulated in the original parent material.

A brief summary of the cradle-to-grave potential for the formation and/or mobilisation of contaminants and ecological effects of generating electricity from fossil fuel is provided in Table 9.3.

9.4.2   The issue of producing electricity from nuclear energy

The harnessing of nuclear energy to generate electricity is a relatively new technology. In December 1951, an experimental breeder reactor generated enough electricity to turn on four light bulbs. Currently nuclear power produces 17% of the world's electricity, with considerable variation among countries in terms of this proportion.

Nuclear reactors vary greatly in detail of engineering design, type of fuel, and nature of the moderator and coolant whether heavy water, light water, or gas. All have the controlled fission of isotopes of uranium (U) or plutonium (Pu) in common (Section 8.5.2). The resulting heat is used to make steam to drive the turbines from which electricity is produced. As discussed in Section 8.5.2, the nuclear fuel cycle is essentially a five-step process with the mining and milling of ore at the "front end", fuel enrichment, use and reprocessing, and the disposal of a range of waste products at the "back end".

The nuclear fuel cycle begins with the mining of uranium ore. The toxicological effects of the extraction and purification are common to those of all metal mining (Sections 9.2.2 and 9.2.4) but with additional concern for the release of radioactivity at each stage of the process. Uranium ores are generally sulphide ores, meaning that the waste is acid generating. The mode of extraction of ore-bearing rocks is determined by their depth and distribution. Strip and pit mining deeply scar the landscape as do the heavy equipment needed to extract the ores from the deposit and mill it in preparation for the enrichment process. A combination of waste rock, effluents, and drainage can exert significant biological, physical, and chemical effects through isotope and metal contamination of surface and subsurface waters, soil, and biota. Aluminium, iron, arsenic, and several other trace metals are common contaminants of uranium mining and mill effluents. Drainage from mines or leachate from tailings may be either acidic, in which case metals are readily mobile, or basic due to the presence of boron, which, in addition to being caustic in solution, is also a plant toxicant. Additionally, the so-called front end waste, namely the waste rock and mill tailings from uranium extraction, is characterised by being large in volume but low in radioactivity. The radioactive mate-

Table 9.3. Contaminants related to the generation of electricity from fossil fuel

Activity	Process	Substance(s)	Potential environmental effect(s)
Removal of coal from ore	Underground or surface mining	Acid mine drainage: sulphuric acid and dissolved metals	Damage (acute) to aquatic ecosystems in receiving areas and long-term degradation of streams; potential for groundwater contamination
Oil and gas extraction	Drilling and extraction	Hydrocarbons released, small amount routinely, large losses in accidents	Contamination, including bioaccumulation and effects (acute and chronic) on ecosystems in receiving areas
Transportation of fuel	Accidental leaks and spills	Large releases of hydrocarbons and other petroleum products released	Contamination of soils and water, often in remote areas including the Arctic, by oil and gas components, with physical and toxic effects (acute and chronic) on ecosystems in receiving areas
Energy production	Combustion of fossil fuel	Release into the atmosphere of SO_2, NO_x, CO_2, trace metals especially mercury, radioactive materials (amounts and types dependent upon control measures, precipitators, scrubbers, etc.)	LRTAP of S and N compounds, leading to acidic deposition; build-up of atmospheric CO_2, deposition of trace metals, and radionuclides
Waste disposal	Handling and long-term disposal of solid waste, including contaminated dust and fly ash from precipitators	Potentially toxic dusts and other solids, containing many inorganic contaminants	Potential contamination of soils, surface water, and groundwater

rial will normally decay to safe levels in less than 50 years. But this waste also contains potentially mobile metals and is acid generating; therefore, tailings ponds and waste piles are very difficult to stabilise and revegetate. The volumes of low-level waste in Ontario alone amounted to $1,380,000 \, m^3$ in 1993. Low-level nuclear waste also includes isotopes of nickel and cobalt from pipes and radioactive material in workers' clothing and tools used by reactor workers. The materials can be stored in shallow landfills. In 1989, the nuclear power industry in the United States produced 52% of the country's low-level waste. To put this into some kind of context, commercial industry in the United States produced 35%, and the medical industry produced less than 0.1% of the country's low-level waste. The U.S. Low Level Waste Policy Act of 1985 required that all states either form a pact with other states or select a disposal site themselves. These disposal sites cause problems best described by commonly used acronyms: LULU (locally unwanted land use), NIMBY (not in my backyard), and NIMTO (not in my term of office) (Ahearne, 1993). In Canada, the stabilisation and vegetation of uranium mill tailings continue to challenge experts, and low-level waste disposal in that country continues to be a subject of much debate. By 2025, it is estimated that the Canadian inventory of low-level radioactive waste will have increased to $1,762,400 \, m^3$, a figure that includes a programme of power plant decommissioning after 2010. From 1986 to 1994, annual low-level waste generation and disposal in the United Kingdom fell from c. 41,000 to c. $21,000 \, m^3$, although high-level waste rose during that time (see following discussion).

Purified uranium ore (UO_2) is the final product received from the mining and milling process for further purification to nuclear grade uranium. Purification is accomplished by either a dry or wet process. The wet process, mainly used by commercial purification facilities, entails the use of anhydrous fluoride volatility, wherein hydrofluoric acid is used to convert uranium to the tetrafluoride form, which is then exposed to fluorine gas under special conditions to produce UF_6. The hexavalent uranium is then melted to release uranium metal, which is then formed into metal rods (for Magnox fuel) or roasted with steam and hydrogen gas to form the ceramic material UO_2, which is pressed in solid pellets to create the fuel rods. The conversion of the purified uranium into fuel for the various types of reactor does not generally involve any risk of environmental contamination. The major remaining concerns involve the reactor, particularly accidents, the disposal of highly radioactive waste from spent fuel, and site decommissioning.

In the routine operation of a nuclear reactor, the day-to-day concerns for safety are focused on the workers and other persons in or in close proximity to the plant. Routine release of radioactive substances is extremely small, and both from the occupational and environmental aspects, regulation and built-in safety devices are generally considered to be satisfactory. Data from Cochran and Tsoulfanidis (1990) indicated that in normal operation, radiation from a coal-fired power plant actually exceeds that from a nuclear plant of similar size (Table 9.4). Of greater ecological significance are the thermal effluents generated by nuclear power plants as a result

Table 9.4. Relative radiation doses from electric power generation[a]

Organ	Maximum individual dose[b], mrem/yr		Population dose[c], man-rem/yr			
	Coal-fired plant[d]	Nuclear plant (PWR)	Coal-fired plant[d]			Nuclear plant (PWR)
			Stack height, 325 ft	Stack height, 650 ft	Stack height, 975 ft	
Whole body	1.9	1.8	21	19	18	13
Bone	18.2	2.7	225	192	180	20
Lungs	1.9	1.2	29	23	21	9
Thyroid	1.9	3.8	21	19	18	12
Kidneys	3.4	1.3	50	43	41	9
Liver	2.4	1.3	29	26	25	10
Spleen	2.7	1.1	34	31	29	8

[a] Comparison of radiation doses to the population from a coal-fired and a nuclear 1,000-MW(e) power plant.
[b] At the plant boundary, to be 500 m from the point of release.
[c] Out to 55 miles for a Midwestern site assumed to have a population of 3.5 million within that radius.
[d] Assume an ash release of 1% and coal containing 1 ppm uranium and 2 ppm thorium.
Cochran and Tsoulfanidis (1990).

of the huge amounts of cooling water needed to condense steam for recycling. Ecological effects of discharges have been well studied and are known to effect shifts in plant and animal species composition, species abundance, mortality, and disease occurrence rates. Accidents at nuclear plants are, however, a major concern for the public (see Section 9.4.4). Small releases of radioactive substances and of deuterium (H_3) from heavy water plants and reactors that use heavy water are of concern, but the major fear is for some kind of catastrophic failure in the core of the reactor. The close calls that have been recorded and the massive release of radioactive material from the Chernobyl reactor (see Section 9.4.4) are certainly of technical concern, yet these accidents and close calls have been due in every instance to human error rather than straight technical failures.

High-level nuclear waste is the end-product of any reactor. Relatively small in volume, this highly radioactive spent fuel from the core takes centuries to decay to safe levels of radioactivity. The waste from the CANDU and similar reactors using natural uranium as fuel has the potential to be processed and used as fuel for other types of reactor designs. At present, high-level waste is stored on-site at the power plants while decisions on the final resting place for this hazardous material remain to be made. The U.S. Congress continues to debate over sites for permanent disposal. The fear of nuclear energy leading to nuclear arms proliferation may not be a grounded one according to its track record, but it is an apparent fear that must be dealt with. All nuclear power plants use either plutonium or uranium, and it is possible for these elements to be used for weapons instead of energy generation. According to the report by the National Research Council, "articulate elites" believe the potential for arms proliferation is a serious threat caused by the use of nuclear energy as a power source (Ahearne, 1993). A manifestation of this fear is clearly apparent at the international level where 1996–7 negotiations regarding the construction of nuclear power generation facilities on the Korean peninsula were influenced by concerns over the generation of weapons-grade nuclear material.

Any nuclear power plant has a limited lifetime. Once a reactor is no longer in operation, the entire plant must be decommissioned. Parts of the plant will be contaminated with radioactive material, and if left to deteriorate, the contents of the more highly radioactive parts may escape into the general environment. Options include the removal of some of the most contaminated materials and treating them with other high-level waste, "mothballing" the entire area, or mothballing without removing anything. In any case, the site requires continued surveillance long after power generation has stopped. The technology and the cost of this stage was apparently not built into the cost estimate for early nuclear powered electricity generating plants. Similar to the issue of the final storage for the high-level waste, the issue of decommissioning nuclear plants remains unresolved. Over the past three decades, Canada has had at least three major studies on the safe disposal of high-level waste and has still not finalised the issue. In the United Kingdom, high-level radioactive waste fell from 100 to $80\,m^3\,y^{-1}$ between 1986 to 1987 but had risen again to $>110\,m^3\,y^{-1}$ by 1994 as a result of increased power generation from nuclear fuel and an increase in fuel reprocessing.

Table 9.5 provides a summary of the major sources of environmental contamination that can result from electricity generation from nuclear power. Radiation damage to humans and to ecosystems is anticipated from environmental exposure to ionising radiation (Chapter 8), so it is obviously prudent to make every effort to avoid such exposure. However, in the case of radiation, perhaps more than for any other potentially harmful substance, the chronic effects of low-level exposure are predicted from models based on extrapolation from high-level exposure. The truth concerning threshold effects is simply not known, and even experts are in disagreement (see Section 8.4). Furthermore, we lack real-world examples of the effects of exposures such as would result from Chernobyl. As discussed in Chapter 8 and Section 9.4.4, the evidence for the degree of adverse effects from releases of radioactive material from Chernobyl is still far from conclusive.

If one considers the overall types and amounts of potentially damaging substances for human health and for ecosystems, in routine operation, nuclear power plants are much "cleaner" than fossil-fuelled power plants. But, as is discussed in Section 9.4.4, this fact has not paved the way for their acceptance by the general public nor by the governments of many countries.

Table 9.5. Contaminants related to the generation of electricity from nuclear energy

Activity	Process	Substance(s)	Potential environmental effect(s)
Removal of uranium from ore	Underground or surface mining	Acid mine drainage of sulphuric acid and dissolved metals and radionuclides – relatively small volumes involved in comparison with coal mining	Damage (acute) to aquatic ecosystems in receiving areas and long-term degradation of streams; potential for groundwater contamination
Uranium extraction	Milling and water processing	Slurry of finely ground, acid-generating material, containing metals and radionuclides	Tailings ponds occupy relatively large areas in the external environment, frequently filling natural lakes; potential for blowing particles containing contaminants and for surface and groundwater contamination; tailings ponds are infertile and require expensive management to become vegetated
Nuclear fission in reactors	Accidental leaks and spills or major (core meltdown) incidents – Rare but potentially devastating	Various radio isotopes including plutonium-239, cobalt-60, strontium-90, cesium-137, tritium (^{3}H)	Contamination of soils and water by radioactive material – can lead to cellular damage and genetic changes
Heavy water production and use (CANDU-type reactors)	Deuterium enrichment process	Release into the atmosphere of H_2S and deuterium	Hydrogen sulphide is extremely toxic to mammals
High-level radioactive waste from nuclear power plant	Long-term disposal of solid waste, including spent reactor fuel and other material from the plant	Material continues radiological decay for thousands of years with production of radioisotopes and heat	Potential contamination of soils, surface water, and groundwater
Decommissioning	Long-term storage and surveillance of the entire infrastructure of the nuclear plant	Similar to high-level waste	Potential for release of radioactive material into the environment if containment is not well maintained

9.4.3   The issue of hydroelectric power

Hydroelectricity generation involves trapping water behind a dam, or utilising an existing fall of water, and allowing the water to fall through a turbine. Hydrosystems can be massive as in the case of the Churchill River diversion in Manitoba, the LaGrande system in Quebec, or the Columbia River system, which flows through British Columbia, Canada, and Washington state, USA, and has a total of 450 major dams within its basin. Many smaller plants successfully generate electricity along lesser river systems (e.g., the Trent-Severn waterway in Ontario, Canada).

The construction of hydroelectric dams has been criticised by environmentalists and conservationists for such reasons as dams are changing land use, generating methane (see Section 9.5.2), affecting wildlife, destroying fish habitat, and disrupting traditional aboriginal ways of life. All these arguments are tenable and need to be considered against the more conventional position that hydroelectric power generation is environmentally benign. However, they do not have direct relevance to the present topic of environmental toxicology.

One aspect of hydroelectric dams does have toxicological implications. It emerged on the scientific scene relatively recently and concerns the bioavailability of mercury in newly impounded sites. Details of mercury toxicology and of the mechanism of the so-called "reservoir effect" are provided in Section 6.4.4. The reader is reminded here that the mechanism involves the influx of existing mercury (natural or anthropogenic) from the previously terrestrial and now flooded area, in combination with an influx of organic matter from the same source. Bacterial activity is stimulated by the organic matter. There is no mercury pollution per se, simply a conversion of existing mercury into methylmercury by bacterial methylation. The resulting methylmercury is bioconcentrated and biomagnified up the food chain, resulting in unacceptable concentrations in fish. It is an elegant illustration of the importance of the chemical form of a metal in determining its biological availability. Until the mid 1970s, the problem had not been uncovered. The socioeconomic effects can be severe. Whitefish from LaGrande system in northern Quebec, Canada, contained up to 1.6 ppm of mercury in the 1980s, rendering them unfit for consumption. This has impacted the Cree Indians, who lost terrestrial habitat for hunting when the river was flooded, but who were expected to benefit from the resulting fishery. Recent measurements indicate that Hg concentrations in fish from LaGrande system are declining, without any intervention.

After the flooding has occurred, there is no practical remedy for this situation because there is no source of mercury to be controlled. It has been speculated that with time methylation may decrease so that older reservoirs may show more acceptable mercury levels in fish. New systems may benefit from preimpoundment treatment of the terrestrial habitat, through harvesting most of the vegetation prior to flooding. However, this solution may increase the tendency for erosion from the banks, which is another frequent complication with dams on rivers.

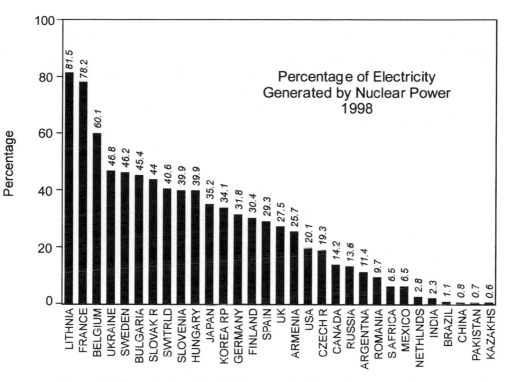

Figure 9.1 Percentage of electricity generated by nuclear power for a range of countries, 1998 (IAEA, 1998).

9.4.4   Socioeconomic considerations

Commitment to nuclear energy varies greatly among countries. In 1998, the United States produced 22% of its electricity from nuclear power plants, placing it eighteenth in the world for percentage of electricity generated from nuclear power plants (Figure 9.1), although in overall terms it still maintained the world's largest nuclear energy production from 107 commissioned plants. Although nuclear energy in the United States grew rapidly in the beginning, it has slowed down and its contribution is far from the predictions made two decades ago for its use. For example, not one reactor has been ordered by an American utility company since 1978. However, by 2001 the U.S. government was reevaluating this in the light of increased energy demands and recommissioning of nuclear power generation facilities again seemed likely.

In Canada, Ontario depends on nuclear power for approximately two thirds of the electrical needs of that province, a proportion equal to that in France and Belgium and higher than that of all other sovereign states. Following a decade of slow growth, by 2001 Ontario was considering re-opening some nuclear power plants. In Canada as a whole, the percentage of electricity produced from nuclear power is 20%, approximately the same as in the United States, whereas in the United Kingdom there has been an increase in the percentage of electricity generated from nuclear power. Public opinion often has a powerful affect on the use of nuclear power as a source of electricity generation. Germany has recently determined to eliminate nuclear power as a source of electricity, despite the fact that nuclear reactors currently supply approximately one third of the country's electricity. Australia currently uses no nuclear power to generate electricity.

The slow growth rate of the nuclear power plants has many causes. The main reasons for decline in the growth rate in the United States were explored by Ahearne (1993). Using information generated by the U.S. National Research Council, he identified a slowed growth in the demand for energy, an increase in cost and construction time for nuclear power plants, nuclear power plants that did not achieve the expected levels of operation, and public opinion that turned away from nuclear energy as a desirable new technology. Nevertheless, following two decades of low oil prices, a reversal of this trend coupled with concern over greenhouse gas emissions has provided new incentives for nuclear power.

The fall in the rate of growth for the demand of energy seen in the 1980s and 1990s was caused by both economical and environmental factors. Two "oil shocks" came in the 1970s when OPEC (Organisation of Petroleum Exporting Countries) expressed its power to raise oil prices. This forced revisions of how developed countries used energy. The rise in oil prices caused a change in attitudes towards energy consumption. A new interest in conservation and efficiency slowed down the rate at which the demand for energy was growing (Ahearne, 1993). To save one unit of energy is less expensive than to supply one unit of energy because it costs less to protect the environment through efficiency than through expensive pollution control technologies. Saving energy reduces the harmful side effects that go along with fossil fuel combustion, including acid rain and increases in carbon dioxide emissions with implications for climate changes (see Section 9.6), and reduces the problem of radioactive waste disposal from nuclear fuel.

Historical increases in cost and construction time of nuclear power plants in the United States appear to be due almost solely to environmental concerns. Fears over safety have been fed by events such as the 1979 partial meltdown at Three Mile Island, near Harrisburg, Pennsylvania, and the 1986 meltdown at Chernobyl, Ukraine (Case Study 8.1). Lack of proper storage and/or disposal of both low- and high-level radioactive wastes and the fear that nuclear power could lead to the proliferation of nuclear arms are additional factors of which the public have become aware, forcing additional safety and environmental protection. Although the effects of the Chernobyl incident are still not all known, a much more complete picture of the effects of the accident has emerged over the last decade. In the United States, the accident at the Three Mile Island plant led to 2 years of deliberation by the U.S. Nuclear Regulatory Commission to decide what regulatory changes would result from the 1979 experience. The new regulations included costly modifications to existing plants as well as to plants under construction. In March 2000, the United States registered its first nuclear electricity generation plant in over a decade.

In Ontario, which uses the CANDU design (a heavy water reactor which uses unenriched uranium), a major commission on nuclear safety published in 1988 identified the most serious problem as the poor performance of the zirconium pressure tubes in the reactor core. Small leaks or even rupture of these tubes have occurred and result in discharge of heavy water. To the date of the report, defective tubes had been safely detected and replaced, but at great cost. The commission considered this as a serious economic problem because the tubes were not meeting their expected lifetime and replacement is extremely costly, but it did not consider that it represented a threat of radioactive exposure to the public. Workers would certainly be at risk were there a major failure in pressure tubes.

By the early 1980s, electric power company owners in the United States were facing uncertainties over the risks posed by nuclear power plants and were already beginning to notice a decline in the energy demand growth rate. These uncertainties were further compounded by management problems at some facilities in what is one of the most highly regulated industries in the world. The Tennessee Valley Authority's nuclear power program, one of the largest in the United States, was shut down by the U.S. Nuclear Regulatory Commission for over 4 years because of management problems. Ontario Hydro (a Crown Corporation with, until recently, complete responsibility for the province's power generation and distribution) identified similar problems with management in the 1990s. Public opposition to nuclear energy has grown over the last 20 years because of the perception that it is unsafe and costly. Along with fears of accidents are concerns over safe permanent disposal for low- and high-level waste.

Notwithstanding these concerns, it is clear that a wide international disparity exists in terms of commitment or opposition to nuclear power. Although the industry seems stalled in Germany and, until recently, the United States, many countries are pressing ahead with ambitious nuclear energy programs. In 1998, 47 nuclear power plants were being constructed worldwide (many of these in India). Between 1995 and 1998, the United Kingdom had shifted its nuclear share of electricity generation from 26 to 36%. This represents an update of the information shown in Figure 9.1. If the terms of the 1997 Kyoto agreement on greenhouse gases are to be complied with, significant reductions in share of electricity generation from fossil fuels must be considered. A future projection of nuclear share of electricity generation of 10% in the United States by the year 2010 would probably be incompatible with emissions targets in that country unless there is a phenomenal rise in the development of alternative energy sources. By 2001 increased energy prices in the western United States had prompted a resurgence of interest in nuclear power generating stations and an evaluation of the options of recommissioning old plants.

The role of fossil fuels in the future is also being reconsidered. The use of coal, oil, and natural gas in the future, like the use of nuclear energy, will be based on their economical and environmental costs. The monetary costs of fossil fuels will continue to rise due to the locations of the source and the rising production costs.

The environmental costs are also likely to continue to rise if more stringent controls on the release of pollutants are indicated.

It is apparent just from looking at the costs associated with the generation of nuclear power and energy generated from fossil fuels that society pays for both sources of energy in a number of ways. These costs include "social costs", which are often not included in the market prices of fossil fuels and nuclear energy. Such costs embrace damage to plant and animal life, human health, corrosion and weathering of buildings, and climate effects. The social cost of fossil fuels is found to be two to five U.S. cents per kilowatt hour. The social cost of nuclear energy has been calculated at five to ten U.S. cents per kilowatt hour. When these penalties are added to market costs, renewable energy sources may become competitive, and eventually cheaper, than conventional energy sources such as fossil fuels and nuclear energy. At present these calculations remain hypothetical.

9.5   Global warming

9.5.1   The issue

The issue of global warming is a widely debated topic. Even though many scientists agree that there has been a slight increase in global surface temperatures, they disagree over the degree to which these increased temperatures are attributable to anthropogenic increases in so-called greenhouse gases or to natural climatic variations. Temperature records over the past century show that there has been an increase in the global mean temperature of about 0.5°C. However, this increase has not been steady. The coolest part of the century was at the beginning with increasing temperatures between the 1920s and the 1940s followed by a period of cooling lasting until the 1970s, when the temperature began to rise again with six of the warmest years on record occurring within the last decade. 1997 was the warmest year since records began.

Ice core samples have shown that the Earth undergoes natural periodic warming and cooling phases. One study on the Vaostock ice core data showed a link between these warming trends and CO_2 concentrations (McElroy, 1994; Sun and Wang, 1996). The CO_2 in the atmosphere traps the sun's heat, leading to an increase in surface temperatures known as the greenhouse effect.

9.5.2   The greenhouse effect

The greenhouse effect occurs when gases such as carbon dioxide allow radiant energy from the sun to penetrate the Earth's atmosphere and warm the surface. However, when the Earth's surface reradiates the energy, the greenhouse gases reflect the energy back to the surface, much in the way that the glass on a greenhouse does. This occurs because CO_2 absorbs sunlight at the near infrared wavelength, resulting in very little absorption. However, it is able to strongly absorb and emit terrestrial radiation, which occurs at about 12–18 µm. The Earth is able to release some of this trapped heat through evaporation. Water absorbs vast amounts of heat as it is evaporated from the Earth's surface and then releases this energy when it condenses, forming into clouds. Much of the energy released as the water condenses is then able to escape into space. Because the evaporation/condensation of water releases some of this trapped energy into space, it would seem that, as temperatures increased, so would evaporation, and the two effects would counter-balance each other. However, as the temperature increases, the efficiency of the Earth to radiate the heat back into space through evaporation decreases. With an increase in air temperature, the concentration of water vapour needed to reach saturation increases. This means that the surface temperature must increase for the same amount of heat to be radiated through evaporation. As the temperature increases and the air saturation point increases, there will be a decreased frequency of precipitation (more time needed to reach saturation) with an increased intensity of the precipitation events.

9.5.3 Substances of concern: Greenhouse gases and their sources

The most important greenhouse gases that humans have influenced in the past century are carbon dioxide, nitrous oxide, methane, ozone, halocarbons, and chlorofluorocarbons.

Carbon dioxide (CO_2) has been the main focus of discussions on greenhouse gases. It is estimated that 700 billion tonnes of CO_2 are cycled through the atmosphere each year from natural sources alone. However, there are also natural processes that are capable of removing this added CO_2 from the atmosphere. Human activity adds approximately 24 billion tonnes of CO_2, primarily from the burning of fossil fuels, to the atmosphere each year (Morrissey and Justus, 1997). Although this amount is small relative to natural sources, it is considered by proponents of global warming to be enough to overburden the natural removal processes. In addition to increased CO_2 production, deforestation and inefficient land use practices have resulted in the reduction of some natural CO_2 consumers. It is estimated that average atmospheric CO_2 concentrations have increased by 30% over the last hundred years (Morrissey and Justus, 1997). At the beginning of the industrial revolution, the CO_2 concentration in the atmosphere was 280 parts per million (ppm) compared with 360 ppm today. Over the next hundred years, the relative contribution of CO_2 to the greenhouse effect is estimated to be approximately 60%.

Methane (CH_4) is produced naturally by ruminants, natural wetlands, reservoirs, aerobic soils, and also termites (Boeckx and Van Cleemput, 1996; Morrissey and Justus, 1997). Anthropogenic sources of methane include cattle raising, rice paddies, trash dumps and losses of natural gas during its production and transport. Anthropogenic sources of methane are estimated to be responsible for at least half of the total atmospheric methane inputs, although there is evidence that these have subsided in recent years. Within the sediments of the deep ocean, vast quantities of methane hydrates have been found. Some scientists have speculated that if the deep oceans were warmed, some of these methane hydrate layers could be melted, releasing large amounts of methane leading to further increases in global surface temperatures (Zimmer, 1997). The relative contribution of methane to the global warming effect over the next hundred years is estimated to be approximately 15%.

Ozone (O_3) is a gas that has been both increased and depleted in the atmosphere as a result of anthropogenic activity. Ozone is a very important gas in the stratosphere that blocks some of the sun's harmful rays from penetrating the Earth's atmosphere. This ozone layer is a naturally occurring shield that has become depleted through the use of chemicals such as chlorofluorocarbons and other aerosols. The loss of stratospheric ozone decreases absorption of solar radiation and allows more heat to escape from the upper atmosphere into space. These effects coupled with the decreased heat inputs from the troposphere (lower atmosphere) have resulted in a net cooling of the stratosphere (Manabe, 1997). Through fossil fuel combustion, a new ozone layer has been created in the troposphere, which blocks heat from escaping the planet's surface. The potential contribution made by ozone to the greenhouse effect is approximately 8%.

The two other key greenhouse gases are nitrous oxide (N_2O) with a potential 4% contribution to the greenhouse effect over the next hundred years, and chlorofluorocarbons with a potential 11% contribution. N_2O primarily emanates from fossil fuel combustion but is also released naturally by biological processes in soils. CFCs, however, seem to mainly exert a cooling influence, largely through their interaction with ozone, and this serves to possibly mitigate against some of the greenhouse warming. Some recent evidence points to a possible positive association with stratospheric ozone depletion, the so-called hole in the ozone layer and global warming.

9.5.4   Global climate models

The Earth is known to undergo periodic warming and cooling cycles, making it difficult to determine any causal relationship between the composition of gases in the atmosphere and changes in temperature. Scientists have developed computer models to help them determine whether the slight warming trend we are currently seeing is the result of natural cycles or caused by anthropogenic inputs of greenhouse gases into the atmosphere. To make accurate assessments of increases or decreases in the Earth's temperature, these models need to consider global atmospheric and ocean circulation patterns. Because of the number of factors that affect global circulation, it takes a tremendous amount of computer power to run these models. As available computer power has increased, models have become more complex, and the results are assumed to be more accurate. Also, the amount of climate data available for models has increased as our technologies for making climatic measurements (e.g., satellite temperature imaging) have improved.

Early three-dimensional models developed in the 1970s were referred to as global circulation models. These models were superseded by coupled ocean-atmosphere global circulation models (CGCM). The current climate models are CGCMs that integrate basic fluid dynamical equations (Hasselmann, 1997). The resolution of these models is usually a few hundred kilometres. At this resolution, clouds and ocean eddies typically cannot be appropriately represented. This situation poses a problem with these models because clouds are of great significance to heat budgets (Karl et al., 1997). Other important feedback parameters to heat budgets include natural phenomena (e.g., snow cover and sea ice) and anthropogenic factors (e.g., the impact of aerosols). For example, sulphate aerosols are believed to have a cooling effect on the atmosphere by reradiating heat and by blocking or reflecting light from the surface. The cooling effect is most prevalent during daylight hours in highly industrialised areas where concentrations of sulphate emissions are high. Stratospheric ozone depletion has, likewise, not been accounted for in these models.

The global warming issue has sharply refocused attention on the economic and environmental costs of energy production worldwide. Currently, fossil fuels provide about 80% of global energy needs, yet there is increasing concern that this degree of commitment to fossil fuel combustion will lead to unacceptably high greenhouse gas emissions. To comply with the terms of the December 1997 Kyoto agreement, several developed countries gave assurances to cut their level of emission of greenhouse gases by agreed percentages, relative to 1990 emission levels. Using 1990 as a reference point, Japan would have to cut its emission of greenhouse gases by 6%, the United States would cut its emissions by 7%, and the European Union, acting as a whole, would cut its emissions by 8%. Some countries, such as Australia, Iceland, and Norway would be allowed an increase over 1990 emissions, although no agreement was reached with respect to emission levels in developing countries. To achieve these goals and still maintain projected increases in electricity production in these countries, some difficult decisions would have to be made concerning alternative energy sources. Included in this equation is nuclear energy. In 2001 the U.S. government disassociated itself from this agreement, thereby leaving its future in doubt.

The last 20 years have seen numerous published texts on global warming and climate change. Although many excellent reports exist and updates abound, readers are directed to Silver and DeFries (1990), which has been recommended as additional reading in Section 9.11.

9.6   Atmospheric pollution

9.6.1   The issue

Most harmful atmospheric pollution is caused by products of fossil fuel combustion either singly or acting in concert with each other and several climatological factors. The geographical scale of the resultant environmental problems has ranged from localised photochemical pollution in urban areas to potentially global issues involving greenhouse gases. Some chemicals, such as NO_x, have been implicated in a variety of largely unrelated, potentially harmful effects. Two effects, photochemical oxidants (smogs) and acid precipitation, are considered in further detail later. These two phenomena have been responsible for most of the air pollution control legislation that has been enacted. Both became recognised as serious environmental problems in the 1950s. By the late 1960s, legislation began to be enacted to first control photochemical oxidants and, later, principal acid rain components. The relevant Acts for the United States, Canada, and the United Kingdom-Europe, respectively, can be found in Tables 12.1, 12.2, and 12.3.

Air pollution continues to be a major problem in the United States despite the 1980 Clean Air Act and its 1990 amendments. Many cities do not comply with its standards even though they have taken small measures to relieve the problem. One of the largest air pollution control agencies in the United States has determined that reducing NO_x emissions is worth up to \$24,500 per tonne and \$14 per tonne for carbon dioxide. Pollution control technologies such as scrubbers may help reduce one pollutant but not all pollutants. The scrubbers trap sulphur (S) but do nothing to control nitrogen (N). Fossil fuel combustion still adds about 5.4 billion tonnes of carbon to the atmosphere each year. That is more than one metric tonne for each human on the Earth. This loading is not shared equally among different countries. For example, 1986 figures quoted by Thurlow (1990) using data from the U.S. Oak Ridge National Laboratory and the U.K. Department of Energy indicated that the United States was responsible for 23% of global CO_2 emissions, EEC countries contributed 13%, and countries comprising the USSR at that time emitted 19% of the global total. China contributed 10%, although this is expected to rise considerably with that country's expanding economy. The United Kingdom's total of 160 million tonnes comprised 3% of the world total and represented an output of about 4 tonnes per head of population. This compared with the EEC average of 2.2 tonnes per head, the U.S. average of 5 tonnes per head and 0.4 tonnes per head in developing countries. On a global basis, electricity production generates 52% of CO_2 emissions from coal burning, 25% of CO_2 emissions from natural gas burning, and 9% of CO_2 emissions from oil burning. On a national basis, Figure 9.2 indicates how the relative proportion of these sources changed in the United Kingdom over a 38-year period between 1950 and 1988. Over this period, total CO_2 emissions in the United Kingdom varied remarkably little in the face of a more than threefold increase in global CO_2 emission from 1,700 MT (as carbon) CO_2 in 1950 to 5,300 MT in 1986. Contributory factors were a decline in coal burning, a higher thermal efficiency in electricity generation, and an increase in nuclear-powered electricity generation.

9.6.2   Substances of concern: photochemical oxidants

The recognition of ozone (O_3) as a greenhouse gas is predated by the recognition of a broad array of environmental problems associated with tropospheric pollution by the gas stemming largely from automobile exhaust. Due to serious air pollution in the city of Los Angeles in the 1940s, the state of California (USA) legislated for air pollution control. Initial efforts were aimed at particulate emissions from stationary sources such as steel mills and refineries. Although these measures achieved a two-thirds reduction in particulates, symptoms such as eye and throat irritation persisted. In 1952, Haagen-Smit demonstrated the formation of oxidis-

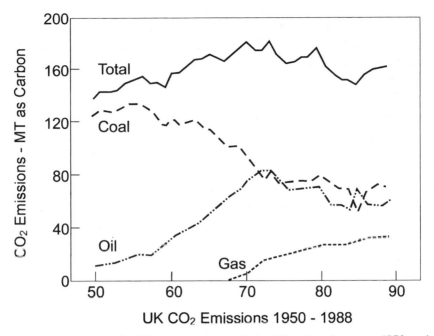

Figure 9.2 Carbon dioxide sources in the United Kingdom between 1950 and 1988 (Thurlow, 1990).

ing products, notably ozone (O_3) in the presence of sunlight, oxides of nitrogen (NO_x), and hydrocarbons (HCs). Both NO_x and atmospheric HCs are primarily produced by processes associated with transportation, including the refining industry, although at least 40% of NO_x emission is from stationary combustion. It should also be noted that NO_x are also released into the environment as a result of agricultural practices (see Section 9.5.3).

Increasing recognition of the role of automobile emissions in this process culminated in the regulation, in the United States, of HCs and carbon monoxide in car exhausts since 1968, and NO_x since 1973. By 1975, NO_x and HC emissions from automobiles were similar in the United States, Canada, Japan, and most European countries. In heavy traffic, the formation of oxidising smog involves a complex series of photochemical interactions between anthropogenically emitted pollutants (NO, HCs) and secondary products (O_3, NO_2, aldehydes, peroxyacetyl nitrate). Concentrations of these compounds exhibit a pronounced diurnal pattern depending on traffic density, sunlight, and atmospheric conditions (Figure 9.3). In a photochemical smog, O_3 is produced through the following coupled reactions:

$$NO_2 + h\nu \rightarrow NO + O \tag{9.1}$$

$$O + O_2 \rightarrow O_3 \tag{9.2}$$

$$NO + O_3 \rightleftarrows NO_2 + O_2 \tag{9.3}$$

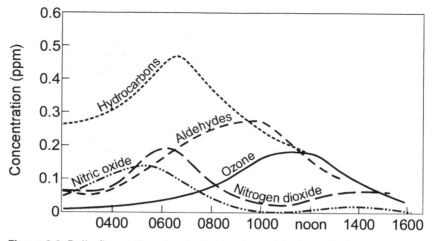

Figure 9.3 Daily fluctuations in photochemical oxidants and related compounds associated with automobile exhaust.

Hydrocarbons and uncharged portions of hydrocarbon molecules such as the peroxy radical, $RO \cdot_2$, participate in a number of different chain reactions involving the oxidation of NO to NO_2.

$$RO \cdot_2 + NO = RO \cdot + NO_2 \tag{9.4}$$

NO_2 is subsequently photolysed to yield an oxygen atom (9.1). This, in turn, combines with O_2 to produce O_3 (9.2). These combined reactions have the net effect of pushing Equation 9.3 to the left with the net production of O_3. Peroxy radicals are formed when a variety of hydrocarbons are attacked by OH radicals, or through OH attack on intermediate breakdown products such as aldehydes and ketones. Other participants in this process are alkoxy radicals (RO) formed as in Equation 9.4. Equation 9.5 illustrates how the ethoxy radical reacts with oxygen to produce a hydroperoxyl radical and an aldehyde (acetaldehyde), another minor product of oxidising smogs.

$$CH_3CH_2O + O_2 \rightarrow HO_2 + CH_3CHO \tag{9.5}$$

Peroxyacetyl radicals are formed as a result of OH attack on acetaldehyde (Equations 9.6 and 9.7) or through the photolysis of molecules containing the CH_3CO group (e.g., biacetyl, Equations 9.7 and 9.8).

$$OH + CH_3CHO \rightarrow H_2O + CH_3CO \tag{9.6}$$

$$CH_3CO + O_2 \rightarrow CH_3COO_2 \tag{9.7}$$

$$CH_3COCOCH_3 \rightarrow CH_3CO + CH_3CO \tag{9.8}$$

Photochemical reaction of peroxyacetyl radicals with NO_2 result in the formation of peroxyacetylnitrate:

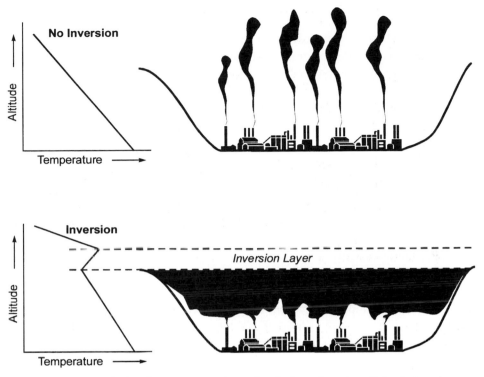

Figure 9.4 Role of temperature inversion in trapping air pollutants in urban areas.

$$CH_3COO_2 + NO_2 \rightarrow CH_3COOO_2NO_2 \tag{9.9}$$

Although peroxyacetylnitrate is a thermally unstable compound that typically has a lifetime of about 40 minutes at room temperature [reaction (9.9) moves to the left], it may reach concentrations between 5 and 50 ppb during a photochemical smog event. At the higher end of this range, PAN is a potent eye irritant. Other homologous compounds such as peroxybenzoyl nitrate may also be formed depending on the availability of different alkyl or aromatic groups.

The net result of the chain reactions occurring in a photochemical smog is the removal of NO from the atmosphere favouring the formation of ozone, O_3, via Equation 9.3. Ozone can, therefore, comprise up to 90% of photochemical smog oxidants with oxides of nitrogen (mainly NO_2) contributing approximately 10% and PAN <1%.

Under normal meteorological conditions, the components of photochemical smogs dissipate towards the end of the day through dilution by fresh air masses or consumption by photochemical reactions including the formation of peroxides, nitrated organics, and other terminal products. In some cities such as Los Angeles, however, high atmospheric pressure and topographical features may combine to produce temperature inversions wherein a layer of warm air overlies a layer of cooler air, trapping pollutants close to the ground (Figure 9.4). Since their characterisation in the United States, photochemical smogs have been described in Australia, southern Ontario, Mexico, South Africa, and several tropical Asian cities. Although not yet regarded as a serious problem in European cities, oxidising smogs have been demonstrated with increasing frequency in hot day summers in southern European cities such as Athens.

9.6.3   The environmental toxicology of photochemical oxidants

OZONE TOXICITY

In understanding ozone toxicity, it is important to make the distinction between ozone in the upper atmosphere, the stratosphere, and ozone in the lower atmosphere, the troposphere. The production of ozone in photochemical smogs represents only a tiny fraction of that produced in the stratosphere through the photochemical oxidation of molecular and atomic oxygen. However, O_3 levels exceeding $2,000 \mu g\, O_3\, m^{-3}$ (approximately 500 ppb) have been recorded under smoggy conditions, about 100 times background levels, and such concentrations cause both human health problems and serious crop damage. Note, too, that the role of NO_x relative to O_3 differs in the troposphere compared with the stratosphere (see Section 9.7.3 and Figure 9.7).

OZONE DAMAGE TO PLANTS

Although plants have a wide range of tolerance to ozone, several important agricultural crops are highly sensitive to it. Acute injury to tobacco is seen after a 2- to 3-hr exposure to 50–60 ppb ozone and a 12% reduction in yield of a variety of crops including potatoes, soybean, corn, snap bean, and tomatoes has been attributed to ozone in the eastern United States. Heck et al. (1986) estimated that the current U.S. maximum allowable O_3 standard (120 ppb for 1 hr) crop losses would represent 2–4% total potential yield [up to $5 billion (1980)].

In most plant species, O_3 causes a characteristic leaf stippling, which coalesces to form pigmented blotches. Plants that rely heavily on leaf appearance for marketability are, therefore, particularly vulnerable. Other symptoms include a reduction in net photosynthesis, reduction in unsaturated fatty acids, increase in respiration and membrane permeability, and reduction in shoot and root growth and in size and number of fruits. At the cellular level, both carbohydrate metabolism and lipid synthesis are inhibited by ozone.

PAN also produces several plant symptoms similar to O_3, and it is likely that the oxidation of sulphydryl (—SH) groups by such strong oxidants results in the inhibition of several enzymes.

HUMAN HEALTH EFFECTS

Because of the increasing incidence of photochemical smogs, many North American cities chronically exceed the North American Air Quality Standard of 120 ppb for 1 hr. Many metropolitan areas are exposed to such ozone levels for 6 hr or more during peak smog periods, and this exposure has prompted considerable concern over both acute and chronic ozone toxicity to humans and other animals.

Experiments with animals have demonstrated epithelial damage along the whole length of the respiratory tract following exposure to >200 ppb for 2 hr.

9.6.4   Substances of concern: Acid precipitation

The phenomenon known as acid precipitation is largely caused by the release into the atmosphere of sulphur dioxide and oxides of nitrogen, NO_x (e.g., nitric oxide, NO, or nitrogen dioxide, NO_2), which become oxidised to sulphuric and nitric acids. In the eastern United States, sulphuric acid comprises about 65% of this acidity; nitric acid, approximately 30%. Other acids contribute about 5% of the acidity. At the end of the 1980s, total global emissions were estimated to be about 180 million metric tonnes of SO_2 (mainly from stationary fuel burning sources) and 75 million metric tonnes of NO_x. However, the gap between these outputs is expected to close as a result of SO_2 emission control and increasing NO_x emission from worldwide automobile use.

The detrimental effects of acid precipitation on aquatic systems were noticed in southern Norway as early as the 1920s. Indeed, historical records of the acidity of precipitation in urban areas go back to the nineteenth century, although it was several decades after the 1920s before the cause of the problem was fully realised. Early signs were a decline in the catch of Atlantic salmon in several of the southernmost Norwegian rivers, several fish kills of adult salmon and brown trout, and increased mortality of eggs and fish larvae in local hatcheries. In the Adirondack State Park in New York State, fishermen in the 1950s began to complain of rapidly diminishing numbers of trout in the park's more than 200 mountain lakes and streams. Surveys conducted in the 1970s found that more than half of the Adirondack Lakes over an elevation of 650 m had pH < 5; of these lakes, 90% contained no fish.

The first North American studies linking lake acidification directly to acid precipitation were conducted in a region in Ontario around the site of several metal smelters near the town of Sudbury. Of the 150 lakes surveyed, 33 were classed as critically acid (pH < 4.5), and 37 classed as endangered (pH 4.5–5.5). Some 200–400 lakes within a 50-mile radius of the smelters contained few or no fish. However, the interpretation of the damage was complicated by the presence of metal pollution, also originating from the smelters.

Whereas the Sudbury studies implicated a specific source for acid fallout, numerous studies throughout the 1970s established a regional pattern of acid precipitation covering almost all of eastern North America, spreading to the south and west with a focus on the U.S. states of Pennsylvania, Ohio, Maryland, and West Virginia and the southern portion of the Canadian province of Ontario (Figure 9.5). In northern Europe, the greatest intensity of acid precipitation was shown to be over southern Scandinavia. By 1990 in Norway, 33,000 km² were affected by acid precipitation with 13,000 km² devoid of fish, and over 2,500 lakes in Sweden were

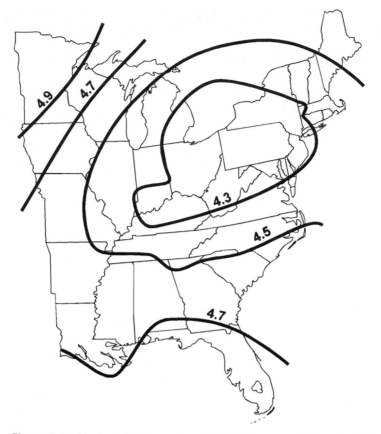

Figure 9.5 pH of rainfall in eastern United States and Canada, 1986.

documented to be affected (c. 4% of the total lake surface area). In Sweden, 70% of the sulphur in the atmosphere in the southern part of the country results from human activity with nearly 80% of this emanating from sources outside Sweden (e.g., Germany and the United Kingdom).

Table 9.6 shows a chronology of several "milestones" in the study of the effects of acid deposition on aquatic and terrestrial ecosystems. The table is modified from Gorham's (1998) paper, which should be consulted for specific references associated with these events and as a comprehensive review of acidification studies as a whole (see Section 9.11).

9.6.5   The environmental toxicology of acid precipitation

The adverse physiological effects of pH on fish have been summarised in Chapter 5. However, it is impossible to consider the effects of acidification of freshwater on fish species without taking into account the effect of low pH on trace metal

Table 9.6. A chronology of some notable advances in the study of acid deposition from the atmosphere to aquatic and terrestrial ecosystems

1852	Linkage between air pollution and acidity of urban precipitation discovered (Manchester, England).
1872	*Air and Rain: The Beginnings of a Chemical Climatology*, R. A. Smith (1872) published.
1911–13	Urban acid deposition and its effects on plants, soil bacteria, and soils studied.
1919	Soil acidification in Austria ascribed to sulphur dioxide in air pollution.
1955	Sweden observed that acid precipitation caused by sulphuric acid occurs distant from urban-industrial sources of air pollution.
1955–58	Lake acidification reported on hard, slowly weathering rocks in the English Lake District.
1958	Further acidification of already acid bog pools in northern England discovered.
1959	Serious damage to the freshwater fishery of southern Norway associated with acid deposition.
1960–63	Lake acidification and a decline in the diversity of aquatic macrophytes near smelters in Ontario, along with damage to plant communities of forests by sulphur dioxide fumigation and soil acidification reported.
1967–68	Studies suggested that acid deposition caused by long-distance transport of air pollution from central Europe and the United Kingdom was injuring aquatic and terrestrial ecosystems in Sweden.
1972	Sweden's case study, "Air Pollution Across National Boundaries: The Impact on the Environment of Sulphur in Air and Precipitation", presented to the United Nations Conference on the Human Environment in Stockholm.
1972	Nitric acid first implicated as a significant component of acid deposition.
1972	Decline in Ontario, Canada, fish populations associated with reproductive failure due to acid deposition.
1976	Effects of an acid pulse during snowmelt on the balance of blood salts implicated in a Norwegian fish kill.
1978	Sulphate adsorption suggested to be a factor mitigating acid deposition on strongly leached and weathered soils.
1979	Increase in free aluminum shown to be toxic to fish in Adirondack lakes subject to acid deposition.
1979	Empirical titration model of surface water acidification proposed, based on >700 Norwegian lakes.
1980	Importance of slope and soil depth in lake acidification by acid deposition recognized.
1980	Linkage between forest decline in Germany and acid deposition proposed.
1980	Mitigation of experimental acid deposition by internal generation of alkalinity due to microbial sulphate reduction in an Ontario lake observed.
1980	Mitigation of acid deposition to bog waters by microbial reduction processes observed.
1980	The utility of diatoms in sediments as palaeoecological indicators of lake acidification reported.

Table 9.6. (*Continued*)

1983	Canadian negotiators with American colleagues on the problem of transboundary acid deposition suggested a critical load of wet sulphate ($20\,kg\,ha^{-1}\,yr^{-1}$) to protect most sensitive surface waters.
1984	Recovery from acidification followed experimentally in an acidified lake in a roofed-over forest catchment.
1985	Loss of successive prey organisms observed to cause starvation of lake trout in an experimentally acidified lake.
1986	Nitrate saturation of forest soils reported.
1986	Acidification of Swedish forest soil profiles over half a century, independent of age and owing to acid deposition, described.
1986	Export of complex colored organic acids from peat bogs shown to acidify Nova Scotian lakes and to add to the effect of acid deposition (with some mitigation by microbial reduction in the bog itself).
1994	Decline of base cations in atmospheric deposition observed to retard the expected decline in acidity owing to controls on emissions of sulphur dioxide.
1995	Atmospheric ammonia and its reaction product with sulphuric acid emphasised as very important agents of ecosystem acidification.
1996	Interactions of acid deposition, ozone depletion, and climatic warming reported to enhance penetration of surface waters by UV-B radiation.

Modified from Gorham (1998).

mobility and toxicity. For example, aluminium (Al) concentrations in acidic lakes in southern Norway, southwest Sweden, and the northeastern United States are 5–10 times higher than in circumneutral lakes in the same area. The comprehensive data sets on water chemistry gathered from affected areas have enabled extensive multivariate analyses to be made. In these analyses, a suite of water chemistry parameters have been entered into multiple regression equations describing fish abundance as the dependent variable. Such an approach indicates that aluminium is an important independent variable explaining fish scarcity in Adirondack lakes, although aluminium apparently makes a much smaller contribution to the overall effect of acid in Scandinavian lakes. Fish species differ greatly in their sensitivity to acid conditions (Figure 9.6), although there is general agreement on the toxicological effect of excess acidity on freshwater fish (see Sections 5.3.2 and 6.2.2). In addition to fish, amphibians have also been shown to be adversely affected by acid conditions with resulting poor reproductive performance and survival.

9.7   Agriculture

9.7.1   The issue

As the human population grows, it must also be maintained. Part of this maintenance consists of creating a resource base large enough for its support. This

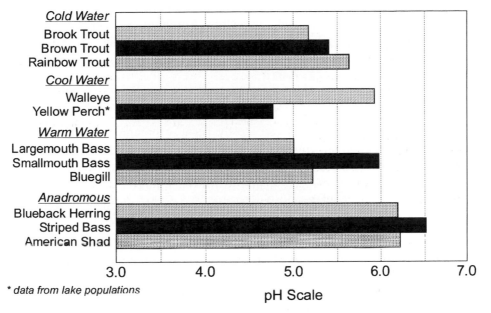

Figure 9.6 Sensitivity of different fish species to pH.

resource base is made possible through such human enterprises as agriculture, industry, fishing, and international commerce (Vitousek et al., 1997). These enterprises are not without ecological consequence, the most notable of which may be the transformation of land surfaces. On a global scale, 39–50% of the world's land has been transformed (Kates et al., 1990; Vitousek et al., 1986). Moreover, often what remains as unmodified has been fragmented to such an extent that ecosystem functioning is similarly disrupted. Because of this, the Earth has seen dramatic decreases in genetic and species diversity leading to current extinction rates of 100 to 1,000 times that before humans dominated Earth's ecosystems (Chapin et al., 1997; Pimm et al., 1995; Lawton and May, 1995).

Land transformation and habitat conversion to yield goods such as food are quantitatively the most significant human alteration of land. Total area of cultivated land worldwide increased 466% from 1700 to 1980 (Matson et al., 1997). Currently, 10–15% of Earth's land is occupied by row crop agriculture, and 6–8% has been converted to pasture (Olson et al., 1983). Such increases in cultivation have led to modifications in structure and functioning of ecosystems, including changes in interactions with the atmosphere, with surrounding land, and with aquatic systems.

Land transformation for agricultural use affects the atmosphere mostly through climatic changes occurring at local and even regional scales. When land is transformed from forest to agriculture or pasture, the surface roughness of the land is decreased, resulting in increasing temperatures and decreasing precipitation. If this process involves burning, the reactive chemistry of the troposphere is altered, producing elevated carbon monoxide concentrations and episodes of air pollution that may travel several thousand kilometres. Agricultural practices contribute nearly 20% to current anthropogenic carbon dioxide emissions, and more substantially to the increasing concentrations of the greenhouse gases, methane, and nitrous oxide (Vitousek et al., 1997).

Agricultural cultivation also negatively affects the surrounding land by changing the soil itself in areas that are heavily farmed. The loss of soil organic matter due to permanent agriculture is very costly because organic substrates provide nutrients, maintain soil structure and water-holding capacity, and reduce erosion (Matson et al., 1997). In temperate regions, 50% of soil carbon is lost during the first 25 years of farming. Organic matter is lost much more rapidly from tropical regions, reaching the same levels of loss within 5 years. Biotic changes in the soils are also substantial because the natural, diverse communities of microbes and invertebrates are significantly different in cultivated areas compared with natural ones. This difference can lead to changes in nutrient cycling and availability, soil structure, biological regulation of decomposition, and other properties regulated by the activities of these organisms.

The rate of expansion of land used for agriculture has slowed dramatically over the last three decades. Remarkably, the food supply has still out-paced the rapid growth of the Earth's population. This growth is due to the intense modifications in management on land already used for agriculture such as the use of new high yielding crops, irrigation, chemical fertilisers, pesticides, and mechanisation. Agriculture has perhaps been perceived by the general public as relatively benign from an environmental perspective. This perception is less than realistic.

Arguably the most dramatic ecological modifications due to agriculture are those affecting the hydrologic cycle and thus Earth's natural water supplies. Globally, humanity uses more than half of the freshwater that is readily accessible. Agricultural use is responsible for approximately 70% of this freshwater use through the impoundment or impedance of natural waterways for irrigation. In fact, only about 2% of the United States' rivers run unimpeded, and 6% of river runoff is lost to evaporation as a consequence of this human manipulation (Postel et al., 1996). Development of irrigated land increased at a rate of 2.2% per year between 1961 and 1973. Globally, 40% of crop production now comes from the 16% of agricultural land that is irrigated (Chapin et al., 1997).

These agricultural alterations of natural waterways have their own effects on surrounding ecosystems. For example, the hydrological cycle may be modified, resulting in further climatic changes. Irrigation increases atmospheric humidity in semiarid areas often increasing precipitation and thunderstorm frequency (Vitousek et al., 1997). Intense irrigation also has effects on semiarid cultivated land, continuously degrading soils through salinisation and waterlogging. In fact, developing countries have experienced reduced yields over approximately 15 million ha due to these problems (Matson et al., 1997). Furthermore, irrigation return flows typically carry more salt and minerals than surface or groundwater due to evaporation and concentration, leading to pollution of natural waters, especially when there is little dilution as in dry climates.

Although increased irrigation itself has many consequences, none is as significant as the effect on the water following farming activity. Agricultural runoff can have significant effects on streams, rivers, lakes, and estuaries through direct contamination of surface and groundwaters by agrochemicals such as fertilisers and pesticides. Contamination by agrochemicals is not the only consequence of agricultural runoff, but it is probably the issue that receives the most attention from the public. Trends of pesticide and fertiliser use in the United States to date provide insight as to how the environment and ecosystems are affected when agricultural runoff containing agrochemicals enters natural waterways.

9.7.2   Substances of concern: Fertilisers

Agricultural practices cause a net removal of nutrients from the soil, reducing productivity. This occurs through the harvesting of crops, depletion of organic matter reducing nutrient release, and loss of nutrients in agricultural runoff. Furthermore, the world's population is growing by approximately 90 million people per year. To meet future food needs, agricultural yields must be increased without expanding land use. Therefore, fertiliser application is necessary.

Fertilisers are generally nitrogen-based, synthetically produced chemicals used to replace nutrients lost from the soil in agricultural fields (see Section 6.10). Nitrogen is a unique element in that its cycle includes a huge reserve of atmospheric nitrogen (N_2), but it must be fixed (microbial process combining nitrogen with oxygen, hydrogen, or carbon usually taking place in the soils) before it can be utilised by most living organisms. Synthetic nitrogen for fertilisers is produced through a fossil-fuel-consuming industrial process that converts abundant atmospheric nitrogen to these available forms. Phosphorus is also used as fertiliser, although not nearly to the same degree. It is primarily obtained through mining of phosphorus-rich deposits in rocks.

In the 1990s, 80 million metric tonnes of nitrogen were produced globally in industrial nitrogen fixation each year. This is an increase from less than 10 million metric tonnes per year in the 1950s and is expected to increase to greater than 135 million metric tonnes per year by the year 2030 (Matson et al., 1997; Vitousek et al., 1997). Additional nitrogen is fixed biologically by the cultivation of legume crops and through the inadvertent fixation of nitrogen during the combustion of fossil fuels; the resulting increase in human-generated fixed nitrogen is approximately 60 million metric tonnes per year (Vitousek et al., 1997). The result is an overall human contribution of fixed nitrogen equal to the natural biological contribution of fixed nitrogen.

Globally, fertiliser use has increased dramatically since the 1960s, although this has happened primarily in developing countries with estimates of a 4% annual increase in nitrogenous fertiliser consumption (Matson et al., 1997; Matthews, 1994). North American and European use has levelled off in recent years, but still represents 17 and 14% of global fertiliser use, respectively. This contribution corresponds to actual U.S. consumption of fertiliser during 1985–86 of 435,000 metric tonnes. In contrast, developing countries now contribute 58% of total global fertiliser use with East Asia using 37% alone, an increase of almost 10% per year. The most widely used nitrogenous fertiliser is urea, approaching 40% of the world's total, 75% of that used in Asia. Ammonium nitrate, used in central Europe, West Asia, and Africa makes up 25% of global consumption (Matthews, 1994).

Fertiliser demand into the twenty-first century is projected to reach 208 million nutrient tonnes of use per year, 86 million nutrient tons used by developed nations and 122 million nutrient tonnes used by the still developing nations by the year 2020. Even these levels are not predicted to meet the goals of future food security as necessary levels are predicted to be 251 million nutrient tonnes of fertiliser per year. Bumb and Baanante (1996) projected a shortfall of 51 million nutrient tonnes if production capacity was not increased by the year 2000.

9.7.3   The environmental toxicology of fertilisers

Fertilisers provide many benefits for improving agricultural production by replacing lost nutrients in the soil. However, these improvements are not without consequence. The primary harmful consequences of fertiliser use include effects on atmospheric constituents, nitrate contamination of surface and groundwaters, and eutrophication of natural waters.

Nitrogenous fertilisers have impacts on atmospheric components such as nitrogen gases that are emitted to the atmosphere. Nitrous oxide is a greenhouse gas (Section 9.5.2), chemically inert in the troposphere but unable to alter Earth's radiation balance (Webb et al., 1997). In the stratosphere, nitrous oxide (N_2O) is reduced to NO, which has the ability to destroy ozone. These gases, known as NO_x gases, then interact chemically in the troposphere to produce ground-level ozone, known to be a health hazard to humans and potentially harmful to vegetation (see Section 9.6.1). Agriculture is a major contributor of N_2O emissions because of the use of chemical fertilisers. In fact, 1–5% of annual N_2O emissions are from terrestrial sources. It has been estimated that up to 3% of the current annual increase of atmospheric nitrous oxide is contributed by fertiliser use alone (Matthews, 1994). Ammonia emission coefficients are also associated with nitrogenous fertilisers as the ammonia volatilises. In 1984, a total of 10–15% of annual ammonia emissions was concentrated in subtropical Asia due to the use of urea as the dominant fertiliser (Matthews, 1994). As these data suggest, the type of fertiliser used has an impact on the level of NO_x emissions. Webb et al. (1997) found that NO emissions after urea and ammonium nitrate additions were significant and up to 20 times higher than from plots treated with sulphur-coated urea. Also, the use of a slow-release form of urea reduced N_2O emissions considerably. Therefore, the investigators concluded that the type of fertiliser had an impact on both NO and N_2O emissions, with the magnitude of emission dependent on soil moisture

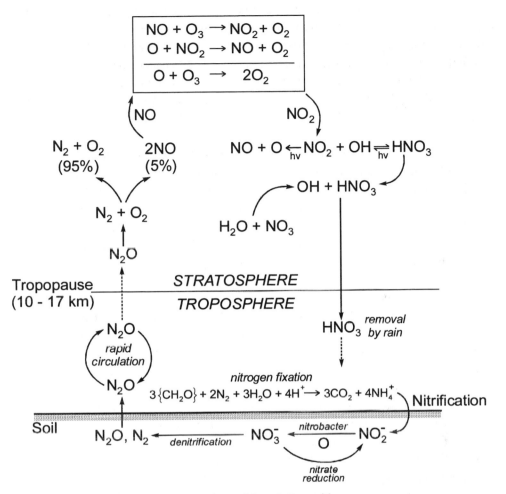

Figure 9.7 The nitrogen cycle and its relation with ozone.

conditions at the time of fertiliser addition. A generalised scheme of the main features of the nitrogen cycle and its effects on stratospheric ozone is shown in Figure 9.7.

Another consequence of nitrogenous fertilisation is increased concentrations of nitrates, nutrients commonly derived from fertilisers, flowing into surface waters and leaching into groundwaters from the soil surface. Nitrate is one of the major ions found naturally in waters with mean surface water concentrations reaching about $1–2\,mg\,L^{-1}$. However, concentrations in individual wells or groundwaters affected by contamination are often found to reach potentially hazardous levels. Nitrates, if ingested, cause two types of adverse health effects: induction of met-hemoglobinemia, or blue-baby syndrome, and the potential formation of carcinogenic nitrosamines. Methemoglobinemia occurs as nitrate is reduced to nitrite in the saliva and gastrointestinal tract and occurs much more often in infants than adults due to the more alkaline conditions in their digestive systems. The nitrite then oxidises haemoglobin to methemoglobin, which cannot act as an oxygen carrier in the blood. Anoxia and death can occur. The established carcinogenicity of many aromatic amines has prompted concern over the health risks associated with nitrated aromatic compounds. Certain dietary components have been shown to react in the gut with high nitrite concentrations to form carcinogenic nitrosamines. Due to these perceived human health hazards, the U.S. EPA has proposed a maximum nitrate concentration limit of $10\,mg\,L^{-1}$ in drinking waters and for the protection of aquatic life. Similar guidelines are in place in Canada and the EEC. Mueller et al. (1996) conducted the most extensive study to date of nutrients in the nation's water resources consisting of 12,000 groundwater and 22,000 surface water samples. They found that nitrate concentrations in rural wells were much higher (12% exceeded the U.S. EPA's drinking water limit) than those in public water supply wells used by cities and towns, where only 1% exceeded the drinking water standard. Additionally, they found that elevated concentrations of nitrates in groundwaters were found primarily in agricultural areas and were highest in shallow groundwater. Regions of the United States most vulnerable to nitrate contamination of groundwater were in the highly agricultural areas of the Northeast, Midwest, and the West Coast.

The contamination of surface waters by nitrates has also followed a pattern related to land use. For example, nitrate concentrations have been shown to be elevated downstream from highly agricultural areas. Overall, nitrates in the major rivers of the United States have increased three- to tenfold since the early 1900s (Chapin et al., 1997). This increase is due both to agricultural use and to improvements made in the wastewater industry, reducing levels of ammonia through tertiary treatment, but transforming it to nitrates (Mueller et al., 1996). However, surface water concentrations rarely reach levels found in groundwaters.

Often described as the primary consequence of pollutants associated with agricultural runoff is nutrient enrichment or eutrophication. Surface runoff containing nitrogen and phosphorus provide substantially higher concentrations of what are normally limiting factors for plant growth. Nutrients present in such high concentrations are considered contaminants. Generally, when water is contaminated with nutrients, the result is increased productivity, decreased biotic diversity, and reduced stability of the ecosystem due to the loss of organisms' ability to adapt to change (see Section 6.10).

Agricultural sources of phosphorus, and in some instances nitrogen, provide significant contributions to the eutrophication process in many freshwater aquatic systems (Matson et al., 1997; Foy and Withers, 1995). This is often manifested by increased algal production. Delong and Brusven (1992) showed that chlorophyll-*a* concentrations in Idaho streams receiving agricultural runoff were at least twice those of undisturbed streams. Algae can interfere with the abstraction of water for drinking water purposes by producing noxious odours, tastes, and toxins and by physically clogging water filters. Algae also have negative aesthetic effects as increased productivity produces scums, reducing water transparency. Generally, increased algal productivity is not sustained and decreases rather rapidly in a few weeks or months. When the algae die and biodegrade, oxygen is consumed, and O_2 in the water may decrease to levels hazardous for fish and aquatic life.

Nutrient additions not only increase phytoplankton growth, but they also change the composition of the community and spatial distribution. By altering the communities in this way, the distribution and storage of nutrients is also altered, resulting in further changes in ecosystem processes. There are also impacts on macrovegetation of different types due to increased nutrients. Commonly, competition between plant species depends on the balance of nutrients with other environmental factors. When fertilisers are added, large shifts in species may occur due to changes in competitive advantages.

Benthic communities will also be affected with increased eutrophication through the loss of macrovegetation due to shading out by increased surface productivity. When macrovegetation decreases, macroinvertebrates also decrease, thereby reducing species diversity and changing spatial composition. Generally, benthic species found in highly eutrophic conditions are chironomids and oligochaetes. And, finally, eutrophication changes fish species composition from salmonid and coregonid species that require low temperatures and high oxygen to warm-water species that are tolerant of eutrophic conditions. Certain fish of this type, such as carp, have feeding habitats that modify the littoral substrate to the point that many submerged macrophytes can be eliminated. This will, again, affect benthic invertebrate compositions and could also lead to increased turbidity, resulting in negative aesthetic effects.

Examples of large freshwater systems that have been adversely affected by eutrophication are the Great Lakes system, Chesapeake Bay (see Case Study 4.5), and the Florida Everglades. An account of the Everglades system is given as a case study at the end of this chapter to illustrate the complexity of eutrophication issues and to describe ongoing restoration measures.

HARMFUL ALGAL BLOOMS AS A CONSEQUENCE OF NUTRIENT RUNOFF
In recent decades, there is evidence that increased nutrient runoff has contributed to the development of harmful algal blooms (HAB) in estuaries and other coastal regions.

Among the scientific community, there is a consensus that the scale and complexity of harmful algal blooms in coastal waters is expanding. Harmful algal blooms are by no means a new phenomenon in coastal systems. In fact, there is evidence that blooms of toxic algal species occurred over two centuries ago (Horner et al., 1997). On the west coast of the United States, the earliest documented case of saxitonin poisoning caused by the toxic dinoflagellate *Alexandrium* was in 1793 when five members of Captain George Vancouver's crew became ill and one died. Historically, blooms were limited to specific regions where conditions for algal growth were ideal during a finite period of the growing season. It has been proposed, however, that the breadth of these bloom events has expanded as a result of increased nutrient input to coastal regions and also through the transport of harmful algal species to previously unaffected regions in the ballast water of ships.[1]

Initially, these blooms were referred to as red tides for those algal species that contained reddish pigmentation. This term has traditionally been applied to any reddish discoloration of the water; however, it is a misnomer because blooms are not tidal events. Contrary to public belief, red tides are not usually harmful. In fact only about 8% of the estimated 3,800 phytoplankton species can reportedly produce red tides, and of these only about a quarter are actually toxic or harmful (Smayda, 1997a). Many of these species never reach densities high enough to be visible as water discolourations; however, they are capable of disrupting marine habitats by producing potent biotoxins, altering trophic dynamics, and overshadowing other aquatic organisms. These effects on coastal ecosystems inevitably infringe upon coastal business such as shellfish farming and fishing due to the noxious effects to seafood of these harmful species. Due to these extensive impacts, the HABs are so named because they are arrays of both microscopic and macroscopic algal species that are capable of a variety of negative effects on humans and aquatic life.

Public health concerns over bloom-related shellfish diseases have become a global issue. The biotoxins produced by many harmful algal species are capable of entering the aquatic food chain, ultimately becoming concentrated in many shellfish and finfish, which may be consumed by humans. The most significant health problems caused by harmful algae include amnesic shellfish poisoning (ASP), ciguatera fish poisoning (CFP), diarrhetic shellfish poisoning (DSP), neurotoxic shellfish poisoning (NSP), and paralytic shellfish poisoning (PSP). Each of these syndromes results from ingestion of seafood contaminated with biotoxins released by different organisms. Some of these illnesses invoke only temporal symptoms, but others are fatal if high toxic concentrations are consumed.

Associated with these health risks are the economic losses incurred as a consequence of restricted fishery harvests. Losses to shellfish culturing alone during a single bloom episode have exceeded $60 million, and over an 18-year period in Japan, fishery farming losses exceeded $100 million (Smayda, 1997b). Economic losses are often the result of the necessity to close down and regulate fisheries following poisoning outbreaks. In addition, many finfish populations are capable

---

[1] Concern over the importation of exotic organisms, some of which have been shown to adversely affect native species, has gained impetus over the last decade. Accidental introduction of the zebra mussel into North America and the jellyfish into the Black Sea, both probably as a result of ballast water discharge, have led to calls by several governments and the International Maritime Organisation (IMO) for more stringent control of ballast water exchange and exploration of methods for biocidal treatment of ballast water. The latter option essentially treats the issue as a form of biological pollution and raises the possibility of clean water legislation being invoked to control such introductions.

of migrating away from harmful algae populations, thus causing economic loss to local fishermen.

In addition to human health effects, HABs may affect ecosystems in a variety of different ways. Effects may be chemical or nonchemical. Harmful algae can cause starvation in grazing populations by being too small to provide adequate nutritional value for the consumers, which, in turn, causes impaired growth and fecundity in the grazing populations. Mechanical damage to the larvae of shellfish and other organisms is caused by collisions with harmful algae during high-density conditions. In this case, death is usually caused by respiratory failure, haemorrhaging, or bacterial infection resulting from particle irritation during these collisions. Additionally, many harmful algal species are capable of secreting cellular polymers that clog and tear gills and cause osmoregulatory failure in fish (Smayda, 1997b).

Among the chemical modes of injury induced by harmful algae, mortality in aquatic species is often caused by anoxic or hypoxic conditions following the decline of large harmful algae blooms (Smayda, 1997b). Many harmful algal species also produce phycotoxins, which can cause mortality either by ingestion (endotoxins), by exposure (exotoxins), or through trophic relationships. Some harmful algal species (particularly *Nociluca*) produce the ammonium ion, which can cause mortality in fish. Allelopathic substances, or secondary metabolites, are also given off by many harmful species causing physiological impairment.

Another form of chemical impact to aquatic species is illustrated by the ambush behaviour of *Pfiesteria piscicida*. This recently discovered behaviour involves chemical intoxication to stun prey (Smayda, 1997b). The dinoflagellate *P. piscicida* has been responsible for recent fish kills and the appearance of lesions on fish from estuaries on the mid-Atlantic coast of the United States. Even though definitive causes of *Pfiesteria* outbreaks remain to be found, increased input of nutrients such as phosphates from agricultural activity have been implicated as causative agents.

A summary of some algal species known to have formed harmful blooms is shown in Table 9.7, together with their mode of action.

9.7.4   Substances of concern: Pesticides

Each year, the chemical industry produces more than 100 million tons of organic chemicals representing some 70,000 different compounds, about 1,000 of which are new (Vitousek et al., 1997). Only a fraction of these are tested for toxicity to humans, fish, and wildlife or other environmental impacts. Synthetic organic chemicals such as pesticides have brought humanity many benefits, such as increased agricultural productivity and decreased incidence of insect-vectored disease. However, several are toxic to humans and other species, and some can persist in the environment for decades. Major pesticide groups have been summarised in Chapter 7.

Table 9.7. Summary of some HABs and their mode of action

Organism	Mode of action
	Dinoflagellates
Alexandrium fundyense, A. catebekkam, A. tamarense	Responsible for earliest records of paralytic shellfish poisoning; release saxitonin (Na⁺ channel blocker); responsible for fish and marine mammal kills and some human deaths; northern United States and Canada.
Gymnodinium breve	Produces brevetoxins (Na⁺ channel activators) causing mortality in fish and marine mammals through respiratory arrest; mainly known from Gulf of Mexico
Prorocentrum lima, Gambierdiscus toxicus, Ostreopsis siamensis	Produce highly potent ciguatoxin; lipid-soluble Na⁺ channel activators causing ciguatera fish poisoning
Pfiesteria piscicida	Cause of major fish kills in estuaries and coastal waters of mid-Atlantic and southeastern U. S. coast; complex life-cycle includes toxic zoospores and amoeboid forms stimulated by inorganic P and induced by chemical stimulus from fish to produce toxins causing ulcerative fish disease and human health problems
Dinophysis spp.	Cause of diarrhetic shellfish poisoning in Asia through production of okadaic acid
	Diatoms
Pseudo-nitzschia spp.	Responsible for fish and bird kills (e.g., brown pelicans, cormorants) and amnesic shellfish poisoning in humans through production of amino acid, domoic acid, which acts as surrogate for neurotransmitter L-glutamic acid
Chaetoceros convolutus, C. concavicornis	Irritation of fish gills in salmonid aquaculture operations causing suffocation through excess mucus production
	Raphidophyte
Heterosigma akashino	Cause of red tides responsible for heavy mortality of cultured fish in Asia and north western Pacific coasts of United States through mechanical gill damage
	Pelagophyceae
Aureococcchus anophageffrens	"Brown tide" picoplankter responsible for reproductive failure in some bivalve spp., possibly through nutritional inadequacy in certain parts of food chain
Noctiluca spp.	Fish kills caused by ammonium ion production in heavy blooms

Current estimations of pesticide use per year in the United States reach approximately 450 million kilograms with 70% of that used purely for agricultural production. Approximately 250 million kilograms is attributed to herbicides which are only applied once a year, but insecticide and fungicide use contribute 150 million and 30 million kilograms, respectively, and in some cases may be applied up to eight times per year. The annual costs for pesticide use reach $4.1 billion. However, farmers save $3 to $5 in crops for every $1 spent on pesticides, making it economically feasible (U.S. DA, 1997).

PESTICIDE OCCURRENCE

Not long ago, it was thought that pesticides would not migrate to groundwater, but this problem is now widespread. As part of a study conducted by U.S. Geological Survey, concentrations of pesticides in surface and groundwaters throughout the United States during 1991–5 were determined. The study included analysis of 85 pesticides in about 5,000 water samples (both surfacewater and ground) in 20 of the nation's major watersheds. The 85 herbicides, insecticides, fungicides, and selected metabolites targeted for study represented 75% of agricultural pesticides used in the United States (U.S. GS, 1997).

Results of the study show that 75 pesticides were detected at least once in both groundwater and surface water, most frequently in streams. At least one pesticide was found in every stream sampled and in 95% of the samples obtained, whereas only 50% of samples collected from wells contained pesticides. The same herbicides were most commonly found in streams and groundwater: atrazine and metolachlor, used for corn and soybeans; prometon, used primarily for urban applications; and simazine, used in both agricultural and nonagricultural settings. Some insecticides were found in streams, but not in groundwater. The most common were diazinon, chlorpyrifos, and carbaryl, all used extensively in urban and agricultural settings.

In agreement, data collected by the U.S. Environmental Protection Agency indicated that of 45,000 wells sampled around the United States for pesticides, 5,500 had harmful levels of at least one pesticide; an additional 5,500 had traces of 73 different pesticides in amounts considered not harmful. Of the 73 pesticides found, some were carcinogenic, some teratogenic, and others mutagenic. All were attributed to normal agricultural use (U.S. GS, 1997). These findings are especially significant as the report also states that more than 50 million people in these counties rely on groundwater for their drinking water source. Of these people, 19 million obtain water from private wells, which are possibly more vulnerable to contamination by pesticides because they are not regulated. None of these pesticides were found at levels considered harmful to humans.

A review compiled by Nowell et al. (1997) lists more than 400 monitoring studies and 140 review articles from the 1960s to 1992 dealing with pesticide contamination in sediments and aquatic biota of U.S. rivers. Forty-one pesticides or their metabolites were detected in sediments, and 68, in aquatic biota. Most detected were the organochlorine insecticides or their metabolites, reflecting their persistence in the environment. The researchers state that pesticide detection frequencies and concentrations reflect a combination of past agricultural use, persistence in the environment, water solubility, and analytical detection limits. National trends showed residues of DDT, chlordane, dieldrin, endrin, and lindane decreasing during the 1970s then levelling-off during the 1980s; toxaphene residues declined nationally during the 1980s.

9.7.5   The environmental toxicology of pesticides

Historically, public perception of pesticides has been heavily influenced by problems associated with persistent chlorinated hydrocarbon insecticides such as DDT, dieldrin, and mirex. These compounds have been banned from use in many countries for several years, but they are still found in trace amounts throughout the environment as already illustrated (Nowell et al., 1997) and in the herring gull monitoring programme (see Section 4.4.3). More recently, pesticide formulation has resulted in the production of less hazardous chemicals with lower application rates necessary to obtain similar pest knock-down, no bioaccumulation in organisms, less persistence in the environment, and less toxicity to nontarget species (Wall et al., 1994).

Nonetheless, nonspecific effects of pesticides on humans, fish, and wildlife will inevitably lead to varying degrees of toxicity to nontarget species. This means that growth of some plant species will be inhibited in water known to contain herbicides that inhibit photosynthesis, and some animal populations will be eliminated where there is contamination by pesticides. The result is altered ecosystem functioning due to the reduction of species diversity and changes in community composition. Probably the most common effect on community composition occurs through alterations in competitive advantages between species that utilise similar resources. A review compiled by Havens and Hanazato (1993) showed changes in the zooplankton community structure in freshwater due to pesticide contamination. The contaminated environment favoured small cladocerans and rotifers, indicating a dominance of the smallest taxa. Reasons given for their enhanced survival were their rapid reproductive rates, physiological tolerance, development with few transitions through sensitive stages, and the great richness of small species in general. The investigators also state that with stresses of pesticide contamination resulting in small zooplankton dominance, the efficiency of carbon and energy transfer from algae to zooplankton is reduced.

A change in competitive relationships was also illustrated by Yasuno (1991) with an additional, interesting twist. When streams were treated with temephos, an organophosphorus insecticide, indigenous invertebrates were destroyed. Niches then left vacant were taken over by exotic species, an event never observed before. The exotics were species belonging to Chironomidae which have short life-cycles and high reproductive rates, possibly explaining their ability to quickly invade open niches. They disappeared when the original fauna were restored. However, if the previously dominant species had not returned, or if the exotics had already established themselves adequately by the time the orginal fauna returned, this restoration might not have occurred.

Another example of a change in community structure can occur when the predator-prey balance is disrupted, as shown by Yasuno (1991). When the density of animal populations decreases due to contamination of an insecticide or other pesticide, it is often followed by a large increase in algal production or an increase of another invertebrate species. For example, when permethrin was applied during an enclosure experiment, a copepod species, *Acanthodiaptomus pacificus*, increased. This was due to the disappearance of its predator, *Chaoborus flavicans*. *Chaoborus* is also a predator on cladoceran species like *Daphnia*. Although *Daphnia* were also acted upon by insecticides, they were affected to a lesser degree than their predator *Chaoborus*. In further enclosure experiments, the investigator found a quicker recovery of the cladoceran species from the insecticide when *Chaoborus* was reduced.

This type of alteration in competition through reductions in diversity can have negative impacts all the way back to the farm. Many farmers plan low-diversity crops, known as a monoculture. Pesticide application in these cases can actually result in a greater loss from an insect pest. This loss occurs when pesticides decrease most of the insects present, but one tolerant one remains and prospers. A similar result may occur with tolerant microbial pathogens. The result can be devastating because the entire crop might be eradicated by one species of insect or pathogen due to lack of competition (Chapin et al., 1997).

The negative effects of fertilisers and pesticides on surface water, drinking water, the atmosphere, and the structure and function of ecosystems must be weighed against their benefits. Failure to use fertilisers to maintain agricultural land can also have adverse effects. For example, increased soil degradation, deforestation, and depletion of the natural resource base can occur as poor farmers undertake cultivation of increasingly marginal land. Similarly, farmers save billions of dollars each year and increase agriculture productivity due to the use of pesticides. Future progress will be allied to better management of farmland and improved enforcement of the existing laws to pave the way for compatibility between good environmental decisions and good business decisions.

9.8   Oil extraction, transportation, and processing

9.8.1   The issue

In a 1985 U.S. National Research Council report it was estimated that between 1.7 and 8.8 million metric tonnes of oil enter the world's oceans each year. Eighty percent or more of oil contamination of the marine environment results from human activities, and more than half of this is related to its transportation. Oil is often obtained from environments far removed from its main markets, and probably no other single commodity involves such a massive transportation effort.

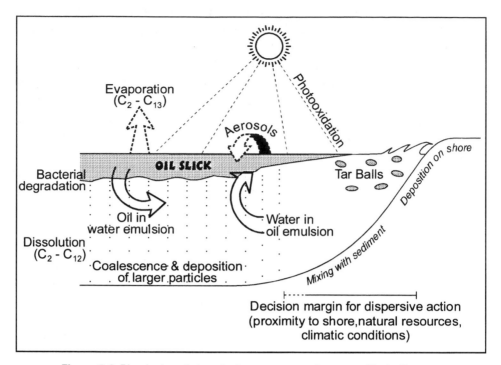

Figure 9.8 Physical and chemical processes acting on spilled oil.

Although tanker accidents comprise only 25–30% of transportation losses, media coverage of massive oil spills elevates such events to prominence in the public consciousness. Dealing with oil spills often represents a balance between protecting natural resources and public amenities. Their unpredictability makes the remedial process a very difficult one. The most important factors influencing oil spill damage are proximity to land, type of shoreline, proximity to valuable natural resources (e.g., fisheries, hatcheries, wildlife refuges), type of oil, wind/sea conditions, and climate. Physicochemical factors in particular play a large part in determining the time course of processes acting on spilled oil (Figure 9.8).

The relative importance of these processes will depend on the initial chemical composition of the spilled oil. Major chemical components of oil are listed in Chapter 7 and will not be repeated here. Heavier hydrocarbon compounds in crude oils are essentially insoluble in water, whereas the smaller molecules, particularly aromatics such as benzene and toluene, are partially soluble. Lighter compounds are also more volatile, and evaporation plays a much more important role in oil dispersion than dissolution. Typically, evaporation is 10 to 100 times more rapid than water solubilisation. Photooxidation, by sunlight, of oil components, especially aromatics, will result in the formation of more polar compounds such as phenols and sulphoxides, although this is a slow process. A more important consequence of photooxidation of aromatic compounds is the formation of surface-active resins that stabilise water-in-oil emulsions resulting in the formation of persistent solid products such as tar-balls.

Under suitable conditions, biodegradation by microorganisms (e.g., bacteria) can play a major role in dispersing oil in the marine environment. Following an oil spill, hydrocarbon-metabolising bacteria may increase from $<10^2 L^{-1}$ to $50 \times 10^6 L^{-1}$. Important factors influencing this process are the concentration of nutrients (nitrates and phosphates), oxygen, and temperature. Even if the concentration of microorganisms at a spill site is initially low, a rapid increase will occur if conditions are favourable, and many bacteria will facultatively use hydrocarbons as an energy source. The most easily biodegraded components are the straight-chain saturated hydrocarbons (n-alkanes) (see Section 7.1).

Biodegradation will only take place where oil interfaces with water. Therefore, stranded oil above the tidal zone will only be degraded extremely slowly. In the sea, the formation of oil droplets by natural or chemical dispersion will increase the biodegradation rate at least 10 to 100 times compared to surface oil due to the increase in area of the oil/water interface. It is very difficult to predict the rate of dispersion of oil in seawater because it is influenced by many factors. Consequently, the overall ecological impact of an oil spill may have more to do with physical conditions and proximity to important natural resources than with the volume of the spill. Case histories of four oil spills are summarised in Table 9.8 and illustrate that no two oil spills are alike.

9.8.2   The environmental toxicology of oil

The toxic effects of oil can be divided into two categories.

- Heavy viscous components may have a blanketing effect on a variety of static filter-feeding organisms causing starvation and oxygen deprivation. The latter effect may be exacerbated by the process of microbial degradation of oil, which may expand and prolong the incidence of anoxic conditions in the benthic environment. Birds coated with oil lose the insulating and water-repellent qualities of their feathers and become water-logged and hypothermic. Preening behaviour may also result in the ingestion of toxic levels of hydrocarbons.

- Systemic toxicity of oil to marine life is generally associated with lighter fractions and is increased where refined products are involved. Of the known components of crude and refined oil, only relatively few are highly toxic to marine organisms. Most of the toxic effects of petroleum can be attributed to the low-molecular-weight alkanes (methane to decane), monocyclic (BTEX – benzene, toluene, ethyl benzene, and xylene) aromatics, two and three ring PAHs, and related heterocyclic compounds (Neff and Anderson, 1981). The large initial mortality of soft-shelled clams and lobsters at the North Cape spill site (Table 9.8)

Table 9.8. Case histories of four oil spills illustrating variability of effects under different circumstances

Exxon Valdez (March 24, 1989)	When tanker grounded on Bligh Reef in poor weather conditions, 38,000 tons (10.8 million gal) of Alaskan crude oil spilled into Prince William Sound, Alaska. A massive physical clean-up involved booms, skimmers, and burning and hot water spraying of beaches. Some experimental use was made of fertiliser to stimulate bacterial biodegradation. Four weeks after the spill, 10% of the oil remained as a slick on the surface of the sound, 35% evaporated, 17% was recovered, 8% was burned, 5% biodegraded, 5% dispersed, and the remainder was on the shoreline. Approximately 30,000 bird deaths among 90 different species including >150 bald eagles were directly attributed to the spill. Of an estimated 10,000 sea otters in Prince William Sound prior to the spill, >1,000 deaths were confirmed. Although several local finfish and shellfish fishing operations were temporarily disrupted, the two principal fisheries, for herring and salmon, showed no signs of degradation, and, in fact, the salmon catch reached a record in 1990, the year following the spill. Subsequent monitoring confirmed robust and rapid recoveries of most wildlife components in the years following the spill. This notwithstanding, the incident resulted in the largest legal settlement in the history of environmental litigation, >1$ billion by Exxon Corporation, and precipitated the enactment in the United States of the Oil Pollution Control Act of 1990.
Braer (January 5, 1993)	When tanker grounded and broke up in >60 knot (>force 8) winds at Garths Ness at the southern tip of the Shetland Islands, 85,000 tons (25 million gal) Norwegian crude oil lost. About 150 tons of dispersant was applied but ineffectively. Violent weather conditions and wave energy physically dispersed oil into $1-70 \mu m$ droplets and mixed to >5M. In near shore areas, up to 30% of oil was forced into sediments. Localised finfish and razorshell mortalities were reported close to wreck, but there were no lasting effects on fisheries. By April 1993, hydrocarbon levels in seawater were at background levels, and fishing ban was lifted. By July, 1993, hydrocarbon tainting of salmon from farms 25 km from site no longer apparent.
North Cape Barge (January 19, 1996)	The *North Cape Barge* broke free from the tug *Scandia*, which caught fire in Long Island Sound, NY, in 60-knot winds and waves 5–7 m high. The barge grounded on Nebraska Shoals, a national wildlife refuge spilling 3,000 tons (840,000 gal) No. 2 fuel oil. High wave energy and the relatively high solubility of the refined oil caused an extraordinarily high degree of mixing. A concentration of 5.5 mg L^{-1} total petroleum hydrocarbons was found 17.8 m below the water surface. Very high mortalities of surf clams and lobsters (500,000 deaths) were reported, and fisheries were closed over 460 km² area for 5 months following spill. Over 400 bird deaths were reported.
Sea Empress (February 15, 1996)	In 40 knot (force 6) winds which blew for 5 days, 72,000 tons (21 million gal) forties crude oil and 350 tons of heavy fuel were spilled at St. Anne's Head near the mouth of Carmarthen Bay, South Wales. Oil mainly drifted southwest, away from shore, and was treated with 450 tons of dispersant sprayed from helicopters (dispersant was restricted to >1 km from shore). Approximately 50% of oil was dispersed, and 40% evaporated. About 200 km of coastline were contaminated, and 6,900 seabirds were killed. By May 21, 1996, levels of contaminants in seawater were declared within safe limits for bathing, and the exclusion order for all species of finfish was lifted on that date.

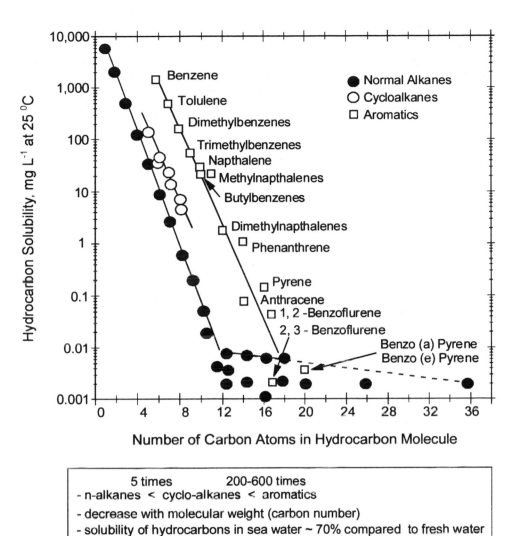

Figure 9.9 Solubilities of petroleum hydrocarbons in water.

was undoubtedly due to the rapid transport of toxic, low-molecular-weight compounds to the benthic environment. The highly volatile BTEX fraction also contributes a narcotic effect, which would lead to lethal stranding of crustaceans and finfish.

Acute toxicity is related, in part, to water solubility, which correlates logarithmically with molecular size within each hydrocarbon group (Figure 9.9). Although this relationship breaks down in normal alkanes larger than 12 carbon atoms, the principal petroleum hydrocarbons sort neatly into three groups. For molecules of similar size, cycloalkane solubilities and aromatic solubilities are respectively 5

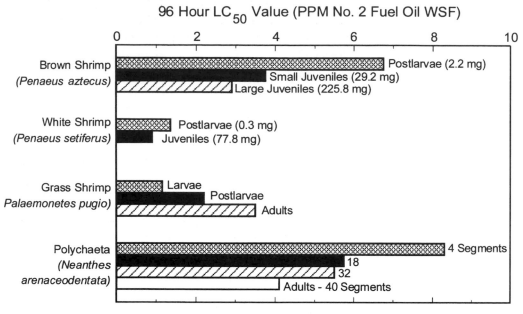

Figure 9.10 Toxicity bioassays for fuel oil (National Academy of Sciences, 1975).

times and 200–600 times those of their *n*-alkane counterparts. The solubility of hydrocarbons in seawater is approximately 70% compared with freshwater.

Many toxicity bioassays have been conducted using No. 2 Fuel Oil as a means of characterising a worst case situation (Figure 9.10) and this remains a standard for many oil-related studies. Toxicity studies on complex oil mixtures remain problematic, however. Considering that major qualitative differences in hydrocarbon components may exist among oils, total hydrocarbon levels are inadequate as a basis for bioassays. Another problem is dose. Within the context of a transient event such as an oil spill, a 96-hr static acute toxicity bioassay probably has only limited relevance, and over the last three decades, much effort has been devoted to defining an appropriate oil-accommodated water sample for use in oil toxicity assessment. CYP1A induction (see Section 7.9.2) has often proven a useful indicator of PAH exposure following oil spills (Stagg et al., 2000) although toxicological implications remain uncertain. Summing of PAH toxic units has been used to predict the toxicity to benthic animals of PAH mixtures to marine sediments (Swartz et al., 1995; see Chapter 5) and some attempts have been made to employ a similar approach to model the toxicity of complex petroleum hydrocarbon mixtures in the aqueous environment. Results thus far have failed to yield satisfactory data as toxicities predicted for several higher molecular-weight PAHs exceed their water solubilities. For lower molecular-weight, more volatile compounds such as BTEX, a critical body residue (CBR) approach (Chapter 3) may help explain their acute toxicity, although complex mixtures such as oils continue to pose problems for toxicologists interested in dose-response relationships. Differing physiocochemi-cal characteristics of oils under very variable environmental conditions mean that oil spill toxicology will remain a substantially empirical process pursued in large part case by case.

9.8.3   Oil spill legislation and control

Although major oil spills receive large media coverage, most releases of oil, whether accidental or intentional, tend to go unnoticed. In U.S. waters alone, it is estimated that there are, on average, 13,000 oil spills each year. Although most of these spills are modest in size, collectively they contribute from 10 million to 14 million gallons of oil annually. It was not until the Exxon *Valdez* spill that the U.S. Congress approached the oil pollution issue with renewed interest, and in 1990, a year after the accident, the Oil Pollution Act became law. This law addressed issues such as the establishment of tougher civil and criminal penalties for oil discharge violations; the development of tighter standards for licensing of vessel, personnel, equipment, and tanks; the granting of rights to federal authorities to enforce and conduct cleanup response; the creation of a federal fund (>$1 billion) to pay for oil spill cleanup; and the infusion of money into research institutes for research and development of new technologies for oil spill response.

A variety of methods are commonly employed during response to oil spills. These methods include mechanical containment of oil by booms, recovery of oil from the water surface by skimming vessels, burning, bioremedial action (e.g., nutrient enhancement, site inoculation with oil-degrading microbes), shoreline removal, cleansing and replacement, use of chemical shoreline cleaners, and use of chemical dispersants.

9.8.4   Use of oil dispersants

The use of chemical dispersants has created a dilemma for the oil industry and much recent effort has been directed to establishing criteria for their use. The first major application of dispersants occurred in 1967 when the tanker *Torrey Canyon* spilled nearly 1 million barrels of crude oil at the western end of the English Channel. At that time "dispersants" were actually synonymous with degreasing agents that had been developed for cleaning tanker compartments and bilges. These detergents contained alkylphenol surfactants and aromatic hydrocarbons such as toluene and benzene. Heavier fractions of the solvent formed stable detergent-oil emulsions instead of breaking up in the water column. The solvent carriers proved more toxic than the oil itself, and their use caused massive mortalities of indi-genous species.

Although the exact composition of currently available dispersants is unknown because of proprietary considerations, most are composed principally of nonionic surfactants. The most common nonionic surfactants are sorbitan monooleate, sulphosuccinate, and polyethylene glycol esters of unsaturated fatty acids. Small amounts of anionic surfactants (5–25%) are generally included in the formulation.

These surfactants are mixed in a solvent carrier of either water, water-miscible hydroxy compounds, or hydrocarbons that help reduce the overall viscosity of the mixture. Dispersant effectiveness trials have concluded that a 20:1 oil-to-dispersant ratio is appropriate for most applications. Assuming an average thickness of 0.1 mm oil in an oil slick, this equates to 7 L of dispersant required for each hectare of oil-slicked water.

9.9   Case study

Case Study 9.1. The Florida Everglades: A case study of eutrophication related to agriculture and restoration

The Florida Everglades system is a 500,000-ha freshwater wetland consisting of two portions: the 350,000-ha water conservation areas (WCAs) and the 150,000-ha Everglades National Park. North of the Everglades is the Everglades Agricultural Area (EAA), a 280,000-ha tract of land south of Lake Okeechobee. The ECAs were created in the 1960s to provide flood control and supply water to the EAA and urban areas by impounding wetlands from the northern and central Everglades. The EAA, with approximately 200,000 cultivated acres, has resulted in substantial increases in nutrient input into the southern Florida ecosystem. Nutrient-rich agricultural drainage from this area is either pumped north to Lake Okeechobee, south to the WCAs or to the east or west coasts of the Florida peninsula. The increased nutrient input and water management practices have resulted in a shift in water quality, plant community dominance, algal community structure, oxygen levels, sediment chemistry, and food web dynamics.

Historically, the Everglades were oligotrophic with particularly low levels of phosphorus. Although some degree of nutrient enrichment of Lake Okeechobee can be traced back to the 1950s, heavy algal blooms in the lake did not prompt serious concerns over eutrophication in the system until the early 1970s. Steady increases in both nitrogen and phosphorus in the lake were recorded in the late 1970s, and in 1982 an effort was made to decrease nutrient input into the lake, reducing nitrogen but not phosphorus levels because of the internal loading from sediments. These efforts resulted in a decrease in the nitrogen-to-phosphorus ratio from 30:1 to 20:1, a condition which favoured nitrogen-fixing cyanobacteria. Blue-green algal blooms increased in the 1980s, dominated by *Anabaena circinalis*. The most visible effect of eutrophication in the Everglades system has been a change in the macrophytic community wherein the historically dominant sawgrass (*Cladium jamaicense* Crantz) communities have been increasingly replaced by cattails (*Typha domingensis* Pers.).

To decrease the incidence of cyanobacterial algal blooms, water quality management strategies were adopted that were designed to minimise phosphorus loading and enhance nitrogen-to-phosphorus ratios. Phytoplankton dynamics in Lake Okeechobee are determined by several factors including temperature, underwater light attenuation, and the supply rate of available phosphorus to the water column. The latter is a complex process involving both external tributary loadings and internal sediment phosphorus flux. Loading reductions have been attained by restricting fertilisation, reducing livestock, and pumping less water from farm land during certain periods. A major phosphorus removal mechanism is through peat accumulation and man-made marshes.

The Everglades Forever Act of 1994 mandates that the waters of the Everglades Protection Area meet Class III water quality standards (i.e., safe for recreation and the maintenance of a well-balanced fish population and wildlife). The plan to achieve this includes the completion of the construction of six storm-water treatment areas encompassing approximately 17,800 ha. The target for phosphorus removal represents an 80% overall load reduction in the Everglades protection area.

9.10 References

Addison, R. F., and Fraser, T. F. (1996) Hepatic mono-oxygenase induction in benthic flatfish sampled near coastal pulp mill in British Columbia, *Marine Environmental Research*, **42**, 273.

Ahearne, J. F. (1993) The future of nuclear power, *American Scientist*, **81**, 24–35.

Antonovics, J., Bradshaw, A. D., and Turner, R. G. (1971) Heavy metal tolerance in plants, *Advances in Ecological Research*, **7**, 1–85.

Beyer, W. N., Pattee, O. H., Sileo, L., Hoffman, D. J., and Mulhern, B. M. (1985) Metal contamination in wildlife living near two zinc smelters, *Environmental Pollution, Series A*, **38**, 63–86.

Boeckx, P., and Van Cleemput, O. (1996) Flux estimates from soil methanogenesis and methanotrophy: Landfills, rice paddies, natural wetlands and aerobic soil, *Environmental Monitoring and Assessment*, **42**, 189–207.

Bothwell, M. L. (1992) Eutrophication of rivers by nutrients in treated Kraft pulp mill effluent, *Water Pollution Research Journal of Canada*, **27**, 447–72.

Bourdier, G., Couture, P., and Amblard, C. (1990) Effects of a pulp and paper mill effluent on the structure and metabolism of periphytic algae in experimental streams, *Aquatic Toxicology*, **18**, 137–62.

Brumley, C. M., Haritos, V. S., Ahokas, J. T., and Holdway, D. A. (1996) Metabolites of chlorinated syringaldehydes in fish bile as biomarker of exposure to bleached eucalypt pulp effluents, *Ecotoxicology and Environmental Safety*, **33**, 253–60.

Buchauer, M. J. (1971) Effects of zinc and cadmium pollution on vegetation and soils. Ph.D. dissertation, Rutgers University, New Brunswick, NJ.

Bumb, B. L., and Baanante, C. A. (1996) The role of fertilizer in sustaining food security and protecting the environment to 2020. International Fertilizer Development Center 2020 Vision Discussion Paper 17, report vol. 18, no. 3. IFDC, Muscle Shoals, AL.

Chapin, F. S., Walker, B. H., Hobbs, R. J., Hooper, D. U., Lawton, J. H., Sala, O. E., and Tilman, D. (1997) Biotic control over the functioning of ecosystems, *Science*, **277**, 500–3.

Cochran, R. G., and Tsoulfanidis, N. (1990) *The Nuclear Fuel Cycle: Analysis and Management*, LeGrange, IL.

Delong, M. D., and Brusven, M. A. (1992) Patterns of periphyton chlorophyll a in an agricultural nonpoint source impacted stream, *Water Resources Bulletin*, **28** (4), 731–41.

Dubé, M. G., and Culp, J. M. (1996) Growth responses of periphyton and chironomides exposed to biologically treated bleached Kraft pulp mill effluent, *Environmental Toxicology and Chemistry*, **15**, 2019–27.

Foy, R. H., and Withers, P. J. A. (1995) The contribution of agricultural phosphorus to eutrophication. Proceedings of the Fertilizer Society, Peterborough.

Franzin, W. G. (1984) Aquatic contamination in the vicinity of the base metal smelter at Flin Flon, Manitoba Canada – A case history. In *Environmental Impacts of Smelters*, ed. Nriagu, J. O., pp. 523–50, Wiley Interscience, New York.

Franzin, W. G., and McFarlane, G. A. (1981) Fallout, distribution and some effects of Zn, Cd, Cu, Pb, and As in aquatic ecosystems near a base metal smelter on Canada's Precambrian Shield. Proc. 7th Annual Aquatic Toxicology Workshop, Technical Report, Canadian Journal of Fisheries and Aquatic Sciences, Montreal, Nov. 5–7.

Franzin, W. G., McFarlane, G. A., and Lutz, A. (1979) Atmospheric fallout in the vicinity of a base metal smelter at Flin Flon, Manitoba, Canada, *Environmental Science and Technology*, **13**, 1513–22.

Freedman, B., and Hutchinson, T. H. (1980) Pollutant inputs from the atmosphere and accumulations in soils and vegetation near a nickel-copper smelter at Sudbury, Ontario, Canada, *Canadian Journal of Botany*, **58**, 108–32.

Gorham, E. (1998) Acid deposition and its ecological effects: A brief history of research, *Environmental Science and Policy*, **1**, 153–66.

Haagen-Smit, A. J. (1952) Chemistry and physiology of Los Angeles smog, *Industrial Chemistry*, **44**, 1342–6.

Harrison, S. E., and Klaverkamp, J. F. (1990) Metal contamination in liver and muscle of northern pike (*Esox Lucius*) and white sucker (*Catostomus commersoni*) and in sediments from lakes near the smelter at Flin Flon, Manitoba, *Environmental Toxicology and Chemistry*, **9**, 941–56.

Hasselmann, K. (1997) Are we seeing global warming? *Science*, **276**, 914–15.

Havens, K. E., and Hanazato, T. (1993) Zooplankton community responses to chemical stressors: A comparison of results from acidification and pesticide contamination research, *Environmental Pollution*, **82**, 277–88.

Heck, W. W., Heagle, A. S., and Shriner, D. S. (1986) Effects on Vegetabio: Native, Crops, Forest. In *Air Pollution*, ed. Stern, A. C., pp. 247–350, Academy of Science, New York.

Horner, R. A., Garrison, D. L., and Plumley, F. G. (1997) Harmful algal blooms and red tide problems on the U.S. west coast, *Limnology and Oceanography*, **42** (5), 1076–88.

Hunter, B. A., and Johnson, M. S. (1982) Food chain relationships of copper and cadmium in contaminated grassland ecosystems. *Oikos*, **38**, 108–17.

Hutchinson, T. C., and Whitby, L. M. (1974) Heavy-metal pollution in the Sudbury mining and smelting region of Canada: I. Soil and vegetation contamination by nickel, copper and other metals, *Environmental Conservation*, **1**, 123–32.

IAEA. (1998) *Nuclear Power Reactors in the World*, April 1998 ed., International Atomic Energy Agency, Vienna.

Jeffries, D. S. (1984) Atmospheric deposition of pollutants in the Sudbury area. In *Environmental Impacts of Smelters*, ed. Nriagu, J. O., pp. 117–54, Wiley Interscience, New York.

Karl, T. R., Nicholls, N., and Gregory, J. (1997) The coming climate, *Scientific American*, **276**, 78–83.

Kates, R. W., Turner, B. L., and Clark, W. C. (1990) *The Earth as Transformed by Human Action*, Cambridge University Press, Cambridge.

Lander, L., Jensen, S., Soderstrom, M., Notini, M., and Rosemarin, A. (1990) Fate and effects of pulp mill chlorophenolic 4,5,6-trichloroguaiacol in a model brackish water ecosystem, *Science of the Total Environment*, **92**, 69–89.

Larsson, D. G. J., Hällman, H., and Förlin, L. (2000) More male fish embryos near a pulp mill, *Environmental Toxicology and Chemistry*, **19**, 2911–17.

Lawton, J. H., and May, R. M. (1995) *Extinction Rates*, Oxford Press, Oxford.

Lindesjoeoe, E., Thulin, J., Bengtsson, B.-E., and Tjaernlund, U. (1994) Abnormalities of a gill cover bone, the operculum, in perch *Perca fluviatilis* from a pulp mill effluent area, *Aquatic Toxicology*, **28**, 189–207.

MacLatchy, D. L., and Van Der Kroak, G. (1994) The plant sterol β-sitosterol decreases reproductive fitness in goldfish, The Second Annual Conference on Environmental Fate and Effects of Bleached Pulp Mill Effluents.

Manabe, S. (1997) Early development in the study of greenhouse warming: The emergence of climate models, *Ambio*, **26**, 47–51.

Matson, P. A., Parton, W. J., Power, A. G., and Swift, M. J. (1997) Agricultural intensification and ecosystem properties, *Science*, **277**, 504–9.

Matusgek, J. E., and Wales, D. L. (1992) Estimated impacts of SO_2 emissions from Sudbury smelters on Ontario's sportfish populations, *Canadian Journal of Fisheries and Aquatic Sciences*, **49** (Supp.), 87–94.

Matthews, E. (1994) Nitrogenous fertilizers: Global distribution of consumption and associated emissions of nitrous oxide and ammonia, *Global Biogeochemical Cycles*, **8**, 411–39.

McElroy, M. B. (1994) Climate of the Earth: An overview, *Environmental Pollution*, **83**, 3–21.

Morrissey, W. A., and Justus, J. R. (1997) *Global Climate Change*. Committee for the National Institute for the Environment, Washington, DC.

Mueller, D. K., Hamilton, P. A., Helsel, D. R., Hitt, K. J., and Ruddy, B. C. (1996) Nutrients in ground water and surface water of the United States. NWQEP Notes, the NCSU Water Quality Group Newsletter, U.S. Geological Survey, Reston, VA. no. 76.

Munkittrick, K. R., Van Der Kraak, G. J., McMaster, M. E., and Portt, C. B. (1992) Reproductive dysfunction and MFO activity in three species of fish exposed to bleached Kraft mill effluent at Jackfish Bay, Lake Superior, *Water Pollution Research Journal of Canada*, **27**, 439–46.

National Research Council. (1985) *Oil in the Sea: Inputs, Fates and Effects*. National Academy Press, Washington, DC.

Neff, J. M., and Anderson, J. W. (1981) *Responses of Marine Animals to Petroleum and Specific Petroleum Hydrocarbons*, London Applied Science Publishers, London.

Nowell, L. H., Dileanis, P. D., and Capel, P. D. (1997) Pesticides in sediments and aquatic biota of United States Rivers: Occurrence, geographic distribution, and trends. Recent National Synthesis Pesticide Reports. U.S. Geological Survey, Reston, VA.

Nriagu, J. O., Wong, H. K., and Coker, R. D. (1982) Deposition and chemistry of pollutant metals in lakes around the smelters at Sudbury, Ontario, *Environmental Science and Technology*, **16**, 551–60.

Oikari, A., Lindstroem-Seppae, P., and Kukkonen, J. (1988) Subchronic metabolic effects and toxicity of a simulated pulp mill effluent on juvenile lake trout, *Salmo trutta* m. *lacustris, Ecotoxicology and Environmental Safety*, **16**, 202–18.

Olson, J. S., Watts, J. A., and Allison, L. J. (1983) *Carbon in Live Vegetation of Major World Ecosystems*, Office of Energy Research, U.S. Department of Energy, Washington, DC.

Owens, J. M., Swanson, S. M., and Birkholz, D. A. (1994) Environmental monitoring of bleached Kraft pulp mill chlorophenolic compounds in a northern Canadian river system, *Chemosphere*, **29**, 89–109.

Perrin, C. J., and Bothwell, M. L. (1992) Chlorate discharges from pulp mills: an examination of effects on river algal communities, *Water Pollution Research Journal of Canada*, **27**, 473–85.

Pimm, S. L., Russell, G. J., Fittleman, J. L., and Brooks, T. M. (1995) The future of diversity, *Science*, **269**, 347–52.

Postel, S. L., Daily, G. C., and Ehrlich, P. R. (1996) Human appropriation of renewable fresh water, *Science*, **271**, 785–92.

Priha, M. (1994) Bioavailability of pulp and paper mill effluent phosphorus, *Water Science and Technology*, **29**, 93–103.

Rabitsch, W. B. (1995a) Metal accumulation in arthropods near a lead/zinc smelter in Arnoldstein, Austria I, *Environmental Pollution*, **90**, 221–37.

(1995b) Metal accumulation in arthropods near a lead/zinc smelter in Arnoldstein, Austria II. Formicidae, *Environmental Pollution*, **90**, 239–45.

(1995c) Metal accumulation in arthropods near a lead/zinc smelter in Arnoldstein, Austria III. Archnida, *Environmental Pollution*, **90**, 249–57.

Ripley, E. A., Redmann, R. E., and Crowder, A. A. (1996) *Environmental Effects of Mining*, St. Lucie Press, Delray Beach, FL.

Rose, G. A., and Parker, G. H. (1983) Metal content of body tissues, diet items, and dung of ruffed grouse near the copper-nickel smelters at Sudbury, Ont, *Canadian Journal of Zoology*, **61**, 505–11.

Sandstrom, O., Karas, P., and Neuman, E. (1991) Pulp mill effluent effects on species distributions and recruitment in Baltic coastal fish, *Finnish Fisheries Research*, **12**, 101–10.

Sileo, L., and Beyer, W. N. (1985) Heavy metals in white tailed deer living near a zinc smelter in Pennsylvania, *Journal of Wildlife Distribution*, **21**, 289–96.

Silver, C. S., and DeFries, R. S. (1990) *One Earth, One Future: Our Changing Global Environment*, National Academy of Sciences, Washington, DC.

Smayda, T. J. (1997a) Harmful algal blooms: Their ecophysiology and general relevance to phytoplankton blooms in the sea, *Limnology and Oceanography*, **42** (5), 1137–53.

(1997b) What is a bloom? A commentary, *Limnology and Oceanography*, **42** (5), 1132–6.

Stagg, R. M., Rusin, J., McPhail, M. E., McIntosh, A. D., Moffat, C. F., and Craft, J. A. (2000) Effects of polycyclic aromatic hydrocarbons on expression of CYPIA in Salmon (*Salmo Salar*) following experimental exposure and after the *Braer* oil spill, *Environmental Toxicology and Chemistry*, **19**, 2797–805.

Sun, L., and Wang, M. (1996) Global warming and global dioxide emission: An empirical study, *Journal of Environmental Management*, **46**, 327–43.

Swartz, R. C., Schults, D. W., Ozretich, R. J., DeWitt, T. H., Redmond, M. S., and Ferrano, S. P. (1995) ΣPAH: A model to predict the toxicity of polynuclear aromatic hydrocarbon mixtures in field-collected sediments, *Environmental Toxicology and Chemistry*, **14**, 1977–87.

Taylor, G. J., and Crowder, A. A. (1983) Accumulations of atmospherically deposited metals in wetland soils of Sudbury, Ontario, *Water, Air and Soil Pollution*, **19**, 29–42.

Thurlow, G. (1990) *Technological Responses to the Greenhouse Effect*, The Watt Committee on Energy, report no. 23, Elsevier, London.

U.S. Department of Agriculture (U.S. DA). (1997) Control of Agricultural Nonpoint Source Pollution. Internet.

U.S. Geological Survey (U.S. GS). (1997) Pesticides in surface and ground water of the United States: Preliminary results of the National Water Quality Assessment Program (NAWQA). Pesticide National Synthesis Project. U.S. Geological Survey, Reston, VA.

Varadaraj, G., and Subramanian, M. A. (1991) Toxic effect of paper and pulp mill effluent on different parameters of bio-energetics in the fingerlings of *Oreochromis mossambicus*, *Environmental Ecology*, **9**, 857–9.

Varadaraj, G., Subramanian, M. A., and Nagarajan, B. (1993) The effect of sublethal concentration of paper and pulp mill effluent on the haemotological parameters of *Oreochromis mossambicus* (Peters), *Journal of Environmental Biology*, **14**, 321–5.

Vitousek, P. M., Ehlich, P. R., Erhlich, A. H., and Matson, P. A. (1986) *Bioscience*, **36**, 368–75.

Vitousek, P. M., Mooney, H. A., Lubchenco, J., and Melillo, J. M. (1997) Human domination of Earth's ecosystems, *Science*, **277**, 494–9.

Wall, G. J., Coote, D. R., DeKimpe, C., Hamill, A. S., and Marks, F. (1994) Great Lakes water quality program overview.

Webb, J., Wagner-Riddel, C., and Thurtell, G. W. (1997) The effect of fertilizer type on nitrogen oxide emissions from turfgrass. Land Management Report.

Whitby, L. M., and Hutchinson, T. C. (1974) Heavy-metal pollution in the Sudbury mining and smelting region of Canada, II. Soil toxicity tests, *Environmental Conservation*, **1**, 191–200.

Yan, N. D., and Welbourn, P. M. (1990) The impoverishment of aquatic communities by smelter activities near Sudbury, Canada. In *The Earth in Transition: Patterns and Processes of Biotic Impoverishment*, ed. Woodwell, G. M., pp. 477–94, Cambridge University Press, Cambridge.

Yasuno, M. (1991) Significance of interactions among organisms in aquatic ecosystems in response to pesticide contamination, *Reviews of Pesticide Toxicology*, **1**, 31–41.

Zimmer, C. (1997) Their game is mud, *Discover*, **19**, 28–9.

9.11 Further reading

Acidic deposition: Gorham, E. 1998. Acid deposition and its ecological effects: A brief history of research, *Environmental Science and Policy*, **1**, 153–66.

Metal mining and smelting: Ripley, E. A., Redmann, R. E., and Crowder, A. A. 1996. *Environmental Effects of Mining*. St. Lucie Press, Delray Beach, FL.

The safety of Ontario's nuclear power reactors: Hare, F. K. 1988. *A Scientific and Technical Review* vol 1, Report to the Minister, Technical Report and Annexes, Ontario Nuclear Safety Review, Toronto.

Global warming: Silver, C. S., and Defries, R. S. 1990. *One Earth, One Future: Our Changing Global Environment*, National Academy of Sciences, Washington DC.

10

○ ○

Risk assessment

10.1 The context and rationale for ecological risk assessment

Until the 1960s, the basis of environmental protection tended to come from retro-spective, or after-the-fact, recognition of damage. Discoveries of abandoned waste material, the occurrence of accidents and spills, and inadvertent exposures of humans and ecosystems to toxic substances, as well as other unplanned events pro-vided warning signs as well as environmental data on which to base future pro-tection. This type of situation is no longer acceptable, although surprises still occur. Even with the application of more systematic approaches, until the mid 1980s, the potential for adverse effects was evaluated largely through tests for impacts, referred to as hazard assessment (e.g., acute toxicity testing, see Section 2.2.2).

Environmental management of toxic substances, indeed all kinds of environ-mental protection, ideally should involve putting into practice the advice of scien-tists and other specialists in the various disciplines that are involved. In spite of advances in our understanding of environmental toxicology, there are no rules or even guidelines for protecting the environment from potentially toxic substances that can be applied across the board: Every case has its own unique sets of condi-tions and problems. It is axiomatic that no situation is completely free of *risk*. Risk is defined as the "chance of an undesired effect, such as injury, disease, or death, resulting from human actions or a natural catastrophe" (CCME, 1996).

Prospective rather than retrospective approaches are now expected, and the plan-ning process should have built into it measures for protecting humans and ecosys-tems. For example, a cradle-to-grave or life-cycle type of management is advocated for potentially toxic substances, and most jurisdictions require environmental impact assessment prior to approval of major projects. Furthermore, the assess-ment of previously contaminated sites (the legacy of past activities) is now required by most jurisdictions in the developed world, and there is a perceived need to follow this with some kind of rehabilitation.

The way in which regulatory bodies handle risk is almost always a compromise between the ideal and the manageable and/or affordable. A reasonable and responsible level of protection rather than an absolute absence of any potentially harmful substance is the goal of most regulatory agencies. This can be understood as reducing the risk of harm to an acceptable level. Purists will argue that there is no acceptable level for certain types of toxic material, and this has led to a "zero discharge" policy, which is addressed in Chapter 12.

Much research in environmental toxicology was not designed to answer the questions that the regulatory agencies are asking. The preceding chapters have provided a great deal of information about various types of toxic substances as well as complex issues usually involving more than one toxic substance or more than one type of perturbation. Uncertainty and controversial opinions among scientists are the rule rather than the exception for many of these examples. Furthermore, subtle effects and real-world studies are fraught with inconsistencies and uncertainties, which make interpretation very difficult. Additional uncertainties arise in the context of extrapolation among studies, the most notable of which is extrapolation from the laboratory to the field. In many studies of the effects of potentially toxic substances there is not sufficient evidence to reject the possibility that harmful effects could occur, or, to put it another way, there are reasons to suspect potential damage. Dealing with these various types of uncertainty, while still using the best available scientific information, is a major goal of the process known as risk assessment (RA). Risk assessment is concerned with estimating the probability of undesired events. The concept is not new: It was developed by actuaries to estimate the risk of insurance claims. It is used, for example, by engineers to estimate risks of system failure such as plane crashes, by economists to estimate business failures, and by fire-fighters to estimate the risk that forest fires would get out of control. Of relevance in the present context, RA is used to estimate the risks of chemicals to human or ecosystem health (Suter, 1990).

Just as concern for the harmful effect of chemicals on human health (classical toxicology) has expanded into concern for the harmful effects of chemicals on other biota in ecosystems (ecotoxicology), so risk assessment has been expanded to deal with biota in ecosystems. Humans are normally an integral part of ecosystems; however, human health RA is usually dealt with independently from ecosystem health, even though the chemicals of concern may be the same in origin and nature. Although not exclusively the concern of decision makers, ecological risk assessment has become increasingly important in regulation. ERA can be used to derive environmental criteria, especially on a site-specific basis, as for example in planning, or to serve as the basis for making remedial decisions. Ecological risk assessment differs philosophically and technically from approaches to regulatory criteria that are based on the quality and quantity (concentration) of chemical, in that it combines information on the degree of toxicity and the magnitude of exposure.

Risk assessment should logically be coupled with decision making concerning the management of risk, and the combined goal of risk assessment and risk management should be to define and achieve acceptable levels of risk (ICME, 1995).

Table 10.1. Definitions of ecological risk assessment

Definition	Reference
The process of assigning magnitudes and probabilities to adverse effects of human activities or natural catastrophes.	Barnthouse and Suter (1986)
A formal set of scientific methods for estimating the probabilities and magnitudes of undesired effects on plants, animals, and ecosystems resulting from events in the environment, including the release of pollutants, physical modification of the environment, and natural disasters.	Fava et al. (1987)
A subcategory of ecological impact assessment that • predicts the probability of adverse effects occurring in an ecosystem or any part of an ecosystem as a result of perturbation; • relates the magnitude of the impact to the perturbation.	Norton et al. (1988)
The process that evaluates the likelihood that adverse ecological effects may occur or are occurring as a result of exposure to one or more stressors. This definition recognises that a risk does not exist unless the stressor • has an inherent ability to cause adverse effects; • cooccurs with or contacts an ecological component long enough and at sufficient intensity to elicit the identified adverse effects(s). ERA may evaluate one or many stressors and ecological components.	U.S. EPA (1992)

Based on CCME (1996).

Various frameworks, approaches, and definitions have been discussed in the proliferating literature on the topic of ecological risk assessment, and the subject can provide a basis for some challenging intellectual discourse. Entire conferences and symposia have been dedicated to the subject over the past decade; frameworks abound, but practical applications are still evolving. There are a number of different definitions of ecological risk assessment, some of which are provided in Table 10.1. It appears that there is no universally accepted definition; nevertheless, the definitions in Table 10.1 are in agreement in terms of the principles of ecological risk assessment. The methodology is still being developed, and in the following section some of the practical aspects are discussed.

10.2 The methodology of ecological risk assessment and risk management

10.2.1 Risk assessment

THE PARADIGM

The traditional risk assessment paradigm consists of four steps (Glickman and Gough, 1990; ICME, 1995; National Research Council, 1994).

1. *Hazard identification* is defined by the National Academy of Sciences, USA, as "the process of determining whether exposure to an agent can cause an increase in the incidence of a health condition (cancer, birth defect, etc.). It involves characterizing the nature and the strength of the evidence of causation". Hazard identification does not provide any information on the extent or magnitude of risk.

2. *Dose-response evaluation*, a fundamental tenet of toxicology, is exemplified by the phrase from Paracelsus that "All substances are poisons: there is none that is not a poison. The right dose differentiates a poison and a remedy".

3. *Exposure assessment*, which quantifies the exposure to the agent of interest, is often calculated for human and environmental exposures on the basis of predictive models, rather than measurements. The population exposure estimate is much more difficult to calculate than the maximum individual risk (MIR) or the maximally exposed individual (MEI).

4. *Risk characterisation*, the final stage of risk assessment, summarises the information from 1, 2, and 3. This normally involves putting the expression or assessment of risk into a form that is useful for decision makers. The result of a risk characterisation is a qualitative or quantitative description of the potential hazards due to a particular exposure.

THE PROCEDURE

The detailed procedure for carrying out an ERA involves a series of prescribed steps. The philosophy and the essential components are very similar among jurisdictions, although details vary. Certain terms have been coined or adopted in the practice of ERA, but the reader is likely to encounter some variation in the definitions and the uses of such terms. This is inevitable in an evolving process. Figure 10.1A illustrates one particular framework for the entire process, as provided by Environment Canada (1996), based on the scheme provided by the U.S. EPA in 1992. In the present section of our text, this model is followed in some detail, not because it is the only or even the best available one, but because it represents a comprehensive coverage of the components of the process and so provides a useful example.

Note that, at all stages, there should be contribution from risk managers and interested parties (sometimes referred to as stakeholders); this is particularly important in the problem formulation stage (top box in Figure 10.1A). Data collection and data generation are expected to be occurring throughout the process, typically in an iterative process, with midcourse adjustments being normal.

In Figure 10.1A, three main boxes are shown – Problem Formulation, Analysis, and Risk Characterisation – for two models. These stages are described in more detail next.

PROBLEM FORMULATION

The problem formulation stage includes the hazard assessment referred to earlier, but it also incorporates additional concepts. Formulation of the problem includes

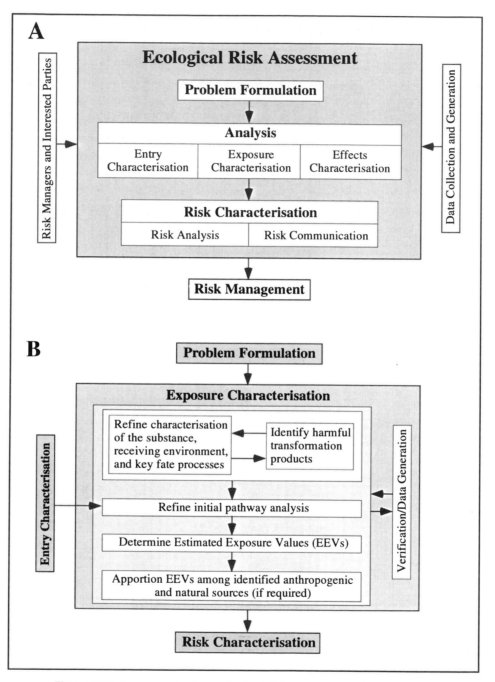

Figure 10.1 Frameworks for ecological risk assessment. A = overall framework; B = detail of exposure characterisation.

- Pathways analysis
- The identification of sensitive receptors

Pathways analysis. Pathways analysis considers the entry of a substance into the environment and the manner in which it will partition among environmental compartments such as air, soil, water, and sediment. Information on the physical/chemical behaviour of the substances under varying environmental conditions (temperature, pH, oxygen tension, redox potential, hydrologic regime, etc.) may be available, or the behaviour may be modelled, based for example on knowledge of molecular structure. The latter is particularly relevant for organic substances (see Section 5.5 and Case Study 10.2).

The rate or amount of release of a substance into the environment normally has to be determined through estimation rather than through direct measurement and may involve consideration of the potential for loss or discharge during manufacture, storage, transportation, and disposal. Natural sources also need to be considered. Geographic areas that are at particular risk through anthropogenic or natural conditions may also be identified through this procedure. Clearly, there are likely to be uncertainties in all of these values.

The identification of sensitive receptors. Data from field or laboratory tests on the effects of a particular substance are used to determine the potential receptors that are likely to be at the greatest risk. In the simplest instance, this would involve selecting the family, genus, or ideally the species for which the lowest LC_{50} had been recorded. In most instances, however, such information for the ecosystems of concern is not available from the literature. Models such as QSARs (see Section 5.5) may be useful to make an initial identification of sensitive organisms.

Confirmation from laboratory testing or field sampling would, however, be needed. The problem formulation step should result in a conceptual model that describes the ways in which a substance behaves in an ecosystem and its possible effects. At this point, it should be possible to identify, in a general manner, the types of ecosystems that may be at risk.

Following problem formulation is the stage of Analysis (second main box in Figure 10.1A).

ANALYSIS
Analysis is composed of three major parts:

- Characterisation of entry
- Characterisation of exposure
- Characterisation of effects

Characterisation of entry. The information on entry that was used in the pathways portion of the problem formulation analysis is verified and refined. In

Table 10.2. Maximum total daily intake of hexachlorobenzene (HCB) for mink[a]

Medium	Maximum concenteration of HCB	Intake rate (medium) day^{-1}	Maximum daily intake of HCB, ng kg body weight^{-1} da^{-1}
Air	0.29 ng m^{-3}	0.55 m^3 day^{-1}	0.16
Water	87 ng L^{-1}	0.1 L day^{-1}	8.7
Diet 1: 100% fish	283 ng g^{-1}	215 g day^{-1}	60,845
Diet 2: 100% birds or mammals	30 ng g^{-1}	158 g day^{-1}	4,740
Total daily intake for air, water, and diet 1			60,854
Total daily intake for air, water, and diet 2			4,749

[a] Estimated for a 1-kg mink (*Mustela vison*) in the region of the St. Clair River, Ontario, Canada. Bioavailability of HCB is assumed to be 100%.
Modified from Environment Canada (1996).

essence, entry characterisation will involve identification of sources and characterisation of releases. The unit of concern may range from a single ecosystem, a region, or an entire country. The unit of concern will not necessarily be one with natural boundaries; indeed, in the case of regulatory use, it is likely to be one of political boundaries.

Normally a table is set up for the specific needs of each assessment. Again, there will inevitably be uncertainties.

Characterisation of exposure. Exposure can be characterised in a number of ways, including direct measurements or estimates of daily intake. Accurate and comprehensive measurement of intake from all sources may be available from controlled experiments, but for real-world examples it is necessary to estimate exposure, sometimes employing models. Table 10.2 illustrates how an estimate can be made of daily exposure for a wild mammal. When exposure is estimated from the concentrations of chemical in the environment, the bioavailable forms of substances rather than the total concentrations should be used in the calculation. However, with the current state of knowledge, this is rarely done. These considerations will be case- and substance-specific.

In many instances, exposure is determined through indirect methods, as for example measurement of body burdens or tissue residues. These methods are typically used for estimating exposure for wildlife or for other biota whose daily intake is unknown. Whole body burden may be less useful than the concentration in a particular tissue: the appropriate tissue will depend upon the chemical of concern.

Obviously, this requires some knowledge of the toxicology of the chemical. Body or tissue burdens are good measures of exposure for chemicals whose forms are not changed through metabolism. Body burdens and tissue residue analysis have been discussed in some detail in Section 4.4.3.

Exposure will generally be quantified as estimated exposure values (EEVs) for each receptor in each area of concern. The expression of exposure will include a time interval (such as a day or a month), the actual time being selected on a case-specific basis, including information such as the nature of the exposure, whether episodic or continuous, and the duration of the life-cycle of the receptor. Statistical treatment of the EEVs is necessary to address both real variability and uncertainty.

Figure 10.1B provides further details of the steps involved in the characterisation of exposure. The process can be complex, and certain steps are depicted as iterative, indicating that the characterisation may need ongoing revisions or refinement, as data become available.

Characterisation of effects. The quantification of effects for the purposes of ERA should lead to a critical toxicity value (CTV). Typically the CTV is based upon sensitive toxicity thresholds such as LOEL or an EC_{10} (Section 2.3.1). The quality of toxicity data as well as analytical procedures are always specified by the regulatory agency or the assessor; some data, particularly those that have been developed for other purposes, will be unacceptable and will be rejected. Lack of suitable data adds to uncertainty and may result in recommendations for further research. Where data are not available, estimates such as those based on QSARS may be used: This normally applies only to organic substances, and even for these substances, data from real tests in the laboratory, or field samplings, are preferred.

Information on toxicity should be taken for a wide range of trophic levels, and this will aid in determining which populations, communities, or ecosystem processes are particularly susceptible to adverse effects from a particular substance. It should be obvious that a chemical that tends to be biomagnified in an ecosystem will be much more ecotoxic than one that is not. At present, this potential is addressed by increasing the safety factor rather than by attempting to make a quantitative model of the biological processes.

Once the CTVs are determined, they are used as inputs to the next phase, Risk Characterisation, the third large box in Figure 10.1A. CTVs are used to calculate the estimated no effects value (ENEV). As the name suggests, this value with a very low probability of causing adverse effects. To estimate this no effects value, the CTV is divided by an application factor (see Section 10.4).

RISK CHARACTERISATION
Risk characterisation can be seen as involving two stages:

- Risk analysis
- Risk communication

Risk analysis. This stage requires a determination of the likelihood and magnitude of adverse effects that would result from exposure to the substance of concern. Risk analysis incorporates the results of the characterisation of entry, exposure, and effects discussed earlier to derive a worst case value. As an example, the first screening tier employs the worst case quotient, which is calculated by dividing the estimated exposure value by the estimated no effects value for each of the end-points. If this worst case quotient is less than unity for all of the end-points available, then the substance is deemed to be nontoxic, and no further risk analysis is required. However, if any of the worst case quotients exceed unity, then higher tiers of risk analysis are required. Details of the various mathematical procedures that can be used to derive other worst case estimates will not be provided here: further reading is provided.

Risk communication. The preceding stages of risk assessment need to be put into a form in which they are scientifically credible for the decision-making process. Communication of risk essentially requires that each of the stages of problem formulation, analysis, and risk characterisation be reported on, with all methodology, data sources, assumptions, and uncertainties. The process should be logical because it is based upon a logical series of interconnected steps.

This process then leads to the practical use of the assessment, in what is called the management of risk.

10.2.2 Risk management

When a significant risk has been identified, it is prudent, indeed necessary, to take action to come to terms with the risk, a process known as risk management. Risk management strategies seek to minimise risks as well as costs. The management of risk can take a number of different forms, including

> Pollution-control technologies
> Pollution monitoring and site remediation
> Limitation of the use of a site, to avoid exposure
> Product reformulation

Ideally, the task of the risk manager is integrated with the risk assessment process, rather than being added later. This is depicted in Figure 10.1A in the vertical box on the left.

10.3 Site-specific risk assessment

"Because risk characterisation is based in significant part on a specific exposure scenario or scenarios, risk characterisation is highly context-specific and cannot be automatically applied from one context or location to another" (ICME, 1995).

In the preceding chapters, as well as in the current one, we have pointed out that the effects of toxic substances are affected by a number of factors. In the context of assessment, then, this frequently leads to the need for individual consideration on a site-by-site basis. Even though a site-specific assessment is likely to involve more effort and resources than, for example, the application of existing water quality criteria and, thus, is likely to cost more than a more generalised or across-the-board approach, it may well prove that an assessment of site-specific risk will be less costly in the long run for the owner of a site or other proponent. The example of the metal-contaminated marine site provided as Case Study 4.1 illustrates this. In that case, the concentration of certain metals in the sediments exceeded the guidelines of several different jurisdictions. Nevertheless, the benthic communities at these sites, based on several different and accepted indices, appeared to be healthy and were not significantly different from those at control sites. In this case, the field data combined with a site-specific assessment of risk would almost certainly obviate the need for any remedial action.

Many risk assessments are indeed site-specific. These are illustrated in Case Studies 10.1, 10.2, and perhaps also 10.3. The procedure for site-specific risk assessment is essentially the same as for the more generalised procedure described in detail earlier. The difference between the two is mainly in the use of the known and documented site conditions for the phases of assessment identified in Section 10.2.1.

The example in Case Study 10.2 is fairly typical. The recent (MOE, 1997) Ontario guideline for use at contaminated sites provides values for generic criteria for most potentially toxic substances in soil and groundwater. These criteria themselves were derived through a risk assessment procedure and already have built into them some safety factors. For any given substance, the criterion varies depending upon the anticipated use of the land or water. If these criteria are exceeded for the expected use of the land or water, then some action may be required. However, according to the guideline, certain results or conditions trigger a site-specific risk assessment. Concentrations of toxic chemicals as well as the existence of sensitive sites or species may be involved in the triggering (see Section 10.5 for more details).

The result of the site-specific assessment may diverge in either direction from the results that were based solely upon generic guideline criteria for the chemicals of concern. For example, it may be demonstrated (as in Case Study 10.2) that certain substances in the groundwater, although at high concentrations, are essentially immobile, so that for nonpotable use, the groundwater does not represent a risk to the receiving system. In this case, the site-specific risk assessment would be more forgiving than the use of generic criteria. On the other hand, if the receiving ecosystem includes a species or a community that is known to be abnormally sensitive to a certain substance and if that substance is likely to leach into the surface water or move in discharged groundwater, then the site-specific risk assessment may prove to be less forgiving than the direct use of criteria. In either case, the site-specific assessment should provide for a more fair and responsible course of action than would the use of generic criteria alone.

Guidelines for contaminated sites are now available for many jurisdictions, but they are all relatively recent and still evolving and as such are liable to be modified or updated. Therefore, the reader is advised to consult the appropriate authority if any further detail is required, or if research is involved.

Reference has been made several times to uncertainty and to safety factors. It is frequently necessary in ERA to account for uncertainty, and some of the available approaches are described next.

10.4 Dealing with uncertainty

Many aspects of environmental toxicology are plagued with uncertainty, and often uncertainty or controversy provides healthy stimuli for research. However, in the area of environmental management or regulation, it is frequently necessary to provide guidance within the limitations of uncertainty or incomplete knowledge of the potential adverse effects of a toxic substance or other perturbation. Rarely is it acceptable to simply recommend further research, when decisions are urgently required. Thus, in ERA we frequently find that some procedure is required to accommodate uncertainty. The usual method is the application of uncertainty or safety factors. As described by Chapman et al. (1998), safety factors are a "conservative approach for dealing with uncertainty related to chemical risks".

According to Chapman et al. (1998), the term "safety factor" includes any means by which known data are extrapolated to deal with situations for which there are no data. Some authorities use other terms for what are essentially subsets of the concept. The terms and respective tools include application factors, acute-to-chronic ratios, and the "precautionary principle". These are described briefly below, based on Chapman et al. (1998).

Uncertainty factors. In common with some other types of safety factors, uncertainty factors (also referred to in the U.S. EPA as assessment factors) were developed for mammalian toxicology. Uncertainty factors involve empirical methods such as adjusting a point estimate (e.g., an EC_{50}) by an arbitrary factor of 10 or 100 to estimate a safe level of a substance. Such an approach will definitely decrease the probability of underestimating risk, but it has little or no relevance to the actual uncertainty. In ecological risk assessment, uncertainty factors tend to be used in risk assessments that use the indicator species approach.

Application factors. Application factors (referred to in Section 10.2.1) exemplify attempts to extrapolate from real toxicity data to the concentration of a substance that would have no harmful effects. Originally derived from the slopes of time-mortality curves, values are now almost always empirically derived; AFs for

Canadian and European regulatory agencies are now standardised at values of 0.1, 0.05, and 0.01.

Acute-to-chronic ratios. These ACRs are essentially a form of AFs, with the acute or lethal end-point being divided by the chronic end-point (see also Section 2.2.4). A conservative value of 40 has been suggested by the European Centre for Ecotoxicology (1993). This value is based on calculated ACRs for a number of substances and a range of species, from which they derived a 90th percentile AR ranging from 4.2 to 22.

The precautionary principle. This principle is an extreme example of the use of a safety factor, which is essentially based on a philosophy of taking protective action "even when there is no scientific evidence to prove a causal link between emissions and effects" (London Conference, 1987), and advocates zero discharge of potentially toxic contaminants. In its basic form, it will result in infinitely large safety factors, but modifications may render this approach useful.

The selection and use of any one of these tools must be recognised as primarily a policy-based, not a science-based, decision. While embracing the real need for protection in the face of uncertainty, empirically derived factors involve the real possibility of overestimating the risk and often lead to unrealistic answers in hazard and risk assessment (Chapman et al. 1998; Whittaker et al. 1998).

10.5 Factors triggering risk assessment

In spite of a widespread recent tendency among regulatory agencies for the use of ERA, this type of assessment is not appropriate for all situations, nor is it always compulsory. According to the CCME (1996), "It must be emphasized that ERA is not necessarily superior to other approaches in the development of remediation strategies".

The following examples provide situations related to contaminated sites for which ERA is appropriate or required and also for the level of complexity of ERA that is needed. The information is based on the Canadian National Contaminated Sites Programme (CCME, 1996). It lists all situations for which the generic criteria may not be adequately protective, as well as other situations for which there is sufficient uncertainty of some kind that it is not possible to judge the applicability of the generic criteria. The guide also points out that "the ERA practitioner is encouraged to consider when an ERA would be inappropriate." For example, if there is an improvement (as may be expected) in the understanding of the risk related to some sites, the need for ERA may decrease.

Triggers for ERA
1. The site is ecologically significant and includes any or all of the following:

- The site contains critical habitat for species of wildlife including migratory birds and fish.
- There are rare, threatened, or endangered species, populations, or ecosystems.
- The land has been designated as a national park or ecological reserve.
- The land is locally or regionally important for fishing, hunting, or trapping.
- There are organisms that are not representative of the data on which criteria are based.
- The criteria values (generic) are based on assumptions that do not apply to the site of concern.
- There are modifying biotic or abiotic modifying factors (e.g., naturally high background of metals) providing conditions under which the criteria cannot realistically be applied.

2. There are unacceptable data gaps. ERA should be considered if any of the following apply:
   - There are one or more chemicals, above background concentrations, and little is known about these.
   - Exposure conditions are very uncertain.
   - Pathways and partitioning of contaminants in the ecosystem are not understood.
   - There is a high degree of uncertainty about levels that have adverse effects (i.e., hazard levels).
   - There are significant data gaps concerning ecological receptors.

3. The site has special characteristics. The ERA might be a practical selection for sites where any of the following apply:
   - The costs of remediation to meet existing criteria are exceedingly high, and priorities must be established to focus remediation efforts.
   - Existing criteria need field testing or improvement.
   - No criteria currently exist for chemicals of concern at the site.
   - The contaminated area is so large that ERA is needed to provide a framework for site investigation and to set remediation priorities.

10.6 Case studies

Case Study 10.1. Risk assessment of the Clark River Superfund site

The so-called Superfund sites in the United States include a number of metal-contaminated terrestrial and aquatic sites, many of them being areas that have been contaminated by mining or smelting. The Superfund (the U.S. Comprehensive Environmental Response, Compensation and Liability Act of 1980) was created with the idea of identifying and rehabilitating such sites, but the massive amounts of effort and funds associated with rehabilitation have prompted the regulatory agency (in this case, the Environmental Protection Agency, U.S. EPA) to require prescreening and assessment of the degree of harm which the sites might exert.

A series of risk assessments has been conducted on metal-contaminated sites, of which the Clark River Superfund site in Western Montana is one. This site includes several different types of habitat, from riparian to completely terrestrial, and the source of contamination was a variety of mining-related activities in the past, as well as aerial deposits from the Anaconda smelter (Pastorok et al., 1995b). Nearby un-affected habitats were also available to provide reference (control) areas. The team identified, for this particular site, arsenic (As), cadmium (Cd), copper (Cu), lead (Pb), and zinc (Zn) as contaminants of concern; the identification was based on prelimi-nary screening of soil contaminants, as well as prior knowledge of the types of activ-ities that had affected the site.

For the riparian plant communities, there were clear gradients of zonation, from nearly bare in the most contaminated sites to relatively normal plant communities. To quantify the effects, original on-site field studies were carried out. Various indices of plant community, as well as trace elements in soils, were measured across the zones. Multiple linear regression was used to relate plant biomass or species rich-ness to trace elements in soils, and the resulting linear regression models were used to estimate plant community effect levels. These estimates were in general higher than expected, based on toxicity values from the literature, by up to an order of mag-nitude. Soil pH was also shown to be an important variable for plant communities, with a threshold of 5.5.

Metal concentrations in plant tissue, invertebrates, and deer mice were also mea-sured (Pastorok et al., 1995a), and were applied to food web exposure models. Esti-mates of site-specific doses allowed direct comparison of exposure estimates to toxicity reference values.

Details of metal concentrations will not be presented here, but note that, even at the most contaminated sites, whole body Cd concentration never exceeded 1.0 for deer mice, whereas tissue concentrations of Cu and Zn for deer mice were higher than those reported in the literature for the same species in other mining/ smelting areas. Nevertheless, bioconcentration factors for (whole body) mice: soil were quite low, only exceeding 1 for Zn at very contaminated sites, whereas bioconcentration factors for As and Pb were rarely above 0.1 and were frequently less than 0.01.

Risk for four wildlife species – deer mice, red fox, white-tailed deer, and American kestrel – were assessed, based on the various food items and toxicity reference values. It was concluded that the metals in the soils and terrestrial food webs of this site posed negligible risk for populations of key ecological receptors, such as foxes, white-tailed deer, and predatory birds. In discussing uncer-tainty, the authors also pointed out that the toxicity reference values are probably overprotective. Further studies on the site (La Tier et al., 1995) demonstrated that there was no risk to reproduction of deer mice related to the elevated elements in the soils.

No route of exposure other than ingestion was included in these assessments.

Case Study 10.2. The Belle Park Island landfill site, Cataraqui Park, Kingston, Ontario: Site-specific risk assessment

In 1974, a municipal landfill site situated on a river that enters Lake Ontario was closed after 22 years of operation. The site had opened in 1952, prior to the introduction of government regulations for waste disposal. Typical of many such municipal landfill sites, no detailed records were kept of the material placed in the site, and the engineering design was typical of many from that time, having no liner or barriers. The landfill operated under a provincial (Ontario) certificate of approval from 1970 until closure. After this, the municipality converted the site to a multiple-use recreational facility, which includes an 18-hole golf course with a clubhouse.

The waterfront and the inner harbour of the municipality were not always developed for residential and recreational use to the extent that they are today. In the past, the area was used for many industrial and other commercial activities, particularly during the nineteenth century and into the first few decades of the present century. All these activities have left chemical and physical legacies, many of which are not well documented, on the land and adjacent water. The extent of risk for human health and ecosystems associated with these sites depends on a number of factors, including land and water use, mobility of contaminants as well as their potential toxicity. A modern approach to such sites is to carry out site-specific risk assessment. Many jurisdictions are producing guidelines on such assessment. Examples of the application of such assessments have already been provided in the text. Remedial action is not automatically required: In many cases, the "do nothing" option is the safest one.

Over time, leachate from the closed landfill was observed at several discharge points around its perimeter, from which points leachate could enter the river system, either directly or via drainage channels. These discharges were highly visible as reddish-brown stains on snow or surface substratum, indicating precipitates of ferric iron compounds. Prior to discharge, the groundwater was anoxic or had very low dissolved oxygen concentrations, so the iron would be in the ferrous form, which is very much more soluble than ferric iron. As the dissolved ferrous iron was exposed to oxygen, it was converted to ferric compounds, which formed the precipitates. These types of stains are commonly observed when anoxic or acidic water rich in iron encounters oxygenated less acidic water, and, although they may be aesthetically unacceptable, they do not of themselves necessarily constitute a risk.

Public concern, stimulated by local activists, led the municipality to investigate the extent of contamination of the environment by the landfill and to address any risks for human health and the aquatic ecosystem related to the site. The Ontario Guideline for Contaminated Sites (MOE, 1997), subsequently referred to as the Guideline, was used, and, based on its protocols, preliminary risk assessments were made for certain chemicals of concern.

Extensive and intensive sampling of groundwater, visible discharges, receiving water and soils, and vegetation were made over a period of 2 years, and 104 chemical parameters were determined repeatedly in these media. The prescribed sampling methods, quality assurance, and quality controls were used. Criteria provided in the

Guideline for nonpotable groundwater and residential/recreational use of soil were applied. Additional criteria were developed for iron in water and in soil and for ammonia in water, since these substances occurred frequently, and no generic criteria were provided.

In the course of the detailed review of all field data, contaminants of concern (COCs) were those that exceeded the criteria for nonpotable groundwater, or soils, or both.

Contaminants of concern were

- Ammonia in water
- Copper and silver in water
- Iron in water and in soil
- Twelve PAHs in water
- Total PCBs in water

Pathways analysis demonstrated that humans were not exposed to any of these COCs; therefore, no human health risk assessment was required.

The nonpotable groundwater criteria have been designed through a risk assessment process to protect most aquatic life. For the present site, groundwater was clearly the source from which the contaminants needed to be traced and assessed, and, in the course of sampling, numerous bore holes had been made across the site. In addition to providing a means of estimating the movement of contaminants across the site and out of the groundwater, the numerous sampling points provided additional validity in the context of risk assessment, since the highest concentrations from a range are chosen to adhere to the worst case concept.

The worst case, or highest concentrations, along with the assessment of effects on the most sensitive organisms, populations, or communities, are some of the built-in safety features of risk assessment.

The procedure used for screening risk assessment was as follows:

1. The most sensitive receptor appropriate for the chemical was selected by reviewing the literature and selecting the combination of the most appropriate and most sensitive aquatic organism with, where possible, the most appropriate chemical conditions of the test. All end-points for receptors that are used in the risk assessment are direct because, with the present state of understanding, indirect (e.g., food chain) effects cannot be incorporated into risk assessment. When only acute toxicity data are available, it is usually necessary to incorporate a safety or uncertainty factor to estimate the chronic toxicity. This may be a factor from 10 to 1,000, depending on the original test, the mode of toxicity, and environmental behaviour of the chemical.
2. A simple, plausible maximum exposure scenario was provided, using the most conservative approach, namely, it was assumed that the receptor was exposed for 100% of the time.
3. From the real data for groundwater, the maximum detected level of the chemical was used.

4. The plausibly maximal on-site exposure for the chosen scenario was provided. The Guideline allows for a ×10 decrease in concentration from groundwater to the point of discharge. To determine exposure in the present exercise, the real measured values of discharge were applied, and a further ×10 dilution was applied as the discharge entered the receiving water. This represents a very conservative value for dilution by the receiving water, since it was shown that the dilution is more likely to be at least several hundred- or thousandfold in reality.

5. The maximum exposure was compared to the appropriate exposure limit of the chemical (i.e., for each contaminant, there was a direct comparison between the calculated worst case exposure and the value that had been identified as hazardous to the most sensitive organism).

If the former were lower than the latter, no further risk assessment would be required at that point in time.

None of the screening risk assessments for any of the PAHs or metals indicated the need for further assessment. For other substances, where the exposure had indicated that the most sensitive organisms might be at risk, further site-specific assessment would be required. This would include an investigation of the open water conditions in the river where the site discharge is received and an assessment of other sources of contamination in the area.

The preliminary conclusion is that, even though the site contains potentially hazardous materials, and even though some of these are detected at rather high concentration in groundwater, based on these screenings and the available data, the site represents no risk for the persistent organic compounds. There may be some risk from some of the inorganic substances, but further investigations, including site-specific biological data, are required.

These studies may contribute to a site rehabilitation plan, should such a plan be required.

Case Study 10.3. An environmental risk assessment for ditallow dimethyl ammonium chloride in the Netherlands

The cationic surfactant compound, ditallow dimethyl ammonium chloride, has been used worldwide as a primary agent in fabric softeners for at least 15 years. The wastewater from laundry that is discharged into sewers will, therefore, contain this substance. The chemical is rather insoluble ($0.52\,pg\,L^{-1}$). Commercial ditallow dimethyl ammonium chloride is synthesised using a procedure that results in a product that contains from 5 to 12% monotallow dimethyl ammonium chloride. The commercial product has rather similar toxicities toward bacteria, algae, invertebrates and fish. Toxicity is affected by organic carbon (ameliorative) and suspended solids (ameliorative).

A risk assessment for the aquatic ecosystem in the Netherlands was made, using a model to estimate the dilution and in-stream removal of ditallow dimethyl ammonium chloride measurements of the substance in surface waters and chronic toxicity studies on an invertebrate and a green alga. The study reported here (Versteeg et al., 1992) is one that followed an earlier risk assessment based on U.S. usage conditions and acute and chronic toxicity tests.

Exposure of organisms in the aquatic system to ditallow dimethyl ammonium chloride was approached by first developing a model to determine the concentrations of ditallow dimethyl ammonium chloride in surface waters. The overall usage (tonnage) and actual per capita waste flow was modelled based on 1990 data. Removal rates in wastewater treatment plants with dilution, in-stream removal, and background ditallow dimethyl ammonium chloride concentrations were incorporated into the model. Based on this model, the 90th percentile river concentration of ditallow dimethyl ammonium chloride below waste wastewater treatment plants was estimated at $0.021 \, mg L^{-1}$, a value that agrees well with the maximum measured concentrations in Dutch surface waters.

The toxicity of commercial ditallow dimethyl ammonium chloride has been reported in a number of studies, with effects levels for acute test ranging from 0.55 to $3.0 \, mg l^{-1}$ for fish, 0.16 to $11.3 \, mg L^{-1}$ for invertebrates, 0.4 to $31.9 \, mg L^{-1}$ for algal assemblages, and $10 \, mg L^{-1}$ for surface water bacterial communities. Pure ditallow dimethyl ammonium chloride was about six times less toxic to fathead minnow than were the commercial products. This was attributed to the fact that commercial ditallow dimethyl ammonium chloride contains monotallow trimethyl ammonium chloride, which may be responsible for part of the toxicity reported for commercial product.

Because of the insolubility of ditallow dimethyl ammonium chloride, many toxicity tests have incorporated a carrier solvent to aid in dosing. To make more realistic assessments of toxicity, activated sludge units were constructed (laboratory scale), ditallow dimethyl ammonium chloride was introduced into the units, and the effluent from these units was used for acute and chronic tests on the invertebrate *Ceriodaphnia dubia* and the green alga *Selenastrum capricornutum*. The tests properly included measurements of the actual concentrations of ditallow dimethyl ammonium chloride and monotallow dimethyl ammonium chloride. The NOEL was $4.53 \, mg L^{-1}$.

For estimates of exposure, the risk assessment used the environmental fate model described earlier with appropriate confirmation using data from monitoring of actual waters, and calculated safety factors, which in this sense are simply the ratios of the no-effect value to the surface water concentration. At the 90th percentile of concentration, the safety factor was 215, whereas at the 50th percentile, the safety factor was 647.

It is always desirable and reasonable to relate the conclusions from the risk assessment to real-world situations. The highest measured value for ditallow dimethyl ammonium chloride in the Netherlands was $0.060 \, mg L^{-1}$. This would have a safety factor of 76.

The study concludes that ditallow dimethyl ammonium chloride "poses an extremely low risk to aquatic life and that its use can be regarded as safe in the Netherlands".

10.7 References

Barnthouse, L. W., and II Suter, G. W. (1986) User's manual for ecological risk assessment. Prepared for U.S. EPA, Office of Research and Development, Oak Ridge National Laboratory, Oak Ridge, TN.

CCME. (1996) A framework for ecological risk assessment: General guidance. Canadian Council of Ministers of the Environment, CCME Subcommittee on Environmental Quality Criteria for Contaminated Sites, The National Contaminated Sites Remediation Program, Winnipeg, Manitoba.

Chapman, P. M., Fairbrother, A., and Brown, D. (1998) A critical review of safety (uncertainty) factors for ecological risk assessment, *Environmental Toxicology and Chemistry*, **17**, 99–108.

Environment Canada. (1996) Ecological risk assessment of priority substances under the Canadian Environmental Protection Act. Guidance Manual Draft 2.0, Environment Canada, Chemical Evaluation Division, Commercial Chemicals Evaluation Branch.

European Centre for Ecotoxicology and Toxicology of Chemicals. (1993) Environmental hazard assessment of substances. Technical Report 51, ECETOC, Brussels, Belgium.

Fava, J. A., Adams, W. J., Larson, R. J., Dixon, G. W., Dixon, K. L., and Bishop, W. E. (1987) Research priorities in environmental risk assessment. Publication of the Society of Environmental Toxicology and Chemistry, Washington, DC.

Glickman, T. G., and Gough, M. (1990) *Readings in Risk*, Resources for the Future, Washington, DC.

ICME. (1995) Risk assessment and risk management of non-ferrous metals: Realizing the benefits and controlling the risks. International Council on Metals and the Environment, Ottawa.

La Tier, A. J., Mulligan, P. M., Pastorek, R. A., and Ginn, T. C. (1995) Bioaccumulation of trace elements and reproductive effects in deer mice (*Peromyscus maniculatus*). Proc. 12th Annual Meeting American Society for Surface Mining and Reclamation, Gillette, Wyoming, June 3–8, 1995, pp. 3–14.

London Conference. (1987) Second International Conference on the Protection of the North Sea. London.

MOE. (1997) Guideline for use at contaminated sites in Ontario, June 1996 and errata to update for Feb. 1997. Queens Printer, Ontario Ministry of the Environment, Toronto.

National Research Council. (1994) *Science and Judgement in Risk Assessment*, National Academy Press, Washington, DC.

Norton, S., McVey, M., Colt, J., Durda, J., and Henger, R. (1998) *Review of Ecological Risk Assessment Methods*. Prepared for U.S. EPR Office of Planning and Evaluation by ECF, Inc., Fairfax, VA.

Pastorok, R. A., LaTier, A. J., Butcher, M. K., and Ginn, T. C. (1995a) Mining-related trace elements in riparian food webs of the upper Clark fork river basin. Proc. 12th Annual Meeting American Society for Surface Mining and Reclamation, Gillette, Wyoming, June 3–8, 1995, pp. 31–51.

Pastorok, R., Ruby, M., Schoof, R., LaTier, A., Mellot, R., and Shields, W. J. (1995b) Constraints on the bioavailability of trace elements to terrestrial fauna at mining and smelting sites. Poster text, Society for Environmental Toxicology and Chemistry Annual Meeting, Vancouver, BC, November 5–9, 1995.

Suter, II, G. W. (1990) Environmental Risk Assessment/Environmental Hazard Assessment: Similarities and Differences. In *Aquatic Toxicology and Risk Assessment*, eds. Landis, W. G., and van der Schalie, W. H., ASTM STP 1096, pp. 5–15, American Society for Testing and Materials.

U.S. EPA. (1992) Framework for ecological risk assessment. Risk assessment forum, February 1992. EPA/630/R-92/001, U.S. EPA, Washington, DC.

Versteeg, D. J., Feijtel, T. C. J., Cowan, C. E., Ward, T. E., and Rapaport, R. (1992) An environmental risk assessment for DTMAC in the Netherlands, *Chemosphere*, **24**, 641–62.

Whittaker, J., Sprenger, J., and DuBois, D. (1998) Assessing the key components of credible risk assessment for contaminated sites, *Environmental Science and Engineering*, **11**, 102–3.

10.8 Further reading

Risk analysis: Methods of calculation for worst case scenarios: Environment Canada. 1996. Ecological Risk Assessments of Priority Substances under the Canadian Environment Protection Act. Guidance Manual Draft 2.0, Environment Canada, Chemical Evaluation Division, Commercial Chemicals Evaluation Branch.

11

○ ○

Recovery, rehabilitation, and reclamation

11.1 The context for site contamination and recovery

The present chapter deals with attempts to improve the condition of some historically contaminated sites, dealing mainly with technical or scientific aspects, but issues of policy also arise. In Chapters 6–8 we described the contamination of air, soil, and water by potentially toxic substances. We also pointed out in several places that, until relatively recently, there was a tacit acceptance of some degree of change in or damage to the environment resulting from industrial, agricultural, and other human activities. In the complex issues of Chapter 9, reference was made to the fact that environmental contamination usually involves a combination of several substances. While recent practice aims to minimise the contamination of the environment or, at the very least, provides for a plan to include the restoration of a site at the close of a project, the legacy of past activities in both developed and less-developed countries is manifest in the form of orphan sites that may be described as historically contaminated.

For the purposes of restoration or clean-up of contaminated sites, it is obviously necessary to have some means of defining what is meant by a contaminated site. As a starting point, we could consider as contaminated any site that has received substances that were either not previously present in that site or, as a result of some activity, are at significantly higher concentrations than were normal for the site. Current attitudes would imply that such sites should be restored to a more acceptable condition, but the definitions of contaminated, restored, and acceptable condition are by no means clear-cut, nor universally agreed upon. Nor is there agreement in terms of policy concerning the restoration of contaminated sites.

It should be appreciated at the outset that it is not going to be possible to bring back all such sites to a "clean" condition. Even if one could define "clean", or know the characteristics of the original conditions of a site, the technical and economic considerations would often be so overwhelming as to render the task essentially impossible. Nevertheless, the clean-up, restoration, or rehabilitation of contaminated sites has been and continues to be attempted, often with acceptable results.

The present chapter looks at various ways in which contaminated sites have been restored or rehabilitated, with some degree of what might be termed recovery. At this stage, we shall refer to any type of recovery as clean-up, without specifying details. If we accept this, then as part of the more manageable, but still overwhelming task, it seems prudent to determine the following:

- Which contaminated sites are likely to result in *exposure* of living organisms, at any level, to potentially harmful substances?
- Which sites that are likely to result in exposure are also likely to represent a real *hazard* to living organisms?
- What is the current and future *use* of a given site?
- What *technical means* are available to remedy any potential risk that has been identified?
- *Who determines* if a site requires clean-up?
- *Who is responsible* for the site if it is determined that it requires clean-up?

11.2 Exposure and hazard

The most usual course leading to a decision concerning clean-up of a site is the identification of, or the suspected or anticipated occurrence of, contaminants on a site. This discovery may occur accidentally (i.e., the discovery of contaminants on a site or their effects is unplanned). Alternatively, the history of the site may be such that contamination is anticipated, and a systematic study may reveal the existence of contaminants. In either case, the public may be instrumental in bringing about further investigations, or the private or public sector may address contamination independently of public concern.

It is reasonable to concede that, in a practical sense, the mere existence of contaminants on a site should not trigger a clean-up. In a proactive mode, it is necessary to show that organisms are likely to be (a) exposed to and (b) affected by the contaminants.

Sometimes there is evidence of overt damage from a contaminated site: If such damage has been demonstrated, the site will already be a candidate for some kind of clean-up, or restricted use, although causal relationships still need to be established and exposure and hazard assessment will still be required. More frequently though, there is less overt evidence of impact. Where contamination has been identified or is even suspected, then, in current thinking, some kind of risk assessment is the most frequent approach for further consideration. Earlier attempts to determine the need for clean-up were often based on concentrations (e.g., the suggested use of a single value guideline for lead in soil to protect young children). Lead is used as an illustration here and in Case Study 11.5, but it exemplifies many of the same problems that are encountered for other contaminants. In the specific case of lead, the use of a single value for clean-up of lead in soil was considered unrealistic for a number of reasons, including the environment of the population at risk, which may include other sources of the same contaminant, the nature of the potential exposure, and the bioavailability of the contaminant (Wixson and Davies, 1993). The so-called trigger values for contaminants in soil should be seen as indicators that further assessment is required, and not as indicators that action needs to be taken in the form of clean-up.

Risk assessment is discussed in Chapter 10. Essentially, the procedure following the identification of one or more contaminants in a site is that the maximum concentrations are compared with some appropriate guideline or trigger values, and contaminants of concern are identified. Then the risk assessment should identify potential migration routes from the site and potential routes of exposure for humans and other biota. Where possible, chemical transformations that are likely to occur in the course of the migration and exposure, and their implications for bioavailability, should be incorporated. Where there is uncertainty, the most conservative approach is to assume 100% availability of the contaminant.

If exposure is likely to occur, then an exposure concentration is calculated, based, as far as possible, on real values obtained in measurements from the site, but models may also be acceptable. Then the risk to the most sensitive organism in the system is evaluated.

If this process indicates that there is no risk to biota, then normally no further action will be required in terms of clean-up. Routine monitoring of the COCs in the site and leaving the site may be required, and land use may be restricted. For example, if the risk assessments of contaminants in an abandoned landfill were based on the fact that groundwater was not being used as a source of drinking water, then obviously the risk assessment would not be relevant if, at a later date, there were a proposal to use the groundwater as a potable source.

11.3 Site use

For most potentially contaminated sites, the land or water is going to be "used" or "managed" in some way. In the preceding example, groundwater is assessed differently depending on its intended use as potable or nonpotable water. For some substances, the criteria for nonpotable water may be as stringent or more stringent than for drinking water: The nonpotable water may impact an aquatic ecosystem where organisms require more protection from a given contaminant than do humans drinking the water.

For land that is to be used for activities such as playgrounds and single-dwelling residences with backyards, or for growing crops or raising animals, the soil conditions would normally need to meet more rigorous conditions than would land that was destined to be paved over, for example, for roads or a parking lot. These uses minimise human contact. But subsequent change in land use may also occur, bringing back the risk of more contact of living organisms with the water or soil. In this context, it is not unusual to find that, following a risk assessment, the use of the land or water is restricted to avoid exposure to humans or other biota. It is current practice in many developed countries to describe the previous use of a site, with any limitations as to its safe use, on the title deed. This protects future buyers from finding out too late that their site is not suitable for their planned use and also protects users from inadvertent exposure. These practices are relatively recent and apply particularly to industrial or commercial sites that are being decommissioned. One might also question whether "old" agricultural sites, when land use is changed (e.g., to residential) should be treated in the same manner. Such an issue would depend on the use patterns and persistence of agrochemicals such as pesticides and, for example, the degree of groundwater contamination (Section 9.7.4).

11.4 Technical approaches

Most methods that have been applied to clean up or rehabilitate contaminated sites involve the expertise of engineering combined with geochemistry, hydrology, ecology, and human and environmental toxicology. Methods can be categorised as

- Removal of the source of contamination;
- Restriction of site use;
- Reconstruction of the site;
- Removal of the contaminated material;
- On-site containment;
- In situ treatment.

These approaches are not mutually exclusive, and often some combination of approaches will be used, as is illustrated in the case studies.

The essence of each of these methods is covered next.

11.4.1 Removal of the source of contamination

Removal of the source of contamination is based on the assumption that the system has not been irreversibly damaged by the contaminant, and that, therefore, cutting off the source will allow the system to recover, ideally without intervention. The most successful of these types of recovery are those with a clearly established cause-effect relationship between the contamination and the receptors. Indeed, control or elimination of the source may have political and economic implications and may be resisted so that the scientific evidence for cause and effect is particularly important. Reference was made in Section 4.7.2 to the whole-system manipulation of nutrient additions in the ELA, an experiment that elegantly demonstrated the key role of phosphorus in eutrophication and paved the way for regulation of the sources of phosphorus entering lakes. An early approach to recovery, resulting from the removal of the source of contamination, is seen in the work on Lake Washington by Edmondson and Lehman (1981). These workers showed that the diversion of the discharge from municipal sewage treatment plants, which had previously enriched the nutrient status of the lake and resulted in eutrophication, permitted some degree of recovery to the previous condition. The diversion began in 1964. After 14 years, the concentration of nutrients in the lake had decreased to a level where the phytoplankton community recovered to something resembling its former state. Total algal biomass decreased, the algal blooms disappeared, and the blue-green bacteria, which are indicators of eutrophication, were no longer dominant.

Some of the St. Lawrence Great Lakes have been affected by nutrient enrichment: Lake Erie has been the most affected. Case Study 11.1 provides more detail of the condition and recovery of Lake Erie. The River Thames in England has also shown the capacity of a water body to recover after the removal or control of the source of contamination (Case Study 11.2). Other examples of recovery after removal of the source(s) of contamination can be seen with the abatement of sulphur dioxide emissions and the resulting deacidification of lakes (Case Study 11.3). The Florida Everglades system (Case Study 9.1) is an example of a large-scale remediation effort currently in progress.

In the Chesapeake Bay system, state agencies in the two bordering states on the eastern seaboard of the United States, Maryland and Virginia, report good progress in meeting year 2000 targets for point source phosphorus removal, and some major sewage treatment facilities, such as the Blue Plains sewage treatment plant serving the city of Washington, D.C., are now well advanced in implementing (more expensive) nitrogen removal by tertiary treatment. Less easily controllable are nonpoint source (e.g., agricultural, atmospheric) inputs, where longer term remediation goals (10–20 years) are proposed. Even so, small-scale reversals of submerged aquatic vegetation decline (see Case Study 4.5) have been coupled with improvements in water clarity in some areas, and a major initiative is in place to improve the oyster population (and the water filtration capacity) by leasing areas for oyster culture.

Most examples of successful recovery following abatement or removal of the source of contamination apply to water bodies. This is because of the nature of aquatic ecosystems such as lakes and especially rivers, from which the water at any point in time is eventually flushed out and replaced. The time period required for replacement will obviously vary considerably depending upon the hydrological characteristics of the system. Furthermore, if the contaminant is persistent and accumulates in the sediments, the recovery may take much longer than anticipated. Even if the water has been replaced, the sediments may remain as an internal source of contamination, until such time as the layer containing the contaminant(s) is buried under more recently deposited sediment.

As an example of poor recovery following removal of the source of contamination, one can cite the terrestrial ecosystem near the Palmerton smelter in

Pennsylvania, where the major contaminants were metals, deposited from emissions. The smelter closed in 1980 after 60 or more years of operation. There had been severe damage to soil flora and fauna, mainly related to high concentrations of zinc (Zn), lead (Pb), and cadmium (Cd) in soils. There was also extensive habitat destruction. Vertebrates were affected by uncontrolled emissions (Beyer et al., 1985). Studies made 7 years after the closure of the smelter showed continuing lack of plant cover within a 5-km radius from the original air pollution source, with poor faunal species richness, attributed in part to habitat destruction, which had not recovered at that point in time (Storm et al., 1993). The longer term recovery of this site would be of interest because it is a terrestrial site and because of the nature of the contaminants. Metals are likely to persist in undisturbed soil, and there are no processes in the terrestrial systems comparable to the flushing of lakes or the burial of sediments.

In conclusion, it appears that successful recovery following the removal of the source is likely to occur (a) for aquatic, dynamic systems; (b) where cause and effect are clearly understood; (c) where the source is a discrete point source or a known type of process, rather than a diffuse source such as long-range atmospheric transport.

Even with all of these conditions, recovery is typically slow and may not be complete.

11.4.2 Restriction of site use

In Section 11.3, we referred to the importance of site use in the context of clean-up and recovery. After a site has been designated as contaminated, the situations may be such that limited or no acceptable clean-up is possible. This may be for technical reasons. Certain types of contaminants and some types of sites are simply not amenable to remediation with current methods. Even if the technology is in place, clean-up may be hampered because of some combination of economic concerns, and the identification of the responsible party (see Section 11.6). Technical solutions may be available but may be prohibitively expensive. The original owner who was responsible for contaminating the site may no longer be traceable, and the municipality inherits the problem. Taxpayers' money has to be spent with regard to priorities. In such cases, the site may be left untreated, the decision having been made that restriction of access and/or use may be the only reasonable alternative at the current time. Such sites may represent a security problem, and often the decision needs to be revisited at a later date.

In other cases, restricted use of a site may be an acceptable solution. Restriction of use would appear to be less problematic in remote areas where land is not at a premium; yet, even if human use of the site can be restricted, there remains concern for the migration of contaminants and for the exposure of wildlife to potentially harmful materials. Ideally, site-specific considerations must always be taken into account.

11.4.3   Reconstruction of the site

Ingenious engineering schemes for physical and biological site reconstruction have been devised for contaminated sites. The approaches range from the physical diversion of flow of surface or groundwater, often by the installation of barriers, through the placing of impermeable clay covers to shed surface water and prevent its infiltration through the soil, to the rebuilding of entire ecosystems. Details of the various engineering constructions are outside the scope of this text. But Case Study 11.4 provides as an example of ecosystem reconstruction. A mine tailings area (see Sections 9.2.4 and 9.2.6), almost devoid of vegetation, acidic and of poor nutrient status, was revegetated following a series of careful plot tests, by establishing a grass cover crop, with the addition of fertiliser and lime. Natural succession followed, with invasion of native flora and fauna, which subsequently became established as part of the "new" ecosystem.

11.4.4   Removal of the contaminated material

Perhaps the most direct approach to "offending material" is to remove it from the place where it is potentially damaging. During the early stages of the U.S. Superfund (see Section 11.8) remedial actions, the most common technological response was excavation of contaminated soils and buried wastes, followed by their placement in a new location. Later, with the amendments to the act in the United States, the U.S. EPA was required to select remedial treatment that "permanently and significantly reduces the volume, toxicity or mobility of hazardous substances, pollutants, and contaminants" (Hrudy and Pollard, 1993). Removal may be appropriate for small volumes of concentrated material, but the concept is fraught with problems. If the material has been determined to be hazardous, then the risk resulting from exposure during the removal process is high for workers, for the immediate ecosystems, and possibly for more distant systems as well. Contaminants may become mobilised during the removal process and may even change their form, becoming more hazardous than before (e.g., reduced, insoluble compounds may be oxidised on exposure to the atmosphere; organic compounds may be volatilised). Even if safe removal can be assured, it is necessary to dispose of the hazardous material subsequently. The selected site, if one can be located, may involve transportation of the material, which introduces new risks of accidents and spills. The removal solution is almost always very costly.

Removal of water by pumping is a somewhat less controversial procedure than the removal of solid material. For contaminated groundwater, in combination with barriers to prevent movement of water off-site, water can be pumped and transported to the most appropriate treatment or disposal operation, which may turn out to be the municipal sewage treatment plant. Depending upon the volumes of water and the nature of the contaminants, whether persistent or not, and whether likely to inhibit the bacterial activity in the STP, pumping and removal, combined with off-site treatment, can be a temporary or permanent solution for contaminated water. But it too is costly and suffers from the same criticism as the solid waste option, namely that the disposal problem is often a major one.

Nevertheless, in spite of all these potential drawbacks, removal of contaminated material has been carried out for certain contaminated sites. Again, site-specific studies are needed. Case Study 11.5 describes the removal of lead-contaminated soils in the city of Toronto.

11.4.5 On-site containment
Containment technologies overlap to some extent with those that we have categorised in Section 11.4.3 as reconstruction of the site. The objective is to contain and immobilise hazardous material on the site, thus obviating the controversial aspects of removal or the need for subsequent treatment, ideally leading to a safe, "walk away" situation. Isolation of hazardous material can be accomplished by covering (capping) the material, or inserting low permeability walls. Solidification or vitrification can also accomplish isolation and containment; for example, for high-level radioactive waste, vitrification is one of the preferred options.

11.4.6 In situ treatment
As expressed by Hrudy and Pollard (1993), "only destruction options provide for the true elimination of the hazardous material". Complete destruction is only feasible for organic contaminants, although inorganic contaminants can, in theory, be rendered less available or less toxic by in situ treatment. Organic contaminants can be destroyed by heat, but the economics of incineration are not favourable, particularly when typically only a very small proportion of the waste is organic, and large volumes of inorganic matter have to be handled. Incineration also tends to be unpopular with the public, who are concerned about air pollution.

The most promising techniques for in situ treatment involve bioremediation. Bioremediation capitalises on the capacity of microorganisms to destroy hazardous contaminants. This method is, of course, the normal function of decomposers in ecological systems: The earliest engineering application of the metabolism of microbial decomposers was in sewage treatment. Today, secondary treatment involves activated sludge, which was invented in the 1800s. More recently, treatments for other types of waste organic materials such as petroleum wastes were engineered on similar microbial principles.

Microbiologists have shown that naturally existing bacteria have the capacity to break down many complex organic substances, including those that are normally persistent in the environment. This phenomenon has been referred to briefly in Section 4.3.2, in the context of mechanisms for tolerance. Strains of bacteria have been isolated from water, sediment, and soils and have the capacity to attack some of the bonds of the organic compounds. After oil spills, yeast and bacteria that can break down fractions of the spilled oil into simpler and often less harmful or harmless products have been isolated from the spill sites (see Section 9.8). The application of this type of capacity to remediate contaminated sites is an attractive one.

There are still considerable limitations to the practical application of microbial metabolism in the bioremediation of contaminated sites. The following conditions are the minimum for a bioremediation process to function effectively.

- The substrate must be amenable to microbial degradation. Complete mineralisation of the substrate may not occur, and the conversion products may themselves be harmful, as in the conversion of 1,1,1-trichlotroethane to vinyl chloride, or may be resistant to further degradation.
- There must be an appropriate and active microbial population. Even though a microorganism may decompose a substance under controlled, laboratory conditions, this does not guarantee that the same type of activity can and will occur at the site. Some decomposers of organic contaminants are poor competitors when confronted by the diverse community of organisms that exist in the field.
- The contaminants must be available to the microbial community. In soil or sediment, contaminants may partition into natural organic matter, which is usually water miscible, but if the contaminants partition into nonmiscible organic waste, they may be rendered inaccessible to microorganisms.
- Waste-specific constraints must be minimised. Weathering processes such as evaporation, photolysis, and hydrolysis, which typically occur in wastes that have been exposed to soil over long periods of time, may decrease the concentration of easily degradable contaminants in favour of refractory residues that resist microbial attack. Waste characteristics such as high salinity or high metal concentrations may inhibit microbial metabolism.
- Environmental conditions must be appropriate. Soil temperature, redox potential, nutrients, humidity, pH, and the availability of electron acceptors all influence microbial activity. Microorganisms have their own range of these factors under which they can operate. In the field, conditions may not be favourable for the microbes of interest (Hrudy and Pollard, 1993).

11.5 Remedial action plans

The remedial action plans (RAPs) for various areas around the St. Lawrence Great Lakes deserve mention at this point. During the 1980s studies were made on various waterfront systems on the Great Lakes, and some 43 sites were designated for remedial action. Even though the specific issues that required remediation varied among sites were typically complex and involved several types of problem, contamination usually figured prominently. The sites were in both the United States and Canada and included urban and some more remote areas, with seven sites on Lake Superior, ten on Lake Michigan, four on Lake Huron, nine on Lake Erie, eight on Lake Ontario, and five on connecting channels including the Niagara River and the St. Lawrence River.

A remedial action plan is being developed for each area; some have already shown marked improvement. For Green Bay, Wisconsin, on Lake Michigan, there were persistent water quality problems in certain parts of the Lower Bay and the lowest seven miles of the Fox River system and these were designated as areas of concern (AOCs). The Lower Green Bay RAP was the first RAP prepared in the Great Lakes basin. It was first adopted in February 1988 and provides a long-range strategy for restoring Lower Green Bay and the Fox River system. Issues included nutrient loading and potentially toxic substances as well as the introduction of exotic species and habitat loss through changes in land use and drainage. The area is rather typical of the parts of the Great Lakes that have undergone industrial and urban development. In 1988, 120 remedial actions were recommended, and 38 of these had been implemented and another 57 had been initiated by 1993. Actions included controlling toxic substance loading from point sources, but nonpoint sources, including contaminated sediments, have not been reduced significantly. Details of the problems – the remedial action plan, the strategy, and the progress – are provided in Lower Green Bay Remedial Action Plan (1993).

Other sites for which RAPs have been developed include the Toronto waterfront and Collingwood Harbour, both on Lake Ontario, Presqu'Ile Bay on Lake Erie, and Cornwall/Massena on the St. Lawrence River (St. Lawrence RAP Team and St. Lawrence Public Advisory Committee, 1994). Remediation plans are multi-faceted, each employing one or more of the six approaches listed in Section 11.4.

11.6 Responsibilities

Some of the early discoveries of contaminated sites carried with them the sense of Caveat Emptor, let the buyer beware, because the current land owner would be responsible were contamination discovered on his or her land; the current owner was often the innocent victim of past contamination. Land purchased with a given use in mind, such as development for housing, or for horticultural or agricultural use, might subsequently be shown to be contaminated. Then the owner would either be responsible for clean-up or would find that his or her investment had lost much of its value. Another buyer of land similarly contaminated might escape detection, simply because there was no systematic assessment of site contamination prior to the 1970s, and discovery often depended on a chance event or the memory of a diligent member of the public.

The ad hoc approach to this type of problem has been replaced relatively rapidly, beginning in the 1980s. The regulatory aspects are dealt with in Section 11.8, but at this point it is useful to point out the main parties who are responsible for clean-up of contamination.

For any change in land use, which will normally require a change in zoning or some type of permit, in developed countries, it is likely that site conditions with respect to contamination will be assessed. If the history of the site is known, then the focus for the type of contaminants and their location may be clear. If contamination is historical, then the current owner will be held responsible, thus the Caveat Emptor warning. The owner may decide to sue if he or she feels that the seller withheld information, but that type of action is not within the present topic.

If the site is being sold in recent time, and it has been used for activities that might have caused contamination, then decommissioning will now include assessment of contamination and clean-up, and clean-up will be the responsibility of the seller. The title deed of the property will normally now show the history of contamination, along with clean-up measures, tests, and the like that have been made.

If a site that is municipally or state or federally owned is suspected of being contaminated, then normally the public authority will be responsible for providing an assessment, which will normally be reviewed by the appropriate regulatory agency. If the site is judged to be contaminated, then the public authority will also be responsible for clean-up, even if there is no change in land use, and even if the site is not changing hands.

The expenses involved with assessment and clean-up, for the private owner or for the municipality, can be extremely high.

The delineation of responsibility or liability for clean-up has been rather clearly defined for the United Kingdom (see Section 11.8).

11.7   Routes for recovery

It can be argued that the "ideal" type of recovery results in the return of a site to its original condition. But how is *original condition* defined? And is the original condition appropriate or needed under the current situation of the land or water? It is clear that some ecosystems are pristine, or that their pristine state is known and desirable; nevertheless, many ecosystems are, in fact, managed, and this appears to be acceptable to most people. The very term *land use* implies a degree of management. The issue of recovery or rehabilitation often becomes one of policy.

Figure 11.1 indicates in a schematic manner some of the basic routes and actions by which rehabilitation can be directed, with a variety of results. The original ecosystem box A on the far left indicates that, at some point in time, an ecosystem, natural or managed or built, existed prior to any contamination. This system may never have been described or understood, and this of itself presents a problem in terms of defining recovery. The contaminated ecosystem box B simply indicates that contamination has occurred. The diagram by-passes the risk assessment stages described earlier: It assumes that the system is changed by contamination. The most benign action then is to do nothing and to let time take its course. This path is illustrated in the top-right series of pathways. Doing nothing may permit a return to the original condition. If the contamination were a single event, in a river reach, for example, self-cleansing would occur as the contaminant washed downstream and the water was replaced with uncontaminated water. With time, the communities that had been affected might recover to their original form, and signs of the

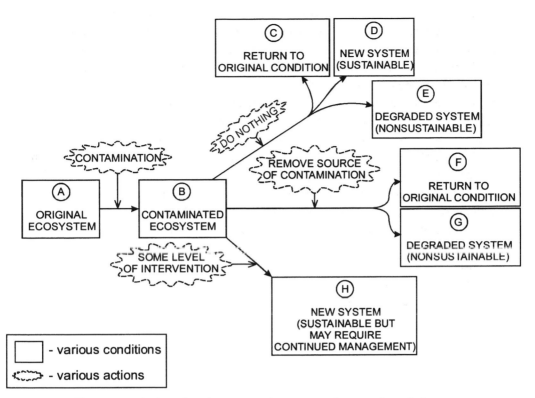

Figure 11.1 Actions for clean-up and recovery of contaminated sites.

impact would be obliterated, as shown in box C. This type of recovery from perturbation happens all the time in natural systems. Some communities are already adapted to unstable conditions. One can think of intertidal communities, desert communities, and estuarine communities where extreme fluctuations are the norm. If recovery to the original condition does not occur, a new viable system may occur with time, as shown in box D. If this is self-sustaining, it may be acceptable; indeed, it may serve the user better than did the original system. An example of this type of recovery can be seen when air pollution thins out the climax forest to such an extent that the shrub layer, formerly kept in check by the shading of the upper storey and its canopy, proliferates and becomes dominant. The tree layer may never recover once the shrub layer has taken over, particularly if the air pollution continues. The third type of outcome of contamination followed by no action may be degradation, box E, when irreversible change has occurred resulting in a system that is not sustainable. For example, the ultimate result of eutrophication can be an anaerobic system with no primary production and nothing but organic matter, the bacterial decomposition of which produces anoxic conditions in which no other organisms can survive. A truly dead system is, in fact, rather rare; even the severely impacted areas of forest around the older point sources of sulphur dioxide and other air pollutants still show remarkable survival of certain plants and animals. But the simplification of systems sacrifices ecological redundancy, which provides a measure of resilience against subsequent insults. The simplified weakened system may succumb to an additional stress, whereas the same stress may not be so damaging to a diverse system.

The second course of action depicted in Figure 11.1 is the removal of the source of contamination, without intervention. The best result of this is complete recovery, box F, but as described in Section 11.4.1, this remedy may not occur over an acceptable period of time, in which case a degraded system, box G, would persist, with or without continued contamination.

The third course of action is described as "some level of intervention", leading to the establishment of a sustainable system but one that is different from the original, box H. Normally, this action implies that a choice has been made, often in the interests of the needs for a particular type of land use, or in the interests of economy. One can think of the construction of a golf course or a park or other recreational facility over the site of a former waste dump, where the original vegetation was forest. Recovery to the original state may be completely impractical; furthermore, it may not be desirable. If the area is urbanised, then an urban park may be an appropriate land use. It should be obvious that the new system is, frequently, not strictly self-sustaining, but it can be sustained with continuing intervention. This type of "recovery" is often the most practical solution for waste dumps in urban areas.

The type of action shown as the lowest arrow in Figure 11.1 is major intervention, which could involve any combination of the approaches described in Section 11.4. With the goal of returning the site to its original condition, box J, the actions would be based on a decision that this was desirable. Possibly the site under consideration might have unique value as a habitat or as a heritage site. Whatever the reason for this type of intervention, the actions would be site-specific.

11.8 Recent regulatory approaches to contaminated sites

During the 1970s, in North America and elsewhere, situations were revealed in which innocent parties had been exposed to hazardous substances in sites where potentially harmful chemicals had been improperly disposed of. The Love Canal in Niagara Falls, New York State, became almost a household name symbolising the mismanagement of hazardous waste (Hrudy and Pollard, 1993). In this case, approximately 20,000 tonnes of waste from a chemical manufacturing plant had been buried in metal drums during the 1930s and 1940s on land that was subsequently sold for a nominal sum to the Board of Education for the construction of a school. Localized health problems associated with chemicals leaking into the groundwater led to the designation of the community as a national disaster area in 1977 with the subsequent evacuation of more than 500 homes (Epstein et al., 1982).

Table 11.1. Estimates of contaminated sites in North America and parts of Europe

Country	Number of contaminated sites[a]	Estimated high-risk sites (where estimates available)
USA	33,000	1,200
Canada	10,000	1,000
UK	100,000	Ne[b]
Denmark	20,000	Ne[b]
Finland	20,000	100–1,000
France	Nd[c]	100 (priority sites)
Germany	100,000	Ne[b]
Netherlands	110,000	6,000
Norway	2,441	61

[a] Based on Smith (1991).
[b] No estimate available.
[c] No data.

The number of contaminated sites is very large, as shown in a 1991 estimate reported by Smith (1991). These numbers are estimates, not all from identical sources, and probably not all based on identical criteria for contamination. One might expect that the estimates were also influenced by the degree of public concern or public sector diligence at the time. For example, some jurisdictions actively looked for waste sites, whereas others relied upon historical and anecdotal information. Certainly the numbers are large by any standards. Of the hundreds of thousands of contaminated sites in Europe and North America, many were also considered to be high risk. Table 11.1 shows these estimates for the United States, Canada, the United Kingdom, and some European countries.

Politicians in North America responded to the public concern for waste sites that grew in the 1970s with two somewhat different approaches. In the United States, the Comprehensive Environmental Response, Compensation and Liability Act of 1980 (CERCLA), better known as Superfund (reauthorised in 1986) was enacted. Under this act, contaminated sites are placed on a national priority list for assessment leading to decisions concerning clean-up. The Superfund approach does not use a common set of contamination criteria to determine the need for clean-up, but rather relies upon existing regulations and guidelines concerning environmental quality. Site-specific risk assessments are performed at Superfund sites, based on a prescribed set of procedures in the form of a manual. The various cases that have been assessed under the Superfund have resulted in a growing collection of published records of decisions, which provide guidance for subsequent cases.

The Superfund between 1980 and 1992 had expended some U.S.$15.2 billion (Miller, 1992). Added to this, various industries and government agencies have undertaken site remediation beyond those recorded on the national priority list, for work including decommissioning and treatment of leaking storage tanks.

Canada's approach to contaminated sites, under provincial jurisdiction, has been guided by the Canadian Council of Ministers of the Environment (CCME, 1991a, 1991b, 1996a, 1996b). Two types of criteria are recognised – assessment and remediation. Assessment criteria provide concentrations of various substances in various media for which site determinations are made. If these are not exceeded, then no action is required for that site. Remediation criteria represent higher concentrations of the various contaminants than assessment criteria, but the former criteria may still be acceptable for a given land use. These remediation criteria are "generic" in that they apply to most situations but are not necessarily appropriate for specific sites, such as those that support rare species or are acknowledged as high-class habitat or wetland. Essentially this apparent double standard is recognition that, realistically, some sites will never be "clean". Remediation criteria are the initial basis for the subsequent site-specific risk assessment, which is common to both jurisdictions. The derivations of the criteria are themselves based on a risk assessment approach.

Several of the larger provinces in Canada have developed their own guidelines for contaminated sites, most of which are quite recent. Ontario published its Guidelines for the Decommissioning and Cleanup of sites in Ontario in 1990 (OMOE, 1990), and in 1997 an extensive Guideline for Contaminated Sites, a detailed manual with criteria for many contaminants in water and soil, with a range of land uses, was published (OMOE, 1997). In addition to a choice of approaches to assessment, the guideline identifies a number of situations where sites are classified as sensitive, and these sites require specific risk assessment, regardless of their conformity to the generic criteria.

In the United Kingdom, there is no specific act like the U.S. Superfund to address contaminated land. The backbone of U.K. legislation on assessment and rehabilitation of contaminated land is Part IIA of the Environmental Protection Act 1990. In October 1999, the U.K. Department of the Environment, Transport and the Regions (DETR) issued a third and probably final draft guidance on the new contaminated land regime (*http://www.detr.gov.uk/consult*). For the preceding 5 years, work on this document was preoccupied with how to define contaminated land. In the United Kingdom, powers for dealing with contaminated land may also reside in different pieces of legislation, such as integrated pollution control (IPC), waste management licensing, or the Environment Agency's new powers to serve "works notices" to deal with water pollution from land.

Responsibility for clean-up appears to be rather well-defined in the United Kingdom. If a company contaminates its own site, then it is responsible, under various pieces of legislation, for clean-up. If a contaminated site is found and the owners are no longer in business, the land is designated as a public liability. Some funds are available to clean up these sites, especially "English Partnerships", a recently formed U.K. government agency that works with central and local government, the Development Agencies, the private sector, and other partners to bring about sustainable regeneration and development in the English regions (*http://www.cnt.org.uk/site/Default.htm*).

Local authorities in the United Kingdom are obliged to draw up lists of contaminated sites in their areas. Designating land as contaminated requires the source of contamination, identification of the target that will be affected, and a means to link source and target. However, the government's Environment Agency will be the enforcing authority for special sites, which include sites where there is acid tar, nuclear sites, military installations, use/disposal of biological weapons, land where substances can escape to water, other hazards, and European Union black list chemicals.

For the European community, there is a new Integrated Pollution Prevention and Control (IPPC), which is the 1996 EC directive on Integrated Pollution Prevention and Control. This is implemented in UK law through the Pollution Prevention and Control Act 1999. IPPC has its own requirements on site remediation.

11.9   Case studies

Case Study 11.1. The Thames Estuary: Compound pollution and recovery

The estuary of the River Thames in England changed from an "open sewer" to a waterway through which salmon move, and this was accomplished between the early 1960s and the late 1970s (Andrews, 1984). This case provides one of the best examples of how the harmful effects of "use" of a water body can be lessened (Moriarty, 1988).

The early history of the River Thames is not atypical of water courses in growing urban areas in the eighteenth and nineteenth centuries. Even though objective scientific documentation of the early condition of the river is lacking, records of the quantities and species of fish that were caught suggest that salmonid fish were still migrating through the river until 1800 (Moriarty, 1988). However, the fish populations, as shown by records of the catches of migratory fish at Taplow, upstream of the City of London (Table 11.2), declined over the period 1774 to 1821. Perhaps, ironically, the single factor that had most harmful effects on water quality was one that improved the human health condition for the City of London. The introduction of the water closet, with collection of human wastes in a sewerage connection, replaced the cesspools of the early nineteenth century, but the increased volumes of waste that were now discharged directly into the Thames, along with the storm water that the sewers were originally designed to carry, resulted in serious pollution of the water. Drinking water was still taken from the Thames and the London cholera outbreak of 1849 resulted in 14,000 deaths (Andrews, 1984).

By the 1860s, trunk sewers discharged large quantities of untreated human waste into the river at ebb tide, theoretically to prevent the sewage from being swept back towards the city on the flooding tide. These developments improved conditions in the City of London, but, within a few years, there were complaints of pollution and

Table 11.2. The River Thames: Catches of migratory fish (salmon) at Taplow, above London

Year	Number of salmon	5-year running average number of salmon
1794	15	
1795	19	
1796	18	
1797	37	
1798	16	21
1799	36	25
1800	29	27
1801	66	37
1802	18	33
1803	20	34
1804	62	39
1805	7	35
1806	12	24
1807	16	23
1808	5	20
1809	8	10
1810	4	9
1811	16	10
1812	18	10
1813	14	12
1814	13	13
1815	4	13
1816	14	13
1817	5	10
1818	4	8
1819	5	6
1820	0	5
1821	2	3

Based on Andrews (1984).

loss of fish for considerable distances downstream of the outfalls. Primary sewage treatment (removals of solids) improved the water quality, and some fish species reappeared. However, as the population of London continued to increase after the First World War, more small sewage works were built, and the oxygen demand of the effluents once again resulted in loss of fish species. Indeed, between 1920 and 1960, the eel (*Anguilla anguilla*) was apparently the only fish species that could be found through much of the tidal Thames (Wheeler, 1969).

Other physical changes and problems of water quality accompanied the loss of oxygen resulting from the burden of organic matter and nutrients that the sewage represented. These included thermal pollution, industrial effluents, and synthetic detergents (surfactants). However, it would appear that no other single factor had as

Table 11.3. Number of fish, crustacean, and jelly fish species in winter samples at West Thurrock Power Station, on the River Thames downstream of Greater London, between 1968 and 1980

Date	Winter sample, number of species (average)	Winter sample, number of species (range)
1968	2	(0–2)
1969	5	(0–11)
1970	4	(0–13)
1971	10	(0–19)
1972		
1973		
1974		
1975	24	(20–27)
1976	27	(22–31)
1977	29	(22–34)
1978	26	(18–36)
1979	28	(21–33)
1980	27	(19–32)

Based on Andrews (1984).

much influence, nor was as amenable to treatment, as the one represented by the overload of sewage effluent.

Gradually, from the third decade of the twentieth century, attempts were made to improve the treatment processes and efficiency of the sewage works. Major improvements occurred in the late 1950s, with secondary treatment (microbial breakdown of wastes), and from 1964 onwards, the estuary was no longer anaerobic (Andrews, 1984). After 1964, marine fish were again reported above the estuary. Andrews (1984) reported two phases of recovery of the ecosystem of the estuary, related to sewage treatment works situated in Greater London, south of London Bridge. The first phase followed the rebuilding of Crossness sewage works in 1964, and the second resulted from the extensions to the Beckton works in 1976.

The recovery was measured by a number of different indices, including species composition of macroinvertebrates, number of individual fish (N), number of fish species (S), fish species diversity (H'), fish evenness (J'), and fish species richness (D).

Table 11.3 shows the numbers of fish, crustacean, and jelly fish species on a per-sample basis for the years 1968 to 1980. The numbers increase between 1968 and 1971 and again in about 1976, after which they appear to stabilise.

Andrews (1984) also presented population dynamics and multivariate statistics for the data, in part to establish some cause-effect relationship between the environment and the population statistics and to attempt to quantify the relative effects of natural and pollution-related influences.

The indices N, S, H', J', and D all proved to be sensitive to seasonal change; for trends over time, the indices H' and D showed the most marked response.

"The successful clean up of the Thames shows that it is possible to redeem even an extremely polluted aquatic system" (Andrews, 1984).

Case Study 11.2. Lake Erie recovery

The five St. Lawrence Great Lakes have been subjected to various types of contamination, ranging from point sources of toxic chemicals to long-range transport of airborne contaminants from diffuse sources. Of the five, Lake Erie has been the most affected by cultural eutrophication. The symptoms in the late 1960s and early 1970s included algal blooms, hypolimnetic oxygen depletion, and the disappearance of certain benthic indicator organisms (Boyce et al., 1987). Extensive loading of phosphorus (P) in the 1950s and 1960s was identified as the principal cause of the eutrophication (IJC, 1986).

Two main management practices were employed to address the excessive loadings of P to Lake Erie. In the United States and in Canada, in the mid 1970s, municipal waste facilities were improved, to bring the municipal discharges into compliance with the $1.0\,mg\,L^{-1}$ P limitation on effluent. At the same time, fisheries management included some stocking of salmonids, following some improvement in water quality (Makarewicz and Bertram, 1991).

Changes in fish communities are another indicator of eutrophication; for example, salmonids cannot tolerate the low concentrations of oxygen that typify the condition, but coarser fish are more tolerant. However, the Lake Erie fish communities had started to change in composition in the nineteenth century, due to habitat deterioration and exploitation, and thus the decline in the Lake Erie fishery was not necessarily a simple indicator of excess P loadings. Nevertheless, following the P abatement programme in the mid 1970s, walleye abundance began to increase and was evident by the early 1980s.

The statistical relationship between phosphorus concentration and chlorophyll-*a* is referred to in Section 6.10. In Lake Erie, the point-source phosphorus load was decreased from approximately 10,000 tonnes yr^{-1} in 1972–3, to 5,700 tonnes yr^{-1} in 1977. It is not, however, axiomatic that the decrease in loading will be reflected in a decrease in total P concentration in the lake water.

Attempts to detect trends over time for a large water body such as Lake Erie encounter a number of problems (El-Shaarawi, 1987). There is high spatial variability, and seasonal cycles in limnological variables vary from year to year. Selection of unsuitable sampling sites also contributes to variability. With these problems in mind, El-Shaarawi (1987) used statistical methods to analyse trends in total phosphorus, chlorophyll-*a*, and hypolimnetic dissolved oxygen for the period 1967–80. The data indicated a decrease in the concentration of P and the concentration of chlorophyll-*a* in all three basins of the lake, over the period studied. The concentrations of hypolimnetic dissolved oxygen were related to water level and hypolimnion temperature as well as to total P. A model for oxygen indicated that, after removal of the effect of water level and temperature, the depletion of hypolimnetic oxygen was related to an increase in total P. The model also showed that for the central basin of Lake Erie, there is always a high chance of anoxia but that this chance is increased with increases in total P. Therefore, according to the model, it is possible by controlling total P loading to improve the anoxic conditions.

There is no question that ambient P levels in the open waters of Lake Erie have decreased since the mid 1970s, but whether trends in biological response parallel these trends in water chemistry has been a matter of debate. The phosphorus reduction plan was designed to decrease the nutrient that was believed to be responsible for the overgrowth of algae in the plankton. Logically then this was expected to be reflected by decreases in chlorophyll-*a* and decreases in phytoplankton biomass.

Several studies have reported decreasing phytoplankton biomass in all three basins of Lake Erie from 1970 to the mid 1980s, and several species of blue-green bacteria that dominate the phytoplankton in eutrophic conditions have shown dramatic reductions over the same period (Makarewicz and Bertram, 1991). Using a classification of trophic status based on the maximal and average phytoplankton biomass, the western basin has changed from eutrophic to mesotrophic, whereas the eastern basin has changed from mesotrophic to oligotrophic over this period (Makarewicz and Bertram, 1991).

Zooplankton can also indicate the trophic status of the lake, and changes from eutrophic indicators to those indicating more oligotrophic conditions have been recorded for Lake Erie from the early to mid 1980s. Comparable data are not available for the earlier years of the 1960s and 1970s, so detailed comparisons cannot be made. Furthermore, some of the recent changes in the composition of the zooplankton community cannot be attributed to nutrient control. The large herbivorous cladoceran *Daphnia pulicaria* was first observed in Lake Erie in 1983 and was dominant in 1984, but not in 1985. The invading species *Bythotrephes cederstroemi* is thought to be responsible for cropping the *Daphnia*. This example points to the multifaceted stresses on the Great Lakes, including not only chemical changes but physical and biological changes as well. The latter may well prove to be bringing about more dramatic and irreversible changes than the effects of toxic chemicals. For example, the nonindigenous zebra mussel (*Dreissena polymorpha*) and the closely related quagga mussel (*Dreissena bugensis*) colonized large portions of the lake subsequent to their introduction into the Great Lakes system in the late 1980s, and by 1993 *D. bugensis* comprised over 80% of the benthic biomass in some areas. This dramatically increased the filtration capacity of the lake resulting in an unprecedented degree of light penetration and significant changes in the benthic cycling of suspended particulates and associated chemical contaminants.

The data for Lake Erie fish show increases in the abundance of the piscivorous walleye (*Schizostedium vitreum*) and decrease in the planktivorous alewife (*Alosa pseaudoharengus*), spottail shiner (*Notropis hudsoniuis*), and emerald shiner (*Notropis atherinoides*). The planktivores form the primary diet of the walleye and salmonid fish in Lake Erie, so their rise may have caused the decline in the planktivores. This, in turn, may have released some of the larger-bodied invertebrates from predation pressure, leading to changes in the zooplankton communities.

Certainly the nutrient status has improved markedly, resulting primarily from decrease in P loadings, which has been justifiably described as a dramatic success

(Makarewicz and Bertram, 1991). Nuisance blooms of algae no longer occur, and crustacean biomass and species composition are indicative of meso- or oligotrophic status, all of which can be attributed to nutrient reduction. The recovery of the walleye fishery and the introduction of a new salmonid fishery have affected the trophic structure of the lake. It is concluded that the return of Lake Erie to its previous biological condition, even with control of P loading, is very unlikely to occur. Although interest focussed for a long time on the eutrophication problem, the stresses on the lake are, in fact, multiple and complex. Cause and effect have not been unequivocally established for either the losses of species nor for the partial recovery.

Case Study 11.3. Deacidification trends in Clearwater Lake near Sudbury, Ontario, 1973–1992

Clearwater Lake is a small oligotrophic water body approximately 15 km southwest of Sudbury, Ontario. Since the first two decades of the twentieth century when taller stacks were erected in Sudbury, airborne emissions from the Sudbury smelters have affected the lake with a combination of acidification and metal contamination. Although detailed historical records of the water chemistry and biota of the lake are lacking, comparisons with similar lakes that are remote from local influences support the idea that changes in chemistry were accompanied by changes in the biota. Notable in the samplings for 1973 and 1975 is the complete absence of fish, the lowered richness and biomass of zooplankton and phytoplankton, which led to shortened food chains, abnormally high biomass of a few species, and increased temporal variability in standing stock (Yan and Welbourn, 1990).

For a number of reasons, the deposition of acidic and metallic contaminants from the Sudbury complex to Clearwater Lake started to decline in the early to mid 1970s. Some of the local smelters closed in 1972; however, new stacks were built. They were so much taller than their predecessors that deposition was much more widely spread, and Clearwater Lake received proportionally less contamination. A series of temporary shutdowns of the smelters in the 1980s, related to the economic conditions as well as to labour disputes, provided a further decrease in added contaminants. Since the mid 1980s, abatement of emissions both locally and regionally has occurred (Bodo and Dillon, 1994).

For Clearwater Lake, nearly 20 years of records from 1973 to 1992 for water chemistry and biota provided an unusual opportunity to track the changes in chemistry and biology related to decreased inputs of contaminants. Thus, this case study is another example of removal of the source of contamination without any other intervention.

The decline of sulphur dioxide emissions over most of eastern North America that began in the 1970s and continued until the mid 1980s was accompanied by improved water quality and recovery of biota in the acid-stressed lakes on the Canadian Shield, including many in Sudbury (Dillon et al., 1986; Gunn and Keller, 1990). There are two related major problems in the locally affected Sudbury lakes of which acidification has arguably received the most attention. Acidification also occurs in susceptible lakes that are distant from Sudbury because of the multiple sources and

Table 11.4. Summary of emissions of sulphur dioxide in the Sudbury area 1960–90

Year	1960	1970	1975	1980	1985	1990
Sulphur dioxide, tonnes $\times 10^6$ yr^{-1}	2.55	2.45	1.45	0.5	0.5	0.7

Based on data from Bodo and Dillon (1994).

Table 11.5. Time trends in the water chemistry of Clearwater Lake

Year	1973	1976	1980	1984	1988	1992
pH	4.3	4.2	4.5	4.6	4.7	5.1
Sulphate (μeq L^{-1})	610	550	460	380	350	300
Nickel (μeq L^{-1})		270	225	200	180	75
Copper (μeq L^{-1})		110	70	50	45	18

Based on data from Bodo and Dillon (1994).

the widely ranging nature of acid deposition in eastern North America. As reported previously, deacidification has been recorded following decreases in the deposition of acidic contaminants.

The other major problem for lakes close to Sudbury is contamination by metals, notably nickel (Ni) and copper (Cu). In affected lakes, the metal contamination and acidification problems cannot be uncoupled; indeed the impoverishment of flora and fauna is almost certainly due to a combination of these stresses. Although metal deposition from Sudbury has also decreased over the past 20 years, metals in lakes are persistent, with reservoirs of metals in sediments, and it appears that recovery of the lake ecosystem from metal contamination may not occur as rapidly as the deacidification.

Data for emissions of sulphur dioxide from Sudbury are summarised in Table 11.4, illustrating that the emissions declined beginning in the 1970s and levelled off by the 1990s. Water quality over the period 1972–92 is shown in Table 11.5. Clearly, the response in terms of water quality is positive and can justifiably be referred to as recovery. Other parameters, not shown in the table, include aluminium (Al) and silicon (Si), concentrations of which have also declined, mirroring the decrease in sulphate. This reduction indicates a deceleration of geochemical weathering as acidic deposition declines (Bodo and Dillon, 1994).

Response by the biota is expected to accompany recovery of the quality of the water. As pH increases and metal concentrations decline, conditions become more favourable for aquatic organisms. However, the recovery process for biota is less well understood and is expected to be complex, not necessarily a direct reversal of the original changes. At the population level, in general, the smaller the organism and the shorter its life cycle, the more rapid will be its return once conditions are suitable. The complete loss of fish that has been described for Clearwater Lake is unlikely to be reversed, unless some refuge supplies recruits, or some deliberate intervention is taken in the form of stocking. Furthermore, major changes in trophic structure referred to in the first paragraph of this case study may mean that a return to the original trophic structure will require a very long time.

Information on the biological status of Clearwater Lake when the most recent chemistry was reported includes chlorophyll-a and Secchi depth. Except for a 1973 high of $1.25\,mg\,L^{-1}$ and a 1991 low of $0.2\,mg\,L^{-1}$, annual median chlorophyll-a concentrations have varied from 0.5 to $0.85\,mg\,L^{-1}$, with a long-term mean of $0.66\,mg\,L^{-1}$ with no appreciable trend (Bodo and Dillon, 1994). Secchi depth similarly showed no trend, with a long-term mean of 8.6 m, except for 1991 when the annual mean was 11.1 m. Dissolved organic carbon showed a small but statistically significant increase in the late 1980s and early 1990s, with a median level of $0.4\,\mu g\,L^{-1}$ in 1981–7, rising to $0.75\,\mu g\,L^{-1}$ over the period 1988–91. Bodo and Dillon (1994) conclude from these and other variables that "in the late 1980s, there was a modest increase in primary biological activity that may have fallen off abruptly in 1991".

Further monitoring of the Clearwater Lake chemistry and biology is clearly desirable. As Bodo and Dillon (1994) comment, "Clearwater Lake ranks foremost among Sudbury area sites for continued surveillance to judge the success of remedial actions implemented in Canada and the U.S. through the coming decade".

Even for this example, which would appear initially to be relatively straightforward, factors other than those related to decreased emissions of sulphur dioxide may influence the water quality of the lake. Drought has been shown to be a major influence that can reverse the expected chemical changes such as decreases in sulphate in water, albeit for relatively short periods. During periods of drought, dry deposition accumulates, while drying and exposure to air induces reoxidation of reduced sulphur that would otherwise be retained normally in saturated zones of the catchment (Keller et al., 1992).

Case Study 11.4. The Inco Mine Tailings reclamation, Sudbury, Canada: Ecosystem reconstruction

One of the legacies of the extraction and beneficiation of metal ores is the material known as mill tailings. Tailings are the waste products from the process of producing a finely ground concentrate that is feed for some types of furnaces. Typically, the tailings are composed of fine particles with little or no organic matter, very low concentrations of plant nutrients, and high mineral content, including metals. Additionally, if the original ore was sulphide, the tailings may be acid-generating. Colonisation of the material by plants is typically very sparse to nonexistent; thus, the tailings deposits remain barren for long periods of time. The bare tailings present a dust control problem, with the dust itself being potentially contaminated. There is also the risk of leaching metals and acidified water into water bodies. These properties, particularly intractable if the tailings are acid-generating, along with the typically large volumes of material, present major problems for the disposal and long-term fate of the material.

The Sudbury, Ontario, region has been and remains one of the world's most important areas for the mining and smelting of nickel and copper and the extraction of 13 other elements. As a result of these practices, many of which were carried out in an uncontrolled manner especially in the early parts of the twentieth century, the Sudbury region is "one of the most ecologically disturbed in Canada" (Peters, 1984). Inco Ltd. initiated its metal mining and processing operations in Sudbury some 90 years ago. Inco and other companies have faced a range of environmental problems including the acidification and metal contamination of water and soil at considerable distances from the smelter. In recent years, various measures have been taken to improve the quality of the Sudbury environment, incorporating abatement of the emissions of sulphur dioxide and revegetation of the natural ecosystems in the urban areas. The topic of the present case study is the rehabilitation of Inco's old tailings disposal area at Copper Cliff in Sudbury.

The tailings area in question, like many of this type, came about through the damming of a valley, followed by the deposition of a slurry of tailings into the resulting "lake". The starting dam was made from waste rock, and then its height was increased using tailings material. The deposition of tailings began in 1930, with the introduction of a new smelter that used the Herreschoff Roaster and a reverbatory furnace. At the time when the present rehabilitation began, the total area occupied by tailings was 600 ha (Peters, 1984).

Early attempts to treat the Copper Cliff tailings disposal area focussed on stabilisation of the surface, since even by the late 1930s, dust control was the immediate concern. Various efforts such as chemical sprays, limestone chips, bituminous sprays, and timed water sprays pumped through an irrigation system were tried, but all of these efforts proved to be either ineffective or uneconomical (Peters, 1984). In the mid 1940s some experimental seedings of various grasses were made, but these failed, apparently because of insufficient attention to factors such as pH control and the selection of suitable species.

In 1957 a series of experimental plots were sown, with the objective of establishing a new ecosystem, a system that would complement as far as possible the natural ecosystems in the area. The climax type for the area had been pine forest with white pine (*Pinus strobus*), red pine (*Pinus resinosa*), and Jack pine (*Pinus banksiana*), but much of the forest had been lost through logging, which began in the 1870s. Subsequently, impacts of mining and smelting referred to earlier included fumigation by sulphur dioxide from the early ground-level roasting of ore, and this also had destroyed vegetation.

As a result of the experiments and review by Inco's agricultural staff, the following principles for revegetation were assembled.

1. The initial establishment of plant communities using available species (note that considerable seed sources are needed to cover large areas, thus availability of seed was a potentially limiting factor; the seed of many suitable wild plants would simply be unavailable). The species should be tolerant of drought, of low soil pH, of poor soil texture, and of lack of organic matter and plant nutrients.

2. The modification of localised microclimates to benefit the establishment of plants.
3. The reestablishment of soil invertebrate and microbial communities to ensure natural organic decomposition, essential to the rebuilding of soil.
4. The reestablishment of nutrient cycles.
5. The establishment of a vegetative habitat suitable for colonisation by wildlife.
6. The establishment of climax plant communities for the area, by manipulating species composition.

The revegetation programme that was eventually established included seeding with a mixture of lawn and agricultural grasses including Canada blue grass (*Poa compressa*), red top (*Agrostis gigantea*), Timothy (*Phleum pratense*), park Kentucky blue grass (*Poa pratensis*), tall fescue (*Festuca arundinacea*), and creeping red fescue (*Festuca rubra*). Prior to seeding, limestone at the rate of 4.4 tonnes ha^{-1} was spread and a further application of limestone at 4.4 tonnes ha^{-1} was spread at seeding time, followed by discing and the addition of 450 kg ha^{-1} of fertiliser. Further fertiliser additions at 392 kg ha were made in the fall. Compared with agricultural practice, these amounts of lime and fertiliser are immense; of course, they reflect the initial poor condition of the substrate.

Various technical devices are being used to prepare and compact the substrate and to provide suitable contours on the site. These procedures have been applied in phases to various parts of the old tailings areas and are being gradually applied to newer areas as these cease to be active in terms of receiving fresh tailings.

Studies of revegetated areas have identified some issues requiring additional remediation, such as high surface temperatures on the waste, requiring a fast-growing companion crop; the formation of an iron pan at varying depths with rather limited penetration of the pan by plant roots, but with root penetration through cracks and root hair penetration into the pan itself in the pan; rather high rates of leaching of plant nutrients during the early years, necessitating the addition of nitrogenous fertiliser; and fixation of phosphorus by the high concentrations of iron oxides in the tailings, limiting the amount of P available for plant growth and necessitating light annual application of additional fertiliser.

Although the original plan had been to establish a sward of grass (i.e., use an agricultural approach) and to harvest the hay, the Inco researchers found that seedling trees soon became established in the grassed areas, paving the way for a return of the reclaimed area to the natural climax vegetation. By the mid 1960s, the area of grass that was being mowed was decreased and then ceased, to avoid cutting of the tree seedlings. By the late 1970s, ten species of volunteer coniferous and deciduous woody species had become established on the reclaimed site, along with dozens of native wild grasses and other herbs and moss species. A large number of bird species were also recorded using the newly established and evolving habitats on the site. The end use for the site is as a wildlife management area.

In summary, some 30 years after the initiation of the project, the following ecological processes have been established on the previously barren site.

1. Soil horizons are now visible in what was a rather homogeneous substrate, not at all comparable to soil.
2. The tailings surface, once stabilised by grass cover, provided for colonisation by native flora and fauna.
3. The voluntary invasion of trees indicates a step towards establishment of the natural climax forest that existed prior to the mining development.
4. The number of bird species nesting and using the site indicates a suitable habitat and adequate food supply for these wildlife species.
5. There is no indication that the soil and other components of the environment are toxic to any species living or reproducing on the Copper Cliff tailings.

Even though this case study can be seen as an individual success story, the experience from the research involved also has wider implications for the construction of "new" ecosystems.

As an adjunct to this case study, the reader is referred to another, more generalised rehabilitation project in the Sudbury area, in which more than 3,000 ha of roadside and other uncultivated land in the Regional Municipality of Sudbury were revegetated. Winterhalder (1996) provided a fascinating account of this landscape renewal. The author also pointed out that even though rehabilitation involved soil treatment, one of the enabling factors for the project was the abatement of ambient sulphur dioxide emissions.

Case Study 11.5. Clean-up of lead-contaminated sites: The Ontario urban clean-up experience

The sources of lead to humans and to ecosystems are numerous: the toxicology of the metal exemplifies the need for a multimedia approach to the study and management of lead exposure. Nevertheless, elevated blood lead, which is the primary exposure indicator for vertebrates, especially for humans, has implicated lead in soil as a major immediate contributor. Whatever its origin, be it airborne lead from automobile exhaust, deposition from smelter emissions, dust from commercial or industrial processes, or flaking paint, lead in soil in urban areas has been seen as a threat to the health of humans, particularly for young children.

The present case study addresses the clean-up of lead-contaminated soil in a particular area of metropolitan Toronto, Ontario, Canada. Emissions from two secondary lead smelters, which processed and essentially recycled lead from batteries, pipes, and other waste materials, had contaminated the soil of residential land (Roberts et al., 1974). Studies revealed concentrations of total lead in surface soil as high as 11,500 mg kg^{-1} (Temple, 1978) in residential areas and up to 51,000 mg kg^{-1} on industrial land (Rinne et al., 1986).

In 1974, the Ontario Environmental Hearing Board was directed by a provincial Order-in-Council to hold public hearings on lead contamination in the metropolitan Toronto area. Soil replacement was recommended by this Hearing Board, for levels at or above 3,000 mg kg^{-1} of total lead on residential properties. Recommended trigger concentrations for lead in the literature, in fact, range from 150 mg kg^{-1} as a level requiring "further investigation" to 600 mg kg^{-1} as an "indication for clean-up" (Moen et al., 1986) and a Toronto working group on lead had recommended a clean-up level of 1,000 mg kg^{-1}. A final guideline of 2,600 mg kg^{-1} was arrived at for the smelter sites, based on a confidence interval for soil lead of ±13% (i.e., 2,600).

Based on this guideline, 67 sites, including frontyards, backyards, and laneways, on 48 residential properties in the vicinity of one of the secondary smelters underwent soil replacement during 1977 and 1978. The top 15 cm of soil was replaced by clean soil (with lead concentration no more than 7 mg kg^{-1}) at a cost of CAN$67,000. There was considerable disruption and damage to gardens and fences for the residents during this period. The company, while not admitting liability, donated a sum for lead research to the University of Toronto. Clean-up would be useless if the problem that contaminated the soil in the first place were not abated. Subsequent to the discovery of the lead in soil problem, the company was in compliance with standards for lead in air.

Following the soil replacement, lead in soil was determined with the following results.

80% of the treated sites had lead <500 mg kg^{-1}.
51% of the treated sites had lead <200 mg kg^{-1}.
9.1% of the treated sites had lead >2,000 mg kg^{-1}.

Four sites had soil lead exceeding 2,600 mg kg^{-1}. Further clean-up was conducted on these properties in 1979.

Retesting in 1985 revealed that soil lead had increased slightly in some of the sites where soil had been replaced. This was interpreted as resulting from contamination of the replaced soil by residual highly contaminated soil and not from deposition of airborne lead (Jones, 1987). In contrast, some sites with no record of soil replacement appeared to show a general decrease in lead over the same 10-year period following replacement (Jones, 1987), but these sites were for the most part in the 500–700 mg kg^{-1} range, so the situation would not strictly be considered as contaminated.

Concentrations of lead in blood at the time of the discovery of the elevated soil lead in the early 1970s were unacceptably high for residents living near the smelters, in comparison with comparable cohorts of urban residents, and a few of the children were hospitalised for treatment. Yet even before the soil removal had taken place, blood leads had begun to decline (Jenkins et al., 1988). This may be related to the careful educational programme that the City of Toronto had carried out for the residents. It included instruction on exposure pathways for lead to humans and the attendant risks for persons living in contaminated areas. Precautions including washing and peeling root vegetables and extra attention to personal hygiene may have resulted in the lower exposure to environmental lead with subsequent lowering of blood lead.

This case study raises some questions about the wisdom of expensive clean-up by soil removal. The situation in terms of soil lead and blood lead is not clear-cut, either before or subsequent to soil replacement. A risk assessment was not carried out at the time of the initial investigation. Possibly in the year 2001 or later, a more conservative approach might have been taken.

11.10 References

Andrews, M. J. (1984) Thames Estuary: Pollution and Recovery. In *Effects of Pollutants at the Ecosystem Level: SCOPE*, eds. Sheehan, P. H., Miller, D. R., Butler, G. C., and Boudeau, P. H., pp. 195–227, John Wiley and Sons, New York.

Beyer, W. N., Pattee, O. H., Sileo, L., Hoffman, D. J., and Mulhern, B. M. (1985) Metal contamination in wildlife living near two zinc smelters, *Environmental Pollution*, **38**, 63–86.

Bodo, B. A., and Dillon, P. J. (1994) De-acidification Trends in Clearwater Lake near Sudbury, Ontario 1973–1992. In *Stochastic and Statistical Methods in Hydrology and Environmental Engineering*, ed. Hipel, K. W., pp. 285–98, Kluwer Academic Publishers, Dordrecht.

Boyce, F. M., Charlton, M. N., Rathke, D., Mortimer, C. H., and Bennett, J. (1987) Lake Erie research: Recent results, remaining gaps, *Journal of Great Lakes Research*, **13**, 826–40.

Canadian Council of Ministers of the Environment (CCME). (1991a) Interim Canadian environmental quality criteria for contaminated sites. Publ. CCME-EPC/CS34, Canadian Council of Ministers of the Environment, Ottawa.

(1991b) National Guidelines for decommissioning industrial sites. Publ. CCME-TS/WM-TRE013E, Canadian Council of Ministers of the Environment, Ottawa.

(1996a) Guidance Manual for Developing Site-Specific Soil Quality Remediation Objectives for Contaminated Sites in Canada. Canadian Council of Ministers of the Environment, Ottawa.

(1996b) A protocol for the derivation of environmental and human health soil quality guidelines. Prepared for the CCME Subcommittee on Environmental Quality Criteria for Contaminated Sites, The National Contaminated Sites Remediation Programme, March 1996, Canadian Council of Ministers of the Environment, Ottawa.

CERCLA. (1980) Comprehensive Environmental Response Compensation and Liability Act. Incudes Superfund Amendments and Reauthorization Act of 1986 (SARA). 42 USC Section 9601 et seq.

Dillon, P. J., Reid, R. A., and Girard, R. (1986) Changes in the chemistry of lakes near Sudbury, Ontario, following reductions of SO_2 emissions, *Water, Air, and Soil Pollution*, **31**, 59–65.

Edmondson, W. T., and Lehman, J. T. (1981) The effect of changes in the nutrient income on the condition of Lake Washington, *Limnology and Oceanography*, **26**, 1–29.

El-Shaarawi, A. H. (1987) Water quality changes in Lake Erie, 1968–1980, *Journal of Great Lakes Research*, **13**, 674–83.

Epstein, S. S., Brown, L. O., and Pope, C. (1982) *Hazardous Waste in America*, Sierra Club Books, San Francisco.

Gunn, M. J., and Keller, W. (1990) Biological recovery of an acid lake after reductions in emissions of sulphur, *Nature*, **345**, 431–3.

Hrudy, S. E., and Pollard, S. J. (1993) The challenge of contaminated sites: Remediation approaches in North America, *Environmental Reviews*, **1**, 55–72.

International Joint Commission (IJC). (1986) A Phosphorus Diet for the Lower Lakes. Executive Summary of the U.S. Task Force Plans for the Phosphorus Load Reduction to Lake Erie, Lake Ontario and Saginal Bay, IJC, Windsor, Ontario.

Jenkins, G., Murray, C., and Thorpe, B. M. (1988) Lead in Soil: The Ontario Situation. In *Lead in Soil: Issues and Guidelines. Environmental Geochemistry and Health Monographs*, eds. Wixson, B. G., and Davies, B. E., Science Reviews Ltd., Northwood, Middlesex, England.

Jones, A. R. (1987) South Riverdale soil lead levels: An explanation for the recontamination of some residential properties in the vicinity of Canada Metal Co. Ltd. Ontario Ministry of the Environment Technical Report, Hazardous Contaminants Branch, Toronto.

Keller, W., Pitibaldo, J. R., and Carbone, J. (1992) Chemical responses of acidic lakes in the Sudbury, Ontario, area to reduce sulphur emissions, *Canadian Journal of Fisheries and Aquatic Sciences*, **49** (Suppl. 1), 25–32.

Lower Green Bay Remedial Action Plan. (1993) Lower Green Bay Remedial Action Plan. 1993 update for the Lower Green Bay and Fox River Area of Concern. Wisconsin Department of Natural Resources, Madison, WI.

Makarewicz, J. C., and Bertram, P. (1991) Evidence for the restoration of the Lake Erie ecosystem: Water quality, oxygen levels, and pelagic function appear to be improving, *Bioscience*, **41**, 216–23.

Miller, S. (1992) Cleanup delays at the largest Superfund sites, *Environmental Science and Technology*, **26**, 658–9.

Ministry of the Environment (MOE). (1990) Guidelines for the decommissioning and cleanup of sites in Ontario. Ontario Ministry of the Environment, Toronto.

(1997) Guidelines for use at contaminated sites in Ontario, June 1996 and errata to update for Feb. 1997. Ontario Ministry of the Environment, Toronto.

Moen, J. E. T., Cornet, J. P., and Evers, C. W. A. (1986) Soil Protection and Remedial Actions: Criteria for Decision Making and Standardization of Requirements. In *Proceeding of First International TNO Conference on Contaminated Soil (Nov. 11–15, Altrecht, Netherlands)*, eds. Assink, J. W., and van den Brink, W. J., pp. 441–8, Martinus Nijhoff Publishers, Dordrecht.

Moriarty, F. (1988) *Ecotoxicology: The Study of Pollutants in Ecosystems*, Academic Press, San Diego.

Peters, T. H. (1984) Rehabilitation of Mine Tailings: A Case of Complete Ecosystem Reconstruction and Revegetation of Industrially Stressed Lands in the Sudbury Area, Ontario, Canada. In *Effects of Pollutants at the Ecosystem Level*, eds. Sheehan, P. J., Miller, D. R., Butler, G. C., and Boudeau, P. H., pp. 403–21, John Wiley and Sons, New York.

Rinne, R. J., Linzon, S. N., and Stokes, P. M. (1986) Clean-up of Lead Contaminated Sites: The Ontario Experience. In *Trace Substances in Environmental Health*, vol. 20, ed. Hemphill, D. P., pp. 308–21, University of Missouri, Columbia.

Roberts, T. M., Gizyn, W., and Hutchinson, T. C. (1974) Lead contamination of air, soil, vegetation and people in the vicinity of secondary lead smelters. In *Trace Substances in Environmental Health*, vol. 8, ed. Hemphill, D. P., pp. 155–66, University of Missouri, Columbia.

Smith, M. A. (1991) Identification, investigation and assessment of contaminated land, *Journal of the Institution of Water and Environmental Management*, **5**, 617–23.

St. Lawrence RAP Team and St. Lawrence (Cornwall) Public Advisory Committee. (1994) Choices for Cleanup: Deciding the Future of a Great River. St. Lawrence River Remedial Action Plan Options Discussion Paper. August 1994. Ontario Ministry of Environment & Energy and Environment Canada. Kingston, Ontario.

Storm, G. L., Yahner, R. H., and Bellis, E. D. (1993) Vertebrate abundance and wildlife habitat suitability near the Palmerton zinc smelters, Pennsylvania, *Archives of Environmental Toxicology and Chemistry*, **25**, 428–37.

Temple, P. J. (1978). Lead content of replaced soil on residential properties near secondary lead industries in Toronto, July 1978. Ontario Ministry of the Environment Report, Phytotoxicology Section, Air Resources Branch, Toronto. Cited in Rinne et al. (1986).

Wheeler, A. (1969) *The Tidal Thames. The History of the River and its Fishes*, Routledge and Kegan Paul, London.

Winterhalder, K. (1996) Environmental degradation and rehabilitation of the landscape around Sudbury, a major mining and smelting area, *Environmental Reviews*, **4**, 185–224.

Wixson, B. G., and Davies, B. E. (1993) Lead in soil: Recommended guidelines. Science Reviews, Society for Environmental Geochemistry and Health, Northwood, Middlesex, England.

Yan, N. D., and Welbourn, P. M. (1990) The impoverishment of aquatic communities by smelter activities near Sudbury, Canada. In *The Earth in Transition. Patterns and Processes of Biotic Impoverishment*, ed. Woodwell, G. M., pp. 477–94, Cambridge University Press, Cambridge.

12

○ ○

Regulatory toxicology

12.1 Introduction

Earlier chapters of this text referred to the identification of toxic substances, to various tests that identify potentially harmful concentrations of these, and to the need for regulation of toxic substances in the environment. As identified by Rand (1995), toxicity data (in his examples, aquatic toxicity data) have a variety of applications in protecting environmental quality. Examples include corporate industrial decisions on product development, manufacture, and commercialisation; registration of products to satisfy regulatory requirements; permitting for discharge of municipal and industrial wastes; environmental or ecological hazard-risk assessments; prosecution and defence of chemical-related activities in environmental legislation; and development of numerical [water and sediment] criteria for the protection of [aquatic] organisms.

This chapter deals with some of the ways in which regulatory bodies, typically governments at various levels or their agencies, attempt to protect the environment from the harmful effects of toxic substances. Environmental law has many facets and is a rapidly growing discipline. Although it is also very young in comparison with many other bodies of law, it is very extensive, and the present text cannot attempt to cover in any detail even those aspects that deal with toxic substances. The main purpose of this chapter is to present some of the principles involved in the regulation of potentially toxic substances, in the context of the protection of the environment as it affects human health and ecosystems. Occupational health is not dealt with except peripherally where the law in that area overlaps with the regulation of toxic substances in the more general environment.

As we discussed earlier, although environmental or ecotoxicology has been built in part on the older discipline of classical or human toxicology, it has distinctive features and problems that require specific approaches not found in classical toxicology. In a similar vein, we might make a distinction between the older and more developed regulatory provisions for protection of health against toxic substances in the workplace (occupational health) and the protection of human and ecosystem health against toxic substances in the more general environment. Such a set of comparisons is not merely an intellectual exercise. The more recent and less "mature" knowledge base and the generally greater uncertainties inherent in the science of environmental toxicology as compared with classical toxicology are equally prominent and significant, one might say equally frustrating, in a regulatory context. The use of risk assessment (see Chapter 10) in environmental assessment is a regulatory response to this uncertainty as well as a way in which to provide margins of safety against potentially harmful substances or events.

The technical aspects of environmental toxicology, which have formed a major component of Chapters 2–9 of this text, are integral components in the processes that lead to the regulation of toxic substances. But economics and nontechnical matters also play significant roles in the regulatory process. In the brief treatment that space permits, we address the basic categories of law that can be applied to the protection of the environment from toxic substances, the principles of environmental protection from a legal standpoint, and voluntary abatement procedures. A few basic definitions of some legal terms are provided, and finally the main statutes relating to toxic substances that currently exist in the United Kingdom, Europe, the United States, and Canada are summarised.

12.2 Possible legal approaches to the regulation of toxic substances

In 1307, the penalty in England for violating air pollution standards was capital punishment (Duncan, 1993). In the twentieth century, until the late 1960s, pollution was considered a minor, if not trivial, offence in England. Similarly, only recently did North Americans begin to consider pollution as a crime. This change of attitude is discussed in Chapter 1: A certain amount of harm to the environment was often seen as a normal and certainly acceptable consequence of some human activities.

The legal bases for addressing environmental issues include components of the common law and many statutes that specifically address the protection of the environment. These include various means by which pollution can be controlled or, more recently, prevented. Certain other statutes that might not immediately be considered as environmental law, such as those relating to transportation, nevertheless, may have provision for environmental protection.

COMMON LAW

The common law is the body of customary law, based upon judicial decisions and embodied in reports of decided cases, which has been administered by the common law courts of England since the Middle Ages. This body of law is the foundation of the legal systems now found also in the United States and in most of the member states of the Commonwealth of Nations. Common law stands in clear contrast to statute law (i.e., the acts of legislative bodies such as Parliament or Congress). In the common law, judges base decisions in part on previously decided cases.

Examples of the common law that have been applied to environmental protection include actions of negligence, nuisance, strict liability (*Rylands v Fletcher*), riparian rights, and trespass. These terms are defined in Section 12.4. Briefly, these bodies of law, developed over centuries, provide the private individual with remedies for injury to his or her own property. Negligence deals with unintended injuries to persons or property; nuisance is an activity or condition that causes an interference with the use and enjoyment of neighbouring privately owned lands, without constituting an actual invasion of the possession of the neighbours; strict liability (but not absolute liability) implies some fault on the part of the keeper of the thing in question, namely that the keeper has brought to, collected, or kept on his land something that is liable to do damage if it escapes; riparian rights embody the legal principle that anyone whose land adjoins a flowing stream or river has the right to the use of that stream in its natural quantity and quality, provided that enough is left for downstream users and the water quality is maintained; trespass is the entry on land (or water) without lawful authority, by the man, his servants, or the thing or animal in question.

Even though these long-standing common laws were not originally used in the context of environmental protection in general, nor specifically for the regulation of toxic substances, they have sufficient flexibility that actions have been brought in such cases. Trespass, for instance, is now commonly alleged in petroleum contamination cases. The judgements are always subject to the broad and flexible interpretation that is characteristic of the common law. For example, in the context of riparian rights, the secondary use of a river or stream for waste disposal would be actionable if the use were "unreasonable." The word *unreasonable* is subject to different interpretations. Some courts have interpreted the "reasonable" standard to mean almost no alteration to the quality or quantity of the water for the user downstream, whereas others have considered local, social, and economic conditions (Charles and VanerZwag, 1993).

The common law is typically used for individual plaintiffs for injuries to their own property. It does not normally give standing to classes of persons nor to non-human entities such as populations of plants and animals, communities, or ecosystems. These are implicated as possessions of the human. The effectiveness of the common law remedies often depends on the resources of the people invoking them. Private plantiffs, for instance, often lack the resources to hire the toxicologists whose testimony is necessary to quantify the damage caused by a toxic release. A good example of the difficulty faced by individuals suing corporations is the lawsuit in *A Civil Action* (Harr, 1996), in which a community sued several corporations for polluting the town water supply.

STATUTORY LAW

A statute, in contrast to the common law, is a formal expression of a rule or set of rules, made in English law by "the King, the Lords and the Commons" (Brittannica, 1965). Statute making or legislation is a more modern process than the establishment of law by custom and judicial interpretation. There is a clear distinction between common and statute law in terms of the way in which each is interpreted. As pointed out earlier, the common law has considerable flexibility. A statute is more formal; typically an "act of Parliament", it usually provides that a particular authority shall have the power to make orders, rules, and regulations in relation to particular matters arising from the Act. The interpreter of the statute has less flexibility than in the common law: He or she must be guided by the precise words used.

Since the 1970s, many statutes have been enacted with the goal of environmental protection. These laws, like many other statutes, have often evolved from experience of common law cases. For example, the U.S. Clean Water Act blends some aspects of the various common law causes of action: As in a trespass case, any addition of a pollutant to the water without permission is a violation; as in a nuisance case, the fact that others have also polluted the water is no defence.

The immediate point to be made here is that there are major differences between the common law and statutory law. Whereas cases that involve the common law can be found in the additional reading material, we concentrate on the more formal processes of setting and administering statutory law and the related regulations and directives. A number of statutes are described in Section 12.5.

12.3 Procedures and policies, including voluntary abatement

In general, the basic philosophy and principles of regulatory toxicology are to protect humans and other living organisms in natural and managed ecosystems from the harmful effects of potentially toxic substances. The means by which this can be done ranges from the direct regulation of individual sources of pollution, through rules for the use of products, and the residues of substances in foods and agricultural feed, to a more general approach to protection of a resource or valued species or whole ecosystem.

12.3.1 Types of approach

Individual sources of pollution such as various industries or commercial endeavours and waste treatment plants are known to release substances variously into air, water, and soil. Most environmental statutes deal with three types of release: release of wastes, accidental releases, and deliberate releases. A fourth category of statutes is more concerned with the protection of a specific component of the ecosystem (e.g., aquatic life in receiving waters). The amounts or concentrations of certain substances are usually regulated, through measuring effluent or discharge. An example of this is the Ontario (Canada) provincial Municipal/Industrial Strategy for Abatement (MISA) programme, which regulates wastes. In some jurisdictions, regulation is done by permit, often at a local (municipal, regional, or state/provincial) level of government rather than at the national or federal level.

Regular monitoring is required and exceedances of the permitted amounts may result in fines or injunctions. The general policies that prevail are: (a) that there will be a certain amount of potentially harmful material entering the environment and (b) that this is acceptable below a certain level, but some risk exists above that level.

More recently a policy known as zero discharge or virtual elimination has been expounded. The U.S. Water Pollution Control Act of 1972 aimed to eliminate the discharge of pollutants into navigable waters by 1985 [CWA 101(a)]. This appears to be the first time that the United States embraced the idea of zero discharge. The meaning of zero discharge in a practical sense is a topic of much debate. At the extreme, it can mean stopping all activities that generate pollutants; more reasonably, it can mean eliminating the effect(s) of undesirable materials, mainly those of synthetic origin and persistent fate. The International Joint Commission for the Great Lakes has a Virtual Elimination Task Force, and the recently enacted Canadian Environmental Protection Act (CEPA) encompasses the concept of virtual elimination for those substances that are all of persistent, bioaccumulable, and toxic (PBT) natures, depending upon the definition of PBT. The European Union (EU) also has embraced the idea of the identification of PBT substances that need to be eliminated.

Specific chemical substances have been banned. For example, DDT is banned in North America except for a few very rare permits that may still be given for one specific task, but even this permitting is being eliminated. The effectiveness of complete banning of DDT has to be questioned; DDT is still manufactured and used in other countries with the full support of the World Health Organisation, which recognises its function in saving lives related to insect-borne diseases such as malaria. The DDT that is discharged in jurisdictions where it is not banned finds its way across international boundaries. For example, it is carried via the atmosphere to the Canadian Arctic, where it accumulates in plant and animal tissues.

Regulations mandating that there be *no* release of certain harmful substances into the environment, and that the substances should be eliminated from the environment are typically applied to extremely toxic and persistent organics and certain metals. The policy of zero discharge may appear to be somewhat idealistic and is clearly at odds with the previous concept of permitting discharges, but with limits. If implemented strictly, zero discharge could result in the need to close down certain industrial or commercial operations. On the other hand, it might accelerate advances in technology that would have overall benefit not only for the environment but also for the industry.

The second category, accidental releases, is normally dealt with differently in the law from routine releases such as those from known industrial or municipal sources. Spills should be rare events, but they can have catastrophic results. Deterrents to careless handling of dangerous materials, for example in the course of transportation, can be achieved through laws. The Merchant Shipping Act in the

United Kingdom has provision for, among other things, regulation of oil pollution in water.

A third category of laws concerning toxic substances in the environment concerns those that are released deliberately. The major categories here are pest control products and fertilisers. These laws are in many ways analogous to those involved in the regulation and control of additives such as drugs and food supplements, but the necessary assessment that underlies the regulation of pest control products is more complex. Originally, if a producer wished to put a pest control product onto the market, the main concern of the regulators was for the efficacy of the product. Now, in developed countries, acts and regulations for pest control products address toxicology, both human and environmental, including mobility and persistence in the environment. Furthermore, there may be restrictions on the persons who can use the products (permits for use and licences for practitioners) as well as the amounts that can be retailed in single packages. Administration and enforcement of these laws and regulations, as with any laws, requires an ongoing monitoring and reporting system. New scientific information and public concern for pest control products mean that the laws may need to be reviewed and updated. Pesticides appear to evoke considerable public concern, possibly because of the dramatic effects on birds of the earlier organochlorine pesticides such as DDT and dieldrin (see Sections 7.2.1 and 9.7.3). Residues of pesticides and other chemicals in food and animal feed are also regulated. Many of the less-developed countries (LDCs) also have laws concerning pest control products, but their effectiveness is often marred by lack of resources to administer the law. Some LDCs, however, both use pesticides that are banned in other countries and have more lax regulations concerning emissions and effluents. This situation can present problems involving transboundary pollution, relative costs of production, and residues in food. For example, in 1990 the U.S. Food and Drug Administration tested about 10,000 food shipments to the United States and found illegal pesticide residues in 4.3% of these (U.S. General Accounting Office, 1992). In the United States, restrictions on imported foods contaminated by unacceptable concentrations of residues of chemicals are legislated under the Imported Food Safety Act of 1998.

A fourth category of regulation protects a specific or general category of receptor, rather than focussing on the process(es) that produce the substance. The ultimate goals of environmental protection are the same as for the earlier examples, and frequently individual toxic substances are identified, such that the target will be the individual polluter, but the route(s) by which the acts have been developed and the general approach can vary. These types of act can be seen as having more flexible application and are useful for a variety of alleged offences. Examples of this type of legislation include the U.K. Environmental Protection Act, the Canadian Environmental Protection and Environmental Assessment Acts, and the U.S. Federal Water Pollution Control Act.

Recent concern for rehabilitation of contaminated sites (see Chapter 11) has resulted in the growth of acts and guidelines for the assessment and possible clean-up of such sites. Arguably the best known of these is the U.S. Comprehensive Environmental Response, Compensation and Liability Act (the Superfund Act) of 1980, reauthorized in 1986, which is discussed at some length in Chapter 11. As pointed out in Chapter 11, the existence of contamination in soil or other substrates does not automatically of itself represent a risk. It may not be feasible, practical, or necessary to clean up all contaminated sites. The primary consideration should be the likelihood of exposure for humans and other biota to the contaminants in such sites. Risk assessment forms the basis of most assessments of contaminated sites. Provincial and other lower level agencies in Canada have produced guidelines for assessment of contaminated sites, using a risk assessment approach (e.g., MOE, 1997; PQ, 1994). The latter are not of themselves enforceable but provide useful guidance in the context of assessing contaminated sites and determining if there is a need for rehabilitation. Decisions concerning rehabilitation normally depend upon the expected use of the land or water resource. These same guidelines have also been developed with the recognition that decommissioning of previously industrial sites cannot reasonably be accomplished without some benchmarks.

The United Kingdom has nothing strictly comparable to the U.S. Superfund Act. The backbone of U.K. legislation on assessment and rehabilitation of contaminated land is Part IIA of the Environmental Protection Act 1990 (U.K. EPA). In October 1999, the U.K. Department of the Environment, Transport and the Regions issued a third and probably final draft guidance on the new contaminated land regime. This draft guidance assumed statutory force from April 2000 under provisions on contaminated land inserted into the U.K. EPA 1990 by section 57 of the U.K. Environment Act 1995. This currently applies only to England because the Scottish Executive and Welsh Assembly will determine how to implement their own legislation.

Powers for dealing with contaminated land in the United Kingdom may also reside in different pieces of legislation, such as integrated pollution control, waste management licensing, or the Environment Agency's new powers to serve "works notices" to deal with water pollution from lands.

In the United Kingdom, if a company contaminates its own site, then it is responsible for clean-up, under various pieces of legislation, including Part IIA of the U.K. EPA 1990. Part II of the U.K. EPA 1990 is called integrated pollution control. There is also the integrated pollution prevention and control, which is the 1996 European Community directive on IPPC (see Table 12.3b) which was implemented in U.K. law through the Pollution Prevention and Control Act 1999. IPPC has its own requirements on site remediation.

More details on rules for contaminated sites in the United Kingdom and the European Community can be found in *http://www.ends.co.uk/report/index.htm*.

12.3.2 Objectives, standards, and related concepts

Regardless of the approach(es) used in law, measures are obviously needed to put potentially harmful environmental contaminants into perspective. Apart from the "zero discharge" approach, all other legal actions will require some numerical or qualitative means by which to measure risk or damage. The following section deals with the most commonly used concepts and terms that need to be understood in this context.

Terms that are frequently used in the assessment and regulation or control of toxic substances in the environment include *objectives*, *criteria*, *guidelines*, and *standards*. Because each of these terms has familiar as well as technical meanings, a brief account of their usual meanings in the present context of regulation of toxic substances follows. The reader is also referred to the definitions in Section 12.4.

OBJECTIVE

An objective is a goal or purpose toward which an environmental control effort is directed. Until recently, selection of objectives involved identifying uses to be protected, mainly implying human purpose. During the 1970s, it was realised that appropriate uses might include more than readily identifiable human uses. In the present climate, objectives, although expressed in a way closely related to human uses (water supply, water-based recreation, damage-free agriculture concerning air quality), also provide for protection and propagation of fish, shellfish, and wildlife (sometimes referred to as the fishable/swimmable goal). Decisions about objectives concerning the quality of air and water resources are neither technical nor legal decisions. In a democratic society, the objectives should express the collective will and perceptions of all people likely to have an interest in the condition of the resource (Lucas, 1993).

CRITERIA

Criteria are compilations or digests of scientific data that are used to decide whether or not a certain quality of air or water is suitable for the chosen objectives. Criteria consist of the supporting scientific data that enable us to answer questions such as "Is this water safe to drink?", "Is this air safe to breathe?", and "Can this lake support a healthy trout fishery?" In contrast to the formulation of objectives, the formulation of criteria is strictly a scientific matter. It involves the gathering of data on the effects of potential pollutants on the uses in question. It uses existing data, normally of specific quality as defined by the authority responsible for the resource, as well as, where necessary, directing experimentation to generate new information (Lucas, 1993).

GUIDELINES

Guidelines are designed to help environmental managers to make decisions on a range of issues that affect environmental quality. In certain jurisdictions, the term *guideline* has a more precise definition. For example in Ontario, Canada, where a Provincial Water Quality Objective is "a numerical and/or narrative limit recommended to protect all forms of aquatic life" and is "established when a defined minimum information base is available" (MOE, 1992). A guideline value for water quality (PWQG) is used when the available data are insufficient for the establishment of an objective. The guideline value is calculated using a prescribed method of applying uncertainty factors determined by the quality and quantity of the data. It should be cautioned that different jurisdictions throughout the world use the terms *objective* and *guideline* under their own definitions.

STANDARDS

The term *standard* in environmental regulation denotes a prescribed numerical value or set of values to which concentrations or amounts actually occurring in the ambient medium (ambient standard) or the discharge or effluent (effluent standard) may be compared.

The formulation of ambient quality standards is the prescription of the quality of air or water that is considered necessary to protect the desired objectives. The process utilises scientific knowledge (supplied by criteria) as well as socioeconomic factors affecting the control of pollution. Ambient standards may set *upper* limits to the mass or concentration of harmful substances or *lower* limits in the concentration of desired substances (e.g., oxygen) or prescribe a range of values (e.g., for pH).

The setting of ambient standards is not entirely a technical decision. Even if one assumes that the criteria provide a very complete technical background for the decision, the standards set inevitably reflect value judgements. This is an important point, and one that has often been misunderstood (Lucas, 1993).

Ambient standard setting has the following phases.

a. *Technical evaluation of ambient standards* involves scientists, economists, and experts in various industries. In addition to the strictly technical evaluation of the proposed standards, two other types of issue need to be considered. First, there is the question of whether the proposed standard makes sense in the real world. If the proposed standard has been determined largely from laboratory studies (i.e., on good technical data), then it is possible to conceive of a situation in which a body of water would be in regular violation of the standards that were formulated to protect fish life and yet be found to support a healthy fish population of the desired type. Second, it has to be asked whether the proposed standard is measurable by methods currently available. A nonmeasurable standard not only cannot be enforced but also actually invites violation and disrespect of the entire regulatory process (Lucas, 1993).

b. *Political choice process of ambient standards* requires the involvement of the public. A regulatory agency is expected to have secured some public input during the earlier stages of the standard-setting process, but once the standard is proposed, provision should be made for further input. This is often in the form of an invitation for the public and interest groups to consider and comment explicitly on proposed standards; to bring forth new

Table 12.1. Public acceptance of environmental risks

Less acceptable	More acceptable
Catastrophic	Subtle
Imminent	Deferred
Permanent	Temporary
Irreversible	Recoverable
Probable	Unlikely
Rare	Common
Obvious	Intangible
Unpredictable	Predictable
Abstruse	Understood
Quantifiable	Nonquantifiable
Global	Local
Involuntary	Voluntary
Consequences of human action	Consequences of natural processes
Innocent party	Polluter at risk

Based on Jacobson (1981).

information that may not have been accessible to those making the technical evaluation; to state preferences concerning matters that are hard to quantify, such as aesthetic or spiritual values; and to give views on the acceptability of risks inherent in the proposed standards and to indicate willingness to pay for necessary measures (see also Section 13.5). A second component of the political choice process is the determination of risk acceptability. The acceptability of a risk may depend not only on the character of the use that is at risk but also on the attributes of the risk itself. Public perception of risk acceptability is dependent on such factors as rarity of the predicted type of damage (i.e., if the damage is alien to experience and, therefore, induces fear, in contrast to commonly experienced damage). The public also tends to be less accepting of pollution coming from an industrial plant owned by outsiders (the concept of an innocent party being at risk) and more accepting of damage from the community's own waste collection system. Table 12.1 lists the types of risk that the public finds acceptable.

c. *Benefit-cost judgements involve* some complex value judgements that go beyond the technical. In essence, since maintenance of standards costs money, and because resources are limited, it is necessary to ask whether the protection of certain uses or the maintenance of a certain quality is "worth it." Issues such as the protection of unusually sensitive individuals arise here. For example, should air quality standards protect not only those susceptible to asthma or emphysema but also those with extreme environmental allergies, who may need to wear special masks to even walk in an urban street? The first is reasonable and attainable; the second may be impossible to achieve or cost too much for the majority to accept.

Ambient standard setting addresses the desired condition of air, water, or other media. An ambient standard alone may be unenforceable because it gives no indication of cause and effect, nor any guidance to either the regulatory body or the regulated party about the necessary courses of action. Effluent standards address the limits on discharges to the resource that will permit that desired state to be attained. But an effluent standard that lacks a basis in an ambient quality standard has no realistic basis. Ideally, both approaches are necessary: They are complements, not alternatives, for regulation.

Like the setting of ambient standards, the effluent standard requires both technical and nontechnical inputs.

   a. *Technical considerations for effluent standards* typically involve mathematical models that relate the ambient quality to inputs of pollutants and to pollutant removal mechanisms. The advent of computers (see Chapters 1 and 4) has greatly enhanced the possibilities for such modelling. The usefulness of models (see Section 4.6) is not that they give the "correct" answers. In the present context, with relatively modest expenditures of extra time and money, models can be used in sensitivity analysis that can provide information about the response of the cost to changes in the input data, the assumptions about pollutant behaviour and fate, or the constraints. Thus, models can generate a number of options for consideration.

     Compliance with effluent standards should be feasible. Like ambient standards, effluent standards should be set at a level that make the monitoring of compliance possible. Where serious reservations exist about the safety of extremely low or even nondetectable amounts of a substance, a very low or even "no discharge" standard may be set. In such a case, the cooperation of the industry in the form of process change and of other government entities in restricting the use or sale of a substance may be sought.

     An example of an unrealistic effluent ban is seen in the proposed limit on mercury (Hg) discharges to the ocean off California, U.S. PL 92-500 (1992), which assumes (or is premised on) an average contribution of only 0.4 mg mercury per person per day in the City of Los Angeles' population of 3 million served by the main municipal treatment plant. The personal use of medicinal substances such as merthiolate and mercurichrome could be significant in this context because achieving this effluent standard would require a legislative ban on these and other commonly used products containing mercury. A ban so widely affecting commerce is politically difficult to achieve. When faced with a dilemma such as this, the regulatory authority might have to rethink its proposed standards, making sure that they have a firm basis in scientific data. If indeed the very stringent standard were indicated, then it may be necessary to set a more lenient standard for an interim period.

An alternative means for controlling substances that may be present in industrial discharges at less-than-detectable concentrations is that of requiring the industry to submit periodic inventories for the questionable substances.

b. *Non-technical considerations* related to effluent standards include those already identified for ambient standards. Because effluent standards are frequently associated with industrial processes, there is an element of bargaining between regulators and regulated. In fact, some see the bargaining process as an integral component of environmental regulation. The bargaining process can also act as an interface between the technology and economics of pollution control, and the economies of the country and the world. Because of the perceived need for confidentiality about financial matters, this may be the first opportunity given to the regulator authority to consider the effects of the proposed effluent standards and prospective expenditures on a particular discharger's competitive position within the country's economy and, if standards for an entire industry are being set, to consider collectively that industry's position in the world market. Finally, the bargaining process can draw attention to the weak points of the scientific and economic information supporting the standards-setting process. Often, bargaining will take place precisely because such weak points exist, and will leave room for manoeuvring (Lucas, 1993).

c. *Technology-based effluent standards* are limits that are designated without reference to any ambient standards, balance of benefits and costs or other guiding principles. They are based on what is "practical", "possible", or "achievable". For the most part, these types of regulations have not worked well in terms of improving water quality, although data are difficult to interpret. It would be necessary to compare systematically the results from technology-based regulation with water-quality standard-based approaches, and this analysis has not been done. Furthermore, in the opinion of some, technology-based standards tend to discourage innovation in treatment processes and encourage reliance on existing technologies even if they fail to produce the desired quality of effluent (Lucas, 1993).

12.3.3 Risk assessment in a regulatory context

Uncertainties are facts of life in environmental toxicology and assessment. The risk assessor and risk manager use techniques that deal with uncertainties; for regulatory as well as other purposes, the risk assessor's technique should attempt to ensure that uncertainties are treated in the same way for all chemicals. The risk assessor can make statements about the *relative risk* that different chemicals represent but cannot realistically make authoritative statements about the *absolute risk* posed by a specific chemical. Authorities that conduct risk assessments for the purposes of regulatory decision making have adopted certain policies, primarily in the context of specific assumptions that should be used to deal with deficiencies or gaps in knowledge concerning the effects of specific chemicals (Rodricks, 1992). For example, the term *guideline* is used by some authorities to denote very specific treatment of a limited set of toxicity data for a given chemical, which leads to a useful but weaker or more limited assessment than a full criterion. This is discussed in Section 12.3.2.

12.3.4 Voluntary systems of regulation

Changes are occurring in environmental management from the situation in which government oversees the private sector completely, to a level of self-regulation by the private sector. Such initiatives as environmental auditing and monitoring and internal reuse and recycling of potentially harmful substances are increasingly being undertaken by companies in the private sector.

Carra (1999) described voluntary approaches and industry/government partnerships to achieve environmental goals as a "quiet revolution" in the field of environmental protection. He suggested that five factors have contributed to this revolution. Even though he discussed this concept in the context of the U.S. EPA, his points have more general relevance. His suggested the following factors:

- The increase in direct and indirect costs and costs of liability insurance resulting from regulations;
- The need of industry to adjust to a fast-paced marketplace by being proactive to keep ahead of regulatory requirements;
- The lack of flexibility in regulations, resulting in missed opportunities for improvements in environmental and public health;
- The increasing difficulty of designing regulations that meet statutory requirements and the growing complexity of the regulatory process;
- The increased expectations of the public for pollution prevention.

Carra suggested that the people are "hungry for information" on the environment and are using such information to participate in discussions with industry on finding solutions. Through the EPA's Toxic Release Inventory (TRI) programme, chief executive officers (CEOs) of large companies were made aware, often for the first time, of data on their own companies' environmental releases. As a result of these data, CEOs have demanded changes within their companies.

In summary, voluntary compliance has a number of advantages: For the company, environmental sustainability is frequently a prerequisite for economic sustainability. A proactive approach not only avoids excessive insurance costs and expensive suits, fines, and injunctions but may also lead to new initiatives in control or other technologies that are economically beneficial of themselves, and the company's image is improved (as seen in various "green" awards for environmental responsibility). Furthermore, it appears that government-imposed environmental regulations have deficiencies (Morelli, 1999). Thus, according to Morelli, we can expect to see a dramatic change in responsibility and accountability, which in fact is already occurring in a number of developed countries.

Examples of the new regulatory model in which government and industry, and possibly public interest groups, are each seen as stakeholders include the U.S. EPA's 33/50 programme, which reduced high-risk industrial chemicals by 33 and 50% by preset target dates; Canada's Accelerated Reduction/Elimination of Toxic Substances (ARET) programme; and the OECD's Ministerial Declaration that called for member countries to take action in reducing exposure to lead by decreasing the risk of major sources of lead exposure. This latter example is, in fact, an international agreement (see Section 12.3.5); the concept of voluntary compliance is one of the principles of environmental protection at the international level, since formal law and its enforcement as understood at the federal or lower levels for one nation do not apply at the international level.

Some detailed examples will serve to illustrate the voluntary or cooperative model in a practical manner.

ARET IN CANADA
The Accelerated Reduction/Elimination of Toxic Substances programme was established in Canada "as a consensus-oriented multi-stakeholder group to use good science combined with common sense" (MAC, 1995). The stakeholders originally included members from several industries and their associations, health and academic associations, environmental nongovernment environmental organisations (ENGOs) and government, both provincial and federal. The ENGOs withdrew from the process, even though the other stakeholders remain.

The objectives of ARET include a systematic selection of candidate substances, based on toxicity (including carcinogenicity) to human and other biota, persistence, and bioaccumulation. Parties that were emitting such substances were challenged in 1994 to put together their own action plans for reducing and eliminating emissions. Industry members of ARET have stated that they believe voluntary programmes can be more effective than regulatory programs because they allow companies to plan more efficiently for the reduction or elimination of emissions.

ARET's challenge in 1994 was to effect a major or total reduction of emissions from the base year 1988 by the year 2000. Within the mining industry in Canada, some fifteen companies are participating in ARET. The ARET goal was for 50% decrease in all emissions by the year 2000. Twelve substances that were emitted by the mining and smelting industries were listed through the substance selection process: arsenic (inorganic), cadmium (respirable and inorganic forms), chromium (hexavalent), cobalt (inorganic, soluble), copper (inorganic salts), cyanides,

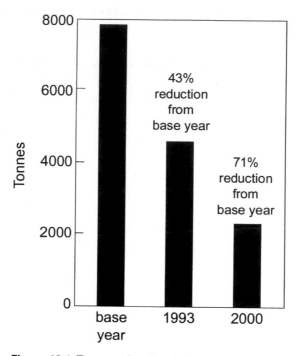

Figure 12.1 Tonnes of major ARET-listed substances released during the year by the mining industry ARET participants.

hydrogen sulphide, lead (all forms except alkyl), mercury (elemental and inorganic), nickel (inorganic, respirable, soluble), silver (soluble inorganic salts), and zinc (inorganic, respirable, soluble). Figure 12.1 shows that a considerable reduction in emissions had occurred by 1993. The ARET goal of 50% reduction by the year 2000 was achieved by the mining companies, as shown in the projection for the year 2000 in Figure 12.1. In fact, the projection shows a decrease of 71% overall by the year 2000.

Of course, individual operations and individual contaminants need to be evaluated to determine the effectiveness of the decreased emissions, particularly on sensitive or previously damaged ecosystems. However, to date this would appear to be a successful programme.

THE EPA 33/50 INITIATIVE

The U.S. Environmental Protection Agency uses a nationwide data base of toxic chemical discharge, the Toxic Release Inventory, as a baseline for measuring improvements in companies across the nation. The TRI comprises over 80,000 reports representing several million tonnes of chemical releases from more than 20,000 manufacturing facilities and 200 federal facilities. In 1988 the agency challenged 1,300 companies that were responsible for the emission of 17 target chemicals to reduce voluntarily their emissions of these chemicals 33% by 1993 and

50% by 1995. According to reported TRI data, targets were met ahead of schedule, with a reduction of ca. 400,000 tonnes of these chemicals.

12.3.5 International considerations: Treaties and informal agreements

The usual understanding of international law is a system of principles and rules that govern relationships between states and other internationally recognised parties. There is no legislature for the enactment of new laws nor any court system to which nations are required to submit for dispute settlement, and there are only limited means of enforcing international law.

Some may question whether international law is law at all. However, according to Muldoon (1993), international law can still be considered law because states regard it as law. The reason for this is partly self-interest, since the protection provided by international law compensates for the need to respect it. International law, which has evolved over many centuries, may be more important now than ever before, particularly in the context of environmental issues, which some see as having had an important role in forcing the development of a more comprehensive framework of laws to govern states.

Even though, as already stated, there is no legislature to enact international law, a definitive statement on the sources of international law can be seen in Article 38 from the Statutes of the International Court of Justice:

> The Court, whose function is to decide in accordance with international law such disputes as are submitted to it, shall apply:
>
> > international conventions, whether general or particular, establishing rules expressly recognised by the contesting states;
> > international custom as evidence of a general practice accepted as law;
> > the general principles of law recognised in civilised nations;
> > judicial decisions and the teachings of the most highly qualified publicists of the various nations, as subsidiary means for the determination of rules of law.

International agreements for environmental protection involve the relatively straightforward concept of the "good neighbour" principle, with bilateral and multilateral agreements and treaties. Bilateral agreements are arguably the most manageable types of international law, but multilateral agreements have also been attempted. A few examples follow.

BILATERAL AGREEMENTS

The boundary waters treaty. There is a relatively long history of shared resource management between the United States and Canada. As early as 1909, the Boundary Waters Treaty was signed between the United States and Great Britain; this related to the boundary waters between the United States and Canada, which included the main inland waterways along which the international boundary "between the United States and the Dominion of Canada passes". It involved the use of waters for "domestic and sanitary purposes", for "navigation, including the service of canals for the purposes of navigation", and "for power and irrigation purposes". The treaty provided for the appointment of a joint commission, known as the International Joint Commission, with three members from each of the two parties. This commission continues to the present day. The major water quality agreement between these parties is the Great Lakes Water Quality Agreement (GLWQA) of 1978. This agreement involves the ecosystem approach to resource management, and, as such, the Great Lakes Ecosystem is defined in the agreement as "the interacting components of air, land, water and living organisms, including humans, within the drainage basin of the St. Lawrence River at or upstream from the point at which this river becomes the international boundary between Canada and the United States".

The IJC considered that the GLWQA was a milestone document, "one of the first international statements that technical, diplomatic, and administrative approaches to resource management need to be considered in terms of holistic ecological concepts" (IJC, 1984).

The trail smelter case: a bilateral convention. In 1908, a dispute arose between Canada and the United States involving emissions from a smelter in Canada that were affecting the adjoining state in the United States. The case went before the IJC, which wrote a report, without invoking its legal powers, but the United States rejected the commission's report. As a result of this case, a bilateral convention was signed by a tribunal in 1965 to the effect that "under the principles of international law, as well as the law of the United States, no state has the right to use or permit the use of its territory in such a manner as to cause injury by fumes in or to the territory of another or the properties therein, when the case is of serious consequence and the injury is established by clear and convincing evidence" (Dinwoode, 1972). This principle is reminiscent of the principles of the common law as applied to persons (see Section 12.4.2).

MULTILATERAL AGREEMENTS

In 1972, a landmark conference was organised by the United Nations and held in Stockholm. At this conference, the Declaration on the Human Environment was formulated. According to its principle 21, "States have, in accordance with the Charter of the United Nations and the principles of international law, the sovereign right to exploit their own resources pursuant to their own environmental policies, and the responsibility to ensure that activities within their own jurisdiction or control do not cause damage to the environment of other States or of areas beyond the limits of national jurisdiction."

Since the Stockholm Conference, a number of multilateral environmental agreements have been drawn up. Some comments on a few of these are discussed here, and in Table 12.2 we have summarised the status of several more from European nations, which deal wholly or in part with environmental contaminants (see *http://www.mem.dk/aarhus-conference/other/status.html*).

Table 12.2. Implementation and impact of multilateral environmental agreements

	WE	CE	EE	CA
Global agreements				
Basel Convention on the control of transboundary movement of hazardous wastes	G	G	F	N/P
Vienna Convention for the protection of ozone layer	G	G	G	F
Framework convention on climate change (FCCC)	F	G	G	N/P
Ramsar Convention on wetlands of international importance especially as waterfowl habitat	F	F	F	P
Regional agreements				
Geneva Convention on long-range transboundary air pollution	G	F	F	N/P
Espoo Convention on environmental impact assessment in a transboundary context	F	F	F	N/P
Helsinki Convention on protection and use of transboundary watercourses and international lakes	G	G	G	N/P
Subregional agreements				
Barcelona Convention for the protection of the Mediterranean Sea against pollution	F	N/A	P	N/A
Helsinki Convention on the protection of the marine environment of the Baltic Sea area	G	G	F	N/A
Convention for the protection of the marine environment of the northeast Atlantic	F	F	N/A	N/A
Bucharest Convention on the protection of the Black Sea against pollution	N/A	P	P	N/A
OsParCom (North Sea and Atlantic Ocean)	F	F	N/A	N/A
Rhine EAP	G	N/A	N/A	N/A
North Sea EAP	F	N/A	N/A	N/A

Legend: WE, Western Europe; CE, Central Europe; EE, Eastern Europe; CA, Central Asia.
G, good: Positive development in introducing and implementing policies, reduction of pressures on the environment.
F, fair: Some positive development, but insufficient progress to deal with the problems. Can also indicate uncertain or mixed development.
P, poor: Negative development or little development in introducing and implementing policies.
N/A, Not applicable (subregional agreements).
N/P, No parties to this multilateral environmental agreement in the subregion.
Note: A policy cannot rate better than fair if there is no improvement in pressures on the environment.

The long-range transboundary air pollution convention.* The Convention on long-range transboundary air pollution (LRTAP) was signed in 1979. There are currently 40 signatories. It is an excellent example of regional environmental management and has been included as a specific case study (Case Study 12.1).

OECD's Ministerial Declaration. The Organisation for Economic Development and Co-operation (OECD)'s Ministerial Declaration called for member countries to take action in reducing exposure to lead by decreasing the risk of major sources of lead exposure.

The Montreal Protocol. In 1987, a conference on depletion of the ozone layer resulted in The Montreal Protocol on Substances that Deplete the Ozone Layer, which was itself an amendment to the earlier Vienna Convention of the Protection of the Ozone Layer (1985). The signatories agreed that, within 6 months of signing, each would maintain its use of the listed ozone-depleting substances to the level of 1986, until July 1993. After this, levels should be decreased to 80% of the 1996 levels by July 1, 1993, and to 50% of the 1986 levels by July 1, 1998.

The Basel Convention. The United Nations, through its environmental program, convened a meeting in Basel in 1989 to address concerns for the traffic and dumping of hazardous wastes into developing countries. This meeting led to the creation of the Basel Convention on the Control of Transboundary Movements of Hazardous Wastes and their Disposal with some 109 parties being involved. According to the convention, "the parties shall co-operate with a view to adopting, as soon as practicable, a protocol setting out appropriate rules and procedures in the field of liability and compensation for damage resulting from the transboundary movement and disposal of hazardous wastes and other wastes."

The first meeting adopted a decision entitled "Liability and Compensation" by which it established an ad hoc working group of legal and technical experts to consider a draft protocol on liability and compensation, possibly including "the establishment of an international fund for compensation for damage resulting from the transboundary movements of hazardous wastes and their disposal." By 1997 it had been ratified by 110 countries and has already had some successes due, in part, to the availability of criminal sanctions (Case Study 12.2).

Kyoto agreement. The United Nations Framework Convention on Climate Change, signed in 1992, came into force as an international treaty in 1994 and has now been ratified by 181 nations. Signatories continue to assemble in periodic conferences, of which the most significant occurred in Kyoto, Japan, in November 1997. This conference resulted in a protocol that stated emissions goals for six gaseous compounds or groups of compounds to be achieved by industrialized nations by the year 2012 (see Section 9.5.2). These goals were based on the 1990 greenhouse gas emissions from these countries, which were Australia, Canada, European Community, Japan, Russian Federation, and the United States. Not all goals represented cutbacks. Some countries, such as Australia, were permitted under the agreement to increase their greenhouse gas emissions, and, even though the European Union as a whole was committed to an 8% reduction, some countries within the union were exempt.

The agreement would only come in effect if ratified by those nations responsible for 55% of the net worldwide CO_2 emissions. Some dissenters claim that the agreement represents an erosion of national sovereignty; others protest the exclusion of developing nations in the agreement. The agreement approves in principle the establishment of mechanisms to assist developing nations achieve "clean development", although specific details are lacking. Trading of emission rights and credits for tree planting (CO_2 sinks) are also principles espoused by the agreement although, again, the protocol is vague on these matters. Proponents of the protocol claim the agreement to be a critical foundation to be strengthened and expanded in subsequent conferences. Critical to the success of the Kyoto Protocol would be the eventual degree of compliance by emerging and very large economies, such as China, which are rapidly expanding their share of greenhouse gas emissions. In view of the 2001 U.S. decision to opt out of the agreement it remains to be seen whether a compromise strategy can be negotiated.

12.4 Definitions

12.4.1 Types of law

Common law is the body of customary law, based upon judicial decisions and embodied in reports of decided cases, that has been administered by the common law courts of England since the Middle Ages. From this experience has evolved the type of legal system now found also in the United States and in most of the member states of the Commonwealth of Nations.

Statute law is often known as "Acts of Parliament or Congress" (which does not include all statutes). A statute is a formal expression of a rule or set of rules, made in English law by "the King, the Lords and the Commons" (Brittannica, 1965). Statute making or legislation is a more modern process than the establishment of law by custom and judicial interpretation (the common law). There is a clear distinction between common and statute law in terms of the way in which each is interpreted. The common law has considerable flexibility. Judges base decisions in part on previous decided cases. A statute is more formal, usually having provision that a particular authority shall have the power to make orders, rules, and regulations in relation to particular matters arising from the act. The interpreter of the statute has less flexibility: He or she must be guided by the precise words used.

12.4.2 The common law

Negligence is broadly defined as failure to take care in view of foresee-
able danger. It is set off against intent on the one hand and mischance
or accident on the other. It deals with unintended injuries to persons
or property (originally arising from horse and buggy collisions); the
action is also employed to recover damages for injuries resulting
from defective performance of professional services. The action for
trespass for physical injuries was displaced by the concept of negligence
in what has been called the most radical revolution in common law
history. It was the courts' response to the industrial revolution. Under
mediaeval tort law, financially weak industries and enterprises, with
their dangerous and imperfect machines could not do business. The
severity of the early law was completely reversed by the negligence
action, under which the burden was placed on the victim to sustain his
case.

Nuisance is a legal term used to denote a human activity or a physical condition that is harmful to others. Private and public nuisance are distinct.
A public nuisance is an offence against the state and is only actionable
by the state, either by way of criminal proceedings or injunction. A
private nuisance is an activity or condition that causes an interference
with the use and enjoyment of neighbouring privately owned lands,
without constituting an actual invasion of the possession of the neighbours. A private nuisance is only actionable by persons who have a
property interest in such land.

Rylands v Fletcher is a case of strict but not absolute liability that has
come to represent a rule in common law. It implies some fault on the
part of the keeper of the thing in question, namely that the keeper has
brought to, collected, or kept on his land something that is liable to do
damage if it escapes. If the defendant has allowed it to escape, there is
no need to prove negligence, except through an act of God or the act of
a stranger, then he or she must answer for the natural and anticipated
consequences.

Riparian rights is the legal principle that anyone whose land adjoins a
flowing stream or river has the right to use water from that source, provided that enough is left for downstream users and the water quality is
maintained. The riparian doctrine is the common law principle governing water use in most of Canada, the eastern United States, and the
United Kingdom.

Trespass is the entry on land without lawful authority, by the man, his servants, or the thing or animal in question. The plaintiff must have possession of the premises. In addition to damages for trespass, the court
may grant an injunction prohibiting trespass.

12.4.3 Some general legal terms

Compliance means conformity to official (legal) requirements.

A regulation is a rule or order having the force of law issued by an executive authority of a government; regulation(s) made under an act; a specific law that legally applies in all relevant situations.

A plaintiff is one who commences an action or lawsuit to obtain remedy for an injury to his or her rights.

A defendant is a person or party sued in a legal action or suit.

12.4.4 Terms used in assessment and regulation of toxic substances

Abatement is the reduction in the degree or intensity of, or eliminating, pollution. Abatement may be voluntary or imposed.

Action levels are regulatory levels (concentrations) recommended by a federal or other authority for (a) pesticide residues in food or feed commodities, or (b) a contaminant concentration, in the Superfund (United States, see CERCLA), high enough to trigger a response under Superfund Act Reauthorization (SARA), or (c) similar use in other regulatory programmes.

A contaminant is any physical, chemical, biological, or radioactive substance or matter that has an adverse effect on air, water, or soil.

Criteria are compilations or digests of scientific data that are used to decide whether or not a certain quality of air or water is suitable for the chosen objectives.

Guidelines are designed to help environmental managers to make decisions on a range of issues that affect environmental quality. In certain jurisdictions, the term guideline has a more precise definition.

An objective is a goal or purpose towards which an environmental control effort is directed. Objectives, although expressed in a way closely related to human uses (water supply, water-based recreation, damage-free agriculture concerning air quality), provide also for protection and propagation of fish, shellfish, and wildlife.

Standards are often thought to be numbers that are enforceable, but this is not strictly true. The term standard in environmental regulation denotes a prescribed numerical value or set of values to which concentrations or amounts actually occurring in the ambient medium or the discharge or effluent may be compared.

12.5 Federal statutes

Tables 12.3a, 12.3b, 12.4, and 12.5 provide a summary list of some of the major federal acts that have been developed for the protection of environmental quality

Table 12.3a. Selected UK statutes affecting toxic substances

Name of statute	Date of proclamation	Major revisions and notes
Atomic Energy Act 1989 (c. 7)	September 1, 1989	
Atomic Energy Authority Act 1995 (c. 37)	Nov. 8, 1995	
Biological Weapons Act 1974 (c. 6)	February 8, 1974	
Chemical Weapons Act 1996 (c. 6)	Section 39 in force April 3, 1996; rest of act in force Sept. 16, 1996	
Clean Air Act 1993 (c. 11)	August 27, 1993	Repealed the Clean Air Act 1956, the Clean Air Act 1968, and the Control of Smoke Pollution Act 1989
Consumer Protection Act 1987 (c. 43)	Part I in force March 1, 1988	
Control of Pollution Act 1974 (c. 40)	Sections in force on a range of dates from December 12, 1974, to January 31, 1985	Control of Pollution (Amendment) Act 1989 (c. 14) – registration of carriers of controlled waste. Many sections repealed by the Environmental Protection Act 1990
Deregulation and Contracting Out Act 1994 (c.40) s. 33	November 3, 1994	Regarding duty of care and the like as respects waste
Environment Act 1995 (c. 25)	Sections came into force on a range of dates from July 28, 1995, to January 1, 1999. Sections 24, 60 (1), (2) not in force	
Environment and Safety Information Act 1988 (c. 30)	April 1, 1989	
Environmental Protection Act 1990 (c. 43)		Repealed the Alkali, and Chemical Works Regulation Act 1906
Finance Act 1996 (c. 8) Sections 39–71, 197(1), (2)(d), (3)–(5), (7), 206, Sch.5		Landfill tax provisions
Food and Environment Protection Act (FEPA) 1985 Part III Pesticides		Repealed the Dumping at Sea Act 1974. Control of Pesticides Regulations under FEPA 1985 passed in 1986, replacing a previous voluntary notification scheme. Pesticides Act 1998 (c. 26) amends FEPA1985 in respect of powers to make

Table 12.3a. (continued)

Name of statute	Date of proclamation	Major revisions and notes
		regulations concerning pesticides and enforcement of provisions relating to the control of pesticides
Food Safety Act 1990 (c. 16)	Sections in force on a range of dates from June 29, 1990 to Jan 1, 1991	
Health and Safety at Work Act 1974 (c. 37)	Sections came into force on range of dates from October 1, 1974 to March 17, 1980	Check dates of sections in force with respect to dates
Merchant Shipping Act 1995 (c. 21) Part VI Prevention of pollution	January 1, 1996	Prevention of Oil Pollution Act 1971 amended Sections 131– 151 Merchant Shipping (Oil Pollution) Act 1971 amended Sections 152–171 The Merchant Shipping (Salvage and Pollution) Act 1994 amended Sections parts of 128, 129, and Sections 152–171
Merchant Shipping and Maritime Security Act 1997		Pollution control measures, carriage of hazardous substances, etc.
Planning (Hazardous Substance) Act 1990 (c. 10)	Some sections in force on March 11, 1992, with remainder on June 1, 1992	
Radioactive Material (Road Transport) Act 1991 (c. 27)	August 27, 1991	
Radioactive Substances Act 1993 (c. 12)	August 27, 1993	
Road Traffic Act 1991 (c.41)	Sections in force on a range of dates from October 1, 1991, to July 5, 1994	Covers vehicle emissions
Road Traffic (National Targets) Reduction Act 1998 (c. 24)	July 2, 1998	
Waste Minimisation Act 1998 (c. 44)	November 19, 1998	
Water Act 1989 (c. 15)	July 6, 1989; some sections September 1, 1989	
Water Industry Act 1991	December 1, 1991	
Water Resources Act 1991 (c. 57)	December 1, 1991	

Modified from *http://www.mem.dk/aarhus-conference/other/status.html.*

Table 12.3b. Selective important EU regulations, directives,[a] and decisions affecting toxic substances in the United Kingdom

Title	Date in force	Revisions and notes
Directive 85/337/EEC environmental impact assessment	July 5, 1988	Extensively amended by Directive 97/11/EC – must be in force by March 14, 1999
Directive 90/313/EEC access to environmental information	December 31, 1992	

Air quality

Title	Date in force	Revisions and notes
Directive 96/62/EC air quality framework	Articles 1–4, 12 and Annexes I–IV on May 21, 1998, the rest depending on Article 4(5) provisions	Sets overall objectives for ambient air quality
Directive 80/779/EEC SO_2 and particulates (air)	August 30, 1982	
Directive 82/884/EEC lead (for lead concentrations in air)	December 31, 1984	*To be replaced by new standards under the Air Framework Directive, which will also cover benzene, ozone, CO, and other atmospheric pollutants.*
Directive 85/203/EEC nitrogen oxides (air)	January 1, 1987	*See note for Directive 82/884/ EEC Lead.*
Directive 92/72/EEC tropospheric ozone pollution	March 21, 1994	
Directive 70/220/EEC motor vehicles		Extensively amended by directives since 1970 to gradually make emissions requirements more stringent
Directive 88/77/EEC heavy-duty diesel engines emissions	July 1, 1988	
Directive 72/306/EEC diesel exhaust smoke	February 20, 1974	
Directive 97/68/EC emissions from nonroad mobile machinery	June 30, 1998	

Table 12.3b. (continued*)*

Title	Date in force	Revisions and notes
Directive 94/63/EC VOC (volatile organic compound) emissions from storage and transport of petrol	December 31, 1995	
Directive 85/210/EEC lead content of petrol	January 1, 1986	Will be replaced by *Directive 98/70/EC on quality of petrol and diesel fuels as of January 1, 2000*
Directive 93/12/EEC sulphur content of liquid fuels	October 1, 1994	Replaced 75/716/EEC

Waste management

Directive 75/442/EEC waste framework	July 25, 1977	Decision 94/3/EC initiated the European Waste Catalogue
Directive 91/689/EEC hazardous waste	December 12, 1993	
Directive 78/176/EEC titanium dioxide waste	February 25, 1978	
Directive 75/439/EEC storage and disposal of waste oils	July 25, 1977	
Directive 96/59/EC PCBs and PCTs	September 24, 1996	
Directive 91/86/EEC batteries	September 18, 1992	
Directive 86/278/EEC sewage sludge	July 4, 1989	
Directive 89/429/EEC (existing installations) and 89/369/EEC (new installations) municipal waste incineration	December 1, 1990 (both)	
Directive 94/67/EEC hazardous waste incineration	December 31, 1994	
Proposal on landfill COM(97)105		
Regulation EEC/259/93 shipment of waste	February 9, 1993	

Table 12.3b. (continued)

Title	Date in force	Revisions and notes
Water quality		
Proposed water framework directive COM(97)49	Proposed date of transposition December 31, 1999	*All marked * will be integrated into this.*
Directive 91/271/EEC urban wastewater treatment*	June 30, 1993	
Directive 75/440/EEC surface water*	July 25, 1977	
Directive 78/659/EEC fish water*	August 14, 1980	
Directive 79/879/EEC shellfish water*	November 10, 1981	
Directive 80/68/EEC groundwater*	January 26, 1982	
Directive 76/464/EEC dangerous substances to the aquatic environment	no date of transposition listed; implementation dates from 1976 to 1978	Includes seven "daughter" directives on particular substances
Directive 76/160/EEC bathing water	February 5, 1978	
Directive 80/778/EEC drinking water	August 30, 1985	*Directive 98/83/EC on drinking water will take effect December 25, 2000*
Directive 91/271/EEC urban wastewater	June 30, 1993	
Directive 91/676/EEC nitrates from agricultural sources	December 31, 1993	
Pesticides		
Directive 91/414/EEC plant protection products	August 19, 1993	Primarily affects agricultural pesticides
Regulation 3600/92/EEC review of active substances (pesticides)	February 1, 1993	
Directive 76/895/EEC pesticide residues on fruits and vegetables	December 9, 1978	
Directive 86/362/EEC pesticide residues on cereals	June 30, 1988	

Table 12.3b. (continued)

Title	Date in force	Revisions and notes
Directive 86/363/EEC pesticide residues on animal foodstuffs	June 30, 1988	
Industrial pollution control and risk management		
Directive 96/61/EEC IPPC (integrated) pollution prevention and control	*October 30, 1999*	
Directive 88/609/EEC emissions from large combustion plants	June 30, 1990	
Directive 84/360/EEC air pollution from industrial plants	June 30, 1987	will be replaced by the IPPC directive in 2007
Directive 96/82/EEC ("Seveso II") on the control of major accident hazards	February 3, 1999	Replaced Directive 82/501/EEC Seveso – control of major accident hazards
Proposed directive on industrial emissions of VOCs solvents COM(96)538	*Proposed date of transposition: April 2001*	
Regulation 1836/93/EEC ecomanagement and audit scheme	July 13, 1993	
Nuclear safety and radiation protection		
Directive 96/29/EURATOM basic safety standards framework directive for health & safety with respect to artificial radiation sources	*May 13, 2000*	*Will replace Directive 80/836/EURATOM*
Directive 97/43/EURATOM radiation protection related to medical exposures	*May 13, 2000*	*Will replace Directive 84/466/EURATOM*
Regulation 1493/93/ EURATOM shipments of radioactive substances	July 9, 1993	
Regulation 737/90/EEC imports of agricultural products following the Chernobyl accident	April 1, 1990	

Table 12.3b. (continued)

Title	Date in force	Revisions and notes
Regulation 3954/87/ EURATOM contaminated foodstuffs in case of a nuclear accident	April 2, 1990	
Directive 98/618/ EURATOM public information on radiological emergencies	November 27, 1991	
Directive 90/641/ EURATOM on radiation protection of outside workers	December 31, 1993	

Chemicals and genetically modified organisms

Title	Date in force	Revisions and notes
67/548/EEC classification, packaging and labelling of dangerous substances	January 1, 1970	
EC Directive 79/831 scheme for assessing risks posed by new chemicals	From September 18, 1981 to September 18, 1983	
Directive 88/379/EEC dangerous preparations	June 7, 1991	Extends Directive 67/548/EEC rules to mixtures of two or more chemical substances
EC Regulation 793/93 evaluation and control of existing substances	June 4, 1993	
Regulation EC/1488/94 risk assessment of existing substances	August 28, 1994	
Directive 90/219/EEC contained use of genetically modified microorganisms	October 23, 1991	
Directive 90/220/EEC deliberate release of genetically modified microorganisms	October 23, 1991	
Directive 94/55/EC transport of dangerous goods by road	January 1, 1997	
Directive 76/769/EEC restrictions on marketing and use of certain dangerous substances and preparations	January 26, 1978	

Table 12.3b. (continued)

Title	Date in force	Revisions and notes
Regulation EEC/2455/92 import and export of dangerous chemicals	November 29, 1992	
Regulation EC/3093/94 ozone-depleting substances	December 23, 1994	
Directive 87/217/EEC asbestos	December 31, 1988	the first substance-oriented directive integrating emissions to air, land, and water
Directive 73/404/EEC detergents	May 22, 1975	

[a] Note that, on the date of commencement of directives and regulations, directives include an article requiring member states to bring into force the laws, regulations, and administrative provisions necessary to comply with the directive by a specified date the date of transposition. Other deadlines for specific actions may be set separately in the articles dealing with those actions. They are known as "implementation dates". The date listed here for directives is the date of transposition. The date listed for regulations is the date the regulation is in force.

in the context of toxic substances, for three major jurisdictions. Although not strictly comparable, a number of common themes can be seen among the three jurisdictions.

The time period over which these laws have been enacted is remarkably short.

12.5.1 The United Kingdom and Europe

Tables 12.3a and 12.3b provide summaries of the main statutes. Note that a characteristic of the European Union law is that there are three major forms of binding law: regulations, directives, and decisions. *Directives* are by far the most common form. They stipulate that member states must enact laws, regulations, and administrative provisions in their own legislation by a certain date but leave states latitude to determine how to go about this.

About 10% of EU law is made up of *regulations* that are, on the other hand, binding on all member states without national legislation being required.

Decisions are individual legislative acts that are binding on the parties to whom they are addressed. They are usually very specific and can refer to individual member states. They are not included in the tables, but their importance should be noted. For example, the monitoring mechanism for EC CO_2 and other greenhouse gas emissions is determined by Decision 93/389/EEC.

Table 12.4. Summary table of the major federal acts for Canada related to the regulation of toxic substances, showing the dates of proclamation as well as any major revisions

Name of Canadian act	Date of proclamation	Major revisions and other notes
Arctic Waters Pollution Prevention Act R.S.C. 1985, c. A-12	R.S.C. 1970, Chap. 2 (1st Supp.) in force August 2, 1972	
Atomic Energy Control Act R.S.C. 1985, c. A-16	October 8, 1946	
Canada Agricultural Products Act R.S.C. 1985, c. 20 (4th supp.)	July 7, 1988	Replaced the Canada Agricultural Products Standards Act – R.S.C. 1970
Canada Labour Code R.S.C. 1985, c. L-2	Canada Labour (Safety) Code proclaimed January 1, 1968	1987 – Amended by the Hazardous Materials Information Review Act
Canada Shipping Act R.S.C. 1985 Chap. S-9	Act itself goes back to at least 1906 Ch. 113.	Main environmental provisions are: Part XV Pollution Prevention and Control Sections 654 to 672 Part XVI Civil Liability and Compensation for Pollution, Sections 673–727 S.C. 1970-71-72, c. 27 Pollution-related amendments in force July 1, 1971
Canada Water Act R.S.C. 1985 c. C-11	Part III and parts of section 2 and part IV necessary to give effect to Part III proclaimed August 1, 1970 The rest proclaimed September 30, 1970	Part III repealed by CEPA (June 30, 1988)
Canadian Environmental Assessment Act S.C. 1992, c. 37	Sections 61–70, 73, 75 and 78–80 brought into force December 22, 1994; remainder brought into force January 19, 1995	
Canadian Environmental Protection Act	In force, except Sections 26–30, June	Acts repealed by the introduction of CEPA

Table 12.4. (continued)

Name of Canadian act	Date of proclamation	Major revisions and other notes
R.S.C. 1985, c. 16 (4th supp.)	30, 1988; Sections 26–30 in force July 1, 1994	include the Environmental Contaminants Act (in force Dec. 2, 1975–June 30, 1988), the Clean Air Act (Nov. 1, 1971–June 30, 1988), Part III of the Canada Water Act, and the Ocean Dumping Control Act (Dec. 13, 1975–June 30, 1988) Bill C-32, House of Commons reported with amendments April 15, 1999
Canadian Food Safety and Inspection Act Bill C-80, 1st session, 36th Parliament	House of Commons First Reading on April 22, 1999	This act would consolidate food inspection and safety legislation and repeal the Canada Agricultural Products Act, the Feeds Act, the Fertilisers Act, the Fish Inspection Act, the Meat Inspection Act, and the Seeds Act
Export and Import Permits Act R.S.C. 1985, Ch. E-19	May 14, 1947	
Feeds Act R.S.C. 1985, c. F-9	Legislation regarding feeds goes back to at least 1909 Precursors – Commercial Feeding Stuffs Act RS 1909, ch. 15 Feeding Stuffs Act RS 1937 ch. 30	
Fertilisers Act R.S.C. 1985, c. F-10	R.S.C. 1906 ch. 132	
Fish Inspection Act R.S.C. 1985, c. F-12	November 1, 1967 Previous Fish Inspection Acts go back to R.S.C. 1927 ch. 72. Prior	

Table 12.4. (continued)

Name of Canadian act	Date of proclamation	Major revisions and other notes
	to that, covered under the broader Inspection and Sale Act	
Fisheries Act R.S.C. 1985, c. F-14 Sections 2, 20–22, 26, 34–43, 66–69, 78–88	1st proclaimed 1868	S.C. 1976–77, c. 35: pollution prevention amendments in force Sept. 1, 1977 S.C. 1991, c. 1: amendments to fish habitat protection measures in force Jan. 17, 1991
Food and Drugs Act R.S.C. 1985, c. F-27	1st Food and Drugs Act RS 1920, ch. 27. This replaced parts of the previous Adulteration Act (which goes back to at least 1906)	
Hazardous Materials Information Review Act R.S.C. 1985, c. 24 (3rd supp.)	October 1, 1987, with caveat "but see Sections 54 and 55"	
Hazardous Products Act R.S.C. 1985, c. H-3	RS 1968–69 ch. 42: sec. 3(2) proclaimed May 1, 1970 October 1, 1987	1987 – amended by the Hazardous Materials Information Review Act
Health of Animals Act R.S.C. 1985, c. H-3.3	January 1, 1991	Repealed the Animal Disease and Protection Act
Manganese-based Fuel Additives Act S.C. 1997, c. 11	June 24, 1997	Amendment to remove MMT from the list of controlled substances as in force of July 20, 1998
Meat Inspection Act RSC 1985 (1st supp.), c. 25	May 16, 1985	
Motor Vehicle Safety Act S.C. 1993, c. 16	April 12, 1995	Repealed original act brought into force July 15, 1971 Sections relevant to toxic substances regarding vehicle emissions standards

Table 12.4. (continued)

Name of Canadian act	Date of proclamation	Major revisions and other notes
Northwest Territories Waters Act S.C. 1992, c. 39	June 15, 1993	
Nuclear Liability Act R.S.C. 1985, c. N-28	October 11, 1976	
Nuclear Safety and Control Act S.C. 1997, c. 9	Not yet in force	
Oceans Act S.C. 1996, c. 31	Sections 1–55 brought into force except Section 53, January 31, 1997	
Pest Control Products Act R.S.C. 1985, c. P-9	November 25, 1972	Legislation on pesticides began in 1920s – Agricultural Economic Poisons Act, S.C. 1927, c. 5. superseded by the Pest Control Products Act S.C. 1939, c. 21 1972 – amendments to cover handling and use of pest control products, inert ingredients, and pesticide merchandising
Pesticide Residue Compensation Act R.S.C. 1985, c. P-10	January 21, 1972	
Plant Protection Act S.C. 1990, c. 22	October 1, 1990	Repealed the Plant Quarantine Act
Radiation Emitting Devices Act R.S.C. 1985, c. R-1	March 1, 1972	
Transportation of Dangerous Goods Act, 1992 S.C. 1992, c. 34	June 23, 1992	Repealed the original Transportation of Dangerous Goods Act (in force November 1, 1980– June 23, 1992)
Yukon Waters Act S.C. 1992, c. 40	June 15, 1993	

Table 12.5. Summary of the major acts for the United States of America related to the regulation of toxic substances, showing the dates of proclamation as well as any major revisions

Name of act with title and section from the United States Code	Date of proclamation	Major revisions and other notes
National Environmental Policy Act 42 U.S.C. 4321 et seq.	January 1970	
Pollution Prevention Act 42 U.S.C. 13101-13109	November 5, 1990	
Clean Air Act, 42 U.S.C. 7401 et seq.	December 31, 1970	Previous legislation began with the Air Pollution Control Act in 1955; Clean Air Act 1963; Air Quality Act 1967
		Major amendments in 1977 (extension of deadlines to meet 1970 goals) and 1990 (recognition of 189 hazardous pollutants, acid rain program, provisions to allow for emissions trading)
		Motor Vehicle Control Act 1960; Motor Vehicle Air Pollution Control Act 1965 (emissions standards were subsequently incorporated into the Clean Air Act)
Federal Water Pollution Control Act (also known as the Clean Water Act) 33 U.S.C. 1251-1387	March 1, 1972	Previous legislation began with the Federal Water Pollution Control Act of 1948
		Amendments in 1977 (fine-tuning), 1981 (Municipal Wastewater Treatment Construction Grants Amendments), and the Water Quality Act of 1987 (nonpoint pollution sources)
Marine Protection, Research, and Sanctuaries Act 33 U.S.C. 1401-1445, 16 U.S.C. 1431-1447f, 33 U.S.C. 2801-2805 (commonly known as the Ocean Dumping Act)	1972	1977 amendment on dumping municipal sewage sludge or industrial wastes; 1992 amendment to allow states more stringent standards

Table 12.5. (continued)

Name of act with title and section from the United States Code	Date of proclamation	Major revisions and other notes
Safe Drinking Water Act, 42 U.S.C. 300f-300j	December 16, 1974	Evolved from the Public Health Act of 1944
		1986 amendments to attempt to accelerate contaminant regulation
		1988 Lead Contamination Control Act (to reduce exposure to lead in drinking water)
		1996 major amendments (extended scope, which contaminants to regulate)
Resource Conservation and Recovery Act 42 U.S.C. 6901 et seq.	October 21, 1976	1939 – first landfill policies (established by Surgeon General)
		Previous legislation: Solid Waste Disposal Act 1965; Resource Recovery Act 1970
		Important amendments in 1976 (comprehensive revision resulting in the RCRA) and in 1984 (Hazardous and Solid Waste Amendments, resulting in expanded requirements)
Hazardous Materials Transportation Act, 49 U.S.C. 5101 et seq.	January 3, 1975	
Emergency Planning and Community Right to Know Act, 42 U.S.C. 11001-11050	October 17, 1986	
Comprehensive Environmental Response, Compensating, and Liability Act 42 U.S.C. 9601 et seq. (CERCLA, Superfund Act)	December 11, 1980	Amended October 17, 1986, by the Superfund Amendments and Reauthorization Act
		Other amendments (both dealing with some of the financial implications of CERCLA): Community

Table 12.5. (continued)

Name of act with title and section from the United States Code	Date of proclamation	Major revisions and other notes
		Environmental Response and Facilitation Act, 1992; Asset Conservation, Lender Liability, and Deposit Insurance Protection Act, 1996
Atomic Energy Act, 42 U.S.C. 2011 et seq.	August 13, 1954	
Energy Reorganization Act, 42 U.S.C. 5801 et seq.	January 19, 1975	Revised federal regulation of the nuclear industry and established the Nuclear Regulatory Agency
Nuclear Waste Policy Act, 42 U.S.C. 10101 et seq.	January 7, 1983	1987 – Nuclear Waste Policy Amendments Act
Low-Level Radioactive Waste Policy Amendments Act	January 15, 1986	
Uranium Mill Tailings Radiation Control Act	November 8, 1978	
Oil Pollution Act, 33 U.S.C. 2701-2761	August 18, 1990	Spills legislation
Federal Insecticide, Fungicide, and Rodenticide Act, 7 U.S.C. 136 et seq. (FIFRA)	October 21, 1972	1st regulated by the 1910 Insecticide Act but for commercial, not environmental or health reasons 1947 – 1st Federal Insecticide, Fungicide, and Rodenticide Act added herbicides and rodenticides; began to address health risks 1972 – complete revision, leading to broader coverage and increased enforcement 1988 – amendments to accelerate reregistration 1996 Food Quality Protection Act – amendments on reregistration, coordination between FIFRA and FFDCA

Table 12.5. (continued)

Name of act with title and section from the United States Code	Date of proclamation	Major revisions and other notes
Federal Food, Drug, and Cosmetic Act, 21 U.S.C. 301 et seq. (FFDCA)	June 25, 1938	Major amendments: 1954 Miller amendment (strengthened requirements for pesticide residues); 1996 Food Quality Protection Act
Food Quality Protection Act	August 3, 1996	Amends both the FIFRA FFDCA
Occupational Health and Safety Act, 29 U.S.C. 651 et seq.	December 29, 1970	1977 amendments – move towards generic standards for categories of toxic substances
Toxic Substances Control Act, 15 U.S.C. 2601 et seq.	October 11, 1976	Amendments to address concerns about asbestos (1986), radon (1988), and lead (1992)
Consumer Product Safety Act, 15 U.S.C. 2051 et seq.	October 27, 1972	Includes substances formerly regulated under the Federal Hazardous Substances Act and the Poison Prevention Packaging Act Amended by the Consumer Product Safety Act Improvement Act (1990)
Chemical and Biological Weapons Control and Warfare Elimination Act, 22 U.S.C. 5601 et seq.s	December 4, 1991	

12.5.2 Canada
Table 12.4 provides a listing of the major federal statutes for Canada.

12.5.3 The United States of America
Table 12.5 provides a listing of the major federal statutes for the United States. The broad terms of the statutes are commonly implemented by regulations that are promulgated after public comment and often in response to lobbying. Additionally, the states have enacted their own statutes: Thus, an air polluter might have violated both federal and state law.

12.6 Case studies

Case Study 12.1. European convention on long-range transboundary air pollution

The original (1979) protocols under this convention were designed to address transboundary acidification and photochemical pollution over an area covering western, central, and part of the eastern European subregions. Initial focus was on acidification, and, in 1994, modifications were made to the original sulphur protocol, which did not provide sufficient protection. The 1998 provisions added persistent organic pollutants and metals to substances covered by the convention with a future focus on nitrogen oxides and related chemicals. The metals protocol emphasises reducing emissions from stationary sources, but it also deals with production technologies and with products containing certain metals.

Under the provision, participating countries are committed to reporting emissions and national strategies. Many have developed action plans or long-term strategies. Most countries have adopted a suite of source and effect-oriented approaches in combination with a system of cost-effective, differentiated obligations to obtain maximum effect at minimum cost.

Emissions of acidifying substances have decreased in all subregions concerned after the first protocols went into force. The decrease is sharpest for sulphur dioxide (SO_2), which is the pollutant principally responsible for acidification. Of the nations with the largest overall SO_2 emissions, reductions in 2000 relative to 1980 were about one third (typically in central and eastern Europe) to two thirds or three quarters (typically in western Europe).

Reductions of the emissions of nitrogen oxides, ammonia, and hydrocarbons are expected to be more difficult to achieve due in part to the diffuse and less manageable nature of the sources. The increased complexity due to multiple pollutants makes it more difficult to find optimal solutions for joint implementation. For example, ways to reduce the combined effects of sulphur *and* nitrogen may differ among countries.

Case Study 12.2. Implementation of the Basel Convention: Turning back waste from Hungary

In late 1995, 100 tonnes of chemical waste were transported from Germany to Hungary by the German company Corrado. The material was held at a customs office in Budapest. In August 1996, the consignment was scheduled to be transferred through Croatia en route to China, its final destination. Although all necessary transfer documents had been signed, there were some signs of corrosion of the containers, and the Croatian authorities denied entry of the material and turned it back to Hungary. However, in accordance with governmental decree 102/1996 on dangerous wastes, which incorporates the provisions of the Basel Convention, the Hungarian environmental authorities did not permit the reentry of the waste material, which remained at the Hungarian-Croatian border for several months.

An investigation and examination of the material was initiated and revealed that the substance was a wood preservative that had been incorrectly labelled. It contained carcinogens, and its date of expiration had passed. It was determined that the material should qualify as a hazardous waste and the Hungarian authorities asked that it should be sent back to Germany under article 9 of the Basel Convention covering the illegal transboundary transfer of hazardous wastes. Under this provision, the exporting country is required to take back the hazardous material within 30 days of notification. Finally, at the end of 1996, the waste was taken back to Germany.

This example may be compared to the case of a German company that sent expired pesticides disguised as humanitarian aid to Albania in September 1991. At that time the Basel Convention did not cover such transboundary shipments, and it took more than 3 years and an international campaign by environmental activists to secure the return of the waste to Germany.

12.7 Questions

1. What are the major differences between the formulation and administration of the common law and statute law? Discuss the relative effectiveness of each in the context of regulation of toxic substances.

2. Give examples of the major types of laws that are designed to protect the quality of the environment from harmful effects of toxic substances. Include examples of those laws that are designed to regulate the individual source; those that are designed to protect the consumer, a resource, a species, or the environment more generally; and those that have other major objectives but that include components of protection from toxic substances.

3. Consider, with examples, the effectiveness of informal or voluntary activities in the protection of the environment from toxic substances.

4. Define criterion, objective, ambient standard, and effluent standard. Explain the relationships among these concepts in the context of the regulation of toxic substances in the environment.

5. What are the advantages and disadvantages of specifying contaminant standards in regulations and making the breach of any standard an offence? Suggest alternative ways in which standards have been/could be used.

6. How should the public be involved in the standard-setting procedure? What is/are the role(s) of special interest or activist groups in this process? Include principles as well as practical suggestions for the process(es) that you advocate.

12.8 References

Brittannica. (1965) *Encyclopedia Brittanica*, William Benton, Chicago.
Carra, J. S. (1999) The role of voluntary programs, *ICME Newsletter*, **5** (4), 1, 6.
Charles, W., and VanerZwag, D. (1993) The Common Law Approach. In *Environmental Law and Policy*, eds. Hughes, E. L., Lucas, A. R., and Tilleman, II, W. A., pp. 89–149, Emond Montgomery, Toronto.

Dinwoode, D. H. (1972) The politics of international pollution control: The Trail Smelter case, *International Journal*, **27**, 219. Cited in Muldoon (1993).

Duncan, L. F. (1993) The Quasi-Criminal Process. In *Environmental Law and Policy*, eds. Hughes, E. L., Lucas, A. R., and Tilleman, II, W. A., pp. 305–63, Emond Montgomery, Toronto.

Harr, J. (1996) *A Civil Action*, Vintage Books, Random House, New York.

IJC. (1984) Second Biennial Report under the Great Lakes Quality Agreement of 1978 to the Governments of the United States and Canada and the States and Provinces of the Great Lakes Basin. International Joint Commission on the Great Lakes, Washington, Ottawa, Windsor.

Jacobson, J. S. (1981) Acid rain and environmental policy, *Journal of Air Pollution Control Association*, **31**, 1071–3.

Lucas, A. R. (1993) Regulatory legislation. In *Environmental Law and Policy*, eds. Hughes, E. L., Lucas, A. R., and Tilleman, II, W. A., pp. 165–209, Emond Montgomery, Toronto.

Mining Association of Canada (MAC). (1995) Voluntary Emissions Reduction. The Mining Industry and the ARET program. The Mining Association of Canada.

Ministry of the Environment (MOE). (1992) Ontario's water quality objective development process. Ontario Ministry of the Environment, Toronto, Ontario.

(1997) Guideline for use at contaminated sites in Ontario, June 1996 and errata to update for Feb. 1997. MOE, Toronto.

Morelli, J. (1999) *Voluntary Environmental Management: The Inevitable Future*. Lewis, Chelsea, MI.

Muldoon, P. (1993) Bilateral and Multilateral Dimensions of International Environmental Law. In *Environmental Law and Policy*, eds. Hughes, E. L., Lucas, A. R., and Tilleman, II, W. A., pp. 509–52, Emond Montgomery, Toronto.

Parti Québecois (PQ). (1994) Politique de rehabilitation des terrains contamines. Government of Quebec.

PL 92-500. (1992) Public Law 92-500. U.S. Federal Water Pollution Control Act.

Rand, G. M. (1995) *Fundamentals of Aquatic Toxicology: Effects, Environmental Fate and Risk Assessment*, Taylor and Francis, Washington, DC.

Rodricks, J. V. (1992) *Calculated Risks: Understanding the Toxicity and Human Health Risks of Chemicals in Our Environment*, p. 256, Cambridge University Press, Cambridge.

U.S. General Accounting Office. (1992) Pesticides: Adulterated imported foods are reaching U.S. grocery shelves. GAO/RCED-92-205, U.S. General Accounting Office, Washington, DC.

12.9 Further reading

Hughes, E. L., A. R. Lucas, and W. A. Tilleman II. 1993. *Environmental Law and Policy*, pp. 89–149, Emond Montgomery, Toronto.

Thompson, G. T., M. L. McConnell, and L. B. Huestis. 1993. *Environmental Law and Business in Canada*, Canada Law Book, Aurora, Canada.

Wildavsky, A. 1995. *But Is It True? A Citizen's Guide to Environmental Health and Safety Issues*, Harvard University Press, Cambridge, MA.

13

○ ○

An overall perspective, or where to from here?

13.1 Introduction

This final chapter presents a brief resume of the rapidly evolving science of environmental toxicology and some opinions about the current state of the art and its future progression. It is hoped that the reader will use some of these ideas in formulating his or her own view of the subject and, perhaps, his or her own contribution to the discipline. The authors' careers have coincided with a period of extraordinarily rapid growth of the subject, involving several paradigm shifts, and as a conclusion to the text it may be useful at this point to consider how the science has matured, particularly over the last two decades, and how it might develop in this new millennium.

There are several areas of investigation that we judge to be particularly important for the progress of environmental toxicology in the near future. Our judgement is based on a perception of the need for intelligent and responsible environmental management and regulation of toxic substances, concomitant with the need for scientific progress. As we have stressed throughout this text, unless there is a sound scientific paradigm, environmental protection is unlikely to be successful or sustained. In essence then, in ecotoxicology, even though the need for protection and rehabilitation tends to direct the science, it is critically important that the science inform the decision making that forms the basis of environmental protection.

The areas of investigation that we highlight in our conclusions then are risk assessment (Section 13.2), which includes the expression of toxic action (Section 13.2.1), bioavailability (Section 13.2.2), pathways of exposure for biota (Section 13.2.3), the question of biological scale (Section 13.3), and genotoxicity (Section 13.4). We also add a short section on the role of society in how the environment is perceived and managed (Section 13.5).

13.2 Updating risk assessment

13.2.1 Expressing toxic action

The need for managers and legislators to apply consistent criteria to cases of environmental contamination and other forms of ecosystem disruption has often led to an overemphasis on simple models as a basis for risk assessment. In fairness, such an approach has often been dictated by the need to make rapid management decisions in response to some ecotoxicological "crisis". Out of necessity, such decisions have frequently been based on the best available information bolstered by chemical analyses of field samples and perhaps some relatively short-term bioassays.

Figures 13.1 and 13.2 describe two early examples of hazard assessments designed to assess maximum acceptable environmental concentrations of PCBs and chlordecone (Kepone) in biota from the marine and estuarine environments. Both were initiated in the face of a pressing need to rapidly define "safe" levels for management purposes. More than 20 years have elapsed since the data for Figures 13.1 and 13.2 were generated, and it is useful to reexamine some aspects of these assessments. This reexamination will provide a vehicle to evaluate how the field has progressed over that time and to project how a similar hazard assessment might be made now and in the future. This examination in no way implies criticism of these early studies, which were very effective at the time in achieving their aims. By the time these two studies were initiated, hazards associated with chlorinated hydrocarbons were already well known. For at least 20 to 30 years, controls have been imposed on the manufacture and use of PCBs and chlorinated pesticides such as chlordecone. By the time of publication of Figures 13.1 and 13.2 in 1984, the U.S. Food and Drug Administration had established action levels for both PCBs ($5.0\,\mathrm{mg\,kg^{-1}}$) and chlordecone ($0.3\,\mathrm{mg\,kg^{-1}}$).

The Maximum Permissible Toxic Concentration (MPTC) isopleths shown in Figures 13.1 and 13.2 were calculated as the PCB/chlordecone concentrations in water that would have been required to reach these action levels, given a series of bioconcentration factors derived from laboratory assays conducted on various organisms over different times. The authors acknowledged the relative crudity of these estimates, as well as the fact that, in the case of the PCBs, part of the MPTC isopleth exceeded the solubility of this group of compounds. A more modern version of this sort of hazard assessment would probably focus on specific PCB congeners, notably the so-called coplanar congeners wherein the two phenyl rings line up in the same plane (see Figure 7.18). This flat configuration confers a similar shape to the highly carcinogenic tetrachlorodibenzodioxin molecule and is characteristic of congeners having no chlorine substitutions at the ortho ($2,2',6,6'$) positions. Two potent inducers of mixed function oxidase enzymes, PCB numbers 126 ($3,3',4,4',5$) and 169 ($3,3',4,4',5,5'$) have been assigned toxic equivalency factors, relative to TCDD, of 0.1 and 0.05, respectively. Other congeners having a single chlorine substitution at an ortho site achieve a configuration that is close to

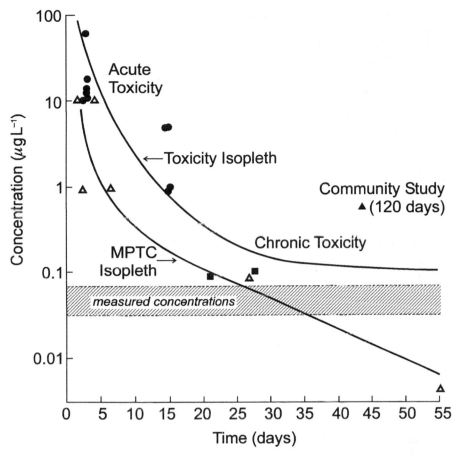

Figure 13.1 Hazard assessment for the PCB Aroclor 1254. Acute data for toxicity isopleth include shell-growth EC_{50} for *Crassostrea virginica* and 96-hr LC_{50}s for crustaceans *Penaeus aztecus* and *Palaemonetes pugio*. Chronic toxicity data include 15-day mortality data for crustacean *Penaeus duorarum* and finfish *Leiostomus xanthurus* and *Lagodon rhomboides*; 21-day LC_{50} and 28-day fecundity assays for sheepshead minnow *Cyprinodon variegatus*; 4-month mixed plankton community test. MPTC = maximum permissible toxic concentration (5 mg kg^{-1} in fish and shellfish) (Gentile and Schimmel, 1984).

coplanar, but they have a much smaller TEF (0.001). Current assessments of hazard associated with PCBs would assume additive toxicity of specific congeners found in tissue and sediments and would express the overall toxic hazard in terms of a total TEF.

In several instances, monitoring known toxic chemicals is now an integral part of the risk assessment process. In some places, we now have the benefit of records for long-term temporal trends in sediments and biota. In the case of a few individual pollutants that have been subject to environmental regulation, declining trends have been noted in sediments and biota (Figure 13.3) and recoveries of birds such as the peregrine falcon and brown pelican and some fish (e.g., lake trout,

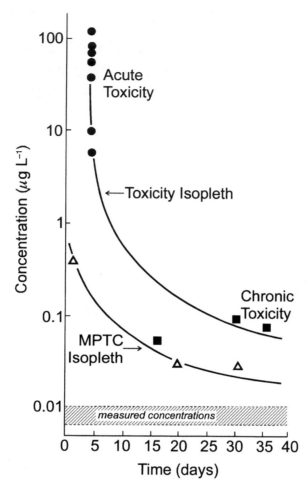

Figure 13.2 Hazard assessment for chlordecone (Kepone). Acute data for toxicity isopleth include 96-hr LC_{50}s for crustaceans *Mysidopsis bahia* and *Callinectes sapidus* and six species of finfish. Chronic data include a 28-day LC_{50} and full life-cycle test for *Cyprinodon variegatus*. MPTC = maximum permissible toxic concentration ($0.3\,mg\,kg^{-1}$) (Gentile and Schimmel, 1984).

walleye, and burbot) are clearly correlated with such chemical declines. Following the restriction of tributyltin usage, recoveries in bivalve settlement have been documented in parts of Europe (Alzieu and Portman, 1984). A significant and fairly recent feature of these studies is that they now include an understanding of the mechanism(s) through which the adverse effects occurred. For example, chemical disruption of sex hormones was responsible for several of the foregoing cases of reproductive failure. Even though estrogen receptors appear to respond to a disparate group of chemicals (Figure 7.21), it seems likely that mimicry of 17β-estradiol may be associated with the existence of a hydroxyl group on the compound in question or through a hydroxylation reaction as the compound is metabolized. It is, therefore, reasonable to assume that future hazard assessments

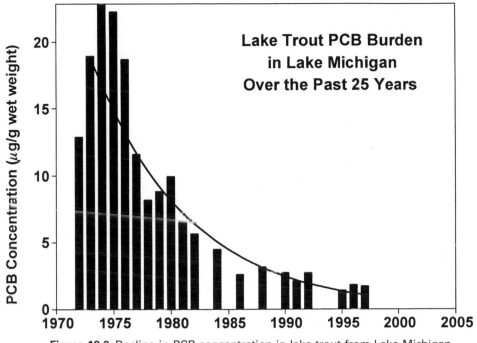

Figure 13.3 Decline in PCB concentration in lake trout from Lake Michigan from 1970 to 1999 (Baker and Stapleton, unpublished).

would contain strong elements of monitoring coupled with some degree of quantitative structure activity analyses (QSARs, see Section 5.5). QSARs would have predictive value not only with respect to the potential for a substance to bioaccumulate, but they may also play a part in assessing the type of toxic response that might be expected.

Guidelines are often provided by regulatory agencies, expressed as concentrations of particular substances in specific media which should not be exceeded, or to which contaminated media should be cleaned, usually for a given use. As the term *guideline* implies, these are not for direct regulatory application; indeed, they are not of universal application, since site-specific assessments are usually recommended. Furthermore, such guidelines should be subjected to continuing scrutiny and be updated as necessary, based upon the more sophisticated approaches outlined in the previous paragraph.

Future hazard assessments will probably expand well beyond the limited suite of chemical compounds that are routinely searched for in most analytical laboratories. Recent refinements in sample preparation and analytical methodologies have pushed reliable detection limits to low concentrations that are unprecedented, and have brought within range of the analyst a range of contaminants, including environmentally discharged pharmaceuticals, that were hitherto not even considered by the environmental toxicologist. Some would argue that we should avoid undue concern over the potential toxicity of an ever-expanding suite of chemical contaminants simply because we can now measure them, but it is instructive to recall that many important environmental contaminants were not recognised as such when they first appeared. PCBs first appeared as unidentifiable peaks during chromatographic screening for pesticides. Likewise, chlordecone (Kepone) in biota and sediments from the James River on the east coast of the United States first came to light in the mid 1970s as an anonymous chromatographic peak, which was recognised, before a positive identification was made, only by its retention time. It is also important to bear in mind that pharmaceuticals were specifically formulated to elicit some biological effect in humans, so even in small concentrations their specific activity might be high.

THE EMPIRICAL APPROACH

The toxicity curves seen in Figures 13.1 and 13.2 reflect an approach still seen in the development of water quality criteria, where acute toxicity data form the upper bounds of what may be tolerated in the environment. For regulatory purposes, this is often considered as a concentration to be tolerated only for a short time or not to be exceeded more than once a year. Chronic toxicity data typically form the basis for the maximum acceptable toxic concentration, which is the backbone of most pollutant discharge legislation. Water quality criteria are ultimately derived statistically from large data sets using regression analysis. A similar strategy has now been proposed for sediments. The approach is an empirical one, which is based on a comparison between sediment contaminant levels and a large historical data base derived from both field and laboratory studies and which has been carefully screened for consistency of analytical rigour. The historical data are divided into an effects range low (ERL) corresponding to contaminant levels that historically resulted in adverse biological effects in 1.9–9.4% of cases involving metals and 5–27% of cases involving organics. Effects range low–effects range medium (ERL-ERM) resulted in adverse effects in 11–47% of cases involving metals and 18–75% of cases involving organics (Long et al., 1995). This empirical approach has resulted in the development of chemical-specific sediment guidelines for 9 metals and 19 organic compounds/groups. Only intended as a management tool, this method of comparative assessment suffers from several drawbacks. It is based on relatively few contaminants and would not directly indicate sediment toxicity due to an unknown pollutant. It is based only on total contaminant levels and does not address the influence of sediment characteristics on contaminant bioavailability.

13.2.2 Bioavailability and uptake pathways as management tools

The question of bioavailability is an extremely important one from a management point of view. Although the regulation of environmental contaminants is based almost exclusively on their total concentration in a particular medium, we have seen in Chapter 5 that numerous extraneous factors can influence the biological availability and toxicity of many chemical pollutants. And in Chapter 6, we discussed the relationship between the chemical form of a metal and its biological availability. For the sake of simplicity, these factors have been dealt with singly in this text, and the term *bioavailability* has been narrowly defined as the form that is assimilable by a living organism. Several useful models have been developed to predict bioavailability of a variety of contaminants. Some examples are shown in Table 13.1. All are imperfect to varying degrees, but it is important to remember that they are only models and can be modified and expanded. One area of expansion of such models recognises that, in reality, chemical uptake by an organism may be integrated among several different media. For example, additive pathway models have been developed for both metals (Thomann et al., 1995) and organic contaminants (Clark et al., 1990), which take into account differential uptake from several media, and the use of nitrogen isotopes has elucidated the passage of organochlorines through a food chain (Kidd et al., 1995). By introducing such concepts into a discussion of bioavailability, we recognise that this might eventually lead to a broader definition of the term than has been traditionally used (e.g., one that encroaches into the area of pathways and vectors).

13.2.3 Pathways/vectors of chemical exposure

In the aquatic environment, organisms may be simultaneously exposed to pore water, sediment, suspended particulates, and living and nonliving food, all of which may constitute a source of the contaminant. In terms of bioavailability, much will therefore depend on how organisms interact with or "sample" these different media. A typical example of this might deal with feeding behaviour. Two organisms occupying apparently similar habitats may feed in entirely different ways and may, therefore, have quite different toxicant exposures. For example, the estuarine clams *Rangia cuneata* and *Macoma balthica* overlap in parts of their respective ranges, although *Rangia* is largely freshwater and *Macoma* is more salinity-tolerant. *Rangia* is a filter feeder, constantly sampling the water column, whereas *Macoma* may switch from filter feeding to deposit feeding as the need arises. Even within the same feeding mode, filter feeders can exhibit food selectivity with potential consequences for contaminant exposure. Decho and Luoma (1996) demonstrated that metal assimilation rose significantly when the bivalve *Macoma balthica* switched from a nondiatom to a diatom diet. Several studies have sought to quantify the relative importance of contaminant uptake food from food and water, often with contradictory results. Dissolved cadmium may be a dominant source under certain experimental conditions (Kemp and Swartz, 1988); however, other studies have demonstrated that food may constitute a significant source of toxicant, but that this may alter according to food availability and type (Reinfelder and Fisher, 1991; Decho and Luoma, 1994; Munger and Hare, 1997). Figure 13.4 shows the results of an experiment wherein copper concentrations in water and a suspended food organism were adjusted independently (Wright, unpublished). The effects of these manipulations on copper uptake by the oyster demonstrated that the contribution of food to metal uptake could be as little as 11% or as much as 50%

Table 13.1. Some methods used to assess bioavailability/toxicity

Bioavailability descriptor	Comments	Reference(s)
Free ionic activity model (some trace metals in water)	Bioavailable form of metal correlates with free ionic activity	Allen et al. (1980); Campbell (1995)
Octanol : water partition coefficient (K_{OW})	Surrogate for membrane permeability in nonpolar organics; correlation with toxicity	Mayer et al. (1991)
Sediment organic carbon (oc)/ tissue lipid concentration (uptake of organics from sediments)	Normalisation factors account for association of organics with sediment oc and organismal lipids	Lake et al. (1987)
AVS/SEM (Trace metals in sediments)	Bioavailable metals $\equiv$ molar fraction of metals exceeding AVS in simultaneous extraction	Di Toro et al. (1990)
Various nonbiological acid/base/oxidising extractants (trace metals)	Extractant used to simulate digested (bioavailable) fraction from solid phase	Luoma (1989); Tessier et al. (1979)
Digestive fluid from GI tract of benthic polychaete (metals and organics)	Digestive fluid itself is used to extract bioavailable fraction from solid phase	Lawrence et al. (1999); Weston and Mayer (1998)
Metal hydroxide (phytoplankton cells)	Log. of M-OH correlates with volume concentration factor	Fisher (1986)
Quantitative structure activity relationships (various groups of organic chemicals and metals)	Molecular properties correlated with bioavailability and/or toxicity; may be applied to K_{OW}/metal hydroxides (above)	Numerous authors (see Table 5.3)
$\delta^{15}N = [(^{15}N/^{14}N_{SAMPLE}$ $^{15}N/^{14}N_{ATMOSPHERIC\ N}) - 1]$ $\times 1,000.$	$\delta^{15}N$ is used as a measure of food chain length that can be correlated with contaminant concentration in biota.	Kidd et al. (1995)

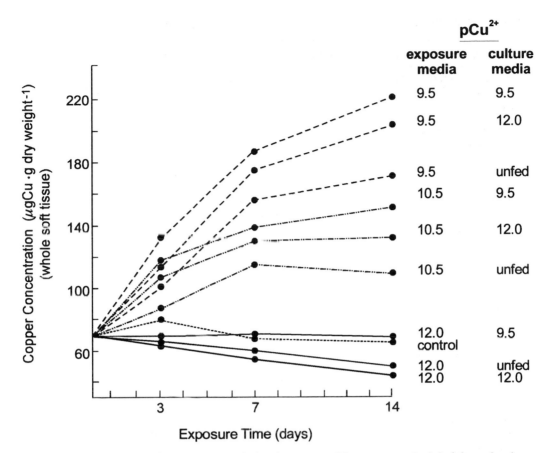

Figure 13.4 Copper accumulation by oysters (*Crassostrea virginica*) from food (*Thalassiosira pseudonana*) and water. Algal cells were grown in media of three different cupric ion concentrations and fed in concentrated suspension into different copper media. Oysters were entrained in media + food in nine different combinations. Copper concentrations in oysters were measured after 7 and 14 days. (Wright, unpublished)

depending on the relative proportions of copper in food and water, respectively. This experiment essentially represented an extension of the one described in Chapter 5 (Figure 5.8), where copper in food and water were adjusted proportionately, and overall copper uptake was related to food availability. Although such results may, at first sight, seem intuitively obvious, they make the point that, rather than devoting an inordinate effort to quantify the relative contribution to contaminant uptake from food versus water in a controlled laboratory setting, more important questions might relate to the multiple uptake pathways (e.g., quality and quantity of food, suspended particle and sediment characteristics) and the spatial and temporal scales governing these. Problems can arise when such scales differ from those forming the bases for physical and chemical characterisation of the ambient medium (e.g., sediment).

13.3 Future paradigm of hazard assessment

We might surmise that a current version of the hazard assessments summarised in Figures 13.1 and 13.2 would incorporate elements of some of the models summarised in Table 13.1. In a more general sense, we can expect future risk assessments and particularly water quality criteria to be much more strongly linked to models of bioavailability and exposure vectors than to computations based on total environmental contaminant levels and historical toxicity data. Existing data bases provide a good foundation for screening purposes, but differences in experimental conditions clearly limit the comparability of many of the data. Indeed, these very differences in physics, chemistry, and biology will provide the basis for the more mechanistic hazard assessment models of the future. We can expect these to be considerably richer in input data and, therefore, much more system-specific than their predecessors, and much more realistic. We have already seen remarkable improvements in fate and transport models over the last two decades, particularly with respect to the roles of atmospheric processes in the dispersion of environmental contaminants.

13.4 The question of biological scale

A strength of the original hazard assessments shown in Figures 13.1 and 13.2 was the use of reproductive data from full life-cycle tests and the inclusion of planktonic community data (Figure 13.1), which was unusual in hazard assessments of that era. Inherent in this was the recognition of higher order effects in the characterisation of toxicity. Such features are common components of current toxicity assessments where an examination of sediment-borne contaminants would probably include effects on benthic community structure and bottom-dwelling fish communities, often as part of a "triad" approach (Chapman, 1986). Although adverse effects of pollutants have been demonstrated at all levels of biological organisation, the linkage between these levels remains difficult to quantify (see Figure 2.1). Biological effects are often much more complicated than the first-order responses demonstrated in the laboratory. Studies of endocrine disrupters have illustrated subcellular events with clear implications for reproductive performance, yet only in a few instances has it been possible to link or translate biochemical effects into significant changes at the population level. Even if such links can be made, the relationships are largely correlative, and we are still lacking satisfactory models that make connections among different levels of biological organisation concerning toxic effects. Some would doubt whether such linkages are indeed possible, so here it is of interest to consider two quotations that straddle this exponential growth period in environmental toxicology:

> *The biochemist is apt to feel that all important biological problems will be solved*
> *at the molecular or submolecular level, while the ecologist feels that the molec-*

*ular biologist is preoccupied with details of machinery whose significance
he does not appreciate. . . . there are a number of levels of biological integration and each level offers unique problems and insights . . . each level finds its
explanations of mechanism in the levels below, and its significance in the levels
above.*
 Bartholemew (1964)

*even though nature could, in principle, be explained in terms of universal basic
laws, in practice our finite mental and computational capacities mean that we
either cannot grasp the ultimate physical explanation of many complex phenomena – or we can't fruitfully link this basic level to higher order phenomena.
. . . additional principles, not evident in the laws governing basic constituents,
are needed to explain higher order phenomena.*
 N. Williams (1997), quoting Thomas Nagel

Bartholemew (1964) adopts a largely reductionist position wherein larger biological units can be understood by analysing their component parts. Thus, the effect
of a toxicant at one level of biological organisation can be explained and predicted
by understanding its action at the next lowest level of organisation. For example,
biochemists can characterise mechanisms of toxic action, but the significance and
perhaps survival value of these mechanisms is seen at higher levels of organisation. Bartholemew's approach is not completely reductionist, however. Rather, he
states that "each level offers unique and important insights". In other words, he
stresses the interdependence of the different levels of organisation.

William's comments on the statement by Thomas Nagel concerning the application of basic principles to universal laws introduce a more extreme, holistic argument. Essentially, Nagel's thesis is that we lack the computational capacity to
provide satisfactory explanations for complex phenomena and are, therefore,
unable to explain "emergent properties" that are found at higher levels of biological organisation. From this point of view, it is, therefore, impossible to use knowledge gained at the biochemical level effectively to explain events at the population
or community level. Taken at face value, such a philosophy would determine that
biochemistry has little to offer the ecotoxicologist. However, we believe that Nagel
is not denying that cellular processes inform events at higher levels. Instead, he is
pointing out that it is impossible to resolve all the questions posed by ecotoxicology simply by studying the subcellular mode of action of a chemical. Williams
further quotes Nagel's call for new principles to explain properties at higher levels
of biological organisation. The development of such principles in the field of
ecotoxicology will require much closer cooperation between toxicologists and
nontoxicologists. This is starting to happen.

In the area of molecular genetic toxicology, as well as at higher biological levels,
we see a convergence with the more basic counterpart disciplines (i.e., those that
are involved with the "normal" molecular biology/physiology/ecology). In other
words, toxicants will be increasingly regarded as yet another environmental
variable to be considered and integrated into demographic, energetic, and other

ecological models. The increasing sophistication of such models in recent years has been a direct result of the increasing engagement of hitherto "basic" physiologists and ecologists in pollution-related matters. It is to be hoped that this fresh infusion of ideas will increase awareness of toxicologists of the variability and stochasticity of undisturbed systems. At the community level, this increased awareness has implications for the restoration and rehabilitation of damaged habitats where knowledge of the timescales for natural change may be important in accurately defining the recovery process.

13.5 Genotoxicity

A relatively unexplored area of environmental toxicology concerns subtle effects on the Darwinian fitness of an organism, which may have only marginal, if any, *direct* effects on reproductive performance. Hypothetical examples might include adverse genetically mediated effects of chemical contaminants on temperature/ salinity tolerance, disease resistance, or a variety of other factors resulting in less robust individuals. Genetically inherited traits could be passed on to subsequent generations, making it extremely difficult to discern the clear involvement of a toxicant in any subsequent loss of fitness of a population. The effect of pollutants on the genetic make-up of populations would, therefore, be a particularly fruitful area of study. In Chapter 4, the phenomenon of tolerance was described in terms of specialists versus generalists and of the survivor effect following moderate to heavy selection pressure. Knowledge of the genetic basis of physiological and biochemical phenotypic variability would be particularly important in evaluating the role of either of these phenomena in survivorship. In a study of marine gastropods, Nevo et al. (1986) demonstrated that highly heterozygous generalists had significantly better survival than congeneric species of low heterozygosity (specialists), following exposure to a range of organic and inorganic pollutants. On the other hand, Guttmann (1994) suggested that lower genetic diversity in fish from polluted sites might result from selection pressures resulting from heavy metal exposure, a hypothesis essentially similar to the survivor phenomenon described by Mulvey and Diamond (1991). However, it is not always necessary to invoke genotoxicity in describing these effects. It is by no means clear that genetic shifts in populations following contaminant exposure are always a result of genetic damage per se. Along with the explosion of growth in molecular genetics and its application to environmental toxicology, there needs to be a renewed appreciation of exactly what we are looking for. Markers for genotoxicity might be useful in determining the effects of certain chemical pollutants; however, for other toxicants, it might be more instructive to follow gene expression associated with their detoxification and metabolic transformation. Although earlier studies of pollutant-induced genetic damage tended to be confined to the DNA molecule itself, there are now many more studies of higher order conditions. Genotoxicity may be manifest in a number of pathological conditions affecting the physiology of organisms. Kurelec (1993) identified a number of symptoms that he collectively identified as the Genotoxic Disease Syndrome. Component disorders included enhanced protein turnover, impaired enzyme function, impaired immune response, decreased scope for growth, decreased fecundity, and production of initiators causing cytotoxic injury. Even though it is often difficult to associate many of these disorders with decreased survival, it is to be expected that such studies will become more widespread, and that more sophisticated linkages of this nature will be made. Lower order events such as damage to the DNA itself may prove even more problematic in this regard due to the efficiency of repair mechanisms, although assays involving the detection of c-K-*Ras* oncogenes hold promise and have been correlated with pathological and physiological changes in liver structure and function (McMahon et al., 1990). It is possible that genetic resistance to toxicants could occur through random mutation, but most recorded instances of pollution-tolerant populations seem to have resulted from chronic exposure to contaminants. Depledge (1994) hypothesised that the progressive loss of genetic diversity often associated with such resistant populations could lead to an increase in homozygosity and potential reductions in fertility and offspring viability resulting from the expression of recessive genes having negative consequences.

In pondering these items, it is instructive to revisit some of the concepts that were presented at the beginning of this book. Figure 1.1 provided a framework for the science of environmental toxicology, yet contained some generalisations that need to be expanded upon. The right-hand box in Figure 1.1 treated toxicological effects (uptake, distribution, etc.) as if we were dealing with a single organism. In reality, the entity represented by this box might equally well be a population, a community, or an ecosystem, and it is at these higher levels of biological organisation that most risk assessments in environmental toxicology are ideally made and remedial actions considered. Therefore, this figure has been reproduced here, in modified form, as Figure 13.5. Such higher order effects are now considered to be covered by the term *ecotoxicology*, which was defined by Forbes and Forbes (1994) as "integrating the ecological and toxicological effects of chemical pollutants on populations, communities, and ecosystems with the fate (transport, transformation, and breakdown) of such pollutants in the environment." This information has typically formed the basis of management decisions that constitute the feedback loop at the bottom of Figure 13.5. Also inserted in the figure is a component addressing the priority placed by society as a whole on environmental management issues.

13.6 Society and the environment

On a more philosophical/abstract level, we draw attention to what Leo Marx (1992) has termed "environmental degradation and the ambiguous social role of science and technology". In his essay, Marx recognises a paradox in the role(s) of science

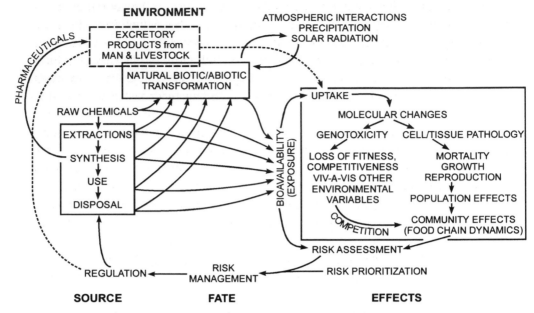

Figure 13.5 Principal components of environmental toxicology showing subject areas of current/future emphasis.

and technology as forging the tools of "civilisation" and at the same time supplying the key to remedying the environmental damage that technical progress has caused. He sees this conflict, and the associated political tensions, as a recent (late twentieth century) phenomenon, but one having its roots in three different philosophical interpretations of the relationship between humans and their environment: progressivism, primitivism, and pastoralism.

Progressivism is well illustrated by the Puritan's Calvinistic view of "the New World" of North America as a wilderness to be tamed with a sacred zeal focussed on turning natural resources into wealth. This religious version of the myth was later to be recast in more secular capitalistic terms, but it still underscored the notion of enlightened settlers in a heroic struggle against a hostile natural environment. Mid-nineteenth century Darwinism as interpreted by Herbert Spencer and many of his contemporaries further reinforced this view of humans as the apex of evolution (civilisation). As Marx points out, social Darwinism metaphysicalised progress. With the increase in scale of technology and the advent of mass production, technical advancement became less of a means to an end and more of an end in itself as the concept of technocratic progress took hold.

Primitivism is seen by Marx as the conceptual opposite of progressivism; it is typified by the desire to escape the excesses of nineteenth century high society in Europe and on the eastern seaboard of North America and to forge westward to the unspoiled frontier of the New World. Bolstered by depictions of "the noble savage" in literature and visual art, the adherents to this viewpoint have tended to gravitate to an articulate elite, perhaps represented by modern day anarchists who see themselves as the standard bearers of aesthetic values in the natural environment. In doing so, they fall into a natural alignment with many of the indigenous peoples of North America and elsewhere whose historical religious values with respect to the ecosystem have contrasted sharply with those of the early European settlers. Ironically, the native American peoples of that continent were often seen by European immigrants as part of the hostile environment that had to be subdued.

Pastoralism in some respects occupies a middle ground between the two foregoing philosophical standpoints, and is exemplified by Thomas Jefferson, the drafter of the American Declaration of Independence, who clearly saw unbridled technology as a threat to the quality of life of the average citizen of his new country. As presented by Marx (1992), pastoralism made several advances in terms of urban planning and the development of national parks in the eighteenth and nineteenth centuries but suffered significant retreats in the face of the accelerating industrial revolution. Nevertheless, we see in this philosophy the predecessor of the conservation movements of the nineteenth century and the reemergence of environmentalism in the 1960s.

At the beginning of the twenty-first century, there is a clear need to reformulate our thinking about the relationship between society and nature. All the preceding viewpoints were initially conceived without any clear evidence of widespread environmental pollution. Well-publicised instances of major contamination events and recent emphases on adverse effects at the population, community, and transboundary ecosystem levels now indicate the pervasiveness of problems. While science and technology have more sophisticated tools than ever before to identify and remediate environmental contamination, global problems will require global solutions and the political will to drive them. Such political decisions can only be made in the light of informed public opinion. This will increasingly involve scientists in the role of communicators and the engagement of politicians, economists, social scientists and philosophers in providing a balanced, realistic framework for ecosystem management.

13.7 References

Allen, H. E., Hall, R. H., and Brisbin, T. D. (1980) Metal speciation: Effects on aquatic toxicity, *Environmental Science and Technology*, **14**, 441–3.

Alzieu, C., and Portman, J. E. (1984) The effect of tributyltin on culture of *C. gigas* and other species, *Proceedings of Annual Shellfish Conference*, **15**, 84–101.

Bartholemew, G. R. (1964) The Roles of Physiology and Behavior in the Maintenance of Homeostasis in the Desert Environment. In *Homeostasis and Feedback Mechanisms*, pp. 7–29, Academic Press, New York.

Campbell, P. G. C. (1995) Interactions between trace metals and aquatic organisms: A critique of the free-ion activity model. In *Metal Speciation and Bioavailability*, eds. Tessier, A., and Turner, D. R., pp. 45–102, John Wiley & Sons, New York.

Chapman, P. M. (1986) Sediment quality criteria from the sediment quality triad: An example, *Environmental Toxicology and Chemistry*, **5**, 957–64.

Clark, K. E., Gobas, F. A. P. C., and Mackay, D. (1990) Model of organic chemical uptake and clearance by fish from food and water, *Environmental Science and Technology*, **24**, 1203–13.

Decho, A. W., and Luoma, S. N. (1994) Humic and fulvic acids: Sink or source in the availability of metals to the marine bivalves *Potomocorbula amurenis* and *Macoma balthica*, *Marine Ecology Progress Series*, **108**, 133–45.

(1996) Flexible digestion strategies and trace metal assimilation in marine bivalves, *Limnology and Oceanography*, **41**, 568–72.

Depledge, M. H. (1994) Genotypic toxicity: Implications for individuals and populations, *Environmental Health Perspectives*, **102**, 101–4.

Di Toro, D. M., Mahoney, J. D., Hansen, D. J., Scott, K. J., Hicks, M. B., Mayr, S. M., and Redmond, M. S. (1990) Toxicity of cadmium in sediments: the role of acid-volatile sulfide, *Environmental Toxicology and Chemistry*, **9**, 1487–502.

Fisher, N. S. (1986) On the reactivity of metals for marine phytoplankton, *Limnology and Oceanography*, **31**, 443–9.

Forbes, V. E., and Forbes, T. L. (1994) *Ecotoxicology in Theory and Practice*, Chapman and Hall, London.

Gentile, J. H., and Schimmel, S. C. (1984) Strategies for utilizing laboratory toxicological information in regulatory decisions. In *Concepts in Marine Pollution Measurements*, ed. White, H. H., p. 743, Maryland Sea Grant, University of Maryland, College Park.

Guttmann, S. I. (1994) Population genetic structure and ecotoxicology, *Environmental Health Perspectives*, **102**, 97–100.

Kemp, P. G., and Swartz, R. C. (1988) Acute toxicity of interstitial and particle-bound cadmium to a marine infaunal amphipod, *Marine Environmental Research*, **26**, 135–53.

Kidd, K. A., Schindler, D. W., Muir, D. C. G., Lockhart, W. L., and Hesslein, R. H. (1995) High concentrations of toxaphene in fishes from a subarctic lake, *Science*, **269**, 240–2.

Kurelec, B. (1993) The genotoxic disease syndrome, *Marine Environmental Research*, **35**, 341–8.

Lake, J. L., Rubinstein, N. I., and Pavignano, S. (1987) Predicting bioaccumulation: Development of a partitioning model for use as a screening tool in regulating ocean disposal of wastes. In *Fate and Effects of Sediment-Bound Chemicals in Aquatic Systems*, eds. Dickson, K. L., Maki, A. W., and Brungs, W. A., pp. 151–66, Pergamon, New York.

Lawrence, A. L., McAloon, K. M., Mason, R. P., and Mayer, L. M. (1999) Intestinal solubilization of particle-associated organic and inorganic mercury as a measure of bioavailability to benthic invertebrates, *Environmental Science and Technology*, **33**, 1871–6.

Long, E. R., MacDonald, D. D., Smith, S. L., and Calder, F. D. (1995) Incidence of adverse biological effects within ranges of chemical concentrations in marine and estuarine sediments, *Environmental Management*, **19**, 81–97.

Luoma, S. N. (1989) Can we determine the biological availability of sediment-bound trace metals?, *Hydrobiologia*, **176/177**, 379–96.

Marx, L. (1992) Environmental degradation and the ambiguous social role of science and technology, *Journal of the History of Biology*, **25**, 449–68.

Mayer, F. L., Marking, L. L., Howe, G. E., Brecken, J. A., Linton, T. K., and Bills, T. D. (1991) Physiocochemical factors affecting toxicity: Relation to bioavailability and exposure duration, *Proceedings of the Society of Environmental and Toxicological Chemistry*, **12**, 141.

McMahon, G., Huber, L. J., Moore, M. N., Stegeman, J. J., and Wogan, G. W. (1990) c-K-*Ras* Oncogenes: Prevalence in livers of winter flounder from Boston Harbor. In *Biomarkers of Environmental Contamination*, eds. McCarthy, J. F., and Shugart, L. R., pp. 229–38, Lewis, Boca Raton, FL.

Mulvey, M., and Diamond, S. A. (1991) Genetic factors and tolerance acquisition in populations exposed to metals and metalloids. In *Metal Ecotoxicology: Concepts and Applications*, eds. Newman, M. C., and McIntosh, A. W., pp. 301–21, Lewis, Chelsea, MI.

Munger, C., and Hare, L. (1997) Relative importance of water and food as cadmium sources to an aquatic insect (*Chaoborus punctipennis*): Implications for predicting Cd bioaccumulation in nature, *Environmental Science and Technology*, **31**, 891–5.

Nevo, E., Noy, R., Lavie, B., Beiles, A., and Muchtar, S. (1986) Genetic diversity and resistance to marine pollution, *Biological Journal of the Linnean Society*, **29**, 139–44.

Reinfelder, J. R., and Fisher, N. W. (1991) The assimilation of elements ingested by marine copepods, *Science*, **251**, 794–6.

Tessier, A., Campbell, P. G. C., and Bisson, M. (1979) Sequential extraction procedure for the speciation of particulate trace metals, *Analytical Chemistry*, **51**, 844–51.

Thomann, R. V., Mahoney, J. D., and Meuller, R. (1995) Steady-state model of biota-sediment accumulation factors for metals in two marine bivalves, *Environmental Toxicology and Chemistry*, **14**, 1989–98.

Weston, D. P., and Mayer, L. M. (1998) Comparison of *in vitro* digestive fluid extraction and traditional *in vivo* approaches as a measure of PAH bioavailability from sediments, *Environmental Toxicology and Chemistry*, **17**, 830–40.

Williams, N. (1997) Biologists cut reductionists approach down to size, *Science*, **277**, 476–7.

Glossary

α **particles** form of ionizing radiation consisting of two neutrons and two protons; carries a +2 charge.

acclimation within the context of a bioassay, the temperature at which organisms are held prior to toxicant exposure. It may or may not differ from the experimental temperature, although organisms are usually acclimated to the experimental temperature for several hours before the initiation of the assay.

acclimatization (acclimation) adaptation of biological function to a change in environmental conditions.

acetylcholinesterase enzyme that catalyses the hydrolysis of the neurotransmitter acetyl choline.

acid deposition the deposition of wet or dry acidic (pH < 5.7) material from the atmosphere onto water or land.

acid mine drainage drainage water from mine workings, waste, and tailings, made acidic by the oxidation of sulphide minerals. The resulting sulphuric acid dissolves iron and other metals from the ore and other geological material. The highly acidic metal-bearing water, often with pH as low as 2.0, drains into surface water bodies.

acid rain wet, acidic (pH < 5.7) material (rain, snow, hail, or fog) that may be deposited from the atmosphere onto water or land.

acid volatile sulfides (AVS) dilute acid (often cold $1\,M$ HCl) extraction of sulfides.

acidosis formation of excess acidity in blood and tissues, either through net production of H^+ ions by metabolic processes or through accumulation of H^+ ions from acid environments.

active transport transmembrane or transepithelial chemical transport against a concentration gradient requiring the expenditure of energy (from ATP).

acute-to-chronic ratio (ACR) the ratio of acute LC_{50} to a measure of chronic toxicity (e.g., MATC).

acute toxicity bioassay cumulative mortality recorded over a period of 96 hr or less (usually 24, 48, or 96 hr).

adduct product of covalent bonding between a xenobiotic and an endogenous molecule such as DNA.

adsorption accumulation of a substance at the boundary of two phases, usually between solid and liquid phases.

advisory a value that is the basis of a warning, rather than a regulation, for the public, concerning the safety of a named product or food item.

anadromous refers to organisms that move from the sea to rivers to breed.

aneuploidy uneven distribution of chromosomes following mitosis.

anoxia, anoxic a condition in which there is no oxygen; synonymous with *anaerobic*.

antagonism less-than-additive effects of two or more substances; normally refers to toxic effects.

aphasia loss or impairment of the ability to use or to comprehend words.

apoptosis a process of programmed cell death.

application factor (AF) multiplier derived from ACR, used to extrapolate from acute to chronic toxicity.

artificial selection selection that occurs when selection pressure is exerted deliberately, under laboratory or other experimental conditions; compare with natural selection, for which the genetic mechanism is the same.

aryl hydrocarbon hydroxylase (AHH) enzyme-catalysing hydroxylation of aryl hydrocarbons such as benzo(a)pyrene; used as a measure of cytochrome P450 activity.

ataxia inability to coordinate voluntary muscular movements.

ATPases adenosine triphosphatases; often involved with active transport of electrolytes across membranes, hence CaATPase. May also transport metabolic precursors and trace metals (esp. in bacteria).

β particles negatively charged electrons formed when an excess of neutrons in the nucleon causes a neutron to be changed into a proton.

B-cells small lymphocytes produced by bone marrow primarily responsible for humoral immunity and production of circulating antibody.

beneficiation the preparation of ore for metallurgical processing, usually by means of crushing and grinding (comminution) and concentration.

Bequerel (Bq) measurement of radioactive emission equivalent to 1 nuclear disintegration per second.

bioaccumulation the tendency of substances to accumulate in the body of exposed organisms with increases over time or with age. See also *bioconcentration*.

bioaccumulation factor (BAF) the ratio of the concentration of a given compound or element in an organism to its concentration in the immediate environment, measured in the field and thus incorporating food chain as well as direct sources. Compare with *bioconcentration factor (BCF)*.

bioassay a measurement, usually quantitative, of the effect of a chemical, physical, or biological action, using the response of a living system, typically a population.

bioavailable, bioavailability see *biologically available*.

biochemical markers biochemical and molecular responses that are a particular category of indicators. Used synonymously with *biomarkers* in the present text, *although biochemical markers* is our preferred term. Some authors have used *biomarkers* in a broader sense, even as synonymous with biological indicators.

bioconcentration the phenomenon whereby a living organism contains higher concentrations of a given substance than the concentration in its immediate source of that substance. See also *bioaccumulation*.

bioconcentration factor (BCF) the ratio of the concentration of a given compound or element in an organism to the concentration of the same in its immediate environment, measured in the laboratory under defined conditions. Compare with *bioaccumulation factor (BAF)*.

biological half-life ($t_{1/2}$) the time required to eliminate from the body 50% of the total body concentration of the named substance.

biological indicator a biological response, at any level of organisation, that is used to provide information concerning the condition of the biological system.

biological oxygen demand (BOD) the amount of oxygen that is consumed by microorganisms in water in a standardized test. BOD is in effect an indicator of the amount of organic matter.

biologically available   in a form that is assimilable by a living organism; synonymous with *bioavailable*.

biomagnification   a food chain or food web phenomenon, whereby a substance or element increases in concentration at successive trophic levels.

biomarkers   see *biochemical markers*.

biominification   the tendency for a substance to diminish in concentration at successively higher trophic levels, a phenomenon exemplified by lead (Pb). Compare with *biomagnification*.

biomonitor   an organism that takes up the contaminant into tissues resulting in concentrations that reflect the exposure of the contaminant in the environment.

bioremediation   application of the phenomenon of degradation or transformation of a toxic substance into benign form(s), by which the organisms can literally clean up a contaminated environment.

biotransformation   chemical change mediated by biological activity.

bioturbation   the action of stirring, irrigating, or otherwise disturbing sediments, by the action of infauna or epifauna.

blood:gas partition coefficient   coefficient describing relative chemical concentrations in blood and adjacent gaseous medium under equilibrium conditions.

bloom, algal bloom   a burst of productivity in phytoplankton, resulting in an increase in standing crop of algae, usually visible as colour and a decrease in transparency of the water.

bottom-top trophic efficiency   that fraction of the carbon fixed by autotrophs that eventually reaches top carnivores.

bottom-up   extrapolation from simple to complex systems (e.g., measurements of rate of maturity and development and of fecundity may contribute functional data on population growth). Analogous measurement of nutrient or contaminant transfer at the bottom of a food chain may allow the assessment of their "flow" to higher trophic levels.

buffering capacity   the ability of a chemical system to neutralize excess acidity.

carbamate pesticides   anticholinesterase pesticide derivatives of carbamic acid (NH_2COOH).

carcinogen   a substance or agent producing cancer.

channels   passages through cell membranes allowing selective movement of electrolytes.

chelate   a complex in which the donor ligand is multidentate.

chelation   a special case of complexation, when two or more donor groups surround the central cation. See also *multidentate*.

chemical speciation   a distribution of the chemical forms in which an element can exist (e.g., the free ion; the chloride, charged or uncharged; the citrate).

chemiluminescence (CL)   property of compounds to emit photons under certain conditions; utilized in measurement of ROI production by the chemiluminescent compound luminol (5'-amino-2,3-dihydro-1,4-phtalazinedione).

chemotactic   orienting or moving in response to a chemical.

chemotactic efficiency   the ability of immunocells such as macrophages to migrate towards a stimulus.

chlorosis   the yellowing or mottling of leaves, which is a symptom of a number of types of abnormal metabolism in vascular plants, typically resulting from some damage to the production or maintenance of chlorophyll by a toxic agent or a pathogen.

cholinesterase   enzyme responsible for the hydrolysis of choline esters. It is inhibited by oganophosphate pesticides; thus, cholinesterase inhibition can be used as a quantitative measure of exposure to this type of pesticide.

chronic toxicity   mortality may still be used as an end-point but may involve longer exposures than 96 hr. In this regard, the term *chronic toxicity* is often used. However in many texts the term *chronic* is conventionally reserved for a group of sublethal bioassays involving graded end-points such as level of biochemical activity, fecundity, or growth.

chronicity index (CI) ratio of acute to chronic expressed as a dose rate (e.g., mg $kg^{-1} day^{-1}$) designed to detect cumulative effects of repeated doses.

cinnabar mercuric sulphide; used as a pigment.

classical toxicology the science of the biological effects of toxic substances, mainly referring to effects on humans.

clastogenisis gene mutations caused by breaks, omissions and translocations of chromosome pieces.

cocarcinogenicity the potentiation of a carcinogen by a noncarcinogen. See *potentiation*.

congeners members of a family of chemicals sharing a common structure but differing in configuration of side chains.

committed equivalent dose unit of radiation risk for humans taking into account the potential for radiation dose to be delivered over long period of time following ingestion.

complete carcinogen chemical able to induce cancer in normal cells having properties of initiation, promotion, and progression.

complexation the combination of an inorganic anion with a positively charged ligand that may be inorganic or organic.

concretion a solid structure within living tissue, usually mineralised, which is not a part of the organism's functional anatomy.

connectance (communities) number of actual interspecific interactions divided by the possible interspecific interactions.

critical body residue (CBR) total mass of toxicant present in organism at time of death.

Curie (Ci) measurement of radioactive emission equivalent to 2.2×10^{12} nuclear disintegrations per minute.

cytochrome P450 family of catalytic haem-based enzymes found in all tissues (especially liver endoplasmic reticulum) that play key roles in metabolism of endogenous compounds such as steroid hormones, fat-soluble vitamins, and fatty acids. Also responsible for detoxication of a wide range of nonpolar xenobiotics. May cause activation of some to carcinogen intermediates. Basic cytochrome P450 action is monooxygenation where a single oxygen atom is incorporated into the substrate.

cytosol intracellular fluid exclusive of nucleus and vesicular inclusions.

degrading reactions chemical changes by which molecules are converted to simpler compounds or elements.

detoxification decrease of the toxic quality of a chemical substance by biological or chemical transformation into one or more less toxic compounds or elements.

dissociation constant (K_a) measure of the degree of dissociation of a weak acid (or base). The negative logarithm (pK_a) is related to pH as follows: $pH - pK_a + \log [(Salt)/(Acid)]$.

dissolved organic carbon (DOC) unspecified organic compounds, naturally occurring or synthetic, collectively determined in water.

diversity indices (communities) measures that combine species richness and evenness with a particular weighting for each.

dose mass of toxic chemical ingested or energy absorbed by an organism.

ecological indicator a biological response that is used to provide information concerning the condition of an ecosystem. See also *biological indicator*.

ecotype a local population of an otherwise widely distributed species that is genetically adapted to a particular set of environmental conditions, such as soils with unusually high metal concentrations.

electrophilic term used to describe molecules with electron-deficient, positively charged atoms capable of sharing electron pairs with electron-rich nucleophiles. See *nucleophilic*.

electrowinning   the removal of metals from a solution by electrolysis.

end-point   a biological process used to quantify response.

endogenous   synthesised by an organism.

entrainment   the process whereby a small amount of material is moved within a large moving matrix, as when soil or dust particles are picked up and carried in a moving air mass.

epidemiological   related to *epidemiology*.

epidemiology   the study of the occurrence, transmission, and control of disease in a population.

epigenetic   category of carcinogens that are not classified as genotoxic.

epiphytic   growing on the surface of a plant.

epoxides   unstable, highly reactive compounds having an oxygen molecule linked to two adjacent carbon atoms in a ring structure. Epoxides are often intermediates or end-products of cytochrome P450 catalysed reactions.

equilibrium partitioning   the proportional distribution of a substance between or among two or more phases, at equilibrium.

ethoxyresorufin-*o*-deethylase (EROD) enzyme-catalysing *o*-deethylation of ethoxyresorufin. Ethoxyresorufin is a three-ringed endogenous compound, not an environmental contaminant, although EROD activity is induced by xenobiotics and is a biomarker.

euryhaline   able to live in waters of a wide range of salinity.

eurytoxic   having tolerance to a wide range of toxicants.

eutrophic   the property of a body of water that has high nutrient loading and is very productive. See also *mesotrophic* and *oligotrophic*.

eutrophication   the process by which a body of water increases in productivity as a response to an increase in the concentration of nutrients.

eutrophication, cultural   eutrophication caused by anthropogenic input of nutrients.

evenness (communities)   variance of the species abundance distribution.

exclusion   a mechanism whereby a substance does not enter the cell or is secreted rapidly.

exogenous   synthesised outside the body of a specific organism.

facilitated diffusion   carrier-mediated molecular transport across a membrane in the direction of the concentration gradient and not requiring energy.

fecundity   reproductive performance, usually measured as number of offspring.

Fick's law   the equation that expresses the rate of movement of a substance across an interface in terms of the diffusion coefficient and the concentration gradient.

flow-through tests   tests performed using metering devices (dosers) designed to deliver an appropriate range of chemical concentrations on a once-through basis.

fluorescence   the emission of electromagnetic radiation, usually as visible light, occurring during the absorption of radiation from some other source.

free ion activity model (FIAM)   a theory stating that the concentration (activity) of the free ion is the best predictor of the bioavailability of a metal.

free radical   a molecule possessing an unshared electron, usually signified by a dot ($\bullet$).

fugacity   a thermodynamic criterion of equilibrium of a substance in solution in two phases. It is closely related to chemical potential and can be regarded as an idealised partial pressure.

γ radiation   electromagnetic photons released from the atomic nucleus of radioisotopes. Their energy is proportional to their wavelength, and highly energetic photons with short wavelengths can pass completely through the human body.

Gaussian curve   curve that reflects a normally distributed population response with early mortalities among more sensitive individuals and prolonged survival of the most resistant organisms.

generalised linear models (GLM)   a method of plotting dose-response without the use of a preconceived model explaining how a test population might respond.

genotoxic   compounds having the potential to alter the genetic code. See *epigenetic*.

genotype   the particular complement of genes present in an individual.

genotypic   related to the *genotype*.

geobotany   the study of the geographical distribution and relationships of plants, including conditions of the soil or other substratum. Also called phytogeography.

glucosuria   decreased renal absorption of glucose.

β-glucuronidase   enzyme occurring in lysosomes and intestinal bacteria, responsible for hydrolysis of glucuronide conjugates.

glucuronyl transferase   enzyme catalysing the formation of conjugates between glucuronic acid and metabolites.

glutamic pyruvic transaminase (GPT)   liver enzyme involved in gluconeo-genesis; used as a nonspecific indicator of toxic liver damage. Damage is inferred from increased serum levels of GPT.

glutathione (GSH)   a tripeptide compound (γ-glutamyl-L-cysteinyl-glycine), which plays two contrasting roles in detoxifi-cation: as an intermediate in phase II metabolism via GST and as an important antioxidant.

glutathione disulfide (GSSG)   oxidized form of glutathione. GSSG^{-} radical can oxidize O_2 to O_2^{-}.

glutathione transferases (GST)   a family of enzymes that function as catalysts for the conjugation of various electrophilic compounds (e.g., organic nitrates, epoxides of PAH) with glutathione.

granuloma   mass of inflamed granulated tissue.

gray (Gy)   measure of radioactive energy absorption of $1\,\mathrm{J\,kg^{-1}}$.

greenhouse effect   warming effect on Earth's surface and lower atmosphere produced by absorption of infrared radiation (heat) by certain gases (greenhouse gases).

Haber-Weiss reaction   chemical reaction: $O_2^{-\bullet} \rightarrow H_2O_2 \rightarrow {}^{\bullet}OH + OH^{-}$.

hazard assessment   determination of the potential for adverse effects.

heat shock proteins (hsp)   group of proteins (also known as *stress proteins*) induced by a variety of stressors, including, but not limited to, salinity and osmotic changes, trace metals, anoxia, heat (and cold) shock, and xenobiotics. They have been found in all organisms examined to date, from prokaryotes to humans. Five groups of hsp have been identified in eukaryotes.

Henderson-Hasselbach equations   equations relating pH to ionized/-unionized forms of weak acids and bases.

Henry's law   the amount of gas dissolved in a liquid is proportional to its partial pressure in the gaseous phase until the point of saturation in the liquid is reached.

histopathology   histological study of the adverse effects of toxicants on tissues and cells.

homeostatic   acting to maintain a stable internal chemical environment.

hormesis   a stimulatory effect of low levels of toxic agent on an organism.

hyperkeratosis   hypertrophy of outer layer of skin with derangement of squamous epithelium.

hypoxia   low oxygen concentrations.

imposex   imposition of male characteristics on females often induced by endocrine-disrupting chemicals.

incipient LC$_{50}$ lowest LC$_{50}$ reached in a time series (i.e., longer exposures to lower toxicant levels produce no further mortality). Also known as *incipient median lethal concentration*.

incipient median lethal concentration   see *incipient LC$_{50}$*.

index of biotic integrity (IBI)   method of quantifying disruptive effects of toxicants on a community. Index may include information on habit and trophic structure, tolerant versus intolerant species and pathological conditions relative to a reference community from a similar habitat. Originally applied to stream communities, IBIs have been modified for other habitats.

inhibition, competitive   inhibitory effect resulting from toxicant competing with a particular substrate for an active site. Typically, this would slow enzymic expression/substrate kinetics.

inhibition, noncompetitive   active site is poisoned in noncompetitive manner (e.g., by denaturation of enzyme).

initiation   causing damage to cellular DNA that results in a mutation.

intercalation   physical insertion of small planar molecules into the molecular structure of large molecules (e.g., intercalation of xenobiotic between base pairs of DNA). Intercalating molecules may be loosely bound or covalently bound. Either case has mutagenic potential.

ionizing radiation   energy released during radioactive decay. Consists principally of α, β, and γ radiation.

isobologram   graphical representation of interaction between two toxicants expressed in terms of equivalent toxic units.

isotope   any one of two or more atoms of an element with the same atomic number and position in the periodic table and the same chemical properties but with differing atomic mass.

isozymes (also isoenzymes)   any of the chemically distinct forms of an enzyme that perform the same biochemical function.

leaching   the process by which dissolved substances are moved, typically by percolating water or other solvent. In metallurgy, leaching has as technical meaning as a process for extracting a soluble metal by a solvent.

lentic   related to still water, typically lakes or ponds.

Leslie matrix model   see stage-based demographic model.

lethal body burden (LBB)   toxicant concentration in the body of an organism at the time of death (same as CBR).

lethal concentration 50 (LC_{50})   the toxicant concentration that kills 50% of the test population at time t. It represents an estimate of that concentration which would cause 50% mortality of the infinitely large population from which the test population was taken.

life-cycle management   the approach that considers all aspects, particularly the potentially harmful aspects, of a substance or process over time, rather than limiting it to a particular point in time. Sometimes referred to as *cradle-to-grave*.

life table   tool for the study of population dynamics. The life table provides a more detailed evaluation of reproductive success through the determination of the intrinsic rate of natural population increase (r).

ligand   a group, ion, or molecule coordinated to a specific atom (typically a metal) in a complex.

limiting nutrient   an essential nutrient that is present in the least supply of all nutrients, relative to the biological or ecological requirements. The limiting nutrient is the one that restricts production. If its supply subsequently exceeds its requirement, then, by definition, some other nutrient will become limiting.

limnocorrals   relatively large enclosures suspended in a water column, typically open to the atmosphere and either closed or open to the sediment, used for experimental manipulation of biotic or abiotic factors, while retaining some of the realism of the natural system.

linear transfer energy (LTE)   measure of ability of ionizing radiation to transfer energy to tissue per unit of exposure.

lotic   related to actively moving water.

lowest observed effect concentration (LOEC)   the lowest toxicant concentration in a bioassay that shows significant difference from the control.

lowest observed effect level (LOEL) same as *lowest observed effect concentration (LOEC)*.

luxury uptake the uptake and storage of a nutrient when it is in excess supply.

macrophages component cells of immune system capable of engulfing foreign particles. They may be motile or nonmotile. Macrophages can secrete soluble factors, cytokines, that assist with phagocytosis.

mass balance modelling a calculation of the concentrations of various chemicals in various media in a specified system.

maximum acceptable toxic concentration (MATC) the geometric mean of two other values: the NOEC and the LOEC.

median effective concentration or EC$_{50}$ the chemical concentration causing a sublethal response in 50% of the test population. In each case, the significance (or otherwise) of the response is measured by comparison with the mean control value but is counted in a quantal way (e.g., an organism is either scored as normal or abnormal). Analogous to the *lethal concentration 50 (LC$_{50}$)*.

median lethal time elapsed time to reach 50% mortality of the test population.

mesocosm the enclosure of a representative segment of a real ecosystem. See also *microcosm*. Microcosms and mesocosms are experimental systems set up to recreate part or all of a particular ecosystem. These range from constructed systems such as simple aquaria, with perhaps only one representative species at each trophic level, to large enclosures (e.g., limno-corrals, test plots) in the external (i.e., real-world) environment, which may be tens of cubic meters in volume.

mesotrophic the condition of an aquatic ecosystem that has moderate supplies of nutrient. It is intermediate between oligotrophic and eutrophic.

metal-tolerant able to live at concentrations of metal in water or soil or other substrate that would normally be considered toxic. The definition is not absolute but rather should be considered in a comparative sense.

metalloids elements such as arsenic and selenium that have some of the same characteristics as metals, but their chemical properties are such that they may form compounds in which they behave as anions *or* cations.

metallothioneins (MTs) small proteins rich in sulphur-containing amino acids (thiols) that effectively bind many trace metals. Originally, the term was restricted to animal proteins, but it now may be used for chemically similar plant proteins, the latter are usually called phytochelatins.

methylation the addition of one or more methyl (CH$_3$) groups, sometimes by microbial or other biological activity, as in the case of the environmental formation of methylmercury from inorganic mercury.

Michaelis constant (K_m) substrate concentration at which enzyme is half saturated. Also used to characterize ionic/molecular pumps where the ion or molecule carried is analogous to the substrate, and the pump is analogous to the enzyme. Low K_m values signify high carrier affinity.

Michaelis-Menten plot plot of enzyme (carrier) substrate complex versus substrate concentration indicating saturable characteristics.

microcosm a small, usually very simple, synthetic ecosystem, typically involving two trophic levels. See also *mesocosm*.

mineralisation transformation into mineral or inorganic form.

mixed function oxidase (MFO) system (cytochrome P450 system) enzyme system responsible for the transformation of nonpolar organics to more polar, water-soluble products.

moving average method method of determining LC$_{50}$ for equal numbers of organisms exposed to a geometric range of toxicant concentrations.

multidentate   an organic ligand with two or more donor groups, which when complexed with a metal, forms a chelate.

multidrug resistance (MDR)   resistance to a variety of unrelated compounds, including xenobiotic chemicals with carcinogenic properties.

multixenobiotic resistance   see *multidrug resistance* (*MDR*).

mutagenicity   potential for altering gene expression.

narcosis   anaesthetic effect produced by narcotic.

natural selection   differential reproduction in nature, resulting from some set of environmental conditions, leading to the increase in frequency of certain genes or gene combinations and to a decrease in the frequency of others.

necrosis   irreversible tissue death often accompanied by inflammation and discoloration of tissue.

neoplasm   relatively autonomous tissue lesion caused by alteration of genetic material of cells.

nitrogen fixers   organisms that can convert gaseous nitrogen into more complex nitrogenous compounds, ultimately organic nitrogen. Known nitrogen fixers are free-living as well as symbiotic with higher plants and include bacteria, both blue-green autotrophs and heterotrophs, as well as some other types of microorganisms.

nitrosamines   compounds, some of which are carcinogens, formed from inorganic nitrogen compounds, such as nitrites, by microbial activity.

no effect concentration (NEC)   operationally defined as the $LC_{0.01}$ obtained by extrapolation from a probit curve at t_{00}. Effectively estimates concentration at which no toxic response is seen.

no observed effect concentration (NOEC)   the highest concentration having a response not significantly different from the control.

no observed effect level (NOEL)   same as no observed effect concentration (*NOEC*).

nonpolar   having no net electrical charge associated with the molecule, which tends to be lipophilic.

normal equivalent deviate (NED)   a measure of the variance associated with the median response of a normally distributed test population. It is equivalent to standard deviation and forms the basis of the probit scale.

nuclear waste, front end   tailings and other waste, especially uranium mill tailings produced in the early stage of the processing of uranium ore; typically large in volume and relatively low in radioactivity.

nuclear waste, high-level   the end-product produced in a nuclear reactor, especially radioactive spent fuel. It is relatively low in volume but highly radioactive and likely to continue decaying and emitting radiation for centuries.

nuclear waste, low-level   radioactive material from the reactor plant, including pipes, radioactive material in workers clothing, and tools used by reactor workers.

nucleophilic   term used to describe molecules containing electron-rich atoms. See *electrophilic*.

octanol:water partition coefficient (K_{ow})   the distribution of a given substance between octanol and water, at equilibrium. Generally, the K_{ow} is a good indicator of the tendency of a substance to move from water to lipid, thus its tendency to move from the aqueous environment into biological membranes.

oligotrophic   a condition of a water body with a low supply of nutrients and low productivity. See also *eutrophic* and *mesotrophic*.

ombrotrophic bog   a wetland that receives no nutrients from ground or surface water so that all inputs are from the atmosphere (i.e., from wet or dry deposition).

oncogene   a gene capable of inducing one or more aspects of the neoplastic phenotype.

ore   naturally occurring mineral containing a potentially valuable metal or metals.

organochlorine a class of compounds in which one or more atoms of chlorine are combined with an organic molecule (e.g., chloroform, carbon tetrachloride, DDT).

organophosphate a phosphate-containing organic pesticide that competes with acetyl choline for the enzyme cholinesterase.

^{32}P post-labelling technique used to identify (DNA) adducts wherein the radioisotope label is added following chromatographic separation.

pathology the study of the nature of diseases, especially the structural and functional changes produced by them.

percentage inhibition concentration (IC$_p$) measure of the degree of response relative to a control (e.g., IC$_{30}$ = 30% inhibition of the measured parameter compared with a control).

periphyton the community of aquatic organisms that grow attached to surfaces, predominantly composed of algae and bacteria.

peroxisomes subcellular organelles involved in lipid, sterol, and purine metabolism including peroxidative detoxification.

persistence the property of a compound by which it remains in its original form in the environment (i.e., is not degraded or transformed). Also refers to the time for which a substance persists.

perturbation in general terms, a disturbance from the regular or normal; used here to indicate any measurable change in the environment, without necessarily implying a harmful result.

phagocytic property of immune system cells to engulf and absorb foreign bodies in tissues and blood, hence phagocytosis.

pharmacology the science of drugs including medical uses and toxicology.

phase I enzymes enzymic components of the mixed function oxidase (MFO) system involved with the catabolism of nonpolar substrates to more polar hydrophilic products.

phase II (conjugating) enzymes anabolic enzymes controlling the conjugation of polar metabolites (usually from phase I reactions) with endogenous polar groups and compounds.

phenotype the physical manifestation of a genetic trait.

phenotypic related to the *phenotype*.

photolytic breaking down in the presence of light.

phytochelatins small proteins rich in sulphur-containing amino acids (thiols) in plants that effectively bind many trace metals. See *metallothioneins (MTs)*.

phytoplankton the community of photosynthetic organisms, mainly algae, that occur freely suspended in the water column.

phytotoxicity the property of being poisonous, toxic, to plants.

phytotoxicology the study of toxic effects on plants.

pinocytosis transmembrane movement of solvents and solutes by the formation of vesicles within the membrane.

pK_a negative logarithm of the acid dissociation constant K_a where $K_a = [(H^+)(A^-)/(HA)]$.

plumbism chronic lead poisoning.

pneumonitis toxicant-induced inflammation of lung tissue.

population distribution vector component of a population projection model that defines the number of organisms at any specific stage of development.

potable in general terms, drinkable; in a regulatory context, refers to conditions that ensure the safety of water for ingestion by humans.

potentiation situation where the toxicity of a combination of compounds exceeds the sum of the toxicities of individual compounds. Special case where a particular compound enhances the toxicity of the other(s).

precautionary principle the philosophy of taking preventive or protective action even when there is no conclusive scientific evidence to prove a causal link between emissions and effects.

principle of conservation of mass   in any chemical change, the quantity of matter at the end of the change is the same as before.

production   the quantity of organic matter that is produced by biological activity, typically expressed per area or volume. Primary production refers to the production of autotrophs; secondary production refers to production by the rest of the trophic chain.

productivity   the rate of production.

progression   stage in carcinogenesis describing transition from production of initiated cells to a biologically malignant cell population.

promotion   the process of increased replication (hyperplasia) of initiated cells leading to the production of a precancerous condition.

proteinuria   decreased renal absorption of protein.

protoporphyrin   precursor of haem. Increased circulating levels are diagnostic of inhibition, by lead, of enzymes involved with haem synthesis. Levels of protoporphyrin in serum erythrocytes increase when lead (Pb) concentrations are elevated.

pseudoreplication   inappropriate clumping of samples from treatments within a laboratory assay or a field sampling programme.

pump   membrane-bound enzyme capable of energy-dependent configurational changes causing translocation of substrate from one side of the membrane to the other.

quantitative structure-activity relationship (QSAR)   relationship between the toxicity of chemicals and their physical structure and properties.

r   intrinsic rate of population increase.

rad.   measure of absorbed radioactivity. Amount of radioactivity causing 1 kg of tissue to absorb 0.01 J of energy.

radioactive decay   process of energy release as a radioisotope reverts to a less energetic state.

radionuclide   radioactive atom of a specific element.

raptor   bird of prey.

reactive oxygen intermediates (ROIs)   highly reactive, cytotoxic radicals (e.g., $O_2^-\bullet$, $OH^\bullet$, H_2O_2), produced by the progressive univalent reduction of molecular oxygen.

receptor   normal body constituents that are chemically altered by a toxicant, resulting in injury and toxicity. Also, receptor molecules capable of binding specific substrates at the initiation of gene activation/metabolism/detoxification sequence (e.g., Ah receptor, RAR, RXR).

rem   measure of destructive radioactive dose. Equivalent of 1 rad of hard X-rays or 0.05 rad of α particles.

resilience   the capacity (of a community or ecosystem) to return to equilibrium following a disturbance.

resistance   in many texts, used synonymously with tolerance (i.e., the ability of an organism to exhibit decreased response to a chemical relative to that shown on a previous occasion). Various authors distinguish resistance from tolerance in several different ways: One is that resistance implies that the magnitude of the chemical change lies outside the normal range, and that negative effects of that stressor will eventually be manifested in the organism. Another considers that resistance is a more general term than tolerance, implying interspecies but not necessarily intraspecies comparisons (i.e., when typical members of entire species can grow without ill effects in the presence of elevated concentrations of a potentially harmful substance).

response   organismal reaction to a toxic dose, quantitative or qualitative.

response, graded   relative response (e.g., percentage reduction in growth), compared with a control.

response, quantal   all-or-none response (e.g., mortality).

risk   the chance of an undesired effect, such as injury, disease, or death, resulting from human actions or a natural catastrophe.

risk assessment determination of the relationship between the exposure of an ecosystem (or part thereof) to a hazard and adverse effects.

risk management action(s) taken by bringing together all the scientific information that is available to assess the magnitude of a risk, making decisions about the need to reduce the assessed risk and, if necessary, designing some action to reduce risk, such as restricted exposure, remediation, and pollution control technology.

roasting in metallurgy, preparation of ore for smelting, by heating the ore to oxidise and drive off sulphur gases.

Scatchard plot a reciprocal plot used to determine the relationship between enzyme (carrier) substrate complex and substrate concentration.

scope for growth (SFG) describes the excess energy available to an organism, after its basic metabolic needs have been met. As such, it represents the energy budget available for somatic and/or germinal growth.

sequester in general, to set apart, isolate; in the present text, to bind up and thus prevent from exerting any effect.

serosal side side of a membrane or epithelium adjacent to the internal body fluid.

Sievert (Sv) measure of destructive radioactive dose equivalent to 100 rem.

sorption from the verb to sorb; the process of holding, by absorption or adsorption.

species richness number of species in an ecosystem.

spectrometry a special case of spectroscopy, where the intensity of electromagnetic radiation is measured using an electronic device (detector).

spectroscopic referring to *spectroscopy*.

spectroscopy the science that deals with the interaction of radiation (UV, visible light, microwave, gamma rays, etc.) with matter.

stage-based demographic models (Leslie matrix models) models that use empirical data from different life stages to estimate probabilities of transition from one developmental stage to the next.

stenotoxic narrow range of tolerance of toxic chemicals.

subacute toxicity toxic effects at concentrations less than the acute LC_{50}.

sulphatases enzymes catalysing the breakdown of sulphate conjugates.

sulphotransferases enzymes catalysing formation of conjugates between xenobiotics and sulphate.

sulphydryl SH group.

surrogate in general use, a substitute; in the context of biological or ecological indicators, the term is used more in the context of an organism that is a representative indicator of the condition of the environment, but that is not necessarily of itself a key component (e.g., "the canary is a surrogate for the human").

static an aquatic assay in which organisms remain in the same medium throughout the test.

static-renewal test test in which some or all of the test medium is replenished periodically.

stress a nonspecific term often used to imply a change in the environment that puts constraints on a biological system. Used in the present text as a synonym for perturbation.

structure activity relationships (SAR) see *quantitative structure-activity relationships*.

synergism more-than-additive effect of two or more substances, typically of toxicants.

T-cells lymphocytes produced by the thymus gland responsible for cellular immunity and hypersensitivity responses (e.g., to bacterial antigens).

teratogenicity potential for causing morphological changes in developing organisms.

thiols molecules containing a sulphydyl group.

threshold concentration   the lowest concentration of a chemical that elicits a toxic effect.

tolerance   the ability to withstand exposure to abnormally high concentrations of contaminants, which would otherwise cause adverse biological effects, and which can be sustained indefinitely.

top-down   holistic approach to ecosystem function, using properties of the ecosystem as a whole, or of the upper trophic level.

toxic   poisonous, harmful in the context of a chemical substance.

toxicants   substances that are toxic or potentially toxic.

toxicity identification estimate (TIE) or **toxicity reduction estimation (TRE)**   a toxicity bioassay method involving the testing and retesting of an effluent before and after a series of chemical extractions and/or pH adjustments designed to isolate and identify the toxic fraction. Also known as *whole effluent toxicity (WET) testing*.

toxicokinetics   the study of the dynamics and partitioning of toxicants within living systems.

transformation, probit   in the probit transformation, percentage mortality is plotted on a probability scale on the *y*-axis versus log chemical concentration on the *x*-axis, hence the logarithmic range of chemical concentrations used for the test.

transport processes   physical changes that alter the place in which a substance is located but which do not result in chemical change.

volume of distribution V_d**; apparent volume of distribution**   hypothetical volume occupied by a substance in the body of an organism, assuming that substance to be evenly distributed.

w_t tissue-specific weighting factors reflecting relative susceptibilities of different body tissues to radiation damage.

W_t **quality (weighting)**   factor used to adjust absorbed radiation dose to effective radiation dose taking into account linear transfer energy. See *linear transfer energy (LTE)*.

whole effluent toxicity (WET) testing   see *toxicity identification estimate (TIE)*.

xenobiotics   refers to compounds not known to occur in nature. Novel or newly synthesised compounds are commonly given the name xenobiotics (from the Greek word *xenos* meaning stranger).

Index

○ ○